Reinforced Concrete Structures

About the Author

David A. Fanella, Ph.D., S.E., P.E., F.ASCE, is Associate Principal, Klein and Hoffman, Inc., a structural and restoration engineering firm. He is the author of numerous technical publications including *Design of Low-Rise Concrete Buildings: Based on the 2009 IBC* and a series of articles on time-saving design methods for reinforced concrete. Dr. Fanella took part in the post-9/11 investigation of World Trade Center Towers 1, 2, and 7, and was the primary author of reports detailing the methods and codes used during the design and construction of the structures.

About the International Code Council

The International Code Council (ICC), a membership association dedicated to building safety, fire prevention, and energy efficiency, develops the codes and standards used to construct residential and commercial buildings, including homes and schools. The mission of ICC is to provide the highest quality codes, standards, products, and services for all concerned with the safety and performance of the built environment. Most U.S. cities, counties, and states choose the International Codes, building safety codes developed by the ICC. The International Codes also serve as the basis for construction of federal properties around the world, and as a reference for many nations outside the United States. The ICC is also dedicated to innovation and sustainability, and Code Council subsidiary, ICC Evaluation Service, issues Evaluation Reports for innovative products and reports of Sustainable Attributes Verification and Evaluation (SAVE).

Headquarters: 500 New Jersey Avenue, NW, 6th Floor, Washington, DC 20001-2070
District Offices: Birmingham, AL; Chicago, IL; Los Angeles, CA
1-888-422-7233
www.iccsafe.org

Reinforced Concrete Structures

Analysis and Design

David A. Fanella

New York   Chicago San Francisco
Lisbon London Madrid Mexico City
Milan New Delhi San Juan
Seoul Singapore Sydney Toronto

The McGraw·Hill Companies

Cataloging-in-Publication Data is on file with the Library of Congress

McGraw-Hill books are available at special quantity discounts to use as premiums and sales promotions, or for use in corporate training programs. To contact a representative please e-mail us at bulksales@mcgraw-hill.com.

Reinforced Concrete Structures: Analysis and Design

1 2 3 4 5 6 7 8 9 0   DOC/DOC   1 9 8 7 6 5 4 3 2 1 0

ISBN 978-0-07-163834-0
MHID 0-07-163834-2

The pages within this book were printed on acid-free paper.

Sponsoring Editor	**Proofreader**
Joy Bramble	Devdutt Sharma
Acquisitions Coordinator	**Indexer**
Alexis Richard	Aptara, Inc.
Editorial Supervisor	**Production Supervisor**
David E. Fogarty	Pamela A. Pelton
Project Manager	**Composition**
Shalini Sharma, Aptara, Inc.	Aptara, Inc.
Copy Editor	**Art Director, Cover**
Aptara, Inc.	Jeff Weeks

Contents

Preface

This book presents subject matter related to the analysis and design of reinforced concrete structural members. The focus is on the design of elements in reinforced concrete buildings where the primary reinforcement is steel reinforcing bars or steel wire reinforcement that is not prestressed.

To safely and economically design reinforced concrete structures, a thorough understanding of the mechanics of reinforced concrete and the design provisions of current codes is essential. The purpose of this book is to present and explain the following in a simple and straightforward manner: (1) the underlying principles of reinforced concrete design; (2) the analysis, design, and detailing requirements in the 2008 edition of *Building Code Requirements for Structural Concrete and Commentary* by the American Concrete Institute (ACI) and the 2009 edition of the *International Building Code* by the International Code Council (ICC). Frequent reference is made to the sections of these documents (especially those in the ACI Building Code), and it is assumed that the reader will have access to them while using this book.

Information on the properties of the materials that constitute reinforced concrete and a basic understanding of the mechanics of reinforced concrete must be acquired prior to exploring code provisions. Design and detailing provisions given in the code change frequently, and it is important to have an understanding of the core elements of reinforced concrete design in order to correctly apply these provisions in practice.

Presented in Chap. 1 are a definition of reinforced concrete and a basic synopsis of the mechanics of reinforced concrete. Typical reinforced concrete members and the fundamental roles they play in buildings are discussed. The main purpose of this discussion is twofold: (1) to introduce the types of concrete elements that are covered in the chapters that follow and (2) to illustrate how all of the members in a structure are assembled. A brief overview of construction documents and the main events that occur in the construction of a cast-in-place concrete building are also covered. Again, the purpose is to make the reader aware of the important topics that are encountered in any building project.

Mechanical properties of concrete and reinforcing steel are summarized in Chap. 2. Basic information on the mechanics of concrete deterioration and failure is provided, which gives insight into the strengths and weaknesses of concrete. Also covered in this chapter are (1) methodologies for proportioning concrete mixtures, (2) durability requirements, and (3) evaluation and acceptance criteria.

General information that is applicable to the analysis and design of any reinforced concrete building is provided in Chap. 3. Included are the loads that must be considered in design and analysis methods pertinent to reinforced concrete structures. Approximate methods of analysis and moment redistribution are also covered.

Chapter 4 contains the general requirements that must be satisfied for strength and serviceability. These requirements form the basis of design of all reinforced concrete members. Concepts of the strength design method of analysis are introduced, including required strength and design strength. Load factors, load combinations, and strength reduction factors are also covered, as are general provisions for deflection control.

General principles and requirements of the strength design method are presented in Chap. 5. The design assumptions of this method and the basic techniques to determine nominal strength of a reinforced concrete section subjected to flexure, axial load, or a combination of both are covered in detail. A thorough understanding of the material presented in this chapter is essential before continuing on to subsequent chapters.

Chapters 6 through 10 contain design and detailing requirements for the following reinforced concrete members: (1) beams and one-way slabs, (2) two-way slabs, (3) columns, (4) walls, and (5) foundations. Each chapter contains techniques on how to size the cross-section, calculate the required amount of reinforcement, and detail the reinforcement. Design procedures and flowcharts provide road maps that guide the reader through the requirements of the code. Also included are numerous design aids and comprehensive worked-out examples that facilitate and demonstrate the proper application of the design provisions. The examples follow the steps of the referenced design procedures and flowcharts and have been formulated using structural layouts that are found in typical concrete buildings. These examples further help the reader to understand how members work together and how loads are transferred through a structure.

Throughout the discussions and in the examples, the practical aspects of reinforced concrete design are stressed at length. These fundamental concepts are presented to familiarize the reader with important aspects of design (other than those that are theoretical) that need to be considered in everyday practice. Emphasis is placed on sizing concrete members on the basis of formwork considerations and detailing reinforcement so that they adequately fit within a section.

The content of this book is geared to both undergraduate and graduate students, as well as to practicing engineers who need to become familiar with current code design requirements or need an update on reinforced concrete design. Engineers studying for licensing exams will also find the material presented here to be very useful.

My sincere thanks to John R. Henry, PE, Principal Staff Engineer, International Code Council, Inc., for review of this text. His insightful suggestions for improvement are most appreciated. I also wish to thank Adugna Fanuel, SE, LEED AP, and Angelo Cicero of Klein and Hoffman, Inc., for reviewing Chap. 2 and producing some of the figures, respectively. Their help was invaluable.

David A. Fanella

Reinforced Concrete Structures

CHAPTER 1

Introduction

1.1 Reinforced Concrete

1.1.1 Definition of Reinforced Concrete

Reinforced concrete is concrete in which reinforcing bars or other types of reinforcement have been integrated to improve one or more properties of the concrete. For many years, it has been utilized as an economical construction material in one form or another in buildings, bridges, and many other types of structures throughout the world. A large part of its worldwide appeal is that the basic constituent materials—cement, sand, aggregate, water, and reinforcing bars—are widely available and that it is possible to construct a structure using local sources of labor and materials.

In addition to being readily obtainable, reinforced concrete has been universally accepted because it can be molded essentially into any shape or form, is inherently rigid, and is inherently fire-resistant. With proper protection of the reinforcement, a reinforced concrete structure can be very durable and can have a long life even under harsh climatic or environmental conditions. Reinforced concrete structures have also demonstrated that they can provide a safe haven from the potentially devastating effects of earthquakes, hurricanes, floods, and tornadoes.

Based on these and other advantages, it is evident that reinforced concrete can provide viable and cost-effective solutions in a variety of applications. This book focuses on the design of reinforced concrete members in building structures.

1.1.2 Mechanics of Reinforced Concrete

Concrete is a brittle, composite material that is strong in compression and weak in tension. Cracking occurs when the concrete tensile stress in a member reaches the tensile strength due to externally applied loads, temperature changes, or shrinkage. Concrete members that do not have any type of reinforcement in them will typically fail very suddenly once the first tension cracks form because there is nothing to prevent the cracks from propagating completely through the member.

Consider the simply supported, unreinforced concrete beam shown in Fig. 1.1 that is subjected to a concentrated load P at midspan. From the strength of materials, the maximum tensile bending stress occurs at the bottom fibers of the beam section at midspan and the maximum compressive bending stress occurs at the top fibers. Because concrete is stronger in compression than in tension, the beam will be able to support the concentrated load and its own weight as long as the maximum bending stress is less than the tensile strength of the concrete in bending (Fig. 1.1a). If the bending stress

1

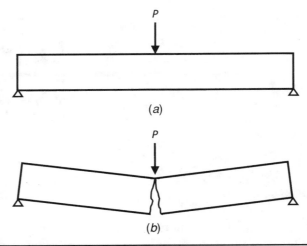

FIGURE 1.1 Response of a simply supported, unreinforced concrete beam due to external loading. (*a*) Bending stress less than the tensile strength of the concrete in bending. (*b*) Bending stress greater than the tensile strength of the concrete in bending.

is equal to or greater than the tensile strength, a crack will form immediately at the bottom fiber of the beam and it will propagate instantaneously to the top fiber, splitting the beam in two (Fig. 1.1*b*).

A different sequence of events would take place if reinforcing bars were present near the bottom of the simply supported beam. Like in the case of an unreinforced concrete beam, a crack will form at the bottom fiber of the reinforced concrete beam at midspan when the bending stress is equal to or greater than the tensile strength of the concrete in bending. However, in contrast to the unreinforced beam, crack propagation will be arrested by the presence of the reinforcement, which has a much greater tensile strength than that of the concrete.

If the magnitude of the concentrated load increases, the crack at midspan will propagate upward in a stable manner and additional cracks will form at other locations along the span where the bending stress exceeds the tensile strength of the concrete (see Fig. 1.2). Assuming that the beam has sufficient shear strength, this process continues until the concrete crushes in compression or until the reinforcement fails in tension. It is shown in subsequent chapters of this book that it is desirable to have the reinforcement fail in tension before the concrete fails in compression.

What is important to remember from this discussion is that reinforcement, which has a tensile strength much greater than that of concrete, is used to counteract tensile

FIGURE 1.2
Response of a simply supported, reinforced concrete beam due to external loading.

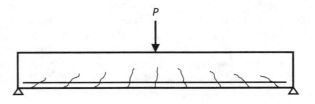

stresses in a reinforced concrete member, and that the reinforcement becomes effective in resisting tension only after cracking occurs. One of the major tasks in designing reinforced concrete members is to determine the required amount and location of reinforcement.

The focus of this book is on the design of reinforced concrete members in building structures where the primary reinforcement is steel reinforcing bars or steel wire reinforcement that is not prestressed. Such reinforcement is commonly referred to as *mild reinforcement* or *nonprestressed reinforcement*; a *nonprestressed concrete member* is a reinforced concrete member that contains this type of reinforcement. A discussion on the material properties of concrete and reinforcing steel is given in Chap. 2.

1.2 Building Codes and Standards

In the United States and throughout the world, the design and construction of building structures is regulated by building codes. The main purpose of a building code is to protect public health, safety, and welfare. Building code provisions are founded on principles that do not unnecessarily increase construction costs; do not restrict the use of new materials, products, or methods of construction; and do not give preferential treatment to particular types or classes of materials, products, or methods of construction.

Many cities, counties, and states in the United States and some international jurisdictions have adopted the *International Building Code* (IBC) for the design and construction of building structures. The provisions of the 2009 edition of the IBC are covered in this book.[1] Chapter 16 of the IBC prescribes minimum nominal loads that must be used in the design of any structure. Chapter 3 contains a summary of these loads as they pertain to the design of reinforced concrete buildings.

Section 1901.2 of the IBC requires that structural concrete be designed and constructed in accordance with the provisions of Chap. 19 of the IBC and the 2008 edition of *Building Code Requirements for Structural Concrete (ACI 318–08) and Commentary.*[2] ACI 318–08 is one of a number of codes and standards that is referenced by the IBC. These documents, which can be found in Chap. 35 of the 2009 IBC, are considered part of the requirements of the IBC to the prescribed extent of each reference (see Section 101.4 of the 2009 IBC). Amendments to ACI 318–08 are given in IBC Section 1908.

Even though it is an American Concrete Institute (ACI) standard, ACI 318 is commonly referred to as the "ACI Code" or the "Code." The ACI Code provides minimum requirements for the design and construction of structural concrete members (see Section 1.1 of that document). The term "structural concrete" refers to all plain and reinforced concrete members used for structural purposes. Section 1.1 also identifies the types of concrete members that are not addressed in the Code and includes general provisions for earthquake resistance.

Throughout this book, section numbers from the 2009 IBC are referenced as illustrated by the following: Section 1901.2 is denoted as IBC 1901.2. Similarly, Section 10.2 of the ACI Code is referenced as ACI 10.2 and Section R10.2 of the Commentary is referenced as ACI R10.2.

It is important to acquire the building code of the local jurisdiction at the onset of any project. Local building authorities may have amended the IBC or other adopted codes, and it is the responsibility of the registered design professional to be aware of such amendments before designing the building.

1.3 Strength and Serviceability

Design philosophies related to reinforced concrete members have changed over the years. Until the early 1960s, the primary design method for reinforced concrete was *working stress design*. In this method, members are proportioned so that the maximum elastic stresses due to service loads are less than or equal to allowable stresses prescribed in the Code.

The *strength design method* was included for the first time in the 1956 edition of the Code, and it became the preferred design method in the 1971 Code. The strength design method requires that both strength and serviceability requirements be satisfied in the design of any reinforced concrete member. In general, reinforced concrete members are proportioned to resist factored load effects and to satisfy requirements for deflection and cracking.

An in-depth discussion on the fundamental requirements of strength and serviceability is given in Chap. 4. Presented are the basic concepts of required strength (including load combinations) and design strength (including strength reduction factors).

Chapter 5 contains the general principles of the strength design method. This method is based on the fundamental conditions of static equilibrium and compatibility of strains. The information presented in this chapter forms the basis for the design of reinforced concrete sections subjected to flexure, axial load, or a combination of both.

1.4 Reinforced Concrete Members in Building Structures

1.4.1 Overview

Structural members in any structure must be designed to safely and economically support the weight of the structure and to resist all of the loads superimposed on the structure. In ordinary buildings, superimposed loads typically consist of live loads due to the inhabitants, dead loads due to items permanently attached to the building, and lateral loads due to wind or earthquakes. Some types of buildings must also be designed for extraordinary loads such as explosions or vehicular impact. Chapter 3 provides a comprehensive discussion on loads.

A typical reinforced concrete building is made up of a variety of different reinforced concrete members. The members work together to support the applicable loads, which are transferred through load paths in the structure to the foundation members. The loads are ultimately supported by the soil or rock adjoining the foundations.

Unlike other typical types of construction commonly used in building structures, such as structural steel and timber, reinforced concrete construction possesses inherent continuity. Cast-in-place reinforced concrete structures are essentially monolithic with reinforcement that extends into adjoining members. As such, reinforced concrete members are analyzed as continuous members in a statically indeterminate structure where bending moments, shear forces, and axial forces are transferred through the joints. Understanding the behavior and response of a reinforced concrete structure is imperative in the proper analysis, design, and detailing of the members in the structure.

Chapters 6 through 10 of this book contain the design and detailing requirements for typical reinforced concrete members found in building structures. A summary of the different member types that are addressed in these chapters is given in the following sections.

It is important to note that the information presented in Chaps. 6 through 10 is applicable to the design of members in structures that are located in areas of low-to-moderate seismic risk. Seismic risk is related to seismic design category (SDC), which is defined in IBC 1613.5.6. In general, SDC is determined on the basis of the level of seismicity and soil type at the site and on the occupancy of the building. Buildings assigned to SDC A and B are located in areas of low seismic risk, whereas buildings assigned to SDC C are located in areas of moderate seismic risk. SDC D, E, and F are assigned to buildings located in areas of high seismic risk.

The provisions in ACI Chap. 21 relate design and detailing requirements to the type of structural member and the SDC. The provisions of ACI Chaps. 1 through 19 and 22 are considered to be adequate for structures assigned to SDC A; no additional requirements need to be satisfied (also see IBC 1908.1.2). ACI Table R21.1.1 gives the sections of ACI Chap. 21 that need to be satisfied as a function of SDC.

Table 1.1 contains a summary of the reinforced concrete members addressed in Chaps. 6 through 10 of this book. Included in the table is the applicability of the information presented in these chapters related to SDC. For example, the design and detailing requirements presented in Chap. 6 are applicable to beams and one-way slabs in buildings assigned to SDC A.

The information presented in Chaps. 6 through 10 is not as limited as it first might appear. In fact, this information forms the basis of design regardless of SDC. For example, the determination of the nominal moment strength of a beam is required in the design of that member no matter what the SDC is for the building. The same is true in regards to the design strength interaction diagram for a column as well as for other important items.

It is also important to point out that satisfying the requirements for SDC B is readily achievable for beams and columns without any special design or detailing. In particular, beams in structures assigned to SDC B must have at least two bars that are continuous at both the top and bottom of the section that must be developed at the face of the supports (ACI 21.2.2). This is usually satisfied in typical beams because these bars are needed over the full length to support the stirrups. In regards to columns, column dimensions in typical buildings are such that the clear height of the column is greater than five times the cross-sectional dimension of the column in the direction of analysis; thus, in many cases, the special shear requirements of ACI 21.3.3 need not be satisfied (ACI 21.2.3).

Chapter	Reinforced Concrete Member(s)	Seismic Design Category (SDC)
6	Beams and one-way slabs	A
7	Two-way slabs	A, B
8	Columns	A
9	Walls	A, B, C
10	Foundations	A, B, C

TABLE 1.1 Applicability of Design and Detailing Requirements in Chaps. 6 through 10

Methods to determine the SDC and comprehensive design and detailing procedures for reinforced concrete members in all SDCs can be found in Ref. 3.

1.4.2   Floor and Roof Systems

Overview

Reinforced concrete structural systems can be formed into virtually any geometry to meet any requirement. Regardless of the geometry, standardized floor and roof systems are available that provide cost-effective solutions in typical situations. The most common types are classified as one-way systems and two-way systems. Examined later are the structural members that make up these types of systems.

It is common for one type of floor or roof system to be specified on one entire level of building; this is primarily done for cost savings. However, there may be cases that warrant a change in framing system. The feasibility of using more than one type of floor or roof system at any given level needs to be investigated carefully.

One-Way Systems

A one-way reinforced concrete floor or roof system consists of members that have the main flexural reinforcement running in one direction. In other words, reactions from supported loads are transferred primarily in one direction. Because they are primarily subjected to the effects from bending (and the accompanying shear), members in one-way systems are commonly referred to as *flexural members*.

Members in a one-way system are usually horizontal but can be provided at a slope if needed. Sloped members are commonly used at the roof level to accommodate drainage requirements.

Illustrated in Fig. 1.3 is a *one-way slab system*. The load that is supported by the slabs is transferred to the beams that span perpendicular to the slabs. The beams, in turn, transfer the loads to the girders, and the girders transfer the loads to the columns. Individual spread footings may carry the column loads to the soil below. It is evident that load transfer between the members of this system occurs in one direction.

Main flexural reinforcement for the one-way slabs is placed in the direction parallel to load transfer, which is the short direction. Similarly, the main flexural reinforcement for the beams and girders is placed parallel to the length of these members. Concrete for the slabs, beams, and girders is cast at the same time after the forms have been set and

FIGURE 1.3 One-way slab system.

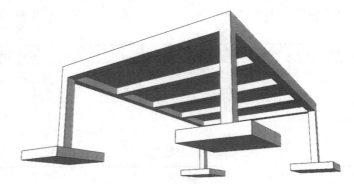

FIGURE 1.4
Standard one-way
joist system.

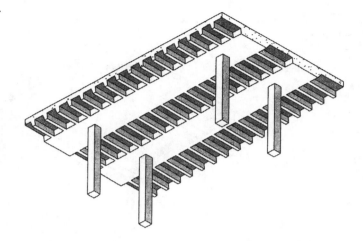

the reinforcement has been placed in the formwork. This concrete is also integrated with columns. In addition, reinforcing bars are extended into adjoining members. Like all cast-in-place systems, this clearly illustrates the monolithic nature of reinforced concrete structural members.

A *standard one-way joist system* is depicted in Fig. 1.4. The one-way slab transfers the load to the joists, which transfer the loads to the column-line beams (or, girders). This system utilizes standard forms where the clear spacing between the ribs is 30 in. or less. Because of its relatively heavy weight and associated costs, this system is not used as often as it was in the past.

Similar to the standard one-way joist system is the *wide-module joist system* shown in Fig. 1.5. The clear spacing of the ribs is typically 53 or 66 in., which, according to the Code, technically makes these members beams instead of joists. Load transfer follows the same path as that of the standard joist system.

FIGURE 1.5 Wide-
module joist
system.

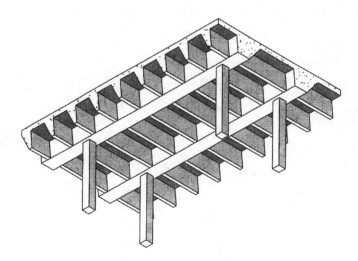

Figure 1.6 Two-way beam supported slab system.

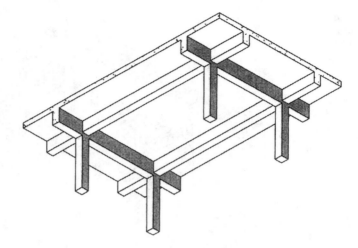

Reinforced concrete stairs are needed as a means of egress in buildings regardless of the number of elevators that are provided. Many different types of stairs are available, and the type of stair utilized generally depends on architectural requirements. Stair systems are typically designed as one-way systems.

Design and detailing requirements for one-way systems (one-way slabs and beams) are given in Chap. 6.

Two-Way Systems

As the name suggests, two-way floor and roof systems transfer the supported loads in two directions. Flexural reinforcement must be provided in both directions.

A *two-way beam supported slab system* is illustrated in Fig. 1.6. The slab transfers the load in two orthogonal directions to the column-line beams, which, in turn, transfer the loads to the columns. Like a standard one-way joist system, this system is not utilized as often as it once was because of cost.

A *flat plate system* is shown in Fig. 1.7. This popular system, which is frequently used in residential buildings, consists of a slab supported by columns. The formwork that is

Figure 1.7 Flat plate system.

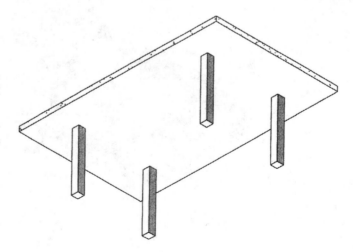

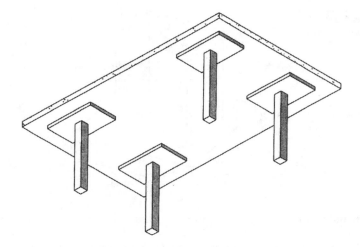

required is the simplest of all floor and roof systems. Because the underside of the slab is flat, it is commonly used as the ceiling of the space below; this results in significant cost savings.

Similar to the flat plate system is the *flat slab system* (Fig. 1.8). Drop panels are provided around the columns to increase moment and shear capacity of the slab. They also help to decrease slab deflection. Column capitals or brackets are sometimes provided at the top of columns.

The two-way system depicted in Fig. 1.9 is referred to as a *two-way joist system* or a *waffle slab system*. This system consists of rows of concrete joists at right angles to each other, which are formed by standard metal domes. Solid concrete heads are provided at the columns for shear strength. Such systems provide a viable solution in cases where heavy loads need to be supported on long spans.

Design and detailing requirements for two-way systems are given in Chap. 7.

1.4.3   Columns

A *column* is a structural member in a building that supports axial loads from the roof and floor members and that transfers the loads to the foundation. Load transfer to columns was illustrated in the previous section for both one-way and two-way systems. Columns

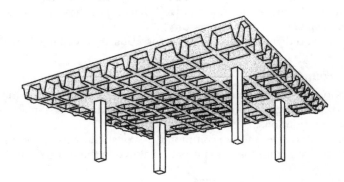

FIGURE 1.10 Walls used to resist the effects from lateral loads.

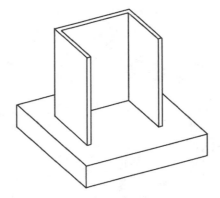

are usually oriented vertically in a building, but any orientation can be provided if needed.

In addition to axial loads, columns may be subjected to bending moments caused by gravity loads or by lateral loads. In general, columns that are part of the structural system that resists lateral loads (i.e., the lateral-force-resisting system) are typically subjected to axial loads, bending moments, and shear forces due to gravity and lateral loads. As such, columns are also referred to as *members subjected to combined axial load and bending*.

Design and detailing requirements for columns are given in Chap. 8.

1.4.4   Walls

In general terms, a *wall* is a member, usually vertical, that is used to enclose or separate spaces in a building or structure. Walls are usually categorized as non–load-bearing and load-bearing: A non–load-bearing wall supports primarily its own weight, whereas a load-bearing wall supports loads from the floor and roof systems. Like columns, load-bearing walls are typically designed for the effects due to axial loads and bending moments, and are referred to as *members subjected to combined axial load and bending*.

Illustrated in Fig. 1.10 are walls that are provided around elevator and stair openings in the core of a building. In addition to supporting tributary gravity loads, they are used alone or in combination with moment frames to resist the effects from wind and earthquakes. Such walls are commonly referred to as *shear walls*.

Basement walls or *foundation walls* resist the effects from gravity loads plus lateral earth pressure that acts perpendicular to the plane of the wall. Illustrated in Fig. 1.11 is a reinforced concrete foundation wall that resists the axial loads from a reinforced concrete wall and lateral soil pressure.

Design and detailing requirements for walls are given in Chap. 9.

1.4.5   Foundations

Foundation systems transfer the loads from the structure above to the soil or rock below the building. There are primarily two types of foundation systems: shallow foundations and deep foundations.

Footings and *mats* are two common types of shallow foundations. A spread footing spreads the load from the superstructure above to the soil so that the stress in the soil

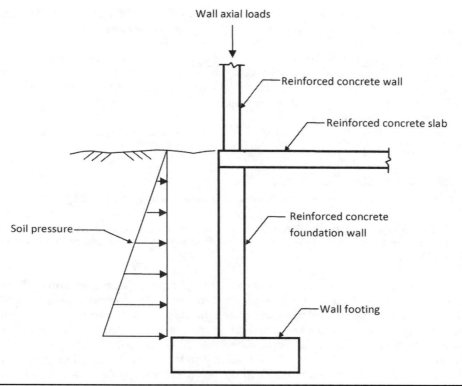

FIGURE 1.11 Reinforced concrete foundation wall.

is less than its allowable bearing capacity. Illustrated in Fig. 1.3 is an isolated spread footing that is supporting a column in a one-way system. A mat foundation is a large concrete slab that supports some or all of the columns and walls in a building. Shown in Fig. 1.10 is a mat foundation beneath the core walls. The loads from the supported members are transferred to the soil, and the mat is designed so that the maximum soil pressure does not exceed the allowable bearing capacity of the soil.

Piles and *drilled piers* are deep foundations that are frequently used to support columns and walls in building structures. Both types of foundation members extend below the strata of poor soil to a level where the soil is adequate to support the loads from the structure above.

Design and detailing requirements for both shallow and deep foundations are given in Chap. 10.

1.5 Drawings and Specifications

1.5.1 Overview
The design and construction of a reinforced concrete building requires input from a variety of design professionals. Structural engineers are responsible for producing structural

drawings and specifications that are used to eventually build the structure. Drawings and specifications, as well as other documents, are referred to as *construction documents* or *contract documents*.

Once the construction documents have been reviewed and approved by the local building authority, a number of important processes are set in motion. One of the first things to move forward is the production of the reinforcing steel *placing drawings* by the reinforcing steel supplier. As the name suggests, these drawings are used in the actual construction of the structure.

Additional information on construction and placing drawings, as well as other pertinent information, is covered in the following sections.

1.5.2   Construction Documents

The following definition of construction documents is given in IBC 202:

> Written, graphic, and pictorial documents prepared or assembled for describing the design, location, and physical characteristics of the elements of a project necessary for obtaining a building permit.

Construction documents consist of calculations, drawings, specifications, and any other data that are needed to indicate compliance with the governing building code. IBC 107 describes the information that must be included in the construction documents, who must prepare them (the registered design professional), and procedures that are used by the building official for approving them.

IBC 1603 and ACI 1.2 contain minimum requirements for construction documents. The following design loads and information must be included in the construction documents (see IBC 1603):

- Floor live load
- Roof live load
- Roof snow load
- Wind design data
- Earthquake design data
- Geotechnical information
- Flood design data
- Special loads

These items are typically listed on the General Notes sheet of the structural drawings. Chapter 3 of this book provides a summary of loads that are typically required in the design of building structures.

For reinforced concrete structures, the following information must also be provided in the construction documents (see ACI 1.2.1):

- Specified compressive strength of all concrete mixtures utilized in the structure at the ages or stages of construction for which each part of the structure is designed.

- Specified strength or grade of all reinforcement utilized in the structure.
- Size and location of all structural members, reinforcement, and anchors.
- Provisions for dimensional changes resulting from creep, shrinkage, and temperature.
- Anchorage length of reinforcement and location and length of lap splices.
- Type and location of mechanical and welded splices of reinforcement.

Structural drawings must show the size, section, and relative locations of all of structural members in a building. The following items are usually included in a typical set of drawings:

- Foundation plans
- Framing plans for all levels at and above ground
- Schedules for the structural members, including foundations, beams, slabs, columns, and walls
- Sections and details

It is important that the structural and architectural drawings be coordinated on a regular basis.

The size of the structural members can be given directly on the plans, or the members can be identified by marks on the plan with the sizes given in applicable schedules (the latter is typically done for beams, columns, walls, and foundations). The same is done for the size, spacing, and length of reinforcing bars. Either method of identification (or both) can be utilized in a project.

Typical details are provided for various types of members utilized in the structure. These details, along with sections cut at various locations in the structure, help in illustrating specific information about the structure.

Specifications are documents that supplement the structural drawings and provide additional information on materials, methods of construction, and quality assurance. Specifications for structural concrete are given in ACI 301.[4] This specification may be referenced or incorporated in its entirety in the construction documents of any reinforced concrete building project together with additional requirements for the specific project. Included in ACI 301 is information on the following:

- Formwork and formwork accessories
- Reinforcement and reinforcement supports
- Concrete mixtures
- Handling, placing, and constructing
- Architectural concrete
- Lightweight concrete
- Mass concrete
- Prestressed concrete
- Shrinkage-compensating concrete

Mandatory and optional requirements checklists are also provided in ACI 301. Although these checklists do not form a part of ACI 301, they assist in selecting and specifying project requirements in the project specifications. The mandatory requirements checklist includes requirements pertaining to specific qualities, procedures, materials, and performance criteria that are not specifically defined in ACI 301. The optional requirements checklist contains a list of actions that are required or available when the specifications are being developed.

A number of master specifications are available that can be utilized in a reinforced concrete building project. One such specification is MasterSpec.[5] Section 033000 of that specification contains comprehensive specifications for cast-in-place concrete. Master specifications can be modified by deleting and inserting text to meet the specific requirements of a project.

1.5.3 Placing Drawings

Once the construction documents have been approved by the local building authority, the documents are used by the reinforcing steel detailer in the preparation of placing drawings and bar lists. Placing drawings are used by the ironworkers at the job site to place (or, install) the reinforcing steel in the formwork. Bar lists are used by the reinforcing steel fabricator to fabricate the reinforcing bars.

When preparing the placing drawings for a specific project, the detailer uses the structural drawings and the specifications to determine the quantity, lengths, bend types, and positioning of the reinforcing bars in all of the members in the structure. The registered design professional reviews and approves the placing drawings once they are complete. Additional information on placing drawings and many other important aspects related to reinforcing steel can be found in Ref. 6.

1.6 Construction of Reinforced Concrete Buildings

1.6.1 Overview

Although each reinforced concrete building is unique, the following sequence of events occurs in the construction of any cast-in-place concrete building with mild reinforcement:

1. Erect formwork
2. Place reinforcement
3. Place concrete
4. Strip forms and provide reshores

This cycle is repeated for each floor of the building. Numerous activities occur within each segment of construction.

It is safe to state that no structure can be built that is perfectly level, plumb, straight, and true. This does not imply that contractors are doing their jobs improperly; rather, it is simply a reality that must be accepted because of the inherent nature of construction.

Fortunately, constructing a "perfect" structure is not necessary. However, some requirements must be established so that the actual structure performs as originally designed. Construction tolerances provide permissible variations in dimensions and locations of the members in a structure. Tolerances are essentially limits within which the work is to be performed. ACI 117 contains comprehensive specifications for tolerances in reinforced concrete construction and materials.[7] This document can be referenced or used in its entirety in the project specifications.

Details of the construction process are covered in the following sections. Included is information that can be incorporated in the preliminary design stages to help in achieving an economical structure.

1.6.2   Formwork Installation

Overview

According to ACI 347,[8] formwork is the total system of support for freshly placed concrete, including the mold or sheathing that contacts the concrete and all supporting members, hardware, and necessary bracing. In essence, formwork is a temporary structure whose main purpose is to support and contain fresh concrete until it can support itself. Concrete buildings require formwork for vertical members (columns and walls) and horizontal members (slabs, beams, and joists). In addition to the weight of fresh concrete, formwork must be designed to support construction loads (workers, material, and equipment) and to resist the effects from wind.

The cost of formwork usually accounts for approximately 50% to 60% of the total cost of the concrete frame. Thus, selecting the proper forming system is crucial to the success of any project. Specifying standard form sizes, repeating the size and shape of concrete members wherever possible, and striving for simple formwork are mandatory in achieving a cost-effective structure.

The type of formwork system to be used is dictated primarily by the structural system that will be utilized in the building. Formwork materials are shipped to the job site and erected. The process of erecting the formwork includes the following:

1. Lifting, positioning, and assembling the various formwork elements.
2. Installing shoring to support the formwork, the weight of the fresh concrete, and the construction loads.

To help ensure that the concrete does not bond to the forms, a form release agent or coating may be applied to the inside of the formwork at this stage. The coating also helps prevent wood formwork from absorbing water from the concrete mixture.

The following is a typical construction sequence in a conventional cast-in-place concrete building:

1. Build formwork for vertical elements
2. Place reinforcement for vertical elements
3. Place concrete for vertical elements
4. Strip formwork for vertical elements
5. Build formwork for horizontal elements
6. Place reinforcement for horizontal elements
7. Place concrete for horizontal elements

The forms for the horizontal elements are stripped after a sufficient amount of time, and the process is repeated in multistory construction. Several levels of reshores are typically required below the newly constructed level to support the weight of the fresh concrete and the construction loads.

Brief descriptions of various types of vertical and horizontal forming systems that are commonly used in reinforced concrete building construction follow. Additional information on these and other types of forming systems can be found in Ref. 9.

Vertical Forming Systems

As noted earlier, the construction of vertical members in the structure (columns and walls) precedes the construction of horizontal members. The following systems are commonly used to form vertical members in a concrete building:

1. Conventional column/wall system
2. Ganged system
3. Jump forms
4. Slipforms
5. Self-raising forms

Slipform and self-raising formwork are classified as crane-independent systems where formwork panels are moved vertically by mainly proprietary mechanisms. Here, we describe the first three systems.

Conventional Column/Wall System   A conventional column forming system consists of plywood sheathing that is nailed together and stiffened by vertical studs. The sides of the forms are held together by clamps that help prevent buckling of the sheathing due to the horizontal pressure imposed by the fresh concrete. Prior to construction of the formwork, a template is made on the floor slab or foundation to accurately locate the position of the column. Round columns are typically formed by steel forms.

In a conventional wall forming system, studs and wales support plywood sheathing that form the wall. Tie rods resist the pressure exerted by the fresh concrete and help maintain the specified thickness of the wall. Wood spreaders can also be used to maintain wall thickness.

Ganged Systems   Ganged systems are large wall form units that consist of aluminum or plywood panels joined together and braced by aluminum or steel frames. Once the system has been assembled on the ground, it is raised into place by a crane. Ganged formwork produces smoother concrete walls that have fewer joints than those constructed with conventional systems.

Jump Forms   Jump forms consist of an upper-framed panel form that is used to form the concrete in a wall member and a supporting structure that is attached to the completed wall below, which carries the entire assembly. Once the concrete in the upper-framed panel form gains sufficient strength, the jump forms are lifted to the next level and the process is repeated.

Horizontal Framing Systems

Wood System   A conventional wood system consists of lumber and/or plywood and is used to form slabs, beams, and foundations. Wood shoring is typically set on wood mudsills. The pieces of this system are made and erected on site.

Metal System   A conventional metal system consists of aluminum joists and stringers that support plywood sheathing. Aluminum or steel scaffolding is commonly used for shoring. Similar to the wood system, this system is also built on site.

Joist-Slab and Dome Forming System   One-way and two-way joist systems are formed by pan forms and dome forms, respectively. These forms come in standard sizes and are made of steel and fiberglass. The forms are nailed to plywood sheathing and are supported by wood or metal shoring.

Flying Form System   A flying form system is a crane-set system that is constructed and assembled as one unit and moved from floor to floor. Flying forms typically consist of sheathing panels that are supported on aluminum joists. These elements, in turn, are supported by steel or aluminum trusses, which have telescoping legs that are used as shoring.

As the name implies, the formwork assembly is flown to the next level by a crane once the concrete on that level has cured and attained sufficient strength.

1.6.3   Reinforcement Installation

Once the formwork has been erected, the required reinforcement is installed in the columns and walls using the placing drawings. For columns, the reinforcement consists of longitudinal bars and transverse reinforcement in the form of either ties or spirals. One or two layers of vertical and horizontal reinforcing bars are provided in walls; in some cases, horizontal ties are required as well. In general, reinforcement is tied together with metal wires into what are commonly referred to as reinforcement cages.

The reinforcement in all of the beams, slabs, and other horizontal structural members is placed and supported in the formwork according to the placing drawings. Inserts for mechanical and electrical equipment and openings for ducts and conduits are some of the typical elements that must be positioned at their proper locations in the formwork as well. Beams require longitudinal and transverse reinforcement, whereas slabs typically require only longitudinal reinforcement at the top and bottom of the section in one or two directions.

Additional information on reinforcement, including recommended practices for placing reinforcing bars, can be found in Ref. 6.

1.6.4   Concrete Placement

Concrete is deposited into the forms after the reinforcement and other construction items, such as ducts and conduits, have been installed at the proper locations. Prior to placement, the concrete is mixed and transported to the job site. Depending on a number of factors, on-site ready-mix plants are sometimes used.

Belt conveyors, buckets and cranes, chutes, drop chutes, and pumping are the most popular methods of transporting concrete to the point where it is needed in the structure. Project size and site constraints are just two of the many factors that dictate which method is the most effective in a particular project.

Once placed, the concrete is consolidated by hand or mechanical vibrators to ensure that the fresh concrete is properly compacted within the forms and around the reinforcement and other embedded items. Proper consolidation also helps in eliminating honeycombs and entrapped air in the mix.

Finishing the exposed concrete surfaces occurs shortly after consolidation. Many attributes can be achieved at this stage, including desired appearance, texture, or wearing qualities.

It is very important to ensure that all newly placed and finished concrete be cured and protected from rapid drying, extreme changes in temperature, and damage from future construction activities. Curing should begin as soon as possible after finishing so that hydration of the cement and strength gain of the concrete continues. Columns and walls are usually cured after the forms are stripped, whereas slabs and beams are cured before and after their formwork is stripped.

Comprehensive information on batching, mixing, transporting, and handling concrete can be found in Ref. 10.

1.6.5 Formwork Removal

Formwork is typically removed (stripped) after the concrete has gained sufficient strength to carry its own weight plus any construction loads. Generally, the formwork is not stripped before the concrete has reached at least 70% of its design compressive strength. Various admixtures are available to accelerate strength development of concrete at an early age so that forms can be stripped sooner (see Chap. 2).

Temporary vertical support is required for the stripped concrete members that have not yet acquired their design strength. Reshores and backshores are two types of shoring that is provided beneath horizontal concrete members after the forms and original shoring has been removed.

Reshores are spaced relatively far apart; this allows the horizontal members to deflect, permitting the forms and original shores to be removed from a large area. Reduced stripping costs are usually realized when using reshores.

Backshores are spaced closer than reshores. The horizontal members are not allowed to deflect, resulting in a small area over which the forms and original shoring can be removed. This permits stripping to occur sooner than if reshores were used.

Reshores and backshores are removed after the structural members have acquired sufficient strength to support all of the required loads. More information on shoring and reshoring of concrete buildings can be found in Ref. 11.

References

1. International Code Council (ICC). 2009. *International Building Code*. ICC, Washington, DC.
2. American Concrete Institute (ACI), Committee 318. 2008. *Building Code Requirements for Structural Concrete and Commentary*, ACI 318–08. ACI, Farmington Hills, MI.
3. Fanella, D. A. 2009. *Design of Low-Rise Reinforced Concrete Buildings Based on 2009 IBC, ASCE/SEI 7–05, ACI 318–08*. International Code Council, Washington, DC.
4. American Concrete Institute (ACI), Committee 301. 2005. *Specifications for Reinforced Concrete*, ACI 301–05. ACI, Farmington Hills, MI.
5. American Institute of Architects (AIA). 2009. *MasterSpec*. Architectural Computer Services, Inc., Salt Lake City, UT.

6. Concrete Reinforcing Steel Institute (CRSI), CRSI Committee on Manual of Standard Practice. 2009. *Manual of Standard Practice*, 28th ed. CRSI, Schaumburg, IL.
7. American Concrete Institute (ACI), Committee 117. 2006. *Specifications for Tolerances for Concrete Construction and Materials and Commentary*, ACI 117–06. ACI, Farmington Hills, MI.
8. American Concrete Institute (ACI), Committee 347. 2004. *Guide to Formwork for Concrete*, ACI 347–04. ACI, Farmington Hills, MI.
9. Hurd, M. K. 2005. *Formwork for Concrete*, SP-4, 7th ed. American Concrete Institute, Farmington Hills, MI.
10. Kosmatka, S., Kerkhoff, B., and Panarese, W. 2002 (rev. 2008). *Design and Control of Concrete Mixtures*, 14th ed. Portland Cement Association (PCA), Skokie, IL.
11. American Concrete Institute (ACI), Committee 347. 2005. *Guide for Shoring/Reshoring of Concrete Multistory Buildings*, ACI 347.2R-05. ACI, Farmington Hills, MI.

CHAPTER 2

Materials

2.1 Introduction

In order to fully comprehend the mechanics of reinforced concrete, an understanding of the material properties of concrete and reinforcing steel is essential. This chapter contains a basic overview of the properties of the constituent parts and how these properties relate to the design and detailing of reinforced concrete members.

In addition to discussing the mechanical properties of concrete, this chapter covers the ACI Code methodologies for proportioning concrete mixtures and how they relate to compressive strength. Also included are the durability requirements and the evaluation and acceptance requirements in Chaps. 4 and 5 of the Code, respectively.

The mechanical properties of both deformed reinforcing bars and welded wire reinforcement are discussed in detail. Information on the various types, sizes, and grades of reinforcement are also provided.

2.2 Concrete

2.2.1 Components of Concrete

The basic components of concrete—cement, water, and aggregates (sand and gravel)—are shown in Fig. 2.1. Cement and water form a paste that fills the space between the aggregates and binds them together. Chapter 3 of the ACI Code contains the minimum requirements for these components and other materials that are commonly used in concrete. Included are references to standards developed by ASTM International.

ASTM International, which was formerly known as the American Society for Testing and Materials (ASTM), oversees the development of technical standards for materials, products, systems, and services. In general, these standards are documents that have been developed and established within the consensus principles of ASTM International and that meet the requirements of its procedures and regulations.

ASTM standards for cementitious materials, aggregates, water, and admixtures are specified in ACI 3.2, 3.3, 3.4, and 3.6, respectively, and are summarized in Table 2.1. For example, portland cement must conform to ASTM C150-05, *Standard Specification for Portland Cement*. Note that "C150" is the serial designation of the standard, and "05" refers to 2005, which in this case is the year that the standard was last revised (otherwise, it is the year of original adoption).

FIGURE 2.1
Concrete
components:
cement, water, fine
aggregate, and
coarse aggregate.
(*Courtesy of the
Portland Cement
Association.*)

FIGURE 2.1
Concrete
components:
cement, water, fine
aggregate, and
coarse aggregate.
(*Courtesy of the
Portland Cement
Association.*)

Cementitious Materials

The eight different types of portland cement referenced in ASTM C150 and their typical applications are summarized in Table 2.2. Type I cement is suitable for use in all types of reinforced concrete structures that do not require the properties of the other cement types. However, Type II cement is sometimes used instead of Type I because of its increased availability, regardless of the need for sulfate resistance or moderate heat generation. Some portland cements may be labeled with more than one type designation. For example, a designation of Type I/II means that the requirements of both Type I and Type II cements have been met. Additional information on cement types can be found in Refs. 1 and 2.

Supplementary cementitious materials such as fly ash, ground-granulated blast-furnace slag, and silica fume are generally added to a concrete mix to enhance one or more properties of the hardened concrete (see Fig. 2.2). Depending on the properties of the materials and the desired effect on the concrete, such materials may be used in addition to or as a partial replacement of cement. Calcined shale, calcined clay, or metakaolin are examples of natural pozzolans that, when added to a concrete mix, contribute to the properties of hardened concrete. In addition to the beneficial effects they have on concrete properties, supplementary cementitious materials are recognized for the potential positive effects they have on energy conservation and the environment.

Aggregates

Fine and coarse aggregates, which typically occupy 60% to 70% of the concrete volume, have a strong influence on the properties of concrete. Fine aggregates usually consist of sand or crushed stone and have diameters smaller than approximately 0.2 in. Coarse aggregates typically have diameters ranging between 0.375 and 1.5 in and consist of gravels, crushed stone, or a combination thereof.

Normal-weight concrete is concrete made with sand, gravel, and crushed stone that conforms to ASTM C33. The density or unit weight of normal-weight concrete is typically between 135 and 160 pcf and is normally taken as 145 or 150 pcf. Expanded shale, clay, and slate are common aggregates used in the production of lightweight concrete, which has a density of 90 to 115 pcf. Sand-lightweight concrete contains fine

Component	ASTM Standard
Cementitious Materials	
Portland cement	ASTM C150-05, *Standard Specification for Portland Cement*
Blended hydraulic cements (excluding Type IS)	ASTM C595-07, *Standard Specification for Blended Hydraulic Cements*
Expansive hydraulic cement	ASTM C845-04, *Standard Specification for Expansive Hydraulic Cement*
Hydraulic cement	ASTM C1157-03, *Standard Performance Specification for Hydraulic Cement*
Fly ash and natural pozzolans	ASTM C618-05, *Standard Specification for Coal Fly Ash and Raw or Calcinated Natural Pozzolan for Use in Concrete*
Ground-granulated blast-furnace slag	ASTM C989-06, *Standard Specification for Ground Granulated Blast-furnace Slag for Use in Concrete and Mortars*
Silica fume	ASTM C1240-05, *Standard Specification for Silica Fume Used in Cementitious Mixtures*
Aggregates	
Normal-weight	ASTM C33-03, *Standard Specification for Concrete Aggregates*
Lightweight	ASTM C330-05, *Standard Specification For Lightweight Aggregates for Structural Concrete*
Water	ASTM C1602/C1602M-06, *Standard Specification for Mixing Water Used in the Production of Hydraulic Cement Concrete*
Admixtures	
Water reduction and setting	ASTM C494/C494M-05a, *Standard Specification for Chemical Admixtures for Concrete*
Flowing concrete	ASTM C1017/C1017M-03, *Standard Specification for Chemical Admixtures for Use in Producing Flowing Concrete*
Air entrainment	ASTM C260-06, *Standard Specification for Air-entraining Admixtures for Concrete*

TABLE 2.1 Summary of ASTM Standards for Concrete Components

aggregates that conform to ASTM C33 and lightweight aggregates that conform to ASTM C330.

Water

In general, water that is drinkable can usually be used for making concrete. Acceptance criteria for water used as mixing water in concrete can be found in ASTM C94/C94M-06, *Standard Specification for Ready-mixed Concrete.*

Cement Type	Application
Type I—normal	General purpose cement commonly used in all types of structures
Type IA—normal, air-entraining	Used in the same structures as Type I where air entrainment is desired
Type II—moderate sulfate resistance	General purpose cement used in structures where protection against moderate sulfate attack is important or where moderate heat of hydration is desired
Type IIA—moderate sulfate resistance, air-entraining	Used in the same structures as Type II where air entrainment is desired
Type III—high early strength	Used in structures where high early strength of the concrete is desired or where structures must be put into service quickly
Type IIIA—high early strength, air-entraining	Used in the same structures as Type III where air entrainment is desired
Type IV—low heat of hydration	Used in structures where a low heat of hydration is desired, such as in massive concrete structures like dams
Type V—high sulfate resistance	Used in structures where high sulfate resistance is desired, such as elements in direct contact with soils or ground waters that have a high sulfate content

TABLE 2.2   Types of Portland Cement and Their Common Applications

Admixtures

Admixtures are ingredients other than cement, aggregates, and water that are added to a concrete mix immediately before or during mixing. Many different types of admixtures are commercially available, and they are typically classified by function.[1,2] Reducing the cost of concrete construction, economically achieving desired properties in concrete, and

FIGURE 2.2
Supplementary cementitious materials. From left to right: fly ash (Class C), metakaolin, silica fume, fly ash (Class F), slag, and calcinated shale. (*Courtesy of the Portland Cement Association.*)

maintaining the quality of concrete during mixing, transporting, placing, and curing are a few major reasons why admixtures are used in concrete. The licensed design professional must approve the use of admixtures that are not identified in ACI 3.6.1 and 3.6.2.

The following are brief descriptions of some common admixtures:

- *Air-entraining admixtures.* These admixtures purposely introduce microscopic air bubbles in concrete to improve its durability when exposed to repeated freeze–thaw cycles. They also increase resistance to scaling due to exposure to deicing chemicals and improve the workability of fresh concrete.

- *Superplasticizers.* These are high-range water reducers that can greatly reduce water demand and cement content without sacrificing workability. High-strength concrete is typically produced with superplasticizers. Using a water reducer can also lead to accelerated strength development of the concrete; this permits formwork to be removed earlier and, thus, overall construction time to be reduced.

- *Corrosion inhibitors.* These are usually used in parking structures, marine structures, and other structures exposed to chlorides, which can cause corrosion of steel reinforcement in concrete. These admixtures chemically arrest the corrosion reaction.

2.2.2   Mechanical Properties of Concrete

Overview

It has been established by numerous experimental investigations that internal microcracks exist in concrete prior to any external loading (see, e.g., the pioneering investigations reported in Refs. 3 through 6). These preexisting cracks, which are usually due to bleeding, shrinkage, and the heat of cement hydration, are typically located on the aggregate–cement paste interface. It is very important to acknowledge the role that preexisting microcracks have in the deterioration and failure of concrete. Experimental investigations have confirmed that the nucleation, growth, interaction, and coalescence of these flaws are the controlling mechanisms that cause macroscopic failure.[5,7]

The mechanical properties of concrete are usually obtained by testing concrete specimens in accordance with ASTM standards. Applicable ASTM standards are referenced in the following sections.

Compressive Strength

Mechanics of Concrete Deterioration and Failure in Compression   The compressive strength of concrete is one of the most important quantities needed in the design of reinforced concrete structural members. The mechanical behavior of concrete in compression is typically acquired by performing tests on concrete cylinders, as shown in Fig. 2.3.

When a concrete cylinder is tested in uniaxial compression in a testing machine that exerts a force at a moderate rate, vertical cracks will typically form at the midheight of the specimen parallel to the direction of the maximum compressive force, as shown in Fig. 2.4. Away from the midheight, the cracks tend to propagate at an angle to the compressive force. This can be attributed to the friction forces that are generated between the steel plates of the testing machine and the ends of the concrete cylinder because of the

FIGURE 2.3 Testing a 6 × 12 in cylinder in compression. (*Courtesy of the Portland Cement Association.*)

differences in the modulus of elasticity and Poisson's ratio of steel and concrete. These friction forces prevent lateral expansion of the cylinder ends and introduce a lateral confining pressure and shear stresses at those locations. Thus, triaxial stresses occur at the ends of the cylinder, and only a small portion of the cylinder near the midheight is actually loaded in pure axial compression.

FIGURE 2.4 Typical failure pattern for a concrete cylinder in uniaxial compression.

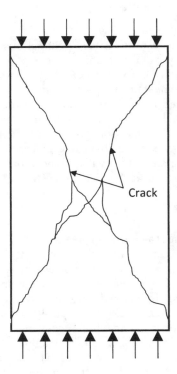

Crack

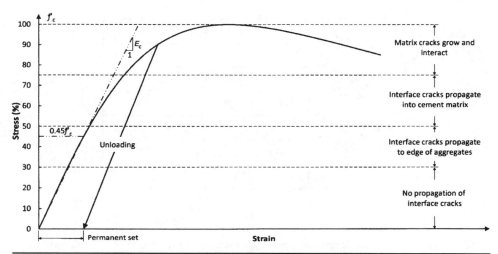

FIGURE 2.5 Generalized stress–strain curve for concrete in compression.

A generalized stress–strain curve for concrete in uniaxial compression is illustrated in Fig. 2.5. The relationship between stress and strain is relatively linear up to approximately 30% of the maximum stress. With an increase in compressive stress, preferentially oriented, preexisting microcracks at the aggregate–cement paste interface (commonly referred to as bond cracks) propagate instantaneously to the edge of the aggregate facet once the shear stress acting on the faces of the cracks reaches a critical value. The cracks are temporarily arrested at the edge of the aggregate because the toughness of the cement paste is greater than that of the interface.

An increasing number of interface cracks become destabilized and increase in length as the compressive stress increases further. This is reflected in a gradual departure from linearity in the stress–strain curve. Once the stress exceeds approximately 50% of the maximum stress, the preferential interface cracks acquire enough energy to overcome the cement paste barrier, and tension cracks form at their tips and extend into the cement paste. These tension cracks kink from the original sliding direction and eventually align themselves in the direction of the axial compressive stress.

The tension cracks in the cement paste propagate in a stable manner with increasing compressive stress. When the applied compressive stress is at approximately 75% of the maximum stress, the tension cracks begin to interact with one another and form a network of internal damage in the concrete. The stress–strain curve becomes even more nonlinear after this occurs.

Shortly after this stage, the network of cracking becomes unstable, and the load-carrying capacity of the uncracked portions of the concrete reaches a maximum value. This maximum stress is referred to as the compressive strength of the concrete and is designated by the notation f_c'.

After f_c' is attained, the concrete can resist only smaller stresses with increasing strains. If a stiff testing machine that can maintain a constant rate of strain is used, it is possible to acquire a descending branch of the stress–strain curve. This indicates that the concrete has not completely failed, even though the maximum stress has been

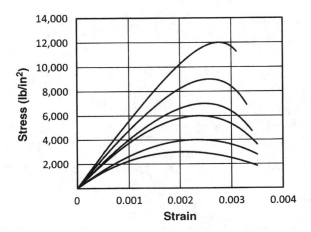

FIGURE 2.6
Stress–strain
curves for
compression tests
on concrete mixes
of varying strength.

attained. In summary, the deterioration of concrete in compression is gradual and does not occur suddenly because of the growth of a single crack.

When load is removed from concrete in the inelastic range, the recovery line is usually not parallel to the original load line (see Fig. 2.5). Thus, the permanent set in the concrete due to the compression load is typically different from the amount of inelastic deformation.

Stress–strain curves of concrete with varying compressive strengths are shown in Fig. 2.6. It is clear that lower-strength concretes have long and relatively flat peaks compared with higher-strength concretes that have sharper peaks. It is also evident that lower-strength concretes are less brittle and fracture at larger maximum strains than higher-strength concretes.

General Requirements   Because concrete is a composite material made of constituent materials whose properties vary, the strength of concrete will vary. It is common for compressive strength test results of cylinders cast from the same batch of concrete to differ, sometimes by relatively large amounts. Thus, the Code utilizes a probabilistic approach to ensure that adequate strength is developed in a structure. In particular, the concrete is to be proportioned so that an average compressive strength f'_{cr} that exceeds the specified compressive strength f'_c is provided. The details of this approach are outlined in the next section.

According to ACI 5.1, compressive strength is to be determined for a concrete mixture on the basis of compression tests of cylinders that have been molded and cured for a specified number of days in accordance with ASTM C31/C31M-06, *Standard Practice for Making and Curing Concrete Test Specimens in the Field*. Requirements for compression tests are given in ASTM C39/C39M-05, *Standard Test Method for Compressive Strength of Cylindrical Concrete Specimens*.

Cylinders that are 6 in in diameter and 12 in in length are commonly used in compression tests. ACI 5.6.3.2 also permits the use of 4 × 8 in cylinders. Regardless of the cylinder size, the procedures for obtaining representative samples of concrete to be used in compression tests are given in ASTM C172-04, *Standard Practice for Sampling Freshly Mixed Concrete*.

It is common practice to test concrete cylinders at 28 days, although it is permitted to specify a larger number of days where warranted (ACI 5.1.3). Consider, for example, a reinforced concrete column that is in the first story of a high-rise building. The specified compressive strength of the concrete for such columns will usually be greater than 6,000 psi (concrete mixtures with compressive strengths equal to or greater than 6,000 psi are generally considered to be high-strength concrete). Depending on the size of the building and the speed of construction, this column will not be fully loaded until a year or more after it has been cast. Thus, a compressive strength based on 56- or 90-day test results would typically be specified in a situation like this.

The specified compressive strength of the concrete f_c' must be indicated in the design drawings and specifications and must be used in the design calculations. The ACI Code requires that the specified compressive strength be equal to or greater than 2,500 psi for concrete structures designed and constructed in accordance with the provisions of the Code (ACI 1.1.1 and 5.1.1). There is no upper limit on the value of f_c' that can be specified except for the maximum values given in specific Code provisions, which will be discussed later.

Proportioning Concrete Mixtures and Required Average Compressive Strength   It is important to understand the methodologies in the Code for proportioning concrete mixtures and how they relate to compressive strength. In general, a concrete mixture must be proportioned to satisfy both the strength and durability requirements of a project. Chapter 5 of the Code describes procedures to produce concrete of adequate strength and prescribes the minimum criteria for mixing and placing concrete. Chapter 4 contains durability requirements for concrete mixtures on the basis of exposure categories. Provisions are included in that chapter for maximum water/cementitious materials ratios and minimum specified compressive strengths as a function of the exposure class.

In addition to strength and durability, the materials for concrete must be proportioned so that concrete is workable. Fresh concrete is generally considered to be workable when it can be easily placed and consolidated in formwork and around reinforcing bars and when it can be easily finished. Concrete should not segregate during transportation and handling, and excessive bleeding should not occur after it has been cast. Additional information on workability can be found in Ref. 2.

Two methods are given in the Code for establishing concrete mixture proportions to satisfy strength requirements. The first method, which is given in ACI 5.3, utilizes laboratory trial batches, compressive strength test records, or both to determine f_{cr}'. In the second method, no trial mixture data or test records that meet the requirements of ACI 5.3 are available, and the mixture proportions are based on other experience or information (ACI 5.4). Details for both methods are discussed next.

Method 1: Proportioning on the basis of field experience or trial mixtures, or both (ACI 5.3)   Three steps are given in ACI 5.3 for selecting a suitable concrete mixture that will satisfy strength requirements:

1. Determine the standard deviation of test records, if available.
2. Determine the required average compressive strength f_{cr}'.
3. Select mixture proportions that will produce an average strength equal to or greater than f_{cr}' and also meet the applicable requirements of ACI Chap. 4 for durability.

Step 1. When a concrete production facility has at least a single group of 30 consecutive strength test records that are not more than 12 months old, the standard deviation s_s can be determined by the requirements of ACI 5.3.1.1. The following equation can be used to determine s_s in this case:

$$s_s = \left[\frac{\sum (x_i - \bar{x})^2}{(n-1)} \right]^{1/2} \qquad (2.1)$$

In this equation, x_i is the result from an individual strength test and $\bar{x}$ is the average (mean) strength value of n strength tests. Note that one test record is the average strength of at least two 6×12 in cylinders or at least three 4×8 in cylinders (see ACI 5.6.2.4 and the discussion given later).

Where two groups of such tests are available, the standard deviation $\bar{s}_s$ is determined by the following equation, which is a statistical average value of standard deviation:

$$\bar{s}_s = \left[\frac{(n_1 - 1)(s_{s1})^2 + (n_2 - 1)(s_{s2})^2}{n_1 + n_2 - 2} \right]^{1/2} \qquad (2.2)$$

The quantities s_{s1} and s_{s2} are the standard deviations from groups 1 and 2, respectively, and the quantities n_1 and n_2 are the number of test results in each group.

If there are less than 30 but at least 15 test results available, the calculated sample standard deviation must be multiplied by the appropriate modification factor given in Table 5.3.1.2 of the Code, which is greater than 1.00. Larger values of s_s result in increased values of f'_{cr}. In essence, these modification factors provide protection against the possibility that the true standard deviation is underestimated because of the smaller number of test results. ACI Table 5.3.2.2 must be used to determine f'_{cr} where less than 15 test results are available.

The only test records that should be used in the calculation of s_s are those obtained from a concrete mixture that was produced with the same general types of components, under similar conditions, and within 1,000 psi of f'_c as the concrete mixture proposed in the project. These requirements are deemed necessary to ensure acceptable concrete. Obviously, test records of a concrete mixture with lightweight aggregate should not be used to calculate s_s for a concrete mixture where normal-weight aggregate is specified. Similarly, test records of a 7,000 psi concrete mixture should not be used to calculate s_s for a proposed 5,000 psi concrete mixture.

Step 2. Once s_s has been established, the equations in ACI Table 5.3.2.1 are used to calculate the required average compressive strength f'_{cr}:

- For $f'_c \le 5,000$ psi, use the larger value of f'_{cr} computed by ACI Eqs. (5-1) and (5-2).

  Equation (5-1): $f'_{cr} = f'_c + 1.34 s_s$
  Equation (5-2): $f'_{cr} = f'_c + 2.33 s_s - 500$

- For $f'_c > 5,000$ psi, use the larger value of f'_{cr} computed by ACI Eqs. (5-1) and (5-3).

  Equation (5-1): $f'_{cr} = f'_c + 1.34 s_s$
  Equation (5-3): $f'_{cr} = 0.9 f'_c + 2.33 s_s$

If a concrete production facility does not have the required information needed to calculate s_s, the required average compressive strength f'_{cr} is determined by the equations given in ACI Table 5.3.2.2:

- For $f'_c < 3{,}000$ psi, $f'_{cr} = f'_c + 1{,}000$.
- For $3{,}000$ psi $\leq f'_c < 5{,}000$ psi, $f'_{cr} = f'_c + 1{,}200$.
- For $f'_c > 5{,}000$ psi, $f'_{cr} = 1.10 f'_c + 700$.

Step 3. Once the required average compressive strength f'_{cr} has been determined, concrete mixture proportions that will produce an average compressive strength equal to or greater than f'_{cr} and that will satisfy the durability requirements of Chap. 4 for the applicable exposure category must be selected. Field strength test records, several strength test records, or trial mixtures can be used to document that concrete strengths are satisfactory (see ACI 5.3.3 for more details).

Method 2: Proportioning without field experience or trial mixtures (ACI 5.4)   The preceding discussion has focused on the requirements for determining concrete mix proportions based on laboratory trial batches or strength test records. In cases where such data are not available, "other experience or information" may be used to proportion a concrete mixture that has a specified compressive strength f'_c that is less than or equal to 5,000 psi, provided the licensed design professional approves the mix (ACI 5.4). The required average compressive strength f'_{cr} determined by this alternative method must be at least 1,200 psi greater than f'_c. It is common for this method to be used in smaller concrete projects where it would be cost-prohibitive to obtain trial mixture data. For concrete strengths greater than 5,000 psi, proportioning on the basis of field experience or trial mixture data is required.

The flowchart shown in ACI Fig. R5.3 contains a summary of the requirements for the selection and documentation of concrete proportions in accordance with ACI 5.3 and 5.4. Utilizing laboratory trial batches or strength test records (ACI 5.3) is the preferred method for selecting concrete mixture proportions (see ACI R5.2.3). Additional information on proportioning mixtures can be found in ACI 211.1, *Standard Practice for Selecting Proportions for Normal, Heavyweight, and Mass Concrete*,[8] ACI 211.2, *Standard Practice for Selecting Proportions Lightweight Concrete*,[9] and Ref. 2.

Evaluation and Acceptance of Concrete   Once a concrete mixture has been selected on the basis of the provisions of ACI 5.3 or 5.4, the provisions of ACI 5.6 are used to determine if the concrete is acceptable or not. In addition to providing the criteria for evaluation and acceptance, ACI 5.6 provides a course of action that must be followed when unsatisfactory strength test results are obtained.

ACI 5.6.1 stresses the importance of using qualified laboratory and field technicians to perform tests and other tasks in the laboratory and at the job site. Laboratory personnel should be certified in accordance with the ACI Laboratory Technician Certification Program or ASTM C1077-07, *Standard Practice for Laboratories Testing Concrete and Concrete Aggregates for Use in Construction and Criteria for Laboratory Evaluation.* Similarly, field technicians should be certified in accordance with the ACI Concrete Field Testing Technician Certification Program or ASTM C1077-07.

Frequency	Minimum Number of Samples for Each Class of Concrete
Per day	Largest of the following: • One • One for each 150 yd^3 of concrete that is placed • One for each 5,000 ft^2 of surface area that is placed for slabs or walls
Per project	• Five from five randomly selected batches where more than five batches of concrete are used • One from each batch where less than five batches of concrete are used

TABLE 2.3   Minimum Number of Samples for Strength Tests where the Total Quantity of a Given Class of Concrete Is Equal to or Greater Than 50 yd^3

Frequency of testing   Because the Code utilizes probabilistic methodologies to establish the strength of concrete, a statistically significant number of samples must be taken to validate strength results.

The minimum number of samples for strength tests is summarized in Table 2.3 in cases where the total quantity of concrete is equal to or greater than 50 yd^3. ACI 5.6.2 prescribes the minimum number of samples that must be taken per day and per project for each class of concrete that is specified in a project. The larger of the two minimum numbers of samples governs.

By examining the second and third criteria for sampling on a per day frequency, it is evident that the third criterion will require more frequent sampling than once for each 150 yd^3 of concrete that is placed where the thickness of a slab or wall is less than approximately 9.75 in. Only one side of a slab or wall should be considered when calculating its surface area.

It is very important that the samples that are taken for a project are done so on a strictly random basis to ensure validity of the statistical analysis. The procedure that must be used for random selection of concrete batches to be tested is provided in ASTM D3665-07, *Standard Practice for Random Sampling of Construction Materials*.

Where the total quantity of a given class of concrete is less than 50 yd^3, strength tests are not required provided that evidence of satisfactory strength is submitted to and approved by the building official (ACI 5.6.2.3).

It was noted previously that the compressive strength of concrete is determined on the basis of the results from compression tests using 6 × 12 in or 4 × 8 in cylinders. A strength test is defined in ACI 5.6.2.4 as the average strength obtained from at least two 6 × 12 in cylinders or three 4 × 8 in cylinders. The cylinders must be made from the same sample of concrete and must be tested at the date specified for the compressive strength f_c'.

At least three 4 × 8 in cylinders must be tested to preserve the confidence level of the average strength results because the results obtained from the smaller cylinders tend to be more variable than those obtained from the 6 × 12 in cylinders. The 4 × 8 in cylinders are generally more popular because they weigh less and require approximately one-half of the testing capacity of the 6 × 12 in cylinders. This last attribute is especially important when high-strength concrete is tested: The smaller cylinders generally do not require high-capacity testing equipment, which is typically not available in most testing laboratories.

For overall consistency, the size and number of concrete cylinders that are used for a strength test should remain constant for each class of concrete that is specified in a project.

Acceptability of strength   Regardless of the method that was used to proportion a concrete mixture, concrete strength of an individual class of concrete is considered to be satisfactory when both of the following requirements are met (ACI 5.6.3.3):

1. Averages of any three consecutive strength tests are equal to or greater than the specified concrete strength f_c'.
2. No individual strength test falls below 500 psi when $f_c' \leq 5,000$ psi or by more than $0.10 f_c'$ when $f_c' > 5,000$ psi.

Requirements and procedures for investigating test results that do not meet these requirements are covered in the next section.

Investigation of low-strength test results   When either of the two requirements of ACI 5.6.3.3 is not met, the average of the concrete tests results needs to be increased. ACI R5.6.3.4 contains steps that can be taken to increase the average test results where the first of these two requirements is not satisfied.

The procedures outlined in ACI 5.6.5 must be followed if it is found (1) that the strength tests of laboratory-cured cylinders fall below f_c' more than the values given in the second of the two requirements of ACI 5.6.3.3 or (2) that tests of field-cured cylinders do not satisfy the strength requirements of ACI 5.6.4.4.

There are many potential reasons for low-strength test results. Some of the most common are the following:

1. *Improper fabrication, handling, and testing of the cylinders.* ASTM standards are not followed and/or uncertified personnel perform the tasks in the field and/or in the laboratory.
2. *Error in concrete production.* The intended concrete mixture was not produced at the concrete production facility according to the specified mixture proportions.
3. *Addition of mixing water at the site.* Water was added to the concrete at the site to achieve a higher slump concrete and/or to "improve" workability.

Regardless of the reason why concrete failed to meet the acceptance criteria, the licensed design professional must ensure that the load-carrying capacity of the structure is not jeopardized. In certain cases, a lower-strength concrete may not be detrimental to the performance of the structure.

Nondestructive testing of the concrete in a structure can help determine whether low-strength concrete is present or not. ACI R5.6.5 lists a number of such tests. The results from nondestructive tests should primarily be used to compare the relative strength of concrete in different portions of a structure rather than to establish the actual strength of concrete.

Drilling and subsequently testing cores taken from the area of a structure with suspect concrete is another method that is permitted to establish concrete strength. A minimum of three cores must be taken and tested in accordance with ASTM C42/C42M-04, *Standard Test Method for Obtaining and Testing Drilled Cores and Sawed Beams of Concrete*, for each strength test that falls below the second of the two values given in ACI 5.6.3.3 (ACI 5.6.5.2). Additional requirements for acquisition and testing are given in ACI 5.6.5.3. According to ACI 5.6.5.4, concrete in the area of a structure where core tests have been performed is deemed structurally adequate when the following two criteria are satisfied: (1) the strength obtained from the average of three cores is equal to or greater than $0.85 f_c'$, and (2) the strength of no single core is less than $0.75 f_c'$.

Strength evaluation of the structure in accordance with Chap. 20 of the Code may be undertaken on a questionable portion of a structure if the criteria for core tests are not satisfied or if for any reason the adequacy of a structure remains in doubt (ACI 5.6.5.5). Load testing procedures and acceptance criteria are contained in that chapter.

Example 2.1   Table 2.4 contains the test records obtained from a concrete production facility for a normal-weight concrete mixture that has a 28-day specified compressive strength f_c' of 4,000 psi. Strength tests were obtained from 6×12 in cylinders. The concrete will be used for the floor slabs in an enclosed building. Assume that the records were established in accordance with the requirements of ACI 5.3.1.1.

Determine the following:

(a) The sample standard deviation s_s for the test records

(b) The required average compressive strength f_{cr}'

(c) The water/cementitious materials ratio to satisfy strength requirements and the applicable durability requirements of ACI Chap. 4

Solution

(a) The results for the individual 6×12 in cylinders are given in columns 3 and 4 of Table 2.4. A minimum of two 6×12 in cylinders are required per ACI 5.6.2.4. The average of the two cylinder tests is given in column 5, which is the strength test result.

The sample standard deviation is determined by Eq. (2.1):

$$s_s = \left[\frac{\sum (x_i - \bar{x})^2}{(n-1)} \right]^{1/2}$$

In this equation, x_i is the result from an individual strength test and $\bar{x}$ is the average (mean) strength value of n strength tests. The mean strength for 30 tests is equal to the sum of the values in column 5 of Table 2.4 (or column 2 of Table 2.5) divided by 30:

$$\bar{x} = \frac{4,845 + \cdots + 4,990}{30} = \frac{156,320}{30} = 5,211 \text{ psi}$$

The data in column 3 of Table 2.5 are obtained by subtracting the 28-day average strength by the mean strength. For example, for test 1, $x_i - \bar{x} = 4,845 - 5,211 = -366$ psi.

Given the data in Table 2.5, the sample standard deviation is

$$s_s = \left[\frac{10,425,290}{30-1} \right]^{1/2} = 600 \text{ psi}$$

On the basis of the data in Table 2.4, the concrete satisfies the acceptance criteria of ACI 5.6.3:

1. The arithmetic average of each set of three consecutive strength tests, which is given in column 6 of Table 2.4, exceeds $f_c' = 4,000$ psi.

2. No single strength test is less than $4,000 - 500 = 3,500$ psi.

Test Number	Mix Code	F28 Test 1 (psi)	F28 Test 2 (psi)	28-day Average (psi)	28-day Average, Three Consecutive Tests (psi)
1	L-1000 4K	5,050	4,640	4,845	—
2	L-1000 4K	4,890	5,170	5,030	—
3	L-1000 4K	5,690	5,670	5,680	5,185
4	L-1000 4K	5,770	5,370	5,570	5,427
5	L-1000 4K	5,990	6,120	6,055	5,768
6	L-1000 4K	5,740	5,770	5,755	5,793
7	L-1000 4K	5,730	5,460	5,595	5,802
8	L-1000 4K	5,290	5,650	5,470	5,607
9	L-1000 4K	5,880	5,920	5,900	5,655
10	L-1000 4K	5,840	5,770	5,805	5,725
11	L-1000 4K	5,050	4,840	4,945	5,550
12	L-1000 4K	5,080	5,110	5,095	5,282
13	L-1000 4K	5,840	5,940	5,890	5,310
14	L-1000 4K	6,010	5,440	5,725	5,570
15	L-1000 4K	5,270	4,990	5,130	5,582
16	L-1000 4K	5,160	4,950	5,055	5,303
17	L-1000 4K	6,570	6,610	6,590	5,592
18	L-1000 4K	4,370	4,270	4,320	5,322
19	L-1000 4K	4,140	4,460	4,300	5,070
20	L-1000 4K	4,660	4,650	4,655	4,425
21	L-1000 4K	4,550	4,600	4,575	4,510
22	L-1000 4K	5,220	5,090	5,155	4,795
23	L-1000 4K	5,580	5,270	5,425	5,052
24	L-1000 4K	5,580	5,930	5,755	5,445
25	L-1000 4K	5,150	5,190	5,170	5,450
26	L-1000 4K	4,390	4,240	4,315	5,080
27	L-1000 4K	4,670	4,760	4,715	4,733
28	L-1000 4K	4,660	4,420	4,540	4,523
29	L-1000 4K	4,220	4,320	4,270	4,508
30	L-1000 4K	4,940	5,040	4,990	4,600

TABLE 2.4   Concrete Strength Test Records for Example 2.1

Test Number	28-day Average (psi)	$x_i - \bar{x}$ (psi)	$(x_i - \bar{x})^2$ (psi)
1	4,845	−366	133,956
2	5,030	−181	32,761
3	5,680	469	219,961
4	5,570	359	128,881
5	6,055	844	712,336
6	5,755	544	295,936
7	5,595	384	147,456
8	5,470	259	67,081
9	5,900	689	474,721
10	5,805	594	352,836
11	4,945	−266	70,756
12	5,095	−116	13,456
13	5,890	679	461,041
14	5,725	514	264,196
15	5,130	−81	6,561
16	5,055	−156	24,336
17	6,590	1,379	1,901,641
18	4320	−891	793,881
19	4,300	−911	829,921
20	4,655	−556	309,136
21	4,575	−636	404,496
22	5,155	−56	3,136
23	5,425	214	45,796
24	5,755	544	295,936
25	5,170	−41	1,681
26	4,315	−896	802,816
27	4,715	−496	246,016
28	4,540	−671	450,241
29	4,270	−941	885,481
30	4,990	−221	48,841
$\sum$	156,320		10,425,290

TABLE 2.5   Data for Calculation of Standard Deviation in Example 2.1

Note that the average of the first set of three consecutive tests is equal to $(4,845 + 5,030 + 5,680)/3 = 5,185$ psi; the average of the second set is equal to $(5,030 + 5,680 + 5,570)/3 = 5,427$ psi; and so on.

(b) Because strength data were used to establish a sample standard deviation, the required average compressive strength f'_{cr} is the larger of the values determined by Eqs. (5-1) and (5-2) where $f'_c \leq 5,000$ psi (see ACI Table 5.3.2.1):

Exposure Class	Maximum w/cm	Minimum f_c' (psi)	Additional Minimum Requirements			Limits on Cementitious Materials
			Air Content			
F0	N/A	2,500	N/A			N/A
			Cementitious materials—types			Calcium chloride admixture
			ASTM C150	ASTM C595	ASTM C1157	
S0	N/A	2,500	No type restriction	No type restriction	No type restriction	No restriction
P0	N/A	2,500	None			
			Maximum water-soluble chloride ion content in concrete, percent by weight of cement for reinforced concrete			Related provisions
C0	N/A	2,500	1.00			None

TABLE 2.6 Requirements for Concrete by Exposure Class for Example 2.1

Equation (5-1): $f_{cr}' = f_c' + 1.34s_s = 4{,}000 + (1.34 \times 600) = 4{,}804 \, \text{psi}$
Equation (5-2): $f_{cr}' = f_c' + 2.33s_s - 500 = 4{,}000 + (2.33 \times 600) - 500 = 4{,}898 \, \text{psi (governs)}$

(c) It was determined in part (b) of this example that the required average compressive strength f_{cr}' is equal to 4,898 psi, which for practical purposes will be rounded up to 5,000 psi. The durability provisions of Chap. 4 of the Code will be examined to determine if a larger compressive strength is required or not.

An exposure class must be assigned on the basis of the severity of the anticipated exposure of the floor slabs in the building (ACI 4.2.1). It is common for structural members that are located inside of a building (not exposed to the elements) to be assigned the following exposure classes (see ACI Table 4.2.1):

- F0—Concrete is not exposed to freeze–thaw cycles.

- S0—Water-soluble sulfate concentration in contact with concrete is low, and sulfate attack is not a concern.

- P0—No specific requirements are needed for permeability.

- C0—Additional protection against the initiation of corrosion of reinforcement is not required.

The applicable requirements for concrete mixtures from ACI Table 4.3.1 are summarized in Table 2.6 for the exposure classes in this example.

It is evident from Table 2.6 that no durability limitations are prescribed on the water/cementitious materials ratio w/cm for the concrete in this example. Also, the minimum compressive strength of 2,500 psi is less than the required average compressive strength of 5,000 psi. Therefore, the concrete production facility can utilize the concrete mixture design that is designated by the mix code L-1000 4K for the concrete slabs.

Comments
Once test results from cylinders cast from the concrete during construction become available, it may be possible to reduce the amount by which the value of f_{cr}' must exceed the value of f_c' [which is

equal to approximately 900 psi; see part (b) of this example] by using a sample standard deviation based on the actual construction data. This reduction will typically produce a more economical concrete mixture.

If the concrete production facility had, for example, only the first 20 strength test records of Table 2.4, the sample standard deviation is calculated by Eq. (2.1) for the 20 records. Because there are less than 30 records, that number would be multiplied by the applicable modification factor given in ACI Table 5.3.1.2, which in this case is 1.08 for 20 test records. The modified sample standard deviation is used in ACI Eqs. (5-1) and (5-2) to determine the required average compressive strength f'_{cr}.

Example 2.2   The concrete in Example 2.1 will be placed in the floor slabs of a 20-story building. A typical floor plate is 63 ft 3 in × 163 ft 0 in, and the slab is 8 in thick. The concrete will be delivered to the site in concrete trucks that have a capacity of 10 yd^3, and one typical floor will be placed in 1 day. Determine the minimum number of 6 × 12 in and 4 × 8 in cylinders that must be cast to satisfy the sampling requirements of ACI 5.6.2.

Solution
Because the concrete will be placed in a 20-story building, the minimum required number of test cylinders in accordance with ACI 5.6.2.1 will be greater than the minimum required in accordance with ACI 5.6.2.2.

The minimum number of samples is the largest of the following:

1. One.

2. One for each 150 yd^3 of concrete that is placed.

   Total volume of concrete per floor $= \dfrac{8}{12} \times 68.33 \times 163 = 7{,}425 \text{ ft}^3 = 275 \text{ yd}^3$

   Minimum required number of samples $= \dfrac{275}{150} = 1.8$

   A minimum of two samples are required for each floor on the basis of this criterion.

3. One for each 5,000 ft^2 of surface area that is placed for slabs or walls

   Total surface area per floor $= 68.33 \times 163 = 11{,}138 \text{ ft}^2$

   Minimum required number of samples $= \dfrac{11{,}138}{5{,}000} = 2.3$

   A minimum of three samples are required for each floor on the basis of this criterion (governs).

On the basis of the strength test requirements of ACI 5.6.2.4, the required number of cylinders per floor is the following:

- For 6 × 12 in cylinders, 2 × 3 = 6 cylinders

- For 4 × 8 in cylinders, 3 × 3 = 9 cylinders

A minimum of 120 of the 6 × 12 in cylinders or 180 of the 4 × 8 in cylinders are required to determine the acceptable strength of the floor slab concrete mixture in this project.

Comments
In a given project, additional cylinders are typically cast and kept in reserve in case any anomalies occur in the test data. Some cylinders are usually tested at 7 days to check, among other things, early strength development of the concrete; this information is used to determine when formwork can be safely removed.

Example 2.3   The strength test data in Table 2.7 were obtained from 4 × 8 in cylinders that were sampled from the concrete cast on site for one of the typical floors of the project outlined in Examples 2.1 and 2.2. The cylinders were tested at 28 days. Determine if the concrete is acceptable in accordance with ACI 5.6.3.

Solution   The average strength values in column 5 of Table 2.7 represent a single test record based on the results from the three sample tests.

Test Number	Sample 1 (psi)	Sample 2 (psi)	Sample 3 (psi)	28-day Average (psi)	28-day Average, Three Consecutive Tests (psi)
1	3,760	3,950	3,875	3,860	—
2	4,425	4,175	4,815	4,470	—
3	4,080	4,220	3,990	4,095	4,142

TABLE 2.7 Strength Test Results from Cylinders Cast on Site in Example 2.3

According to ACI 5.6.3.3, strength level is considered satisfactory when both of the following criteria are satisfied:

- Every arithmetic average of any three consecutive strength tests is equal to or greater than f_c'.

This criterion is satisfied because the average of the three sets of strength tests is equal to 4,142 psi, which is greater than 4,000 psi.

- No strength test falls below f_c' by more than 500 psi when f_c' is less than 5,000 psi.

The test record for test 1 is equal to 3,860 psi, which falls below the specified compressive strength of 4,000 psi by 140 psi. Because this is less than 500 psi, this criterion is satisfied.

If this criterion is not satisfied, the procedure in ACI 5.6.5 must be followed to investigate the low-strength test results.

Modulus of Elasticity

The modulus of elasticity of concrete E_c (Young's modulus) is used in the design of concrete members, including design for deflections and of slender columns. The empirical equation provided in ACI 8.5.1 gives an approximate value of E_c:

$$E_c = w_c^{1.5} 33 \sqrt{f_c'} \tag{2.3}$$

In this equation, w_c is the unit weight of normal-weight concrete or the equilibrium density of lightweight concrete, which must be between 90 and 160 pcf. The compressive strength f_c' has the units of pounds per square inch.

The equation is derived from experimental data and is based on the secant modulus of elasticity, which is defined as the slope of the straight line connecting the point of zero stress and the stress at approximately $0.45 f_c'$ (see Fig. 2.5).

For normal-weight concrete with $w_c = 145$ pcf,

$$E_c = 57,000 \sqrt{f_c'} \tag{2.4}$$

The typical range of E_c for normal-weight concrete is 2,000,000 to 6,000,000 psi. In addition to the compressive strength of concrete, other factors effect E_c, which are discussed next.

The porosity of cement paste has a direct influence on the value of E_c. The porosity of the paste increases as the water/cement ratio increases; thus, both the strength and the modulus of elasticity of cement paste decrease as the water/cement ratio increases. This is accounted for in the equation by expressing E_c as a function of f_c': An increase in the water/cement ratio translates into a lower compressive strength, which results in a lower value of E_c.

The magnitude of E_c is also dependent on the amount and type of aggregate that is used in a concrete mixture. It is evident from the mechanics of concrete deterioration and failure in compression that the elastic modulus is partially dependent on the progressive microcracking at the aggregate–cement paste interface. Thus, the shape, texture, and total amount of aggregate have an influence on E_c. For example, concrete mixtures produced with rounded gravel have values of E_c that are greater than those produced with crushed stone. Similarly, concrete mixtures that have larger maximum aggregate sizes have greater values of E_c than those that have smaller maximum aggregate sizes.

Normal-weight aggregates have modulus of elasticity values that are many times greater than those of cement paste, whereas lightweight aggregates have values that are closer to those of cement paste. Consequently, values of E_c are greater for normal-weight concrete that contains larger amounts of normal-weight aggregate. The amount of lightweight aggregate in a mixture generally has little effect on E_c. Values of E_c for concrete mixtures with lightweight aggregate are typically 40% to 80% of those for concrete mixtures with normal-weight aggregate.

The empirical equation does not capture the influence that aggregates have on the value of E_c. Consequently, measured values of the elastic modulus can range between 80% and 120% of the values determined by Eq. (2.3) or (2.4).

For concrete mixtures with aggregates that have relatively low values of the modulus of elasticity, the provided equation will overestimate E_c. It is important to understand the possible implications of overestimating E_c. Consider a reinforced concrete beam in a building. The calculated deflection of the beam using the value of E_c from the empirical equation would most probably be less than the actual deflection. This may not be an issue in typical situations, but if the beam is to support a glass partition wall that is sensitive to deflections, this could lead to problems. Where deflections are critical, it is recommended to use measured values of E_c. Methods for determining E_c are given in ASTM C469-02, *Standard Test Method for Static Modulus of Elasticity and Poisson's Ratio of Concrete in Compression*.

Tensile Strength

Mechanics of Concrete Deterioration and Failure in Tension
The behavior of concrete in tension is significantly different from its behavior in compression. Consider a plain concrete specimen that is subjected to a uniform tensile force. The preexisting cracks at the aggregate–cement paste interface that are oriented perpendicular to the direction of the force propagate instantaneously to the edge of the aggregate facet once the stress at the tips of the cracks reaches a critical value. Because the toughness of the paste is greater than that of the interface, the cracks are arrested at the edge of the aggregate.

With an increase in tensile force, an increasing number of cracks become unstable. Exactly which cracks become destabilized at a given tensile force depends on their orientation with respect to the direction of the force and their initial length.

As the tensile force increases further, a point at which the preferentially oriented cracks overcome the energy barrier at the facet edge is reached. The microcracks propagate into the cement paste in an unstable manner. For all practical purposes, failure of the entire specimen is imminent at this point because there is essentially no mechanism that can prevent the cracks from splitting the concrete perpendicular to the applied tensile force.

Unlike concrete in compression, the deterioration of concrete in tension is not gradual, and failure occurs suddenly because of the growth of a single crack.

The tensile strength of concrete is significantly smaller than the compressive strength primarily because of the ease with which the cracks can propagate under tensile loads. It is commonly assumed that the tensile strength is equal to 10% of the compressive strength. More information on tensile tests and the relationship between the tensile strength and the compressive strength is given later.

The tensile strength of concrete is usually not considered in the design of structural members because it is small, and it is generally taken equal to zero. As was discussed in Chap. 1, reinforcement is utilized to resist tensile stresses in concrete.

Tension Tests   To date, there is no standard test by ASTM to determine the direct tensile strength of concrete. ASTM D2936-08, *Standard Test Method for Direct Tensile Strength of Intact Rock Core Specimens*, covers the determination of the direct tensile strength of intact cylindrical rock specimens, but it does not specifically address concrete specimens. Specimens with enlarged ends that resemble a "dog bone" and cylindrical or prismatic specimens with end plates glued to the concrete have been used by various researchers to obtain direct tensile strengths. However, issues with respect to specimen preparation and test setup have made the results difficult to reproduce. As a result, indirect tests have been standardized for estimating the tensile strength of concrete. The two most common ones are discussed next.

Splitting tension test   In a splitting tension test, a standard cylinder or a drilled core is tested on its side as illustrated in Fig. 2.7a. Methods for determining the splitting tensile strength of concrete are given in ASTM C496/C496M-04, *Standard Test Method for Splitting Tensile Strength of Cylindrical Concrete Specimens*.

The compression force applied from a testing machine along the diameter of the specimen introduces very high lateral compressive stresses near the points of load application and a nearly uniform lateral tensile stress over approximately the middle

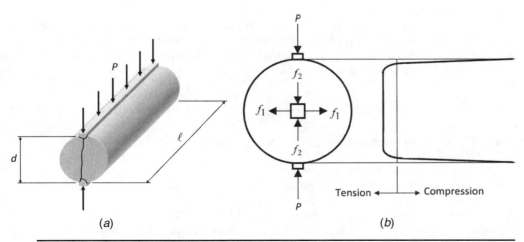

Figure 2.7   Splitting tension test: (a) test procedure and (b) lateral stress distribution on vertical diameter of cylinder.

FIGURE 2.8 Flexural
test of concrete by
third-point loading
method.

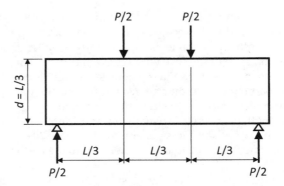

two-thirds of the specimen (Fig. 2.7*b*). Because the concrete is stronger in compression than in tension, the specimen will split along the diameter of the specimen before it crushes at its ends. The stress at which splitting occurs is defined as the splitting tensile strength of concrete *T* and can be calculated by Eq. (1) of ASTM C496/C496M:

$$T = \frac{2P}{\pi \ell d} \tag{2.5}$$

where P = is maximum applied load indicated by the testing machine
  ℓ = length of specimen
  d = diameter of specimen

The value of the tensile strength obtained from split cylinder tests is approximately 15% greater than that from direct tension tests.

Flexural test In a flexural test, a plain concrete beam is loaded at its third points, as illustrated in Fig. 2.8. Methods for determining the flexural strength of concrete are given in ASTM C78-02, *Standard Test Method for Flexural Strength of Concrete (Using Simple Beam with Third-point Loading)*.

The point loads introduce tensile stresses at the bottom surface of the specimen and compressive stresses at the top surface. The tensile strength *R* (or modulus of rupture) of concrete is calculated by Eq. (2) of ASTM C78 if fracture initiates in the tension surface within the middle third of the span length:

$$R = \frac{PL}{bd^2} \tag{2.6}$$

where P = maximum applied load indicated by the testing machine
  L = span length
  b = average width of the specimen
  d = average depth of the specimen

If fracture occurs in the tension face outside of the middle third of the span length by not more than 5% of the span length, *R* is determined by Eq. (3) of ASTM C78:

$$R = \frac{3Pa}{bd^2} \tag{2.7}$$

where a = average distance between the line of fracture and the nearest support measured on the tension surface of the beam.

Results of a test are to be discarded where fracture occurs outside of the middle third of the span length by more than 5% of the span length.

ASTM C293-02, *Standard Test Method for Flexural Strength of Concrete (Using Simple Beam with Center-point Loading)*, can also be used to determine the flexural strength. A single concentrated load is applied to a plain concrete beam at midspan. The tensile strength is calculated by Eq. (2) of ASTM C293:

$$R = \frac{3PL}{2bd^2} \tag{2.8}$$

where all terms have been defined previously.

The value of tensile strength obtained from flexural tests is approximately 50% greater than that from direct tension tests. This can be attributed to the assumption that the flexural stress varies linearly over the depth of the cross-section [see Eqs. (2.7) and (2.8)]; the actual distribution of flexural stress is nonlinear, especially at the surfaces farthest from the neutral axis.

Relationship Between Tensile and Compressive Strengths   As was noted at the beginning of this section, the tensile strength of concrete is commonly taken as 10% of the compressive strength. However, test results have revealed that the ratio of the tensile strength to the compressive strength decreases as the compressive strength increases.

It can be shown that the modulus of rupture is approximately proportional to the square root of the compressive strength. ACI Eq. (9-10) defines the modulus of rupture f_r that is to be used when calculating deflections:

$$f_r = 7.5\lambda\sqrt{f_c'} \tag{2.9}$$

In this equation, f_c' has the units of pounds per square inch and λ is a modification factor that reflects the reduced mechanical properties of lightweight concrete (see ACI 8.6.1):

- $\lambda = 0.85$ for sand-lightweight concrete.
- $\lambda = 0.75$ for all-lightweight concrete.
- $\lambda = f_{ct}/6.7\sqrt{f_c'} \leq 1.0$, where the average splitting tensile strength of lightweight concrete f_{ct} has been determined by tests [Eq. (1) in ASTM C496/C496M]. Note that $6.7\sqrt{f_c'}$ is the average splitting tensile strength of normal-weight concrete.
- $\lambda = 1.0$ for normal-weight concrete.

It is permitted to use linear interpolation to determine λ in cases where a concrete mixture contains normal-weight fine aggregate and a blend of lightweight and normal-weight coarse aggregates. The interpolation shall be between 0.85 and 1.0 on the basis of the volumetric fractions of the aggregates.

A lower value of f_r is used in strength calculations; this will be discussed in subsequent chapters of this book.

Strength Under Combined Stress

Reinforced concrete structural members are rarely subjected to a single type of stress. Most members must be designed to resist a combination of compressive, tensile, and shear stresses that all act at the same time. For example, a beam in a building must be designed to resist the combined effects due to flexural (compressive and tensile) and shear stresses and sometimes due to axial (compressive and tensile) stresses.

Regardless of its complexity, a state of combined stress can be resolved into three principal stresses that are oriented perpendicular to each other. The state of stress is uniaxial when two of the principal stresses are zero and is biaxial when one of the principal stresses is zero. A triaxial state of stress occurs when all three stresses are nonzero. Both biaxial and triaxial states of stress are examined next for plain concrete members.

Biaxial Stress

Biaxial tension   In this case, both an axial tensile stress and a lateral tensile stress are applied to concrete, as illustrated in Fig. 2.9a. Assume that the axial stress is greater than the lateral stress. Like in the case of uniaxial tension, the first microcrack to increase in size is the one of maximum length that is oriented perpendicular to the direction of maximum tensile stress. Once this critical crack overcomes the energy barrier at the edge of the aggregate, the crack propagates through the matrix in an unstable manner and failure occurs.

Some experimental results have shown that the biaxial tensile strength of concrete is equal to the uniaxial tensile strength,[10] whereas others have reported an increase of approximately 18% where the applied lateral stress is equal to 50% of the applied axial stress.[11]

Biaxial compression-tension   Where a lateral tensile stress is present with an axial compressive stress, destabilization of the preferentially oriented interface cracks occurs sooner than it does for uniaxial compression (see Fig. 2.9b). The kinked cracks propagate into the matrix in a stable manner until they acquire a certain length. With the presence

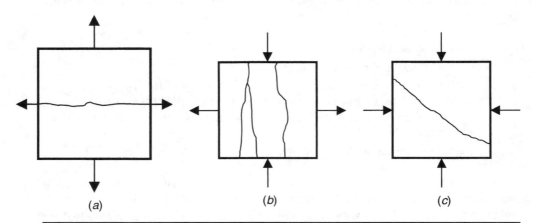

(a)   (b)   (c)

Figure 2.9   Concrete specimen subjected to biaxial stresses: (a) biaxial tension, (b) biaxial compression-tension, and (c) biaxial compression.

of lateral tensile stresses, these cracks become destabilized, and failure of the specimen occurs relatively quickly by splitting perpendicular to the direction of the tensile stress.

The experimental results reported in Ref. 10 show that concrete subjected to a state of biaxial compression-tension fails sooner than it would if subjected to only uniaxial compression. This is also shown to be true for concrete subjected to an axial tensile stress and a compressive lateral stress.

Biaxial compression   A concrete specimen subjected to biaxial compression is illustrated in Fig. 2.9c. In general, lateral confining (compression) stresses inhibit the damage processes: Crack propagation along the interface and into the cement matrix occurs later than it does for uniaxial compression, and the period of stable crack growth in the matrix is longer. In short, the concrete behaves in a more ductile manner, and it fails at a larger stress than it would if subjected to only uniaxial compression.[10]

Where the applied lateral compressive stress is equal to or less than approximately 5% of the applied axial compressive stress, the mode of failure is similar to that of uniaxial compression. For larger lateral confining pressures, the specimen typically splits at an angle of approximately 60 degrees from the horizontal in a shearing mode of failure. Where the lateral stress is almost equal to the axial stress, a strength increase of approximately 20% is attained.

Triaxial Stress   Concrete specimens subjected to triaxial compressive stresses generally have greater strength and exhibit more ductility than those subjected to uniaxial compressive stress. For relatively low lateral compressive stresses, the mode of failure is similar to that for uniaxial and biaxial compressive stresses: The specimen eventually splits parallel to the longitudinal axis of the member. A shearing mode of failure is evident with relatively large lateral compressive stresses.

Triaxial tests on concrete cylinders show that concrete strength increases by approximately a factor of 5 where the specimen is subjected to a constant lateral stress that is approximately equal to the unconfined compressive strength of the concrete.[12] The tests also show that confined concrete exhibits significantly more ductility with long and relatively flat descending branches of the stress–strain curve. For the case where the lateral stress is approximately equal to the unconfined compressive strength, the strain at the peak of the curve is equal to approximately 10 times that of the unconfined concrete.

ASTM C801-98, *Standard Test Method for Determining the Mechanical Properties of Hardened Concrete Under Triaxial Loads*, covers the procedures for testing hardened concrete subjected to triaxial stresses. It was withdrawn in 2004 without a replacement.

Reinforcement is typically used to confine concrete to increase its strength and ductility. Confining concrete in key structural elements of a building is especially important in areas of high seismic risk.

The behavior of concrete subjected to uniaxial compression and lateral tension in two orthogonal directions is similar to that for biaxial compression-tension.

Poisson's Ratio

Poisson's ratio, which is the ratio of the transverse strain to the axial strain of an axially loaded member, can be determined by direct strain measurements when a concrete specimen is tested in compression. Methods for determining Poisson's ratio are given in ASTM C469.

Poisson's ratio generally falls in the range of 0.15 to 0.20 for both normal-weight and lightweight concretes, and it remains approximately constant under sustained loads. It is commonly taken as 0.20.

At a stress of approximately 50% of f'_c, which generally corresponds to the onset of microcracks propagating into the cement matrix, there is an increase in Poisson's ratio; this is evident by an increase in the ratio of the lateral strain ε_3 to the longitudinal strain ε_1. This trend continues as cracking increases with increased stress. Volumetric strain (defined as the ratio of the change in volume of a body to the deformation to its original volume, which is equal to $\varepsilon_1 + 2\varepsilon_3$) also increases at this stage. This translates into an increase in concrete volume.

Volume Changes

Shrinkage, creep, and thermal expansion are the three main types of volume changes that can occur in concrete members. These types of volume changes can cause strains and cracking in a concrete member, which can have a direct influence on strength and serviceability.

Shrinkage Shrinkage is defined as the decrease of hardened concrete volume with time. The decrease in volume can be attributed to changes in moisture content and chemical changes, which occur without the presence of external loading on the concrete. The main types of shrinkage are the following:

1. *Drying shrinkage.* Drying shrinkage is due to moisture loss in concrete that is exposed to the environment and is permitted to dry. Any workable concrete mixture contains more water than is needed for hydration. The excess water—commonly referred to as free water—evaporates with time, which leads to gradual shortening of a concrete member. This is the predominant type of shrinkage for concrete that is not high-strength concrete and is not exposed to a carbon dioxide–rich environment.

2. *Autogenous shrinkage.* Autogenous shrinkage is due to the hydration reactions taking place inside the cement matrix. It is typically neglected except for high-strength concrete mixtures where the water/cement ratio is less than 0.40.[13]

3. *Carbonation shrinkage.* Carbonation shrinkage is caused by the reaction of calcium hydroxide in the cement matrix with carbon dioxide in the atmosphere (e.g., the atmosphere that can be present in a parking garage). This type of shrinkage can be of the same order of magnitude as that of drying shrinkage under certain environmental conditions.

Drying shrinkage is covered in the following discussion.

Ambient temperature, relative humidity, aggregate type, and concrete member size and shape are variables that influence shrinkage. Larger amounts of shrinkage occur in environments where the relative humidity is 40% or less, and virtually no shrinkage occurs where the relative humidity is 100%. Shrinkage can partially be recovered by rewetting the concrete, but complete recovery cannot occur.

Shrinkage takes place primarily in the cement paste and not in the aggregates. As a result, the cement paste tends to pull away from the aggregates, which causes tension in

FIGURE 2.10
Shrinkage strain of
concrete over time.

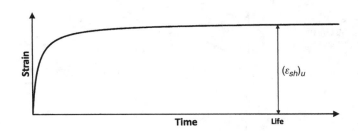

the paste and compression in the aggregates; this leads to microcracks at the aggregate–cement paste interface prior to any external loading. Because aggregates tend to restrain shrinkage, concrete with a larger volume fraction of aggregates will shrink less than that with a smaller volume fraction.

As water evaporates out of concrete, the exterior portion of a concrete member shrinks more rapidly than the interior portion. This induces tensile and compressive stresses in the exterior and interior portions, respectively. Less shrinkage occurs in members with larger volume-to-surface-area ratios because there is a larger interior portion of relatively moist concrete that restrains shrinkage. More information on the factors that affect drying shrinkage can be found in Refs. 14 and 15.

Strains caused by shrinkage in concrete under constant environmental conditions ε_{sh} increase with time as shown in Fig. 2.10. Theoretically, shrinkage will continue without end. It is usually assumed that the ultimate shrinkage strain $(\varepsilon_{sh})_u$ occurs at the estimated life of a structure. It is evident from the figure that most shrinkage occurs soon after a concrete member is exposed to the environment. Generally, approximately 50% of the total shrinkage occurs within a month of exposure, and approximately 90% occurs within a year.

Reference 15 recommends using the following equation to predict shrinkage strain as a function of time for moist-cured, plain concrete at any time t after 7 days:

$$\varepsilon_{sh}(t) = \frac{t}{35 + t}(\varepsilon_{sh})_u \gamma_\lambda \gamma_{vs} \tag{2.10}$$

where
$$t = \text{age of concrete in days}$$
$$(\varepsilon_{sh})_u = \text{ultimate shrinkage strain}$$
$$= 780 \times 10^{-6} \text{ in/in}$$
$$\gamma_\lambda = \text{correction factor that accounts for relative humidity}$$
$$= 1.40 - 0.01\lambda \text{ for } 40 \leq \lambda \leq 80$$
$$= 3.00 - 0.03\lambda \text{ for } 80 \leq \lambda \leq 100$$
$$\lambda = \text{relative humidity expressed as a percentage}$$
$$\gamma_{vs} = \text{correction factor that accounts for volume/surface ratio } (v/s) \text{ of}$$
concrete member
$$= 1.2e^{-0.12(v/s)} \text{ with } (v/s) \text{ in inches}$$

Additional correction factors can be found in Ref. 15, but for practical purposes, the two factors listed earlier should be sufficient to obtain reasonable results. Reference 16 contains additional models and numerous references that can be used to predict shrinkage strains.

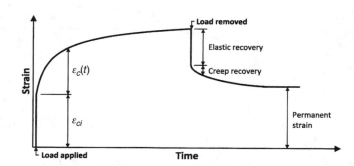

Figure 2.11
Creep strain of
concrete over time.

Methods to determine the measurement of length change for volumetric expansion or contraction of mortar or concrete due to various causes other than applied force or temperature change can be found in ASTM C157/C157M-08, *Standard Test Method for Length Change of Hardened Hydraulic-cement Mortar and Concrete.*

It is important to minimize and control shrinkage. When shrinkage strain is not controlled, cracks that can increase in size with time may form. Not only are these cracks unsightly, but they could eventually become harmful. Reinforcement plays an important role in minimizing the effects of shrinkage.

Creep   Creep is defined as the time-dependent increase in strain in excess of the elastic strain that is induced in a concrete member that is subjected to a sustained external load. Strain caused by creep is unlike strain caused by shrinkage because the latter occurs independent of load. Figure 2.11 illustrates the increase in creep strain with time for plain concrete.

In addition to time, the following factors have an influence on creep: relative humidity, aggregate type, mix proportions, age of concrete at loading, temperature, and the size and shape of a member. Like shrinkage, the aggregates in a concrete mixture restrain creep. More information on how these factors affect creep can be found in numerous references, including Refs. 14 through 16.

Creep strain is obtained by subtracting the elastic (instantaneous) strain from the total measured strain in a loaded specimen. The creep coefficient $v(t)$ relates creep strain $\varepsilon_c(t)$ to the initial elastic strain ε_{ci}:

$$v(t) = \frac{\varepsilon_c(t)}{\varepsilon_{ci}} \tag{2.11}$$

Creep strain approaches a maximum value of ε_{cu} over the life of a structure. The ultimate creep coefficient v_u is defined on the basis of ε_{cu} and the constant elastic strain ε_{ci}:

$$v_u = \frac{\varepsilon_{cu}}{\varepsilon_{ci}} \tag{2.12}$$

Reference 15 recommends the following equation to determine $v(t)$ for moist-cured, plain concrete:

$$v(t) = \frac{t^{0.6}}{10 + t^{0.6}} v_u \gamma_{\ell a} \gamma_\lambda \gamma_{vs} \tag{2.13}$$

where t = time after loading in days

v_u = ultimate creep coefficient = 2.35

$\gamma_{\ell a}$ = correction coefficient that accounts for age of concrete at loading

= $1.25(t_{\ell a})^{-0.118}$

$t_{\ell a}$ = age of concrete at time of loading in days (greater than 7 days)

γ_λ = correction factor that accounts for relative humidity

= $1.27 - 0.0067\lambda$ for $\lambda > 40$

λ = relative humidity expressed as a percentage

γ_{vs} = correction factor that accounts for volume/surface ratio (v/s) of concrete member

= $2[1 + 1.13e^{-0.54(v/s)}]/3$ with (v/s) in inches

If concrete is unloaded, essentially all of the elastic strain is recovered immediately, and a portion of the total creep strain is recovered over time (see Fig. 2.11). A relatively large portion of the total creep strain is irreversible; this translates to permanent deformation (strain) in the member.

A test method that measures the load-induced time-dependent creep strain at selected ages for concrete under an arbitrary set of controlled environmental conditions is given in ASTM C512-02, *Standard Test Method for Creep of Concrete in Compression*.

In reinforced concrete members, creep strains are distributed between concrete and reinforcement, resulting in lower creep strains in the concrete. Nevertheless, the effects of creep on deflections of structural members must be considered, and methods to include the effects of creep in the design process are covered in subsequent chapters of this book.

Thermal Expansion Concrete expands when subjected to increasing temperatures and contracts when subjected to decreasing temperatures. The following factors have an influence on the coefficient of thermal expansion of concrete: (1) type and amount of aggregate, (2) moisture content, (3) mixture proportions, (4) cement type, and (5) age. The first two factors typically have the greatest affect.

Reference 15 contains equations to predict the coefficient of thermal expansion based on the degree of saturation of the concrete and the average thermal coefficient of the aggregate. In general, the coefficient is within the range of 5.5 to $6.5 \times 10^{-6}/°F$ for normal-weight concrete and 4.0 to $6.0 \times 10^{-6}/°F$ for lightweight concrete. A value of $5.5 \times 10^{-6}/°F$ is typically used when calculating stresses induced by changes in temperature.

The test method in ASTM C531-00(2005), *Standard Test Method for Linear Shrinkage and Coefficient of Thermal Expansion of Chemical-resistant Mortars, Grouts, Monolithic Surfacings, and Polymer Concretes*, can be used to determine the coefficient of thermal expansion for concrete.

Properties When Exposed to Temperature Effects

In general, the compressive strength and modulus of elasticity of concrete decrease as external temperature increases. The degree to which these properties are affected depends primarily on the type of aggregate present in the concrete mixture.

Up to approximately 800°F, the strength of concrete in compression is approximately 90% of the compressive strength regardless of the type of aggregate in the mixture.[17] Above 800°F, the strength of concrete made with siliceous aggregates (such as quartzite, granite, and sandstone) drops off dramatically: At approximately 1,000°F, the strength is approximately $0.50 f_c'$, and at temperatures of 1,200°F and 1,600°F, it decreases to $0.30 f_c'$ and $0.20 f_c'$, respectively. In contrast, the strength of concrete made with carbonate aggregates (such as limestone) and with sand-lightweight aggregates is approximately $0.80 f_c'$ at 1200°F. The strength eventually drops to $0.20 f_c'$ at a temperature of 1,600°F. At higher temperatures, the volume of concrete increases rapidly, which leads to spalling at its outermost surfaces.

The modulus of elasticity decreases approximately linearly with an increase in temperature, and the magnitude of the modulus varies at most by approximately 10% for all types of concrete regardless of aggregate type. At 1,200°F, the modulus is at approximately 30% of its initial value.

The preceding discussion is valid for concrete strengths below 8,000 psi. Information on temperature effects on high-strength concrete is given in the next section.

The strength of hardened concrete tends to increase when subjected to colder temperatures.

2.2.3   High-Strength Concrete

ACI Committee 363 defines high-strength concrete as a concrete mixture with a specified compressive strength equal to or greater than 6,000 psi.[18,19]

Components

The components used in the production of high-strength concrete are essentially the same as those for normal-strength concrete. Chemical and mineral admixtures are essential in the creation of high-strength mixtures. Polymers, epoxies, and artificial normal-weight and heavyweight aggregates have been utilized to produce high-strength concrete mixtures as well.

Guidelines on selection of materials and concrete mixture proportions can be found in Ref. 18.

Mechanical Properties

A comprehensive summary of the mechanical properties of high-strength concrete is given in Chap. 5 of Ref. 18 and in Chap. 5 of Ref. 20. When loaded in uniaxial compression, high-strength concrete behaves similar to normal-strength concrete. The shape of a compressive stress–strain curve for high-strength concrete has a more linear ascending branch and a steeper descending branch compared with normal-strength concrete (see Fig. 2.6). At the matrix level, there is greater bond strength at the aggregate–cement paste interface, and there is significantly less microcracking. As such, the network of cracking in high-strength concrete becomes unstable at approximately $0.90 f_c'$, compared with $0.75 f_c'$ for normal-strength concrete. A brittle type of failure subsequently occurs by splitting through the aggregates and paste parallel to the direction of loading. For higher-strength mixtures, failure is sudden and explosive. The compressive strength of high-strength concrete is generally specified at 56 or 90 days.

As the compressive strength increases, test results are more sensitive to testing conditions. Thus, quality control and testing for high-strength concrete are more critical than for normal-strength concrete. Information on quality assurance and quality control, testing, and evaluation of compressive strength results, including statistical concepts and strength evaluation, can be found in Ref. 19.

Modulus of elasticity	$E_c = 40{,}000\sqrt{f_c'} + 1.0 \times 10^6$ psi
Poisson's ratio	0.20–0.28
Modulus of rupture	$11.7\sqrt{f_c'}$ psi for $3{,}000 < f_c' < 12{,}000$ psi
Tensile splitting strength	$7.4\sqrt{f_c'}$ psi for $3{,}000 < f_c' < 12{,}000$ psi

TABLE 2.8 Recommended Mechanical Properties of High-Strength Concrete

Recommended properties of high-strength concrete given in Ref. 18 are summarized in Table 2.8.

Volume Changes

According to Ref. 18, shrinkage in high-strength concrete is approximately proportional to the percentage of water by volume in a concrete mixture and is unaffected by changes in the water/cement ratio. Experimental studies have shown that shrinkage of high-strength concrete is similar to that of normal-strength concrete.

Figure 5.9 in Ref. 18 shows that the creep coefficient is less for high-strength concrete loaded at the same age. As is found in normal-strength concrete, creep decreases as the age at loading increases, and it increases with larger water/cement ratios.

Temperature Effects

Experimental studies on full-scale concrete columns have shown that the fire resistance of concrete with specified compressive strengths equal to or greater than 8,000 psi is smaller than that of lower-strength concrete.[21] When subjected to high temperatures, extremely high water vapor pressure builds up inside higher-strength concrete because of its high density (low permeability). Because there is virtually no means to relieve this pressure, the concrete spalls, and failure occurs shortly thereafter.

2.2.4 High-Performance Concrete

High-performance concrete is typically defined as concrete that meets specific combinations of performance and uniformity requirements that cannot always be achieved when using conventional ingredients and normal mixing, placing, and curing methods. A high-performance concrete mixture is designed to develop certain characteristics for a particular application and environment and contains carefully selected high-quality ingredients. The final product is batched, mixed, and finished to the highest industry standards.

The following are some of the properties and characteristics that may be required for high-performance concrete:

- High strength
- High early strength
- High modulus of elasticity
- High durability
- Low permeability
- Resistance to chemicals
- Resistance to frost
- Ease of placement

High-strength concrete is a common type of high-performance concrete. However, achieving high strength is not always necessary. For example, a normal-strength concrete with high durability and low permeability may be specified to satisfy the performance requirements of a parking structure that is exposed to the environment.

It is important to work with a concrete production facility during the early stages of a project to establish a concrete mixture design that will satisfy all of the required performance criteria.

2.2.5   Fiber-Reinforced Concrete

Fiber-reinforced concrete (FRC) is a composite material consisting of cement, water, aggregate, and discontinuous fibers that are dispersed throughout the mix.[22] The fibers are typically steel, polypropylene, or glass. In structural applications, fiber reinforcement is usually used in a role ancillary to the main steel reinforcement and is added to concrete mixes mainly to improve durability and crack control. Information on the mechanical properties of FRC is available in Ref. 23.

Steel fiber–reinforced concrete is permitted to be used under certain conditions in beams as an alternative to shear reinforcement (see ACI 3.5.1 and 11.4.6). The steel fibers must conform to ASTM A820/A820M-06, *Standard Specification for Steel Fibers for Fiber-reinforced Concrete*. The three conditions in ACI 5.6.6.2 must be satisfied in order for FRC to be deemed acceptable. Strength testing of FRC must satisfy the same requirements as those for concrete without fibers (ACI 5.6.1), and testing must follow the method in ASTM C1609/C1609M-07, *Standard Test Method for Flexural Performance of Fiber-reinforced Concrete (Using Beam with Third-point Loading)*.

2.3   Reinforcement

2.3.1   Overview

Reinforcement is utilized in concrete members to resist primarily tensile forces caused by externally applied loads or volume changes. The most common types of reinforcement are reinforcing bars, prestressing steel, and wire reinforcement made of steel.

ACI 3.5 contains the material requirements for steel reinforcement. The different types of reinforcement and the corresponding ASTM standards are summarized in Table 2.9. Note that all of the referenced ASTM standards are combined standards; that is, the metric (M) designation is included in the official designation of the standard. The referenced ACI sections are also given in the table; these sections contain additional requirements for some types of reinforcement.

Galvanized reinforcing bars, epoxy-coated bars and wires, and stainless-steel bars and wires are commonly used in parking structures, bridges, and other structures in highly corrosive environments where corrosion resistance of reinforcement is of particular concern.

Prestressing steel generally consists of wires, bars, strands, or bundles of such elements. The steel is stressed under high-tension forces either before the concrete is cast (pretensioned) or after the concrete is cast and has hardened (posttensioned).

One type of reinforcement that is not addressed in the Code is fiber-reinforced polymer (FRP) reinforcing bars. These bars are made of composite materials that consist of high-strength fibers embedded in a resin matrix. Fibers provide strength and stiffness

Type		ASTM Standard	ACI Section[*]
Deformed Reinforcement			
Reinforcing bars	Carbon steel	ASTM A615/A615M-07, Standard Specification for Deformed and Plain Carbon Steel Bars for Concrete Reinforcement	3.5.3.1(a)
	Low-alloy steel	ASTM A706/A706M-06a, Standard Specification for Low-alloy Steel Deformed and Plain Bars for Concrete Reinforcement	3.5.3.1(b)
	Stainless steel	ASTM A955/A955M-07a, Standard Specification for Deformed and Plain Stainless-Steel Bars for Concrete Reinforcement	3.5.3.1(c)
	Rail steel and axle steel	ASTM A996/A996M-06a, Standard Specification for Rail-Steel and Axle-Steel Deformed Bars for Concrete Reinforcement	3.5.3.1(d)
	Low-carbon chromium	ASTM A1035/A1035M-07, Standard Specification for Deformed and Plain, Low-carbon, Chromium, Steel Bars for Concrete Reinforcement	3.5.3.3
	Bar mats	ASTM A184/A184M-06, Standard Specification for Welded Deformed Steel Bar Mats for Concrete Reinforcement	3.5.3.4
	Galvanized	ASTM A767/A767M-05, Standard Specification for Zinc-coated (Galvanized) Steel Bars for Concrete Reinforcement	3.5.3.8
	Epoxy-coated	ASTM A775/A775M-07a, Standard Specification for Epoxy-coated Steel Reinforcing Bars	3.5.3.8
		ASTM A934/A934M-07, Standard Specification for Epoxy-coated Prefabricated Steel Reinforcing Bars	
Wire reinforcement	Deformed	ASTM A496/A496M-07, Standard Specification for Steel Wire, Deformed, for Concrete Reinforcement	3.5.3.5
	Welded plain	ASTM A185/A185M-07, Standard Specification for Steel Welded Wire Reinforcement, Plain, for Concrete	3.5.3.6

TABLE 2.9 Material Requirements for Steel Reinforcement (*continued*)

Type		ASTM Standard	ACI Section*
	Welded deformed	ASTM A497/A497M-07, Standard Specification for Steel Welded Wire Reinforcement, Deformed, for Concrete	3.5.3.7
	Epoxy-coated wires and welded wires	ASTM A884/A884M-06, Standard Specification for Epoxy-coated Steel Wire and Welded Wire Reinforcement	3.5.3.9
	Deformed stainless-steel wire and deformed stainless-steel welded wire	ASTM A1022/A1022M-07, Standard Specification for Deformed and Plain Stainless Steel Wire and Welded Wire for Concrete Reinforcement	3.5.3.10
Plain Reinforcement			
Plain bars		ASTM A615/A615M-07, Standard Specification for Deformed and Plain Carbon Steel Bars for Concrete Reinforcement	3.5.4.1
		ASTM A706/A706M-06a, Standard Specification for Low-alloy Steel Deformed and Plain Bars for Concrete Reinforcement	
Plain wire		ASTM A82/A82M-07, Standard Specification for Steel Wire, Plain, for Concrete Reinforcement	3.5.4.2
Headed Shear Stud Reinforcement			
Headed studs and headed stud assemblies		ASTM A1044/A1044M-05, Standard Specification for Steel Stud Assemblies for Shear Reinforcement of Concrete	3.5.5.1
Prestressing Steel			
Wire		ASTM A421/A421M-05, Standard Specification for Uncoated Stress-relieved Steel Wire for Prestressed Concrete	3.5.6.1(a)
Low-relaxation wire		ASTM A421/A421M-05, Standard Specification for Uncoated Stress-relieved Steel Wire for Prestressed Concrete	3.5.6.1(b)
Strand		ASTM A416/A416M-06, Standard Specification for Steel Strand, Uncoated Seven-wire for Prestressed Concrete	3.5.6.1(c)

TABLE 2.9 Material Requirements for Steel Reinforcement (*continued*)

Type	ASTM Standard	ACI Section*
High-strength bar	ASTM A722/A722M-07, Standard Specification for Uncoated High-strength Steel Bars for Prestressing Concrete	3.5.6.1(d)
Headed deformed bars	ASTM A970/A970M-06, Standard Specification for Headed Steel Bars for Concrete Reinforcement	3.5.9

* See the referenced ACI sections for additional requirements.

TABLE 2.9 Material Requirements for Steel Reinforcement (*continued*)

to the composite and generally carry most of the applied loads in tension. The matrix acts to bond and protect the fibers and transfers forces from fiber to fiber through shear. The most common fibers are aramid, carbon, and glass. This type of reinforcement is generally used in highly corrosive environments or in structures that house magnetic resonance imaging units or other equipment sensitive to electromagnetic fields. Additional information on the properties of FRP reinforcing bars is given in Ref. 24, and a specification for FRP reinforcing bars can be found in Ref. 25.

ACI 3.5.7 contains material requirements for structural steel, steel pipe, and steel tubing that are used in composite compression members (members constructed of concrete and steel with or without longitudinal reinforcing bars; see ACI 10.13). Structural steel wide-flange sections are encased in concrete, whereas steel pipe or tubing usually encases a concrete core.

Material requirements for steel discontinuous fiber reinforcement are given in ACI 3.5.8. Also included in this section are limitations on the length-to-diameter ratio of the fibers.

The focus of this book is on the design of reinforced concrete structural members with nonprestressed reinforcement, such as deformed reinforcing bars and welded wire reinforcement. Numerous references are available for the design of concrete members utilizing prestressed reinforcement.

2.3.2 Welding of Reinforcing Bars

Where welding of reinforcing bars is permitted, it must be performed in compliance with the requirements of the *Structural Welding Code—Reinforcing Steel*, ANSI/AWS D1.4.[26] ACI R3.5.2 provides information on the weldability of reinforcing steel and provides guidance on welding to existing reinforcing bars where no mill test reports on the existing steel are available.

Welded splices of reinforcement in special moment frames and in special structural walls are permitted only at specific locations within these members (ACI 21.1.7). Such splices are restricted at locations where it is anticipated that the member will yield because the tension stresses developed in the reinforcement in these regions caused by seismic excitations can easily exceed the strength requirements of ACI 12.4.3.4 for a fully developed weld splice.

FIGURE 2.12
Deformed
reinforcing bars.
(*Courtesy of the
Concrete
Reinforcing Steel
Institute.*)

2.3.3   Deformed Reinforcement

Deformed Reinforcing Bars

General   Deformed reinforcing bars are circular rods with deformations rolled into the surface (see Fig. 2.12). The purpose of the deformations (commonly referred to as ribs) is to enhance the bond between the concrete and the bar (in order for a reinforced concrete member to perform as designed, it is essential that a strong bond exist between reinforcing bars and concrete). Reinforcing bars are placed at judicious locations in the formwork before concrete is cast around them, and they generally do not undergo any significant amount of stress until the structural member is subjected to external loads.

Carbon-steel reinforcing bars conforming to the requirements of ASTM A615 are the most commonly specified type of reinforcing bar and can be used in a wide variety of applications where there are no special performance requirements. ASTM A706 low-alloy bars are specified in situations where enhanced weldability and ductility are needed. ACI 21.1.5 requires that reinforcement in special moment frames, special structural walls, and coupling beams in structures located in areas of high seismic risk comply with ASTM A706 (ASTM A615 bars that satisfy the special tensile strength and yield strength requirements of ACI 21.1.5.2 may also be used in such cases).

Stainless-steel, galvanized, and epoxy-coated reinforcing bars are usually used in applications where high corrosion resistance is needed. The physical and mechanical properties of stainless-steel bars conforming to ASTM A955 are the same as those for carbon-steel bars conforming to ASTM A615 with some exceptions (see the next section). Stainless-steel bars are also used where low magnetic permeability is required.

Galvanized (zinc-coated) bars are obtained by dipping ASTM A615, ASTM A706, or ASTM 996 bars in a molten bath of zinc in accordance with ASTM A767. Reinforcing bars are usually galvanized after fabrication.

Epoxy-coated bars are manufactured in one of two ways. In the first method, a protective epoxy coating is applied by the electrostatic spray method to ASTM A615, ASTM A706, or ASTM 996 bars (ASTM A775). The bars are usually fabricated after application of the epoxy coating. In the second method, ASTM A615, ASTM A706, or ASTM 996 bars are prefabricated and then coated with a protective fusion-bonded epoxy coating by electrostatic spray or other suitable method (ASTM A934).

Low-carbon chromium bars conforming to ASTM A1035 are permitted to be used only as spiral reinforcement in accordance with ACI 10.9.3 or transverse reinforcement in columns of special moment frames in accordance with ACI 21.6.4 (ACI 3.5.3.3). These limitations are imposed because the chromium steel used to manufacture reinforcing bars has low ductility and a relatively large minimum yield strength of 80,000 psi.

Bar Sizes and Grades   Bar sizes are designated by numbers, which range from No. 3 (No. 10) to No. 18 (No. 57). The numbers inside the parentheses following the inch-pound designations are the soft metric bar size designations. More information on soft metric bars is given later.

Designating bar sizes by numbers instead of diameters is necessary for deformed bars because of the surface deformations. Requirements for the height, spacing, and gap of deformations are given in the ASTM standards. The cross-section of a deformed reinforcing bar is illustrated in Fig. 2.13. It is evident from the figure that the overall bar diameter is greater than the nominal bar diameter. In general, nominal dimensions of a deformed bar are equivalent to those of a round bar without deformations that has the same weight per foot as the deformed bar.

For bars up to and including No. 8 bars, the nominal diameter is equal to the bar number multiplied by one-eighth of an inch. For example, the nominal diameter of a No. 3 bar is equal to $3 \times (1/8) = 3/8$ in. For Nos. 9, 10, 11, and 18 bars, the nominal diameter is greater than the bar number times one-eighth of an inch, and for No. 14 bars, the nominal diameter is smaller.

The nominal cross-sectional area of a reinforcing bar is equal to $\pi d_b^2/4$, where d_b is the nominal diameter of the bar. Thus, for a No. 3 bar, the nominal cross-sectional area is equal to $\pi \times 0.375^2/4 = 0.11$ in^2. Nominal diameters and cross-sectional areas

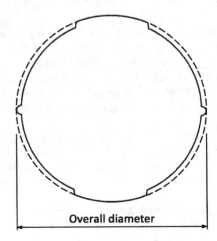

FIGURE 2.13 Overall reinforcing bar diameter.

Overall diameter

ASTM Designation	Bar Numbers	Grade	Minimum Yield Strength (psi)	Minimum Tensile Strength (psi)
A615	3, 4, 5, 6	40	40,000	60,000
	3, 4, 5, 6, 7, 8, 9, 10, 11, 14, 18	60	60,000	90,000
	3, 4, 5, 6, 7, 8, 9, 10, 11, 14, 18	75	75,000	100,000
A706	3, 4, 5, 6, 7, 8, 9, 10, 11, 14, 18	60	60,000	80,000
A955	3, 4, 5, 6	40	40,000	70,000
	3, 4, 5, 6, 7, 8, 9, 10, 11, 14, 18	60	60,000	90,000
	6, 7, 8, 9, 10, 11, 14, 18	75	75,000	100,000

TABLE 2.10   Requirements for ASTM Deformed Reinforcing Bars

for Nos. 3 through 18 reinforcing bars are given in ACI Appendix E and in Table A.1 of Appendix A of this book.

Reinforcing bars with minimum yield strengths of 40, 50, 60, and 75 ksi are available. The corresponding grades are designated as Grades 40, 50, 60, and 75, respectively. Grade 60 bars are used in many common applications, and Grade 75 bars are utilized primarily as longitudinal reinforcement in concrete columns that support relatively large loads such as those in a high-rise building. Note that rail-steel and axle-steel deformed bars are available in Grade 50 (ASTM A996).

A maximum yield strength of 80,000 psi is permitted to be used in design calculations for reinforced concrete structural members (ACI 9.4). However, for spiral reinforcement (ACI 10.9.3) and for confinement reinforcement in special moment frames and special structural walls (ACI 21.1.5.4), a yield strength up to 100,000 psi can be used. ACI R9.4 contains other sections of the Code that limit the yield strength of reinforcement.

Information on sizes, grades, and tensile properties of ASTM A615, ASTM A706, and ASTM A955 deformed reinforcing bars is given in Table 2.10. Additional tensile and bending requirements are given in the ASTM standards for all bar types. Note that not all bar sizes are available in all grades. It is prudent to verify the availability of bar sizes and grades with local reinforcement suppliers at the onset of a project. A list of U.S. manufacturers of Grades 60 and 75 reinforcing bars can be found in Appendix A of Ref. 27.

ASTM specifications require that identification marks be rolled onto the surface of one side of a reinforcing bar, as shown in Fig. 2.14 for Grade 60 bars (see Ref. 27 for marks on bars of other grades). The marks—mill designation of the producer, bar size, type of steel, and grade—facilitate bar identification in the field as the structure is being built.

The first identification mark, which is usually a letter, identifies the mill that produced the reinforcing bar. Mill identification marks for U.S. manufacturers of Grades 60 and 75 reinforcing bars are given in Appendix A of Ref. 27.

FIGURE 2.14
Identification marks
for Grade 60
reinforcing bars.
(*Courtesy of the
Concrete
Reinforcing Steel
Institute.*)

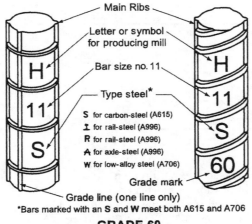

Main Ribs

Letter or symbol
for producing mill

Bar size no. 11

Type steel*

S for carbon-steel (A615)
I for rail-steel (A996)
R for rail-steel (A996)
A for axle-steel (A996)
W for low-alloy steel (A706)

Grade mark

Grade line (one line only)
*Bars marked with an **S** and **W** meet both A615 and A706

GRADE 60

The second identification mark is the bar size. In the United States, virtually all reinforcing bars that are currently produced are soft metric bars that meet the metric requirements of ASTM specifications. It is important to note that the dimensions of soft metric bars are the same as those of inch-pound bars. Metric designations, which are equal to the nominal diameter of an inch-pound bar converted to the nearest millimeter, are essentially a relabeling of inch-pound designations. For example, a No. 3 bar has a nominal diameter of 0.375 in or, equivalently, 9.5 mm, which is rounded up to 10 mm. Thus, a No. 10 bar (metric) is equivalent to a No. 3 bar (inch-pound). Inch-pound designations and the corresponding metric designations are given in Table 2.11. Inch-pound designations for bar size are used exclusively in this book.

The third mark identifies the type of steel that was used in the production of the reinforcing bar. For example, an identification mark of "S" means that the bar was

Inch-pound Designation (No.)	Metric designation (No.)
3	10
4	13
5	16
6	19
7	22
8	25
9	29
10	32
11	36
14	43
18	57

TABLE 2.11 Inch-pound and Metric Reinforcing Bar Designations

ASTM Designation		Identification Mark
A615		S
A706		W
A996	Rail	I
		R
	Axle	A

TABLE **2.12**   Identification Marks for Steel Type

produced of steel that meets the requirements of ASTM A615. A summary of ASTM designations and identification marks is given in Table 2.12.

The fourth identification mark is the minimum yield strength (grade) designation, which is required for Grades 60 (420) and 75 (520) bars only; Grades 40 (280) and 50 (350) bars are required to have only the first three identification marks. The numbers in parentheses following the inch-pound grade designations are the corresponding metric grade designations in megapascals. The minimum yield strength of metric bars is slightly greater than the corresponding minimum yield strength of inch-pound bars. For example, the equivalent inch-pound yield strength of Grade 420 bars is equal to 420/(6.895 Mpa/ksi) = 60.9 ksi, which is greater than the yield strength of Grade 60 bars (60 ksi).

ASTM specifications provide two options for identifying Grade 420 or 520 soft metric reinforcing bars. In the first option, the first digit in the grade number is rolled onto the surface of the bar. An identification mark of "4" corresponds to Grade 420 (60) bars, and a mark of "5" corresponds to Grade 520 (75) bars. In the second option, one additional longitudinal rib (grade line) is rolled onto the bar to designate Grade 420 (60) bars, and two additional longitudinal ribs are rolled onto the bar to designate Grade 520 (75) bars. Inch-pound designations for the minimum yield strength are used exclusively in this book.

Mechanical Properties   Reinforcing steel is manufactured under strict quality control conditions, and as noted previously, the final product must have properties that satisfy applicable ASTM standards. Unlike concrete, there is relatively little variability between the actual properties of reinforcing steel and the specified properties, primarily because of the characteristics of the material. Some basic mechanical properties of reinforcing steel that are pertinent to the design of reinforced concrete structural members are examined later.

Tensile stress–strain curve   Idealized tensile stress–strain curves for Grades 40, 60, and 75 reinforcing bars are given in Fig. 2.15. It is clear from the figure that there are essentially two distinct types of behavior that are dependent on the properties of the steel and the manufacturing process.

In the first type, the stress–strain curve has three distinct parts: (1) an initial linear elastic part up to the yield strength f_y, (2) a relatively flat yield plateau up to the onset of strain hardening, and (3) a strain-hardening part. In the second type, there are two distinct parts: (1) an initial linear elastic part up to the proportional limit and (2) a strain-hardening part.

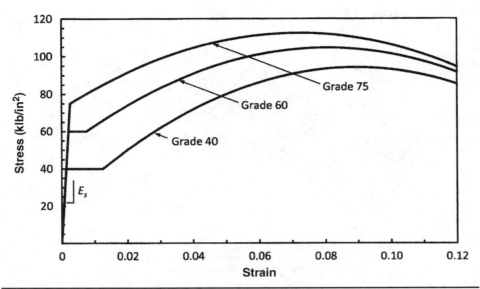

FIGURE 2.15   Idealized tensile stress–strain curves for reinforcing steel.

The initial elastic part of the stress–strain curve is essentially the same for all grades of reinforcing steel. Stress is directly proportional to strain, and the modulus of elasticity E_s is equal to 29,000 ksi (ACI 8.5.2).

The stress–strain curve for low-carbon Grade 40 steel exhibits a relatively long yield plateau where strain continues to increase at constant stress. The yield strength of the steel is established at this plateau. Grade 60 steel will usually have a yield plateau that is shorter in length than Grade 40 steel (or it may have no plateau at all), whereas Grade 75 steel typically enters strain hardening without exhibiting distinct yielding. In cases where $f_y > 60$ ksi, ACI 3.5.3.2 requires that the yield strength of the steel be taken as the stress that corresponds to a strain of 0.0035.

In the strain-hardening part of the stress–strain curve, stress increases with increasing strain. The tensile strength of the steel is reached at the top of the curve, and after this point, stress decreases with increasing strain until fracture occurs.

When comparing the strength of reinforcing steel with that of plain concrete, it is evident that the yield strength of reinforcing steel is at least 10 times the compressive strength of normal-strength concrete and approximately 100 times the tensile strength of concrete.

Thermal expansion   The average value of the thermal-expansion coefficient of reinforcing steel ($6.5 \times 10^{-6}/°F$) is relatively close to that for concrete ($5.5 \times 10^{-6}/°F$). Thus, differential movement between the reinforcing bars and the surrounding concrete is relatively small because of changes in temperature, and detrimental effects such as cracking or loss of bond are usually negligible. Having thermal expansion coefficients that are compatible is important in order for the reinforced concrete member to perform as designed.

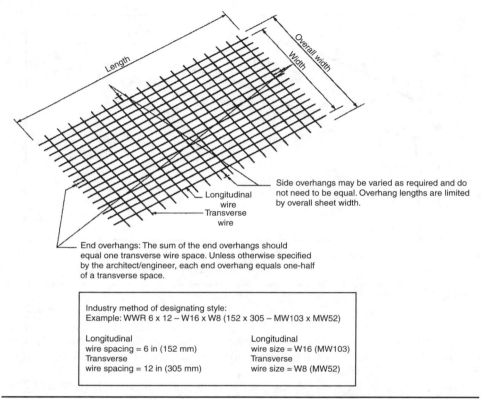

Side overhangs may be varied as required and do not need to be equal. Overhang lengths are limited by overall sheet width.

Longitudinal wire

Transverse wire

End overhangs: The sum of the end overhangs should equal one transverse wire space. Unless otherwise specified by the architect/engineer, each end overhang equals one-half of a transverse space.

Industry method of designating style:
Example: WWR 6 x 12 – W16 x W8 (152 x 305 – MW103 x MW52)

Longitudinal
wire spacing = 6 in (152 mm)
Transverse
wire spacing = 12 in (305 mm)

Longitudinal
wire size = W16 (MW103)
Transverse
wire size = W8 (MW52)

FIGURE 2.16   Style identification of WWR. (*Courtesy of the Concrete Reinforcing Steel Institute.*)

Temperature effects   Similar to concrete, reinforcing bars lose strength when subjected to high temperatures. Figure 2.11 of Ref. 28 shows yield and tensile degradation with respect to temperature. The yield and tensile strengths are a little over 70% of their original values at a temperature of 750°F and are less than 20% at 1,400°F.

Providing sufficient concrete cover protects reinforcing bars from the adverse effects of high temperatures.

Welded Wire Reinforcement

General   Welded wire reinforcement (WWR) consists of high-strength cold-drawn or cold-rolled wires that are arranged in a square or rectangular pattern and are welded together at their intersections (see Fig. 2.16). Like reinforcing bars, the WWR sheets are placed before the concrete is cast, and no significant stress occurs in the WWR until the structural member is subjected to external loads. In the case of plain WWR, bond is developed with the concrete at each welded intersection of wires. Both the welded intersections and wire deformations are utilized in deformed WWR to provide bond and anchorage.

Style Designations   The spacing and size of wires in WWR are identified by style designations. The first part of the designation corresponds to the spacing of the longitudinal and transverse wires, and the second part corresponds to the size of the longitudinal and transverse wires. Plain wire is denoted by the letter "W" and deformed wire by the letter "D." The equivalent metric designations are "MW" and "MD," respectively. The number following the letter gives the cross-sectional area of the wire in hundredths of a square inch. For example, WWR $6 \times 12 - W16 \times W8$ corresponds to the following (see Fig. 2.16):

- Longitudinal wire spacing = 6 in
- Transverse wire spacing = 12 in
- Longitudinal plain wire size = W16 (0.16 in^2)
- Transverse plain wire size = W8 (0.08 in^2)

The equivalent metric style designation is WWR $152 \times 305 - MW103 \times MW52$, where wire spacing is in millimeters and wire area in square millimeters. Note that the terms "longitudinal" and "transverse" are related to the manufacturing process of the WWR and do not refer to the position of the wires in a concrete member. A welded deformed wire style is designated in the same way except that "D" is substituted for "W." Like deformed reinforcing bars, the nominal cross-sectional area of a deformed wire is determined from the weight per foot of wire rather than from the diameter.

The cross-sectional areas for common styles of WWR sheets are given in ACI Appendix E and in Table A.2 of Appendix A of this book.

Mechanical Properties   WWR must conform to the ASTM standards noted in Table 2.9 and to the applicable requirements of ACI 3.5.3. The specified minimum yield strength and minimum tensile strength are 70,000 and 80,000 psi, respectively, for welded deformed wire and 65,000 and 75,000 psi, respectively, for welded plain wire. The minimum shear strength of the welds is also provided in the ASTM standards.

The tensile stress–strain curve of WWR is similar to that for Grade 75 reinforcing bars. Because there is no definite yield plateau, the yield strength is taken as the stress that corresponds to a strain of 0.0035 (ACI 3.5.3.6 and 3.5.3.7). Additional information on the properties of WWR can be found in Ref. 29.

References

1. Kosmatka, S., Kerkhoff, B., and Panarese, W. 2002 (Rev. 2008). *Design and Control of Concrete Mixtures*, 14th ed. Portland Cement Association (PCA), Skokie, IL.
2. Neville, G. B. 2008. *Concrete Manual—Concrete Quality and Field Practices.* International Code Council (ICC), Washington, DC.
3. Dhir, R. H., and Sangha, M. 1974. Development and propagation of microcracks in concrete. *Materiaux et Constructions* 37:17-23.
4. Hsu, T. T. C., Slate, F. O., Sturman, G. M., and Winter, G. 1963. Microcracking of plain concrete and the shape of the stress–strain curve. *ACI Journal Proceedings* 60(2):209-224.
5. Robinson, G. S. 1968. Methods of detecting the formation and propagation of microcracks in concrete. In: Brooks, A. E., and Newman, K., eds. *The Structure of Concrete.* Cement and Concrete Association, London, pp. 131-145.
6. Slate, F. O., and Olsefski, S. 1974. X-rays for study of internal structure and microcracking of concrete. *ACI Journal Proceedings* 60(5):575-588.
7. Diederich, U., Schneider, U., and Terrien, M. 1983. Formation and propagation of cracks and acoustic emission. In: Wittmann, F. H., ed. *Fracture Mechanics of Concrete.* Elsevier, Amsterdam, The Netherlands.

8. American Concrete Institute (ACI) Committee 211. 1991. (Reapproved 2002). *Standard Practice for Selecting Proportions for Normal, Heavyweight, and Mass Concrete*, ACI 211.1. ACI, Farmington Hills, MI.

9. American Concrete Institute (ACI), Committee 211. 1998. (Reapproved 2004). *Standard Practice for Selecting Proportions Lightweight Concrete*, ACI 211.2. ACI, Farmington Hills, MI.

10. Kupfer, H., Hilsdorf, H., and Rusch, H. 1968. Behavior of concrete under biaxial stresses. *Journal of the American Concrete Institute* 66(8):656-666.

11. Tasuji, M. E., Slate, F. O., and Nilson, A. H. 1978. Stress–strain response and fracture of concrete in biaxial loading. *Journal of the American Concrete Institute* 75(7):306-312.

12. Richart, F. E., Brandtzaeg, A., and Brown, R. L. 1928. *A Study of the Failure of Concrete Under Combined Compressive Stresses*, Bulletin 185. University of Illinois Engineering Experiment Station, Urbana, IL, 104 p.

13. Tazawa, E., ed. 1999. *Autogenous Shrinkage of Concrete*. E&FN Spon, London, 424 p.

14. American Concrete Institute (ACI), Committee 209. 2005. *Report on Factors Affecting Shrinkage and Creep of Hardened Concrete*, ACI 209.1R-05. ACI, Farmington Hills, MI.

15. American Concrete Institute (ACI), Committee 209. 1992. (Reapproved 2008). *Prediction of Creep, Shrinkage, and Temperature Effects in Concrete Structures*, ACI 209R-92. ACI, Farmington Hills, MI.

16. American Concrete Institute (ACI), Committee 209. 2008. *Guide for Modeling and Calculating Shrinkage and Creep in Hardened Concrete*, ACI 209.2R-08. ACI, Farmington Hills, MI.

17. American Concrete Institute (ACI), Committee 216. 1981. *Guide for Determining the Fire Endurance of Concrete Elements*, ACI 216R-81. ACI, Farmington Hills, MI. Also in *Concrete International* 3(2):13-47.

18. American Concrete Institute (ACI), Committee 363. 1992. (Reapproved 1997). *Report on High-strength Concrete*, ACI 363R-92. ACI, Farmington Hills, MI.

19. American Concrete Institute (ACI), Committee 363. 1998. *Guide to Quality Control and Testing of High-strength Concrete*, ACI 363.2R-98. ACI, Farmington Hills, MI.

20. Portland Cement Association (PCA). 1994. *High-strength Concrete*, EB114.01T. PCA, Skokie, IL.

21. Kodur, V. K. R., McGrath, R., Leroux, P., and Latour, J. C. 2004. *Fire Endurance Experiments on High-strength Concrete Columns*, Research and Development Bulletin RD138. Portland Cement Association (PCA), Skokie, IL.

22. American Concrete Institute (ACI), Committee 544. 2008. *Guide for Specifying, Proportioning, and Production of Fiber-reinforced Concrete*, ACI 544.3R-08. ACI, Farmington Hills, MI.

23. American Concrete Institute (ACI), Committee 544. 1996. (Reapproved 2002). *Report on Fiber Reinforced Concrete*, ACI 544.1R-96. ACI, Farmington Hills, MI.

24. American Concrete Institute (ACI), Committee 440. 2006. *Guide for the Design and Construction of Structural Concrete Reinforced with FRP Bars*, ACI 440.1R-06. ACI, Farmington Hills, MI.

25. American Concrete Institute (ACI), Committee 440. 2008. *Specification for Construction with Fiber-Reinforced Polymer Reinforcing Bars*, ACI 440.5-08. ACI, Farmington Hills, MI.

26. American Welding Society (AWS). 2005. *Structural Welding Code—Reinforcing Steel*, AWS D1.4/D1.4M. AWS, Miami, FL.

27. Concrete Reinforcing Steel Institute (CRSI), Committee on Manual of Standard Practice. 2009. *Manual of Standard Practice*, 28th edn. CRSI, Schaumburg, IL.

28. American Concrete Institute (ACI), Joint ACI/TMS Committee 216. 2007. *Code Requirements for Determining Fire Resistance of Concrete and Masonry Assemblies*, ACI 216.1-07/TMS-0216-07. ACI, Farmington Hills, MI.

29. Wire Reinforcement Institute (WRI). 2008. *Structural Welded Wire Reinforcement Manual of Standard Practice*. WRI, Hartford, CT.

Problems

2.1. The strength test results for a normal-weight concrete mixture with a specified compressive strength f_c' of 7,000 psi are given in Table 2.13. Strength tests were obtained from 6×12 in cylinders.

Test Number	Mix Code	F28 Test 1 (psi)	F28 Test 2 (psi)	28-day Average (psi)	28-day Average, Three Consecutive Tests (psi)
1	L-401 7K	7,580	7,940	7,760	—
2	L-401 7K	7,580	7,740	7,660	—
3	L-401 7K	6,670	6,880	6,775	7,398
4	L-401 7K	7,330	7,480	7,405	7,280
5	L-401 7K	7,710	7,600	7,655	7,278
6	L-401 7K	7,460	7,280	7,370	7,477
7	L-401 7K	7,650	7,150	7,400	7,475
8	L-401 7K	7,530	7,230	7,380	7,383
9	L-401 7K	7,930	7,620	7,775	7,518
10	L-401 7K	6,970	7,030	7,000	7,385
11	L-401 7K	7,400	7,030	7,215	7,330
12	L-401 7K	7,400	7,340	7,370	7,195
13	L-401 7K	8,200	7,030	7,615	7,400
14	L-401 7K	7,740	8,020	7,880	7,622
15	L-401 7K	7,620	8,030	7,825	7,773
16	L-401 7K	7,460	7,610	7,535	7,747
17	L-401 7K	7,370	7,380	7,375	7,578
18	L-401 7K	7,280	7,770	7,525	7,478
19	L-401 7K	8,230	7,520	7,875	7,592
20	L-401 7K	7,760	7,350	7,555	7,652
21	L-401 7K	7,860	7,940	7,900	7,777
22	L-401 7K	7,760	7,720	7,740	7,732
23	L-401 7K	7,600	7,670	7,635	7,758
24	L-401 7K	6,720	6,620	6,670	7,348
25	L-401 7K	6,150	6,380	6,265	6,857
26	L-401 7K	7,990	7,710	7,850	6,928
27	L-401 7K	7,900	7,460	7,680	7,265
28	L-401 7K	8,200	8,460	8,330	7,953
29	L-401 7K	8,030	7,820	7,925	7,978
30	L-401 7K	7,030	7,210	7,120	7,792

TABLE 2.13   Strength Test Results of Problem 2.1

    1. Determine (a) the sample standard deviation s_s and (b) the required average compressive strength f'_{cr}.

    2. Determine if the concrete is acceptable in accordance with ACI requirements.

2.2.   Repeat Problem 2.1 using only the first 15 test records given in Table 2.13.

2.3.   Determine the following quantities for concrete that is exposed to freeze–thaw cycles and is in continuous contact with moisture and is furthermore exposed to deicing chemicals: (a) the maximum water/cementitious materials ratio, (b) the minimum specified compressive strength, and (c) any additional requirements based on Chap. 4 of the ACI Code.

2.4.   Determine the following quantities for concrete that is exposed to moisture but not to external sources of chlorides: (a) the maximum water/cementitious materials ratio, (b) the minimum specified compressive strength, and (c) any additional requirements based on Chap. 4 of the ACI Code.

2.5.   Determine the modulus of elasticity E_c for a lightweight concrete mixture with a unit weight of 110 pcf and f'_c equal to (a) 3,000, (b) 4,000, and (c) 5,000 psi.

2.6.   Determine the modulus of rupture that is to be used when calculating deflections for sand-lightweight concrete with $f'_c = 4,000$ psi.

2.7.   Determine the shrinkage strain of an unrestrained, plain concrete member with $f'_c = 5,000$ psi as a function of the volume/surface ratio at 2 years after the concrete has been placed in an environment with a constant relative humidity of 70%. Plot the results.

2.8.   Determine the creep coefficient of an unrestrained, plain concrete member with $f'_c = 5,000$ psi as a function of the volume/surface ratio at 2 years after the concrete has been loaded in an environment with a constant relative humidity of 70%. The age of the concrete at time of loading is 60 days. Plot the results.

2.9.   Determine the following quantities for a concrete mixture with a specified compressive strength of 10,000 psi: (a) the modulus of elasticity, (b) the modulus of rupture, and (c) the tensile splitting strength.

2.10.   Identify (a) the bar size, (b) the type of steel, and (c) the grade of steel in both inch-pound and metric units for a steel reinforcing bar with the following markings: C, 29, W, 4.

CHAPTER 3

General Considerations for Analysis and Design

3.1 Introduction

Three major steps are typically undertaken in the analysis and design of any building or structure: (1) determine nominal loads; (2) perform a structural analysis; and (3) design the structural members. The procedures and methods within each step depend on a number of factors, including the type of material used in construction.

Presented in this chapter is the basic information that is needed to analyze reinforced concrete structural members (steps 1 and 2). Nominal loads are the minimum loads that must be used in the design of a structure and are categorized as permanent loads (such as the weight of the structure) and variable loads (including live, snow, wind, and earthquake loads). The magnitudes of nominal loads are specified in building codes (e.g., Chap. 16 of the IBC), and it is important to establish the governing code at the onset of any project. A summary of nominal loads commonly encountered in the design of reinforced concrete structures is given in Section 3.2.

An analysis of the structure is performed using the nominal loads determined in the first step and the load combinations given in Chap. 9 of the Code. General requirements for the analysis and design of concrete structures, which can be found in Chap. 8 of the ACI Code, are covered in Section 3.3. Included is an approximate analysis method that can be used to determine bending moments and shear forces in continuous beams and one-way slabs.

Structural members are proportioned for adequate strength on the basis of the load effects from the structural analysis. Members are designed in accordance with the strength design method, which requires the use of the load factors and strength reduction factors given in Chap. 9 of the Code (ACI 8.1.1). General requirements of the strength design method are summarized in Chaps. 4 and 5, and design methods for specific structural members are given in Chaps. 6 through 10 of this book.

3.2 Loading

3.2.1 Introduction

Applicable nominal loads on a structure are determined from the general building code under which the project is to be designed and constructed. Chap. 16 of the IBC[1]

Notation	Load	Code Section
D	Dead load	IBC 1606
D_i	Weight of ice	Chap. 10 of ASCE/SEI 7
E	Combined effect of horizontal and vertical earthquake-induced forces as defined in ASCE/SEI 12.4.2	IBC 1613 and ASCE/SEI 12.4.2
E_m	Maximum seismic load effect of horizontal and vertical forces as set forth in ASCE/SEI 12.4.3	IBC 1613 and ASCE/SEI 12.4.3
F	Load due to fluids with well-defined pressures and maximum heights	—
F_a	Flood load	IBC 1612
H	Load due to lateral earth pressures, ground water pressure, or pressure of bulk materials	IBC 1610 (soil lateral loads)
L	Live load, except roof live load, including any permitted live load reduction	IBC 1607
L_r	Roof live load including any permitted live load reduction	IBC 1607
R	Rain load	IBC 1611
S	Snow load	IBC 1608
T	Self-straining force arising from contraction or expansion resulting from temperature change, shrinkage, moisture change, creep in component materials, movement due to differential settlement, or combinations thereof	—
W	Load due to wind pressure	IBC 1609
W_i	Wind-on-ice load	Chap. 10 of ASCE/SEI 7

TABLE 3.1 Summary of Loads Addressed in the IBC and ASCE/SEI 7

contains the minimum magnitudes of some nominal loads and references ASCE/SEI 7[2] for others. For a specific project, the governing local building code should be consulted for any variances from the IBC or ASCE/SEI 7.

It is common for nominal loads to be referred to as service loads. These loads are multiplied by load factors in the strength design method, which is the required design method for reinforced concrete members (ACI 8.1.1). The exception is the earthquake load effect E: It is defined to be a strength-level load where the load factor is equal to 1. Additional information on service-level and strength-level loads and on load combinations is given in Chap. 4 of this book.

Table 3.1 contains a list of loads from the IBC and ASCE/SEI 7. A brief discussion on some of the more commonly encountered loads in the design of reinforced concrete buildings follows the table. Comprehensive information on the determination of structural loads can be found in Ref. 3.

3.2.2 Dead Loads

Nominal dead loads D are the actual weights of construction materials and fixed service equipment that are attached to or supported by the building or structure. Specific examples of such loads are listed under the definition of "dead load" in IBC 1602.

Dead loads are considered to be permanent loads because their magnitude remains essentially constant over time. Variable loads such as live loads and wind loads are not permanent loads.

Superimposed dead loads are permanent loads other than the weights of the structural members and include the following: floor finishes and/or topping; walls; ceilings; heating, ventilating, and air-conditioning (HVAC) and other service equipment; fixed partitions; and cladding.

It is not uncommon for the weights of materials and service equipment (such as plumbing stacks and risers, HVAC equipment, elevators and elevator machinery, fire protection systems, and similar fixed equipment) not to be known during step 1 of the analysis and design process. Minimum design dead loads for various types of common construction components, including ceilings, roof and wall coverings, floor fill, floors and floor finishes, frame partitions, and frame walls, are provided in ASCE/SEI Table C3-1, and minimum densities for common construction materials are given in ASCE/SEI Table C3-2. The weights in these tables are meant to be used as a guide when estimating dead loads. Actual weights of construction materials and equipment can be greater than tabulated values, so it is always prudent to verify the weights with manufacturers or other similar resources. In cases where information on dead load is unavailable, values of dead loads used in design must be approved by the building official (IBC 1606.2).

Determining the dead load of reinforced concrete members is typically straightforward. In general, the total dead load is obtained by multiplying the volume of the member by the unit weight of concrete. Thus, for an 18×18 in reinforced concrete column that is 10 ft long and is made of concrete with a unit weight equal to 150 pcf, the total dead load $D = (18 \times 18/144) \times 10 \times 150 = 3,375$ lb.

The dead load of beams and one-way slabs (i.e., slabs that bend in primarily one direction) is usually expressed in pounds per linear foot of the member length. Consider a 20-in-wide and 24-in-deep reinforced concrete beam with a unit weight of 150 pcf. In this case, the dead load $D = (20 \times 24/144) \times 150 = 500$ lb per foot of beam length. For two-way reinforced concrete slabs (i.e., slabs that bend in two directions), the dead load is commonly expressed in pounds per square foot: $D =$ slab thickness $\times$ unit weight of concrete.

3.2.3   Live Loads

General

Live loads are transient in nature and vary in magnitude over the life of a structure. These loads are produced by the use and occupancy of a building or structure and do not include construction loads, environmental loads (such as wind loads, snow loads, rain loads, earthquake loads, and flood loads), or dead loads (IBC 1602).

IBC Table 1607.1 contains nominal design values of uniformly distributed and concentrated live loads L_o as a function of occupancy or use. The occupancy description listed in the table is not necessarily group-specific (occupancy groups are defined in IBC Chap. 3). For example, an office building with a Business Group B classification may also have storage areas that may warrant live loads of 125 or 250 psf depending on the type of storage, which are greater than the prescribed office live loads. Structural members are designed on the basis of the maximum effects due to application of either a uniform load or a concentrated load and need not be designed for the effects of both loads applied at the same time. The building official must approve live loads that are not specifically listed in the table.

Partitions that can be relocated (i.e., those types that are not permanently attached to the structure) are considered to be live loads in office and other buildings. A live load equal to at least 15 psf must be included for movable partitions if the nominal uniform floor live load is equal to or less than 80 psf.

IBC Table 1607.1 prescribes a minimum roof live load of 20 psf for typical roof structures; larger live loads are required for roofs used as gardens or places of assembly.

ASCE Table 4-1 also contains minimum uniform and concentrated live loads, and some of these values differ from those in IBC Table 1607.1. ASCE Tables C4-1 and C4-2 can be used as a guide in establishing live loads for some commonly encountered occupancies.

Reduction in Live Loads

Because live loads are transient in nature, the probability that a structural member will be subjected to the full effects from nominal live loads decreases as the area supported by the member increases. The minimum uniformly distributed live loads L_o in IBC Table 1607.1 and uniform live loads of special-purpose roofs are permitted to be reduced under certain circumstances in accordance with the methods in IBC 1607.9.1 or 1607.9.2. The general method of live load reduction in IBC 1607.9.1 is also given in ASCE/SEI 4.8. Reduction of roof loads must conform to IBC 1607.11.2.

General Method of Live Load Reduction   IBC Eq. (16-22) can be used to obtain a reduced live load L for members that support an area $K_{LL} A_T \geq 400$ ft^2:

$$L = L_o \left(0.25 + \frac{15}{\sqrt{K_{LL} A_T}} \right) \tag{3.1}$$

In this equation, K_{LL} is the live load element factor given in IBC Table 1607.9.1 and A_T is the tributary area supported by the member in square feet. The use of this equation is subject to the limitations of IBC 1607.9.1.1 through 1607.9.1.4.

The live load element factor K_{LL} converts the tributary area A_T into an influence area, which is considered to be the adjacent floor area from which the member derives its load. In other words,

$$K_{LL} = \text{influence area/tributary area} \tag{3.2}$$

Consider interior column B3 depicted in Fig. 3.1. The influence area for this column is equal to the area of the four bays adjacent to the column:

$$\text{Influence area} = (\ell_A + \ell_B)(\ell_2 + \ell_3)$$

The tributary area A_T supported by this column is equal to the product of the tributary widths in both directions:

$$A_T = \left(\frac{\ell_A}{2} + \frac{\ell_B}{2} \right) \left(\frac{\ell_2}{2} + \frac{\ell_3}{2} \right)$$

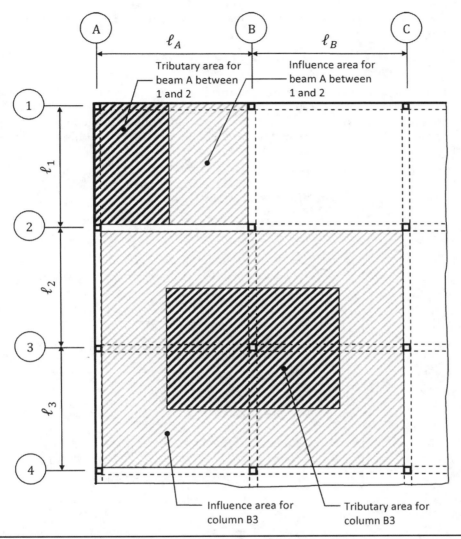

FIGURE 3.1   Influence areas and tributary areas for columns and beams.

Note that the tributary width in this case is equal to the sum of one-half of the span lengths on both sides of the column.

Using Eq. (3.2), it is evident that the live load element factor K_{LL} is equal to 4 for this interior column; this matches the value of K_{LL} given in IBC Table 1607.9.1. It can be shown that K_{LL} is also equal to 4 for any exterior column (other than corner columns) without cantilever slabs.

Now, consider the spandrel beam on line A between lines 1 and 2. The influence area for this beam is equal to the area of the bay adjacent to the beam:

$$\text{Influence area} = \ell_A \ell_1$$

The tributary area A_T supported by this beam is equal to the tributary width times the length of the beam:

$$A_T = \left(\frac{\ell_A}{2}\right)\ell_1$$

Thus, $K_{LL} = 2$ for an edge beam without a cantilever slab. Values of K_{LL} for other beams and other members can be derived in a similar fashion. ASCE/SEI Fig. C4-1 illustrates influence areas and tributary areas for a structure with regular bay spacing.

The general method of live load reduction outlined in IBC 1607.9.1 is summarized in Fig. 3.2.

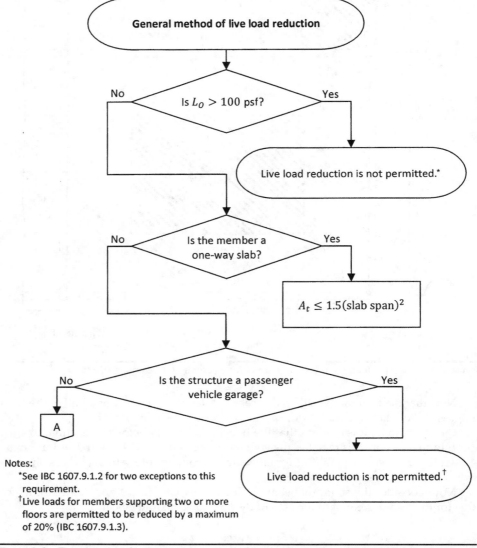

Notes:
*See IBC 1607.9.1.2 for two exceptions to this requirement.
†Live loads for members supporting two or more floors are permitted to be reduced by a maximum of 20% (IBC 1607.9.1.3).

FIGURE 3.2 The procedure for the general method of live load reduction of IBC 1607.9.1. (*Continued*)

Live Load Reduction for Roofs   IBC 1607.11.2 permits nominal roof live loads of 20 psf on flat, pitched, and curved roofs to be reduced in accordance with IBC Eq. (16-25):

$$L_r = L_o R_1 R_2 \tag{3.3}$$

In this equation, L_r is the reduced roof live load per square foot of horizontal roof projection and R_1 and R_2 are reduction factors based on the tributary area A_t of the

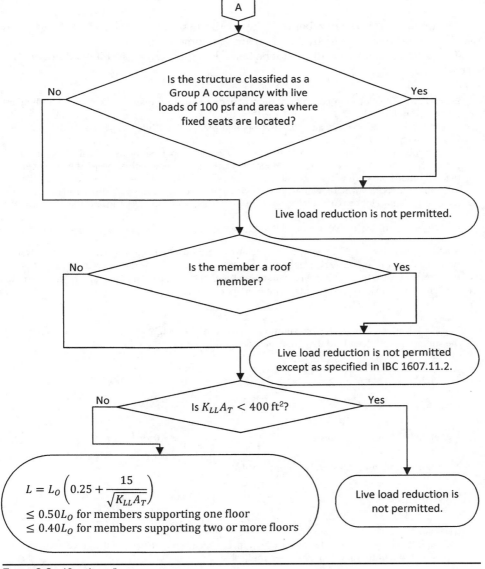

FIGURE 3.2   (*Continued*)

member being considered and the slope of the roof, respectively:

$$
R_1 = \begin{cases} 1 \text{ for } A_t \leq 200 \text{ ft}^2 & \text{IBC Eq. (16-26)} \\ 1.2 - 0.001 A_t \text{ for } 200 \text{ ft}^2 < A_t < 600 \text{ ft}^2 & \text{IBC Eq. (16-27)} \\ 0.6 \text{ for } A_t \geq 600 \text{ ft}^2 & \text{IBC Eq. (16-28)} \end{cases}
$$

$$
R_2 = \begin{cases} 1 \text{ for } F \leq 4 & \text{IBC Eq. (16-29)} \\ 1.2 - 0.05F \text{ for } 4 < F < 12 & \text{IBC Eq. (16-30)} \\ 0.6 \text{ for } F \geq 12 & \text{IBC Eq. (16-31)} \end{cases}
$$

The quantity F is the number of inches of rise per foot for a sloped roof and is the rise-to-span ratio multiplied by 32 for an arch or dome.

The reduced roof live load cannot be taken less than 12 psf, and it need not exceed 20 psf.

Example 3.1   The elevation and typical floor plan of a five-story reinforced concrete building are illustrated in Fig. 3.3.

All beams are 22 in wide and 20.5 in deep, and the wide-module joists are $16 + 4\frac{1}{2} \times 7 + 53$, which weigh 84 psf.

The roof is an ordinary flat roof (slope of 1/2 on 12) that is not used as a place of public assembly or for any special purposes. Floor level 5 is light storage with a specified live load of 125 psf. All other floors are office occupancy. Assume a 10 psf superimposed dead load on all levels including the roof. Normal-weight concrete with a unit weight of 150 pcf is used for all structural members. Neglect lobby/corridor loads on the typical floors, and neglect rain and snow loads on the roof.

Determine the total axial dead and live loads at the base of column B2 in the first story.

Solution   This example illustrates how to perform a column load rundown. This type of analysis is typically done during the early stages of a project to obtain the loads that need to be supported by the foundations so that preliminary foundation types and sizes can be obtained. The results of the load rundown are also used to obtain preliminary columns' sizes.

Dead Loads
The axial dead load supported by column B2 consists of the weight of the structural members and the superimposed dead load that are tributary to this column:

- Weight of joists $= \dfrac{84}{1,000} \times \dfrac{25 + 22}{2} \times 20 = 39.5$ kips

- Weight of beams $= \dfrac{22 \times 20.5}{144} \times 20 \times \dfrac{150}{1,000} = 9.4$ kips

- Weight of column (typical story) $= \dfrac{22 \times 22}{144} \times 10 \times \dfrac{150}{1,000} = 5.0$ kips

- Weight of column (first story) $= \dfrac{22 \times 22}{144} \times 12 \times \dfrac{150}{1,000} = 6.1$ kips

- Superimposed dead load $= \dfrac{10}{1,000} \times \dfrac{25 + 22}{2} \times 20 = 4.7$ kips

Live Loads
IBC Table 1607.1 is used to determine the nominal live loads on the basis of the given occupancies. Live load reductions are taken wherever applicable.

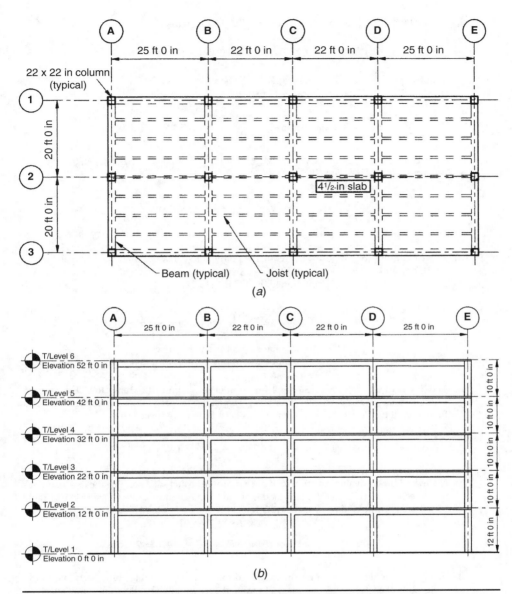

FIGURE 3.3 The building of Example 3.1: (a) typical plan and (b) elevation.

- *Roof:* $L_o = 20$ psf

  Reduced live load is determined by IBC Eq. (16-25):

  $$L_r = L_o R_1 R_2$$

  The tributary area $A_t = \dfrac{25 + 22}{2} \times 20 = 470 \, \text{ft}^2$

Because 200 ft^2 < A_t < 600 ft^2, R_1 is determined by IBC Eq. (16-27):

$$R_1 = 1.2 - 0.001 A_t = 1.2 - (0.001 \times 470) = 0.73$$

A roof slope of 1/2 on 12 means that $F = 1/2$; because $F < 4$, $R_2 = 1$.
Thus, $L_r = 20 \times 0.73 \times 1 = 15$ psf > 12 psf.

$$\text{Axial live load} = \frac{15}{1,000} \times 470 = 7.1 \text{ kips}$$

- *Level 5*: Because Level 5 is storage with a live load of 125 psf, which exceeds 100 psf, the live load is not permitted to be reduced (IBC 1607.9.1.2).

$$\text{Axial live load} = \frac{125}{1,000} \times 470 = 58.8 \text{ kips}$$

- *Typical floors*:

    (a) *Reducible*. From IBC Table 1607.1, reducible nominal live load for an office occupancy = 50 psf (lobby and corridor loads are neglected per the problem statement).
    Reduced live load L is determined by IBC Eq. (16-22):

$$L = L_o \left(0.25 + \frac{15}{\sqrt{K_{LL} A_T}} \right) = 50 \left(0.25 + \frac{15}{\sqrt{K_{LL} A_T}} \right)$$

$$\geq 0.50 L_o \text{ for members supporting one floor}$$
$$\geq 0.40 L_o \text{ for members supporting two or more floors}$$

The live load element factor $K_{LL} = 4$ for an interior column (IBC Table 1607.9.1), and the tributary area A_T at a particular floor level is equal to the sum of the tributary areas for that floor and all the floors above it where the live load can be reduced. Thus,

$$\text{Axial live load (reducible)} = \frac{50}{1,000} \left(0.25 + \frac{15}{\sqrt{4 A_T}} \right) \times 470 = 23.5 \left(0.25 + \frac{15}{\sqrt{4 A_T}} \right) \text{ kips}$$

    (b) *Nonreducible*. A movable partition load of 15 psf must be included because the live load does not exceed 80 psf (IBC 1607.5). This load is not reducible because only the loads in IBC Table 1607.1 are permitted to be reduced (IBC 1607.9).

$$\text{Axial live load (nonreducible)} = \frac{15}{1,000} \times 470 = 7.1 \text{ kips}$$

It is usually convenient to organize the axial load information in a table; this clearly shows the appropriate loads at each story. A summary of the axial dead and live loads on column B2 is given in Table 3.2. In the table, the reduction multiplier is equal to $[0.25 + (15/\sqrt{4 A_T})]$.

At the base of column B2 in the first story, the total axial dead load is equal to the sum of the dead loads in the second column of the table, which is 294.1 kips. Similarly, the total live load is equal to 123.7 kips.

The last column in the table gives the cumulative axial dead plus live loads in each story. At the base of column B2 in the first story, the total axial load is equal to 417.8 kips. As mentioned previously, this load can be used to determine the size of the foundation that supports this column. This load, once factored by the appropriate load factors given in Chap. 9 of the Code, is used to determine the size of the column and the required amount of longitudinal reinforcement. Design of columns is given in Chap. 8 of this book.

Story	Dead Load (kips)	Live Load					Dead + Live Load (kips)	Cumulative Dead + Live Load (kips)
		Nonreducible (kips)	$K_{LL}A_T$ (ft^2)	Reduction Multiplier	Reducible (kips)	Total Live Load (kips)		
5	58.6	—	—	—	7.1	7.1	65.7	65.7
4	58.6	58.8	—	—	—	58.8	117.4	183.1
3	58.6	7.1	1,880	0.60	14.1	21.2	79.8	262.9
2	58.6	7.1	3,760	0.50	11.8	18.9	77.5	340.4
1	59.7	7.1	5,640	0.45	10.6	17.7	77.4	417.8

TABLE 3.2   Summary of Axial Dead and Live Loads on Column B2

Example 3.2   Given the five-story reinforced concrete building illustrated in Fig. 3.3 and the design information given in Example 3.1, determine the dead and live loads along the span of the beam on column line B between 1 and 2 at a typical floor level.

Solution   A one-way, wide-module joist floor system is utilized in this building. The joists support their own weight plus the superimposed dead and live loads along their span lengths. Because the joists are supported by the beams, the reactions from the joists are transferred to the beams. Thus, the beams support the dead and live load reactions from the joists plus their own weight and transfer these loads to the columns (see Fig. 3.4).

For purposes of analysis, the dead and live loads supported by the beams can be assumed to be uniform along their span lengths. The tributary width that is supported by the beams along line B (or any of the other interior beams) is equal to one-half of the transverse span on either side of the column line, which in this case is $(25/2) + (22/2) = 23.5$ ft. For an edge beam, the tributary width is equal to $25/2 = 12.5$ ft.

Dead Loads
- Weight of joists $= \dfrac{84}{1{,}000} \times 23.5 = 2.0$ kips/ft
- Weight of beam $= \dfrac{22 \times 20.5}{144} \times \dfrac{150}{1{,}000} = 0.5$ kips/ft

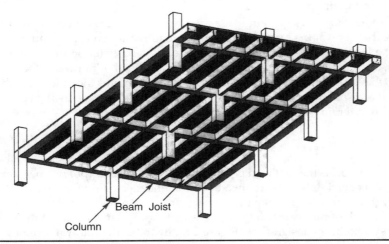

FIGURE 3.4   A one-way, wide-module joist system.

- Superimposed dead load $= \dfrac{10}{1,000} \times 23.5 = 0.3$ kips/ft (conservatively rounded up from 0.235 kips/ft)
- Total dead load $= 2.8$ kips/ft

Live Loads
- *Reducible live load*: From IBC Table 1607.1, reducible nominal live load for an office occupancy $=$ 50 psf (lobby and corridor loads are neglected per the problem statement).

  Reduced live load L is determined by IBC Eq. (16-22):

$$L = L_o \left(0.25 + \frac{15}{\sqrt{K_{LL} A_T}} \right) = 50 \left(0.25 + \frac{15}{\sqrt{K_{LL} A_T}} \right)$$

$$\geq 0.50 L_o \text{ for members supporting one floor}$$

  The live load element factor $K_{LL} = 2$ for an interior beam (IBC Table 1607.9.1), and the tributary area is equal to $23.5 \times 20 = 470$ ft^2. Thus,

$$\text{Live load (reducible)} = \frac{50}{1,000} \left(0.25 + \frac{15}{\sqrt{2 A_T}} \right) \times 23.5 = 1.18 \left(0.25 + \frac{15}{\sqrt{2 \times 470}} \right)$$

$$= 1.18 \times 0.74 = 0.9 \text{ kips/ft}$$

- *Nonreducible live load*: A movable partition load of 15 psf must be included because the live load does not exceed 80 psf (IBC 1607.5). This load is not reducible because only the loads in IBC Table 1607.1 are permitted to be reduced (IBC 1607.9).

$$\text{Live load (nonreducible)} = \frac{15}{1,000} \times 23.5 = 0.4 \text{ kips/ft}$$

- *Total live load*: The total live load is equal to 1.3 kips/ft.

The design of beams is covered in Chap. 6 of this book.

3.2.4   Rain Loads

The requirements for design rain loads are given in IBC 1611. Roofs equipped with hardware that control the rate of drainage are required to have a secondary drainage system at a higher elevation that limits accumulation of water on the roof above that elevation. Such roofs must be designed to sustain the load of rainwater that will accumulate to the elevation of the secondary drainage system plus the uniform load caused by water that rises above the inlet of the secondary drainage system at its design flow.

The nominal rain load R is determined by IBC Eq. (16-35):

$$R = 5.2(d_s + d_h) \tag{3.4}$$

In this equation, 5.2 is the unit load per inch depth of rainwater (pounds per square foot per inch); d_s is the depth of the rainwater on the undeflected roof up to the inlet of the secondary drainage system when the primary drainage system is blocked; and d_h is the additional depth of rainwater on the undeflected roof above the inlet of the secondary drainage system at its design flow. Figure 3.5 illustrates the rainwater depths for the case of perimeter scuppers as the secondary drainage system.

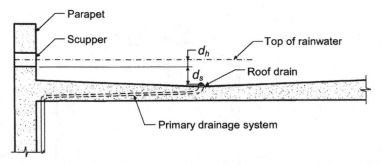

Figure 3.5 Example of rainwater depths for perimeter scuppers.

The nominal rain load R represents the weight of accumulated rainwater on the roof, assuming that the primary roof drainage is blocked. The primary roof drainage system can include, for example, roof drains, leaders, conductors, and horizontal storm drains and is designed for the 100-year hourly rainfall rate indicated in IBC Fig. 1611.1 as well as the area of the roof that it drains. Secondary drainage systems can occur at the perimeter of the roof (scuppers) or at the interior (drains).

3.2.5 Snow Loads

In accordance with IBC 1608.1, design snow loads S are to be determined by the provisions of Chap. 7 of ASCE/SEI 7, which are based on over 40 years of ground snow load data.

The first step in determining S is obtaining the ground snow load p_g from ASCE/SEI Fig. 7-1 or IBC Fig. 1608.2 for the conterminous United States and from ASCE/SEI Table 7-1 or IBC Table 1608.2 for locations in Alaska. Once p_g is established, a flat-roof snow load p_f is obtained by ASCE/SEI Eq. (7-1):

$$p_f = 0.7 C_e C_t I p_g \qquad (3.5)$$

The flat-roof snow load applies to roofs with a slope of 5 degrees or less. The quantities C_e, C_t, and I are related to the roof exposure, roof thermal condition, and occupancy category of the structure, respectively, and are determined from ASCE/SEI Tables 7-2 to 7-4. The occupancy category of the structure is determined from IBC Table 1604.5.

Design snow loads S for all structures are based on the sloped-roof snow load p_s, which is determined by ASCE/SEI Eq. (7-2):

$$p_s = C_s p_f \qquad (3.6)$$

The factor C_s depends on the slope and temperature of the roof, the presence or absence of obstructions, and the degree of slipperiness of the roof surface. ASCE/SEI Fig. 7-2 contains graphs of C_s for various conditions, and equations for this factor can be found in ASCE/SEI C7.4.

The partial-loading provisions of ASCE/SEI 7.5 must be considered for continuous or cantilevered roof framing systems and all other roof systems where removal of

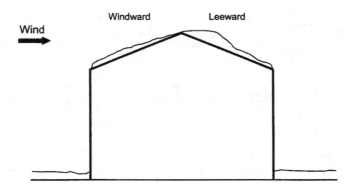

FIGURE 3.6
Unbalanced snow
loads on a gable
roof due to wind.

snow load on one span (e.g., by wind or thermal effects) causes an increase in stress or deflection in an adjacent span. Only the three load cases given in ASCE/SEI Fig. 7-4 need to be investigated.

Wind and sunlight are the main causes for unbalanced snow loads on sloped roofs. Unbalanced loads are unlike partial loads where snow is removed on one portion of the roof and is not added to another portion. For example, wind tends to reduce the snow load on the windward portion and increase the snow load on the leeward portion (see Fig. 3.6). Provisions for unbalanced snow loads are given in ASCE/SEI 7.6.1 for hip and gable roofs, in ASCE/SEI 7.6.2 for curved roofs, in ASCE/SEI 7.6.3 for multiple-folded plate, sawtooth, and barrel vault roofs, and in ASCE/SEI 7.6.4 for dome roofs.

Snow drifts can occur on lower roofs of a building because of the following:

1. Wind depositing snow from higher portions of the same building or an adjacent building or terrain feature (such as a hill) to a lower roof
2. Wind depositing snow from the windward portion of a lower roof to the portion of a lower roof adjacent to a taller part of the building

The first type of drift is called a leeward drift, and the second type is called a windward drift. Both types of drifts are illustrated in Fig. 3.7, which is adapted from ASCE/SEI Fig. 7-7. Loads from drifting snow are superimposed on balanced snow loads, as shown in ASCE/SEI Fig. 7-8. The provisions of ASCE/SEI 7.8 can be used to

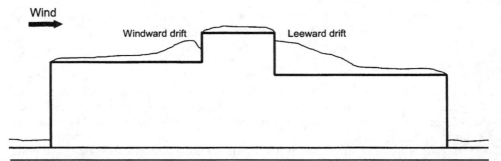

FIGURE 3.7   Windward and leeward snow drifts.

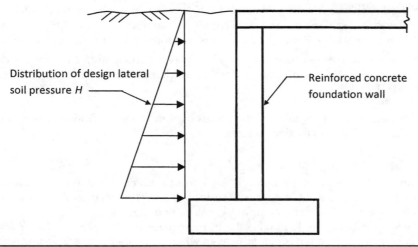

FIGURE 3.8 The distribution of active soil pressure.

determine drift loads on roof projections (such as mechanical equipment) and parapet walls.

3.2.6 Soil Lateral Loads and Hydrostatic Pressure

Foundation walls of a building or structure and retaining walls must be designed to resist the lateral loads caused by the adjacent soil. A geotechnical investigation is usually undertaken to determine the magnitude of the soil pressure. In cases where the results of such an investigation are not available, the lateral soil loads in IBC Table 1610.1 are to be used. The design lateral soil load H depends on the type of soil and the boundary conditions at the top of the wall. Walls that are restricted to move at the top are to be designed for the at-rest pressures tabulated in IBC Table 1610.1, whereas walls that are free to deflect and rotate at the top are to be designed for the active pressures in that table. Figure 3.8 illustrates the distribution of active soil pressure over the height of a reinforced concrete foundation wall.

In addition to lateral pressures from soil, walls must be designed to resist the effects of hydrostatic pressure due to undrained backfill (unless a drainage system is installed) and to any surcharge loads that can result from sloping backfills or from driveways or parking spaces that are close to a wall. Submerged or saturated soil pressures include the weight of the buoyant soil plus the hydrostatic pressure.

3.2.7 Flood Loads

IBC 1612.4 requires that buildings and structures located in flood hazard areas be designed to resist flood loads determined by the provisions of ASCE/SEI Chap. 5. Floodwaters can create the following loads:

- *Hydrostatic loads*: loads caused when stagnant or slowly moving water comes into contact with a building or building component. Lateral hydrostatic pressure is equal to zero at the surface of the water and increases linearly to $\gamma_s d_s$ at the stillwater depth d_s, where γ_s is the unit weight of water.

- *Hydrodynamic loads*: loads caused when water moving at a moderate to high velocity above the ground level comes into contact with a building or building component. ASCE/SEI 5.4.3 contains methods to determine such loads.
- *Wave loads*: loads caused when water waves propagating over the surface of the water strike a building. Methods to determine wave loads are given in ASCE/SEI 5.4.4.

References 3 and 4 contain comprehensive information on how to determine the nominal loads F_a caused by floodwaters.

3.2.8 Self-Straining Loads

According to IBC 1602, self-straining loads T arise from contraction or expansion of a member due to changes in temperature, shrinkage, moisture change, creep, movement caused by differential settlement, or any combination thereof.

When a member is free to move, self-straining loads due to changes in temperature, creep, and shrinkage do not occur. However, when a member is restrained or partially restrained—which typically occurs in a cast-in-place reinforced concrete structure—internal loads will develop.

Chapter 2 of this book provides some general information on how to determine strains due to shrinkage and creep of plain concrete.

3.2.9 Wind Loads

Wind loads on buildings and structures are to be determined in accordance with ASCE/SEI Chap. 6 or by the alternate all-heights method of IBC 1609.6 (IBC 1609.1.1). Note that the basic wind speed, the exposure category, and the type of opening protection required may be determined by IBC 1609 or ASCE/SEI 7 because the provisions in both documents are essentially the same.

Wind forces are applied to a building in the form of pressures that act normal to the surfaces of the building. Positive wind pressure acts toward the surface and is commonly referred to as just pressure. Negative wind pressure, which is also called suction, acts away from the surface. Positive pressure acts on the windward wall of a building, and negative pressure acts on the leeward wall, the sidewalls, and the leeward portion of the roof (see Fig. 3.9). Either positive pressure or negative pressure acts on the windward portion of the roof, depending on the slope of the roof (flatter roofs will be subjected to negative pressure, whereas more sloped roofs will be subjected to positive pressure). Note that the wind pressure on the windward face varies with respect to height and that the pressures on all other surfaces are assumed to be constant.

Method 1 (Simplified Method) and Method 2 (Analytical Procedure) of ASCE/SEI Chap. 6 and the Alternate All-heights Method of IBC 1609.6 are static methods for estimating wind pressures. The magnitude of wind pressure on a structure depends on its size, openness, importance, and location, as well as on the height above ground level. Wind gust and local extreme pressures at various locations on a building are also accounted for.

Although static methods are generally accurate for regularly shaped buildings that deflect primarily in the direction of the wind, such methods cannot be used for many tall, slender structures that respond dynamically to wind forces. Dynamic analyses or Method 3 (Wind Tunnel Procedure) of ASCE/SEI Chap. 6 must be used in these situations.

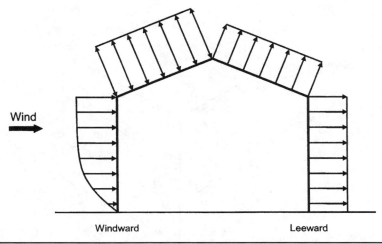

Wind

Windward   Leeward

FIGURE 3.9   The distribution of wind pressures on a building with a gable or hip roof.

Figure 3.10 illustrates how wind forces propagate through a concrete building. The general response of the building can be summarized as follows:

1. The windward wall, which is supported laterally by the roof and floor slabs acting as rigid diaphragms, receives the wind pressures and transfers the resulting forces to the diaphragms.

2. The wind loads are transferred from the diaphragms to the elements of the main wind force–resisting system (MWFRS), which is an assemblage of structural members in a building that are assigned to provide resistance and stability for the entire building. In this case, the walls are the MWFRS. The diaphragms essentially act as beams that are supported by the walls that are parallel to the direction of the wind.

3. The walls transfer the wind forces to the foundations.

The same sequence of events would occur if moment frames, or any other type of lateral force–resisting system, were used instead of or in conjunction with walls.

Only those walls that are parallel to the direction of wind are assumed to be part of the MWFRS. The windward wall, which initially receives the wind, is considered to be a component of the building for analysis in that direction, and the wind pressure on such elements is determined differently than that on the MWFRS. The wind pressure on cladding attached to the building would be determined in a fashion similar to that for components. The IBC and ASCE/SEI 7 refer to elements on the exterior envelope of the building that are not considered MWFRS as *components and cladding*.

For purposes of analysis, the structure is assumed to remain elastic under the design wind forces. Because the wind can occur in any direction, the corresponding critical effects must be considered in design.

The equations given in ASCE/SEI 6.5.12 (Method 2, Analytical Procedure) can be used to determine design wind pressures on the MWFRS and the components and cladding of low-rise buildings and buildings of all heights. Simplified procedures are

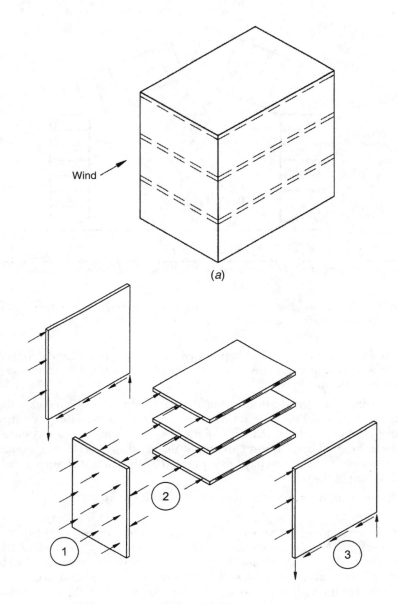

(a)

1: Wind pressure on windward wall.
2: Diaphragms receive wind loads from the windward wall and distribute them to the walls parallel to the wind pressure.
3: Walls parallel to the wind pressure receive the wind loads from the diaphragms and transfer them to the foundations.

(b)

Figure 3.10 The propagation of wind forces in a concrete building: (a) wind load on overall building and (b) distribution of wind load to elements of the MWFRS.

given in ASCE/SEI 6.4 (Method 1) and IBC 1609.6 (Alternate All-heights Method) for low-rise, regularly shaped buildings that meet specific sets of conditions.

Wind forces are determined by multiplying the design wind pressures by an appropriate area. In terms of the MWFRS, it is clear that the magnitude of the wind force is directly proportional to the area of the windward wall perpendicular to the direction of the wind: The greater the area, the greater the wind force.

3.2.10 Earthquake Loads

According to IBC 1613.1, the effects of earthquake motion on structures and their components are to be determined in accordance with ASCE/SEI 7, excluding Chap. 14 and Appendix 11A. The design spectral accelerations, which are proportional to the magnitude of the earthquake forces, can be determined by either IBC 1613 or ASCE/SEI Chap. 11 because the provisions are the same.

The forces that a building must resist during a seismic event are caused by ground motion: As the base of a building moves with the ground, the inertia of the building mass resists this movement, which causes the building to distort. This distortion wave travels along the height of the building. With continued shaking, the building undergoes a series of complex oscillations.

In general, an analysis that considers the acceleration of every mass particle in a building is necessary to determine the inertia forces due to earthquake motion. Such an analysis is usually very complex because there are an inordinate number of degrees of freedom even in small buildings. Because the floor and roof elements (horizontal elements) in a building are relatively heavy compared with the columns and walls (vertical elements), it is reasonable to assume that the mass of the structure is concentrated at the floor and roof levels. This is commonly referred to as a lumped-mass idealization of a structure. For purposes of analysis, the mass of the horizontal and vertical elements of the structure and the mass associated with all other dead loads tributary to a floor level are assumed to be concentrated at the center of mass at that level, and the seismic force is assumed to act through that point. Utilizing a discrete number of masses results in a simpler analysis.

The manner in which a building responds to an earthquake depends on its mass, stiffness, and strength (the strength of the structure generally plays a role beyond the stage of elastic response). There are as many natural modes of vibration as there are degrees of freedom. The seismic response of short, stiff buildings is dominated by the first (fundamental) mode of vibration where all of the masses move in the same direction in response to the ground motion (see Fig. 3.11). Higher modes of vibration contribute significantly to the response of tall, flexible buildings.

Horizontal earthquake forces are transmitted through the diaphragms to walls, frames, or a combination thereof. Collector elements are needed to transfer the diaphragm force where a wall or frame does not extend along the entire edge of the diaphragm. The forces are subsequently transferred to the earth at the base of the foundations. Well-defined load paths such as the one described earlier must be present in every building or structure.

The Equivalent Lateral Force Procedure of ASCE/SEI 12.8 can be used to determine earthquake forces for structures that meet the conditions in ASCE/SEI 12.6-1. It applies to essentially regular buildings that respond to earthquake forces in primarily the first mode. The base shear is computed as a function of the seismicity and soil conditions at

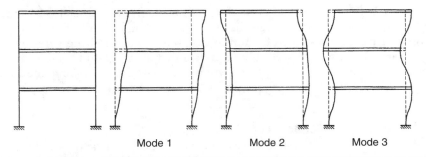

Mode 1   Mode 2   Mode 3

FIGURE 3.11   Modes of vibration for idealized structures.

the site, the type of seismic force–resisting system, and the period of the building (which is related to its stiffness). This force is distributed over the height of the building as a set of equivalent static forces that are applied at each floor level. The magnitude of the force at a given level is based on the height above ground and the weight assigned to that level. This force, in turn, is distributed to the elements of the seismic force–resisting system based on relative stiffness because the diaphragms in a reinforced concrete building are typically rigid.

The simplified alternative method of ASCE/SEI 12.14 can be used to determine design earthquake forces for bearing wall and building frame systems that meet the conditions of ASCE/SEI 12.14.1.1.

For certain types of irregular buildings or for all other structures that cannot be analyzed by the Equivalent Lateral Force Procedure (which are identified in ASCE/SEI Table 12.6.1), more sophisticated analyses are required to determine the design earthquake effects.

The earthquake forces determined by the methods given in ASCE/SEI 7 will be less than the elastic response inertia forces that will be induced in a structure during an actual earthquake. It is expected that structures will undergo relatively large deformations when subjected to a major earthquake. The use of code-prescribed design forces implies that critical regions of certain members will have sufficient inelastic deformability to enable the structure to survive the earthquake without collapsing. As such, these critical regions must be detailed properly to perform in this manner.

The basic difference between wind and earthquake forces exists in the manner in which they are induced in a structure and the way the structure responds. Wind forces are proportional to the exposed surface of a structure, whereas earthquake forces are essentially inertia forces that are generated from the mass of the structure. Structures subjected to the effects of wind are assumed to remain elastic, whereas structures subjected to the effects from a design earthquake are expected to have some level of inelastic deformation depending on the type of structural system.

3.3   Methods of Analysis

3.3.1   Introduction

Once the nominal loads have been established from the applicable building code, an analysis of the structure is performed using these nominal loads and the load

FIGURE 3.12
Vertical load
propagation in a
reinforced concrete
frame.

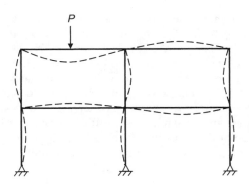

combinations given in Chap. 9 of the Code. It was discussed previously that in the strength design method, nominal loads are multiplied by appropriate load factors, and the members are designed for the effects from critical load combinations. More information on the strength design method is given in Chap. 4 of this book.

According to ACI 8.3.1, an elastic analysis is used to determine bending moments, shear forces, and axial forces in reinforced concrete members due to the combinations of vertical and lateral loads. The assumptions that are given in ACI 8.7 through 8.11 may be used in such an analysis and are discussed later. The purpose of these assumptions is to help simplify the analysis of reinforced concrete structures.

Unlike structural steel or timber structures where the majority, if not all, of the individual members are joined by simple connections that do no transfer any significant bending moments, cast-in-place, reinforced concrete structures are by and large monolithic with reinforcement that extends into adjoining members. Thus, bending moments, shear forces, and axial forces are transferred through the joints. Steel and timber members are commonly analyzed as simply supported members in statically determinate structures, whereas cast-in-place, reinforced concrete members are almost exclusively analyzed as continuous members in statically indeterminate structures.

Consider the two-story reinforced concrete frame illustrated in Fig. 3.12. When a concentrated load P is applied to one bay at the roof level, the effects of the load spread through the rest of the frame because of the continuity at the joints.

A similar scenario is shown in Fig. 3.13, where horizontal wind loads W act on the two-story frame.

Classical methods of analysis, such as the slope-deflection and moment distribution methods, are suited to analyze regular frame buildings of limited size and height.

FIGURE 3.13
Horizontal load
propagation in a
reinforced concrete
frame.

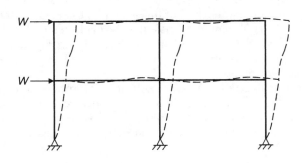

Numerous computer programs that can analyze structures of any size and complexity including interaction between the various structural members and second-order effects are available. When using any computer software, it is essential that the user understands all aspects of the assumptions and limitations of the software.

An approximate method of analysis that can be used to quickly calculate bending moments and shear forces for reinforced beams and one-way slabs in specific types of frames is given at the end of this chapter. For lateral loads, simplified techniques such as the portal method produce adequate results for symmetric frames that meet the limitations of the method. Such simplified methods can also be utilized to obtain preliminary member sizes and to check output from computer software.

3.3.2 Analysis Assumptions

Stiffness

The first analysis assumption is given in ACI 8.7, and it has to do with member stiffness. Any reasonable set of assumptions can be used when computing flexural and torsional stiffnesses of reinforced concrete structural members such as beams, columns, and walls, and these assumptions must be used consistently throughout the analysis.

The stiffnesses used in an analysis for strength design should represent the stiffnesses of the members immediately before failure. Accordingly, flexural and torsional stiffnesses must reflect the degree of cracking and inelastic action that occurs along the length of each member before yielding. Determining these quantities in even a relatively simple frame is very complex, and such a procedure is not efficient for use in a design office. ACI R8.7 contains a number of simplifications that can be used in determining member stiffnesses in some specific applications.

Braced frames are structures that have bracing elements such as walls that inhibit lateral deflection of a building. According to ACI 10.10.1, a story in a structure is assumed to be braced against sidesway when the bracing elements have a total lateral stiffness equal to or greater than 12 times the gross stiffness of the columns in the story. In such cases, one of two assumptions is usually made for member stiffness: (1) the gross flexural stiffness values $E_c I$ are used for all members in the structure, or (2) one-half of the gross flexural stiffness value $E_c I$ of the beam stem is used for beams, and the gross flexural stiffness value $E_c I$ is used for columns.

For frames that are free to sway, such as moment frames, it is especially important to utilize the proper stiffnesses because a second-order analysis may be required. Guidance for the choice of flexural stiffness $E_c I$ in such cases is given in ACI R10.10.4.

Whether it is necessary to consider torsional stiffness in the analysis of a structure depends on the type of torsion present. In the case of equilibrium torsion, where torsion is required to maintain equilibrium of the structure, torsional stiffness should be considered in the analysis. Torsional stiffness may be neglected in the case of compatibility torsion where members twist to maintain deformation compatibility, which is found in a typical continuous system. The design of reinforced concrete beams for torsion is covered in Chap. 6.

For structures subjected to lateral loads, it is important to utilize an appropriate set of stiffness values for the members in the structure so that realistic estimates of lateral deflections are obtained. A nonlinear analysis is required to fully capture the actual behavior of a reinforced concrete structure subjected to lateral loads. In lieu of such an analysis, which is generally complex, the requirements of ACI 8.8 can be used to estimate stiffness. The purpose of these requirements is to simplify the overall procedure: Instead of a complicated nonlinear analysis, a linear analysis of the structure is permitted where reduced member stiffnesses that take into account cracking and other nonlinear behavior of the members are used.

ACI 8.8.2 contains two options for member stiffness that can be employed in the analysis of structures subjected to factored lateral loads. The first of the two options permits the use of the section properties given in ACI 10.10.4.1(a). These properties, which are covered in more detail in Chap. 8, provide lower-bound values for stability analysis of reinforced concrete buildings subjected to gravity and wind loads. Gross stiffnesses of columns, walls, beams, and slabs are multiplied by stiffness reduction factors and are utilized in the analysis along with the full values of the modulus of elasticity and area. This option permits the same structural model to be used to determine lateral deflections and slenderness effects in the columns. The second option permits the use of 50% of the stiffness values based on the gross section properties of the members. Like in the first option, this approximate method produces results that correlate well with experimental results and more rigorous analytical results. References 5 and 6 contain additional information on approximate stiffnesses.

Two-way slab systems without beams, which are commonly referred to as flat plates, can be analyzed for factored lateral loads, using a linear analysis with column stiffness determined by one of the two options of ACI 8.8.2 discussed earlier and slab stiffness determined "by a model that is in substantial agreement with results of comprehensive tests and analysis" (ACI 8.8.3). References 7 through 9 contain acceptable models that can be used in such situations. In general, only a portion of the slab is effective across its full width in resisting the effects from lateral loads. In the effective beam width model of Ref. 8, the actual slab is replaced by a flexural element that has the same thickness as the slab and an effective beam width b_e that is a fraction of the actual transverse slab width. The following equation can be used to determine b_e for an interior column frame:

$$b_e = 2c_1 + \frac{\ell_1}{3} \tag{3.7}$$

In this equation, c_1 and ℓ_1 are the dimension of the column and the center-to-center span length in the direction of analysis, respectively (see Fig. 3.14). For an exterior frame, b_e equals one-half of the value obtained by Eq. (3.7). To account for cracking, the stiffness is usually set equal to one-half to one-quarter of the gross (uncracked) stiffness based on the slab thickness and effective beam width.

For reinforced concrete structures subjected to service lateral loads, ACI 8.8.1 permits the use of a linear analysis where the member stiffnesses are taken as 1.4 times those established by ACI 8.8.2 or 8.8.3. It has been demonstrated that the 1.4 factor is adequate to model effective section properties for structures subjected to service-level lateral loads.

Span Length

The second analysis assumption has to do with span length. In a typical elastic frame analysis, a structure is usually idealized as a simple line diagram where the dimensions are based on the centerlines of the vertical and horizontal members. The usual

FIGURE 3.14 The
effective width of
slab for use in
stiffness
calculations for
slab-column frames.

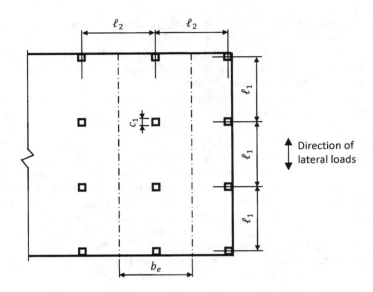

assumption is that the members are prismatic with constant moments of inertia between centerlines. This assumption is not strictly correct. For example, a beam that intersects a column in a monolithic frame is prismatic up to the face of the column. From the face to the centerline of the column, the depth and the corresponding moment of inertia of the beam are significantly larger than those in the span. The same is applicable to columns. Thus, actual variations in member depth should be considered in analysis in order for it to be strictly correct. In general, such an analysis is difficult and time-consuming. ACI 8.9 provides assumptions that help simplify the analysis.

According to ACI 8.9.2, the span length that is to be used in the analysis is the distance between the centerlines of the supports when determining bending moments in frames or similar types of continuous construction. This essentially implies that the reactions are concentrated at the axes of the columns. For the flat-plate system depicted in Fig. 3.14, the bending moments in the slabs must be determined using the span lengths ℓ_1 and ℓ_2.

In the case of beams built integrally with supports, which is typical in cast-in-place, reinforced concrete frame construction, it is permitted to design the beams for bending on the basis of the reduced bending moments at the faces of the supports (ACI 8.9.3). This simplification is based on the presence of a significantly large beam depth from the face to the centerline of the supporting column compared with that in the span. In other words, the critical section for negative bending moments occurs at the face of the support. An acceptable method of reducing bending moments at support centers to those at support faces can be found in Ref. 10. Most computer softwares for the design of reinforced concrete structures automatically take this reduction into account.

Columns

The third analysis assumption deals with columns. Columns in a frame are to be designed for the most critical combinations of factored axial loads and bending moments due to the applied loads (ACI 8.10.1). The design of columns is covered in Chap. 8.

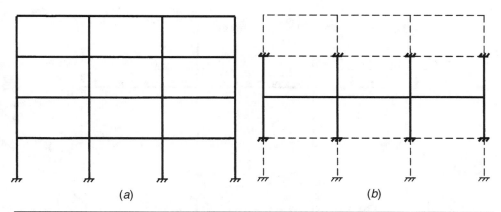

Figure 3.15 The arrangement of live load: (*a*) example frame and (*b*) simplified assumptions for modeling the frame.

When determining gravity load bending moments in columns, it is permitted to assume that the far ends of the column are fixed. More information on this simplification is given next in the assumption on the arrangement of live load.

Arrangement of Live Load

The arrangement of live load on a structure is covered in the last of the four analysis assumptions. There are two assumptions provided in ACI 8.11 that are applicable to gravity load analysis and not to lateral load analysis.

The first assumption has to do with modeling of the frame. The Code permits the use of a model that is limited to the horizontal and vertical framing members at the level at which the far ends of the column are assumed to be fixed. For example, consider the reinforced concrete frame depicted in Fig. 3.15*a*. If we were interested in designing the beams on the second floor level, the live load need only be applied at that level, and the columns above and below that floor level are assumed to be fixed (see Fig. 3.15*b*).

The second assumption considers the arrangement of live load. In typical situations, the arrangement of live load on a structure that will cause critical reactions is not always readily apparent. The engineer is expected to establish the most demanding sets of design forces by investigating the effects of live load placed in various patterns. It is permitted by the Code to assume that the arrangement of live load is limited to combinations of (a) factored dead load on all spans with full factored live load on two adjacent spans and (b) factored dead load on all spans with full factored live load on alternate spans. These loading patterns will be investigated for the three-span frame shown in Fig. 3.16*a*.

In the first loading pattern, the live load is applied on the exterior spans (Fig. 3.16*a*). This pattern produces the maximum negative gravity moment at support A or D and the maximum positive gravity moment in span AB or CD.

In the second loading pattern, live load is applied on adjacent spans (Fig. 3.16*b*). This pattern produces the maximum negative gravity moment at support B (or at support C if spans BC and CD are loaded with the live load).

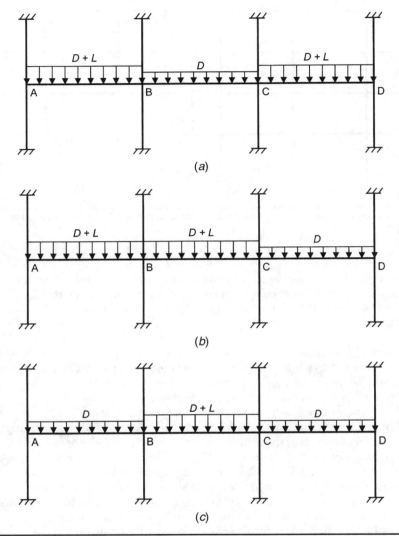

Figure 3.16 The live load patterns for a three-span frame: (*a*) the loading pattern for negative bending moment at support A or D and the positive bending moment in span AB or CD, (*b*) the loading pattern for negative bending moment at support B, and (*c*) the loading pattern for positive bending moment in span BC.

Finally, in the third loading pattern, live load is applied to the interior span only (Fig. 3.16*c*). This pattern produces the maximum positive gravity moment in span BC. Similar scenarios can be developed for other situations.

3.3.3   Approximate Method of Analysis

As an alternative to the frame analysis procedures that have been covered earlier in the chapter, ACI 8.3.3 permits an approximate method of analysis to determine bending moments and shear forces in continuous beams and one-way slabs.

The approximate analysis method can be utilized when all of the following conditions are met:

1. The structure has two or more spans.

2. The spans are approximately equal, with the larger of the two adjacent spans not greater than the shorter one by more than 20%.

3. Loads are uniformly distributed.

4. The unfactored (service) live load L does not exceed 3 times the unfactored (service) dead load D.

5. Members are prismatic; that is, they have a uniform cross-section throughout the span.

Bending moments and shear forces along the span are illustrated in Fig. 3.17 for various support conditions.

The bending moments of ACI 8.3.3 are based on the maximum points of a moment envelope at the critical sections. The moment envelope is obtained by applying live loads on all, alternate, or adjacent spans, considering the five limitations of the method noted previously. This approximate method of analysis gives reasonably conservative values for continuous structures that meet the prescribed conditions.

The quantity w_u is the factored uniformly distributed load along the span length, which is determined by multiplying the service dead and live loads by the load factors given in ACI 9.2. More information on load factors and required strength is given in Chap. 4. For beams, w_u is the uniformly distributed load per length of beam in pounds per linear foot, whereas for one-way slabs, w_u is the uniformly distributed load per unit area of slab in pounds per square foot. It is common to analyze a one-way slab using a 1-ft-wide design strip.

Negative bending moments and shear forces are at the faces of the supports. When calculating negative bending moments, the average of the adjacent clear span lengths $\ell_{n,avg}$ must be used.

Example 3.3 Given the five-story reinforced concrete building illustrated in Fig. 3.3 and the design information given in Examples 3.1 and 3.2, determine the factored bending moments and shear forces in the beam along column line B.

Solution In lieu of using a more rigorous analysis to determine bending moments and shear forces, check if the approximate method of ACI 8.3.3 can be used:

1. There are two spans.

2. Spans are equal.

3. Loads are uniformly distributed (see Example 3.2).

4. Unfactored live load L from Example 3.2 $= 1.3$ kips/ft; unfactored dead load D from Example 3.2 $= 2.8$ kips/ft; $L/D = 0.46 < 3$.

5. Beams have constant cross-section over the entire length.

Because all five conditions are satisfied, the approximate method can be used.

The total factored load w_u is the factored uniformly distributed load along the span length, which is determined by multiplying the service dead and live loads by the load factors in ACI 9.2. In this case, ACI Eq. (9-2) produces the critical effects on the beam:

$$w_u = 1.2w_D + 1.6w_L = (1.2 \times 2.8) + (1.6 \times 1.3) = 5.4 \text{ kips/ft}$$

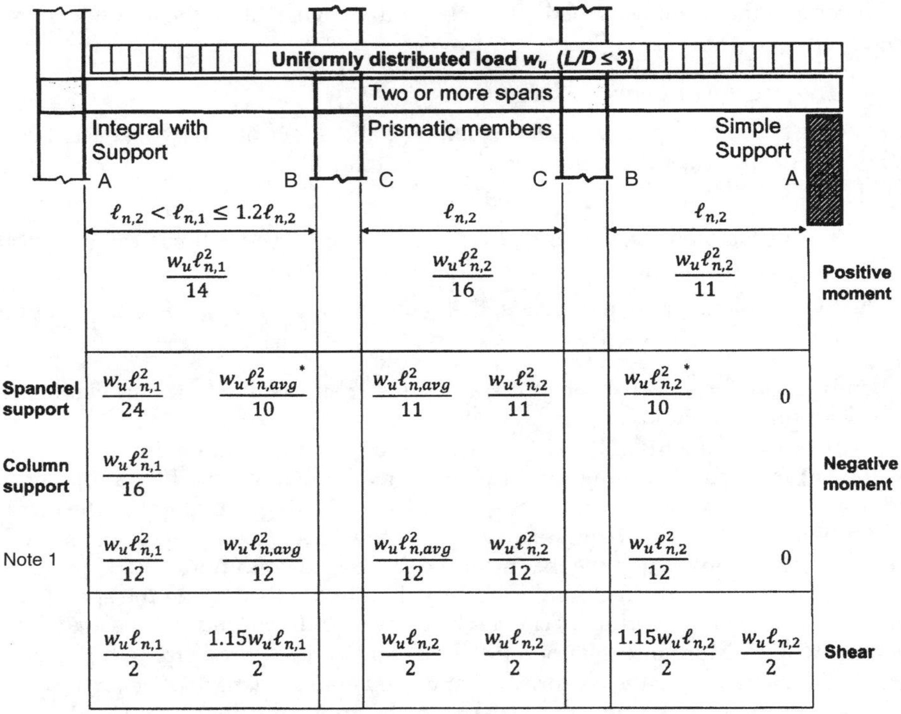

	Integral with Support		Prismatic members (Two or more spans)		Simple Support		
	A	B	C	C	B	A	
	$\ell_{n,2} < \ell_{n,1} \le 1.2\ell_{n,2}$		$\ell_{n,2}$		$\ell_{n,2}$		
	$\dfrac{w_u\ell_{n,1}^2}{14}$		$\dfrac{w_u\ell_{n,2}^2}{16}$		$\dfrac{w_u\ell_{n,2}^2}{11}$		**Positive moment**
Spandrel support	$\dfrac{w_u\ell_{n,1}^2}{24}$	$\dfrac{w_u\ell_{n,avg}^2}{10}$*	$\dfrac{w_u\ell_{n,avg}^2}{11}$	$\dfrac{w_u\ell_{n,2}^2}{11}$	$\dfrac{w_u\ell_{n,2}^2}{10}$*	0	
Column support	$\dfrac{w_u\ell_{n,1}^2}{16}$						**Negative moment**
Note 1	$\dfrac{w_u\ell_{n,1}^2}{12}$	$\dfrac{w_u\ell_{n,avg}^2}{12}$	$\dfrac{w_u\ell_{n,avg}^2}{12}$	$\dfrac{w_u\ell_{n,2}^2}{12}$	$\dfrac{w_u\ell_{n,2}^2}{12}$	0	
	$\dfrac{w_u\ell_{n,1}}{2}$	$\dfrac{1.15w_u\ell_{n,1}}{2}$	$\dfrac{w_u\ell_{n,2}}{2}$	$\dfrac{w_u\ell_{n,2}}{2}$	$\dfrac{1.15w_u\ell_{n,2}}{2}$	$\dfrac{w_u\ell_{n,2}}{2}$	**Shear**

Uniformly distributed load w_u $(L/D \le 3)$

* For two-span condition, first interior negative moment $= w_u\ell_n^2/9$

$\ell_{n,avg} = (\ell_{n,1} + \ell_{n,2})/2$

Note 1: Applicable to slabs with spans equal to or less than 10 ft and beams where the ratio of the sum of column stiffness to beam stiffness is greater than 8 at each end of the span.

A: Interior face of exterior support
B: Exterior face of first interior support
C: Other faces of interior supports

FIGURE 3.17 The approximate bending moments and shear forces for continuous beams and one-way slabs in accordance with ACI 8.3.3.

Because there are only two spans, each span is an end span (there are no interior spans). The clear span length ℓ_n for both spans is equal to the distance between the faces of the columns: $\ell_n = 20 - (22/12) = 18.17$ ft.

A summary of the factored bending moments and shear forces is given in Table 3.3.

At the interior face of the external supports, the applicable negative bending moment is equal to $w_u\ell_n^2/16$ because the beams frame into columns at lines 1 and 3. Also, at the exterior face of the interior support, the applicable negative bending moment is equal to $w_u\ell_n^2/9$ because there are only two spans. Note that the location of the positive bending moment is referenced as "midspan" in Table 3.3. This means that this bending moment is located within the middle portion of the span length and not necessarily at the actual midspan of the beam.

Location	Bending Moment (ft kips)	Shear Force (kips)
Interior face of exterior support	$\dfrac{w_u\ell_n^2}{16} = \dfrac{-5.4 \times 18.17^2}{16} = -111.4$	$\dfrac{w_u\ell_n}{2} = \dfrac{5.4 \times 18.17}{2} = 49.1$
Midspan	$\dfrac{w_u\ell_n^2}{14} = \dfrac{5.4 \times 18.17^2}{14} = 127.3$	—
Exterior face of first interior support	$\dfrac{w_u\ell_n^2}{9} = \dfrac{-5.4 \times 18.17^2}{9} = -198.1$	$\dfrac{1.15w_u\ell_n}{2} = \dfrac{1.15 \times 5.4 \times 18.17}{2} = 56.4$

TABLE 3.3   Summary of Bending Moments and Shear Forces for the Beams Along Column Line B

Example 3.4   Given the five-story reinforced concrete building illustrated in Fig. 3.3 and the design information given in Examples 3.1 and 3.2, determine the factored bending moments and shear forces in the interior wide-module joists.

Solution   Wide-module joists are essentially beams that are constructed using standardized pan forms. The designation $16 + 4^1/_2 \times 7 + 53$ in Example 3.1 means the following (see Fig. 3.18):

- Rib depth = 16 in
- Slab thickness = $4^1/_2$ in
- Rib width = 7 in
- Pan width = 53 in

Each joist supports the load over its length and a tributary width equal to $53 + 7 = 60$ in = 5 ft. The dead loads supported by the joists are as follows:

- Weight of joists = $\dfrac{84}{1,000} \times 5 = 0.42$ kips/ft

- Superimposed dead load = $\dfrac{10}{1,000} \times 5 = 0.05$ kips/ft

- Total dead load = 0.47 kips/ft

The live loads supported by the joists are as follows:

- *Reducible live load*: From IBC Table 1607.1, reducible nominal live load for an office occupancy = 50 psf (lobby and corridor loads are neglected per the problem statement).

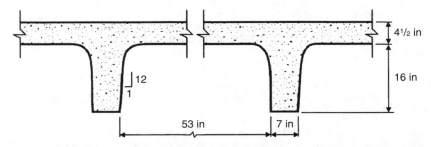

FIGURE 3.18   The dimensions of wide-module joists specified in Figure 3.3.

Reduced live load L is determined by IBC Eq. (16-22):

$$L = L_o \left(0.25 + \frac{15}{\sqrt{K_{LL} A_T}} \right) = 50 \left(0.25 + \frac{15}{\sqrt{K_{LL} A_T}} \right)$$

$$\geq 0.50 L_o \text{ for members supporting one floor}$$

The live load element factor $K_{LL} = 2$ for an interior beam (IBC Table 1607.9.1), and the smallest tributary area is equal to $22 \times 5 = 110 \text{ ft}^2$. Because $K_{LL} A_T = 2 \times 110 = 220 \text{ ft}^2 < 400 \text{ ft}^2$, this live load cannot be reduced.

$$\text{Live load} = \frac{50}{1,000} \times 5 = 0.25 \text{ kips/ft}$$

- *Nonreducible live load*: A movable partition load of 15 psf must be included because the live load does not exceed 80 psf (IBC 1607.5). This load is not reducible because only the loads in IBC Table 1607.1 are permitted to be reduced (IBC 1607.9).

$$\text{Live load (nonreducible)} = \frac{15}{1,000} \times 5 = 0.08 \text{ kips/ft}$$

- *Total live load*: The total live load is equal to 0.33 kips/ft.

As in Example 3.4, check if the approximate method of ACI 8.3.3 can be used:

1. There are more than two spans.
2. Ratio of adjacent span lengths $= 25/22 = 1.14 < 2$.
3. Loads are uniformly distributed (see Example 3.2).
4. Unfactored live load $L = 0.33$ kips/ft; unfactored dead load $D = 0.47$ kips/ft; $L/D = 0.70 < 3$.
5. Joists have constant cross-section over the entire length.

Because all five conditions are satisfied, the approximate method can be used. As in Example 3.3, the total factored load w_u is determined by ACI Eq. (9-2):

$$w_u = 1.2 w_D + 1.6 w_L = (1.2 \times 0.47) + (1.6 \times 0.33) = 1.1 \text{ kips/ft}$$

The clear spans of the joists are equal to the following:

- End spans: $\ell_n = 25 - (22/12) = 23.17$ ft
- Interior spans: $\ell_n = 22 - (22/12) = 20.17$ ft

A summary of the factored bending moments and shear forces is given in Table 3.4.

Two negative moments are calculated at the interior face of exterior supports because the joists along column line 2 frame into columns and the other interior joists frame into beams. In accordance with ACI 8.3.3, the average of the clear span lengths is used to determine the bending moment at the exterior face of the first interior supports. The magnitude of the bending moment at this location (51.7 ft kips) is greater than the magnitude of the bending moment on the other side of the first interior column (47.0 ft kips). It will be shown later that the larger of these two bending moments is used to determine the required amount of negative reinforcement at this location.

Figure 3.19 provides a summary of the factored bending moments and shear forces along the span of the joists.

Example 3.5   Given the five-story reinforced concrete building illustrated in Fig. 3.3 and the design information given in Examples 3.1 and 3.2, determine the factored bending moments and shear forces in the one-way slabs.

Solution   The slabs are supported by the ribs of the wide-module joists and bend primarily in one direction, that is, in the direction perpendicular to the supports. One-way slabs are commonly analyzed and designed using a 1-ft-wide design strip.

Location		Bending Moment (ft kips)	Shear Force (kips)
End span	Interior face of exterior support — Column support	$\dfrac{w_u\ell_n^2}{16} = \dfrac{-1.1 \times 23.17^2}{16}$ $= -36.9$	$\dfrac{w_u\ell_n}{2} = \dfrac{1.1 \times 23.17}{2} = 12.7$
	Interior face of exterior support — Spandrel support	$\dfrac{w_u\ell_n^2}{24} = \dfrac{-1.1 \times 23.17^2}{24}$ $= -24.6$	
	Midspan	$\dfrac{w_u\ell_n^2}{14} = \dfrac{1.1 \times 23.17^2}{14} = 42.2$	—
	Exterior face of first interior support	$\dfrac{w_u\ell_{n,avg}^2}{10} = \dfrac{-1.1 \times 21.67^2}{10}$ $= -51.7$	$\dfrac{1.15 w_u\ell_n}{2} = \dfrac{1.15 \times 1.1 \times 23.17}{2}$ $= 14.7$
Interior span	Interior face of first interior support	$\dfrac{w_u\ell_{n,avg}^2}{11} = \dfrac{-1.1 \times 21.67^2}{11}$ $= -47.0$	$\dfrac{w_u\ell_n}{2} = \dfrac{1.1 \times 20.17}{2} = 11.1$
	Midspan	$\dfrac{w_u\ell_n^2}{16} = \dfrac{1.1 \times 20.17^2}{16} = 28.0$	—
	Interior face of interior support	$\dfrac{w_u\ell_n^2}{11} = \dfrac{-1.1 \times 20.17^2}{11}$ $= -40.7$	$\dfrac{w_u\ell_n}{2} = \dfrac{1.1 \times 20.17}{2} = 11.1$

TABLE 3.4 Summary of Bending Moments and Shear Forces for the Wide-module Joists

The dead loads supported by the one-way slab are as follows:

- Weight of slab $= \dfrac{4.5}{12} \times 1 \times \dfrac{150}{1,000} = 0.06$ kips/ft per foot width of slab
- Superimposed dead load $= \dfrac{10}{1,000} \times 1 = 0.01$ kips/ft per foot width of slab
- Total dead load $= 0.07$ kips/ft per foot width of slab

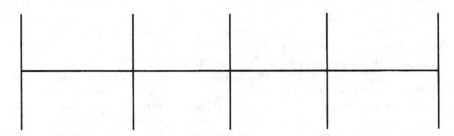

M (ft kips)	−36.9	−51.7 −47.0	−40.7 −40.7	−47.0 −51.7	−36.9
	42.2	28.0	28.0	42.2	
V (kips)	12.7	14.7 11.1	11.1 11.1	11.1 14.7	12.7

FIGURE 3.19 Summary of factored bending moments and shear forces along the span of the wide-module joists.

According to IBC 1607.9.1.1, the tributary area A_T for one-way slabs shall not exceed the area defined by the span length times a width normal to the span length of 1.5 times the span length. In this case, $A_T = 1.5 \times 5^2 = 37.5 \text{ ft}^2$; this is also equal to the influence area because $K_{LL} = 1$ in IBC Table 1607.9.1. It is evident that no live load reduction can be taken on the slab because $K_{LL}A_T < 400 \text{ ft}^2$. Therefore,

$$\text{Live load} = \frac{50 + 15}{1,000} \times 1 = 0.07 \text{ kips/ft per foot width of slab}$$

It can be shown that all five conditions of ACI 8.3.3 are satisfied, so the approximate method can be used to determine the bending moments and shear forces.

As in the previous examples, the total factored load w_u is determined by ACI Eq. (9-2):

$$w_u = 1.2w_D + 1.6w_L = (1.2 \times 0.07) + (1.6 \times 0.07) = 0.2 \text{ kips/ft}$$

Because the center-to-spacing distance between the wide-module joists is only slightly greater than the actual clear span between the joist ribs, conservatively use $\ell_n = 5$ ft.

Because the slabs have spans that are less than 10 ft, the negative bending moment at the faces of all of the supports is

$$\frac{w_u \ell_n^2}{12} = \frac{-0.2 \times 5^2}{12} = -0.42 \text{ ft kips per foot width of slab}$$

For simplicity, use the positive bending moment in the end span for all of the spans:

$$\frac{w_u \ell_n^2}{14} = \frac{0.2 \times 5^2}{14} = 0.36 \text{ ft kips per foot width of slab}$$

Similarly, use the shear force at the face of the first interior support for all of the spans:

$$\frac{1.15 w_u \ell_n}{2} = \frac{1.15 \times 0.2 \times 5}{2} = 0.6 \text{ kips per foot width of slab}$$

3.4 Moment Redistribution

ACI 8.4 permits bending moments calculated by elastic theory at supports of continuous flexural members to be increased or decreased, except where the moments have been computed using the approximate coefficients in ACI 8.3.3. It is customary to reduce the negative moments at the supports and then to increase the positive moment in the span.

The maximum percentage of redistribution is the lesser of 1,000 times the net tensile strain in the reinforcement ε_t or 20%. Methods on how to calculate ε_t are given in Chap. 5.

Moment redistribution is dependent on adequate ductility in plastic hinge regions, which develop at points of maximum moment and which cause a shift in the elastic bending moment diagram. Thus, redistribution of negative moments is only permitted where ε_t is equal to or greater than 0.0075 at the section in which the moment is reduced. Adjustments of the negative moments at the supports are made for each loading configuration, taking into account pattern live loading. It is important to ensure that static equilibrium is maintained at all joints before and after moment redistribution. Thus, a decrease in the negative moments at the supports warrants an increase in the positive moment in the span under consideration.

Figure 3.20 illustrates moment redistribution for a span of a continuous beam. The bending moment diagram determined by analytical methods before redistribution is

FIGURE 3.20
Redistribution of
bending moments in
accordance with ACI
8.4.

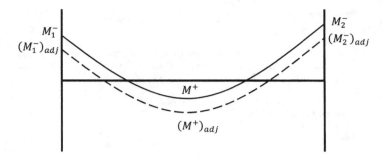

shown along with the maximum negative and positive moments at the faces of the supports and in the span, respectively.

The adjusted bending moment diagram is also shown after redistribution of the negative moments. In this case, it was decided to decrease the negative moments at the supports and to increase the positive moment using statics.

Adjusted negative bending moments at the faces of the supports can be determined by the following equation:

$$(M_i^-)_{adj} = \left(1 - \frac{A\%}{100}\right) M_i^- \tag{3.8}$$

In this equation, $A\% = 1{,}000\varepsilon_t\% \leq 20\%$, which is the permitted percentage adjustment given in ACI 8.4.1.

The required flexural reinforcement at the critical sections can be obtained for these redistributed bending moments using the general principles of the strength design method, which is covered in Chaps. 5 and 6.

References

1. International Code Council (ICC). 2009. *International Building Code*. ICC, Washington, DC.
2. Structural Engineering Institute (SEI) of the American Society of Civil Engineers (ASCE). 2006. *Minimum Design Loads for Buildings and Other Structures, Including Supplements Nos. 1 and 2*, ASCE/SEI 7-05. ASCE/SEI, Reston, VA.
3. Fanella, D. 2009. *Structural Load Determination Under 2009 IBC and ASCE/SEI 7-05*, International Code Council (ICC), Washington, DC.
4. Federal Emergency Management Agency (FEMA). 2000. *Coastal Construction Manual*, 3rd ed. FEMA, Washington, DC.
5. Moehle, J. P. 1992. Displacement-based design of RC structures subjected to earthquakes. *Earthquake Spectra* 8(3):403-428.
6. Lepage, A. 1998. Nonlinear drift of multistory RC structures during earthquakes. Paper presented at Sixth National Conference on Earthquake Engineering, Seattle, WA.
7. Vanderbilt, M. D., and Corley, W. G. 1983. Frame analysis of concrete buildings. *Concrete International* 5(12):33-43.
8. Hwang, S., and Moehle, J. P. 2000. Models for laterally loaded slab-column frames. *ACI Structural Journal* 97(2):345-353.
9. Dovich, L. M., and Wight, J. K. 2005. Effective slab width model for seismic analysis of flat slab frames. *ACI Structural Journal* 102(6):868-875.
10. Portland Cement Association (PCA). 1959. *Continuity in Concrete Building Frames*. PCA, Skokie, IL.

Problems

3.1. For a one-story reinforced concrete building, determine the total axial dead and live loads at the base of an interior column that supports a tributary area of 625 ft². The slope of the roof is 30 degrees, and the load data are as follows:

> Dead load (including weight of column) = 90 psf
> Superimposed dead load = 5 psf
> Roof live load = 20 psf

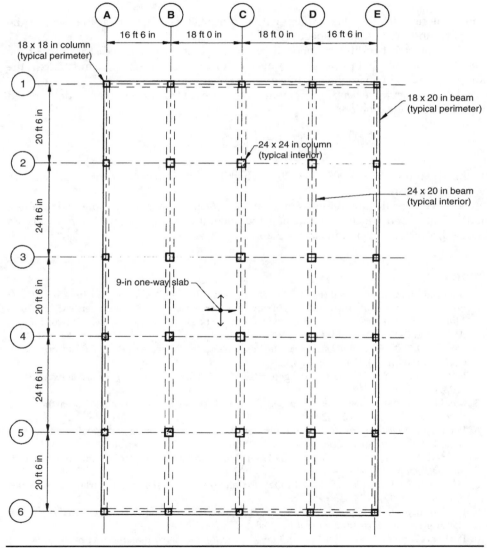

FIGURE 3.21 The floor plan of the reinforced concrete hotel of Problem 3.6.

3.2. Given the five-story building and load data of Example 3.1, determine the total axial dead and live loads at the base of columns (a) A1, (b) B1, and (c) A2.

3.3. Given the five-story building and load data of Example 3.1, determine the total axial dead and live loads at the base of column B2 assuming all floors are typical office occupancy.

3.4. Given the five-story building and load data of Example 3.1, determine the total axial dead and live loads at the base of column B2 assuming that the storage occupancy is on level 4 and that all other floors are typical office occupancy.

3.5. Given the five-story building and load data of Example 3.1, determine the dead and live loads along the span of the beam on column line A between 1 and 2 on (a) a typical floor with office occupancy, (b) the floor with storage occupancy, and (c) the roof.

3.6. The typical floor plan of a reinforced concrete hotel is illustrated in Fig. 3.21.

Determine the factored bending moments and shear forces (a) in the beam along column line A, (b) in the beam along column line D, and (c) in the one-way slab between A and E.

Assume the following: normal-weight concrete with a density of 150 pcf, a superimposed dead load of 5 psf, and the load combination given in ACI Eq. (9-2) produce the critical effects on the members.

CHAPTER 4

General Requirements for Strength and Serviceability

4.1 Introduction

This chapter contains the fundamental requirements for strength and serviceability that form the basis of design of all reinforced concrete members. The basic concepts of the strength design method are presented, as are the general provisions for deflection control.

Throughout the years, there have been basically two design philosophies for reinforced concrete members: working stress design and limit state design (referred to as the strength design method).

From the early 1900s until the early 1960s, working stress design was the primary design method for reinforced concrete. In working stress design, members are proportioned so that maximum elastic stresses due to unfactored loads (also identified as service or working loads) are equal to or less than the allowable stresses prescribed in the Code.

Limit state design involves identifying applicable limit states and determining acceptable levels of safety against occurrences of each limit state. In general, a limit state is a set of performance criteria that must be met when a structure is subjected to loads. The two fundamental limit states are as follows: (1) ultimate limit states, which correspond to the loads that cause failure (strength and stability), and (2) serviceability limit states, which correspond to the criteria that govern the service life of a structure (e.g., deflection and crack width). A structure or structural member is said to have reached a limit state (or, equivalently, is said to have "failed") when it is unable to carry out one or more of the required performance criteria. The statistical methods used to determine the level of safety required in the design process are discussed later.

The 1956 edition of the ACI Code was the first to include provisions for the "ultimate strength design method." The strength method was essentially established as the preferred design method in the 1971 Code, although an updated form of the working stress design method—referred to as the alternate design method—was still permitted to be used. It was not until 2002 that the alternate design method was officially deleted from the Code (however, it has been mentioned in Commentary Section R1.1 in every edition of the Code since then). A more comprehensive history of the evolution of the strength design method can be found in Chap. 5 of Ref. 1.

The gradual elimination of the working stress design method from the ACI Code primarily had to do with the following shortcomings of the method[2]:

- Inability to correctly account for the variability of loads and member resistances
- Inconsistent factor of safety in member design

Chapter 9 of the Code contains the basic requirements for proportioning reinforced concrete members to resist load effects. The main limit states for reinforced concrete structures are strength (ACI 9.1.1) and serviceability (ACI 9.1.2), and both must be considered in the design process.

The basic requirement for strength design is

$$\text{Design strength} \geq \text{Required strength} \tag{4.1}$$

or

$$\phi(\text{nominal strength}) \geq U \tag{4.2}$$

In general, the design strength of a member, which is equal to the applicable strength reduction factor ϕ times the nominal strength of the member, must be equal to or greater than the required strength. The required strength, which is represented by the symbol U, is determined by multiplying service load effects by code-prescribed load factors. Design and required strengths are discussed in more detail in the following sections.

Prior to the 2002 edition of the Code, the load factor combinations in Chap. 9 of the Code were different from those in ASCE/SEI 7, with the latter being the combinations that were used in the design of just about all of the other structural materials. A significant change occurred in the 2002 edition of the Code: The load combinations that were introduced in Chap. 9 matched those in ASCE/SEI 7-02. This change helped in simplifying the overall design process, especially in structures utilizing more than one type of material. Revised strength reduction factors accompanied the revised load combinations in order to provide a consistent factor of safety in design. Details of the statistical analysis that was used to calibrate the resistance factors to the ASCE/SEI load factors can be found in Refs. 3 and 4.

The load combinations in the 2008 edition of the Code match those in ASCE/SEI 7-05 and, for the most part, the load combinations in IBC 1605.2.1. Additional information on the ACI load combinations is given in Section 4.2.

ACI 9.1.3 permits the use of the load factor combinations and strength reduction factors contained in Appendix C of the Code in lieu of those in Chap. 9. These load and strength reduction factors appeared in various forms in Chap. 9 of the Code from the 1960s until 2002, when they were revised and moved to Appendix C. When designing a reinforced concrete member, it is important to use the load factor combinations in conjunction with the corresponding strength reduction factors; in other words, it is not permitted to use the load factor combinations of Chap. 9 with the strength reduction factors of Appendix C.

4.2 Required Strength

4.2.1 ACI Load Combinations

The required strength U is obtained by multiplying service-level (nominal) load effects caused by the nominal loads prescribed in the governing building code by the

ACI Equation Number	Load Combination
9-1	$U = 1.4(D + F)$
9-2	$U = 1.2(D + F + T) + 1.6(L + H) + 0.5(L_r$ or S or $R)$
9-3	$U = 1.2D + 1.6(L_r$ or S or $R) + (1.0L$ or $0.8W)$
9-4	$U = 1.2D + 1.6W + 1.0L + 0.5(L_r$ or S or $R)$
9-5	$U = 1.2D + 1.0E + 1.0L + 0.2S$
9-6	$U = 0.9D + 1.6W + 1.6H$
9-7	$U = 0.9D + 1.0E + 1.6H$

TABLE 4.1 ACI Load Factor Combinations

appropriate load factors given in ACI 9.2. Determination of nominal loads is covered in Chap. 3 of this book.

The load combinations given in ACI 9.2.1 are given in Table 4.1, with the nominal load notation defined in Table 3.1. As noted in Section 4.1, these load combinations are the same as those in ASCE/SEI 2.3.2 and are also the same as those in IBC 1605.2.1 with the following exceptions:

- The variable f_1 that is present in IBC Eqs. (16-3) to (16-5) is not found in the corresponding ACI Eqs. (9-3) to (9-5). Instead, the load factor on the live load L in the ACI combinations is equal to 1.0 with the exception that the load factor on L is permitted to be equal to 0.5 for all occupancies where the live load is equal to or less than 100 psf, except for parking garages or areas occupied as places of public assembly [see ACI 9.2.1(a)]. This exception makes these load combinations in the Code the same as those in the IBC (see IBC 1605.2.1 for the definition of f_1). Note that this load modification factor on L should not be confused with the live load reduction factor presented in Chap. 3; the live load reduction factor is based on the loaded area of a member, and it can be used in combination with the 0.5 load modification factor specified in ACI 9.2.1(a).

- The variable f_2 that is present in IBC Eq. (16-5), which is defined in IBC 1605.2.1, is not found in the corresponding ACI Eq. (9-5). Instead, a load factor of 0.2 is applied to S in the ACI combination. More information on this factor can be found in the third exception in ASCE/SEI 2.3.2, which states that in combinations 2, 4, and 5 [which are the same as ACI Eqs. (9-2), (9-4), and (9-5)], S shall be taken as either the flat-roof snow load p_f or the sloped-roof snow load p_s. This essentially means that the balanced snow load defined in ASCE/SEI 7.3 for flat roofs and in ASCE/SEI 7.4 for sloped roofs can be used in ACI Eqs. (9-2), (9-4), and (9-5). Drift loads and unbalanced snow loads are covered by ACI Eq. (9-3).

Load factors are typically greater than or less than 1.0. Earthquake load effects are an exception to this: A load factor of 1.0 is used to determine the maximum effect because an earthquake load is considered a strength-level load. If the governing building code has provisions where the earthquake effects E are based on service-level earthquake loads, then $1.4E$ must be substituted for $1.0E$ in ACI Eqs. (9-5) and (9-7) [ACI 9.2.1(c)].

This situation will occur where the governing building code is one of the following legacy codes:

- Building Officials and Code Administrators International (BOCA) National Building Code, all editions prior to 1993[5]
- Standard Building Code, all editions prior to 1994[6]
- Uniform Building Code, all editions prior to 1997[7]

Service-level earthquake loads also appear in all editions of ASCE/SEI 7 prior to 1993.

The magnitude of the load factor assigned to each nominal load effect is influenced primarily by the following:

1. *The degree of accuracy to which the load effect can be determined.* Dead loads can vary for a variety of reasons, including the following: member sizes in a structure can be constructed differently than those identified on the construction documents, and the density of the concrete can be different from that specified in the project specifications. However, dead loads are more accurately determined than variable loads. Also, assumptions made in the analysis of a structure—such as stiffness and span length (see Chap. 3)—can result in calculated load effects that are different from those in the actual structure.

2. *The variation that might be expected in the load during the life of a structure.* Variable loads can vary significantly over the life span of a structure. Changes in the magnitudes of such loads can occur over relatively short time intervals.

3. *The probability that different load types will occur at the same time.* The probability that the maximum effects of different variable loads will occur simultaneously on a structure or a structural member is relatively low.

Load combinations are constructed by adding to the dead load D one or more of the variable loads at its maximum value, which is typically indicated by a load factor of 1.6 (except for earthquake load effects where the maximum load factor is 1.0, as was discussed previously). Variable loads are assigned a higher load factor than dead loads D and weights and pressures of liquids with well-defined densities and controllable maximum heights F because they are far less predictable than D or F. Also included in the combinations are variable loads with load factors less than 1.0 that take into account the probability of these loads acting at the same time as the other loads in the combination.

Prior to the 2003 IBC and ASCE/SEI 7-02, the maximum load factor assigned to wind load effects W was 1.3. A wind directionality factor of 0.85 for buildings was explicitly introduced into the wind load provisions of the 2003 IBC and ASCE/SEI 7-02, and it appears in all subsequent editions of both documents. If wind forces are calculated using this directionality factor, the appropriate load factor on W is equal to the previous factor of 1.3 divided by 0.85, which is equal to 1.53 (this factor was rounded up to 1.6 for use in the load combinations). If wind forces are determined from sources where the wind directionality factor is not explicitly considered, such as in editions of the IBC and ASCE/SEI 7 prior to 2003 and 2002, respectively, the previous wind load factor of 1.3 is permitted to be used [ACI 9.1.2(b)].

For structures located in a flood zone and for structures subjected to forces from atmospheric ice loads, the loads and appropriate load combinations are to be determined by the applicable provisions of ASCE/SEI 7 (ACI 9.2.4). Methods to determine nominal loads F_a caused by floodwaters are given in Refs. 3 and 4 of Chap. 3. Once these loads are determined, they are utilized in the following load combinations:

- In V Zones or Coastal A Zones

  ACI Eq. (9-4): $U = 1.2D + 1.6W + 2.0F_a + 1.0L + 0.5(L_r \text{ or } S \text{ or } R)$
  ACI Eq. (9-6): $U = 0.9D + 1.6W + 2.0F_a + 1.6H$

- In noncoastal A Zones

  ACI Eq. (9-4): $U = 1.2D + 0.8W + 1.0F_a + 1.0L + 0.5(L_r \text{ or } S \text{ or } R)$
  ACI Eq. (9-6): $U = 0.9D + 0.8W + 1.0F_a + 1.6H$

Definitions of V Zones, Coastal A Zones, and noncoastal A Zones are given in Refs. 3 and 4 of Chap. 3.

Atmospheric ice and wind-on-ice loads are determined by the provisions of Chap. 10 of ASCE/SEI 7. The weight of ice D_i and the wind-on-ice load W_i are combined with other loads in the following load combinations:

- ACI Eq. (9-2): $U = 1.2(D + F + T) + 1.6(L + H) + 0.2D_i + 0.5S$
- ACI Eq. (9-4): $U = 1.2D + 1.0L + D_i + W_i + 0.5S$
- ACI Eq. (9-6): $U = 0.9D + D_i + W_i + 1.6H$

The subscript u is used to denote the required strength. For example, M_u is the factored bending moment at a section of a reinforced concrete member that has been determined by the applicable load combination(s).

4.2.2 Critical Load Effects

In general, all of the load combinations presented in the previous section must be investigated when designing structural members. The critical load effects obtained by the load combinations are used in determining the size of a member and the required amount of reinforcement. Design methods for various types of structural members are given in subsequent chapters of this book.

The load combinations in Eqs. (9-1) through (9-5) will typically produce the most critical effects in flexural members (i.e., members subjected primarily to bending) where the effects due to dead loads and those due to variable loads are additive. For flexural members subjected to only dead load D effects and floor live load L effects where $D/L < 8$, ACI Eq. (9-2) will produce the critical effects on the member. ACI Eq. (9-1) will produce critical effects in cases where $D/L > 8$. Examples 3.3 through 3.5 illustrate the calculation of factored bending moments and shear forces for flexural members subjected to D and L.

Which of the seven load combinations will produce the most critical effects for members subjected to bending and axial forces (compression or tension) is usually not obvious, so all of them must be investigated.

Certain types of variable loads, such as wind and earthquake loads, act in more than one direction on a building or structure, and the appropriate sign of the variable load must be considered in the load combinations. ACI Eqs. (9-4) and (9-5) are to be used where gravity load effects and lateral load effects are additive, whereas ACI Eqs. (9-6) and (9-7) are applicable where a dead load effect D counteracts a horizontal load effect W or E. Example 4.1 illustrates the determination of load combinations with a horizontal variable load.

It is possible that the critical effects will occur when one or more of the variable loads are set equal to zero. ACI 9.2.1 requires this type of investigation to be performed in every situation. ACI 9.2.1(d) further requires that the load factor on H be set equal to zero in Eqs. (9-6) and (9-7) in situations where it counteracts the effects due to W or E. Because the loads due to lateral earth pressure, groundwater pressure, or pressure of bulk materials are likely to change over time, they are not to be considered as part of the overall resistance to the effects from wind or earthquakes.

Example 4.1   The one-story reinforced concrete frame depicted in Fig. 4.1 is subjected to the loads shown in the figure. The columns are 12 in wide and 12 in deep, and the beam is 12 in wide and 18 in deep. Table 4.2 contains a summary of the bending moments and shear forces in beam BC due to the nominal loads.

Determine the factored load combinations for beam BC.

Solution   Considering the applied loads on this frame, all of the load combinations in ACI 9.2.1 must be investigated except for those in Eqs. (9-5) and (9-7) that contain earthquake effects E.

It is evident that the critical effects on this beam will be obtained by not taking one or more of the variable loads equal to zero in the applicable load combinations. It is also evident by examining Eqs. (9-2) to (9-4) that critical effects will be obtained by using those due to the rain load R because this load produces greater factored bending moments and shear forces than those produced by the roof live load L_r or snow load S. Because critical effects are used to design a member, only the load combinations that contain R are investigated.

The effects of the wind load W must be investigated for sidesway to the right (wind blowing from left to right, as shown in Fig. 4.1) and for sidesway to the left (wind blowing from right to left).

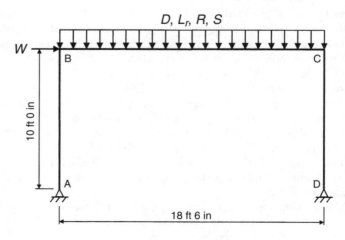

FIGURE 4.1
One-story reinforced concrete frame of Example 4.1.

	Bending Moment (ft kips)		Shear Force (kips)
	Supports	**Midspan**	
Dead load D	−32.0	74.9	23.1
Roof live load L_r	−4.3	9.9	3.1
Rain R	−14.1	33.0	10.2
Snow S	−5.4	12.6	3.9
Wind W	±10.0	—	±1.1

TABLE 4.2   Nominal Bending Moments and Shear Forces for Beam BC in Example 4.1

A summary of the factored load combinations for beam BC is given in Table 4.3. In Eqs. (9-2) to (9-4) and (9-6), the "plus" sign preceding the load factor on W refers to sidesway to the right and the "minus" sign refers to sidesway to the left. For example, in Eq. (9-3), $1.2D + 1.6R + 0.8W$ corresponds to sidesway to the right and $1.2D + 1.6R − 0.8W$ corresponds to sidesway to the left.

A review of Table 4.3 yields the following maximum factored bending moments and shear force, which are obtained by Eq. (9-3):

- Negative factored bending moment M_u at support B or C = −69.0 ft kips
- Positive factored bending moment M_u at midspan = 142.7 ft kips
- Factored shear force V_u at support B or C = 44.9 kips

Because of symmetry, the maximum bending moments and shear forces occur at joints B and C for sidesway to the left and sidesway to the right, respectively. The size of the beam and the required amount of reinforcement are determined for these maximum effects.

As expected, Eq. (9-6) did not yield the critical effects for this flexural member.

Example 4.2   For the frame shown in Fig. 4.1, determine the factored load combinations for column AB using the nominal load data in Table 4.4.

Solution   As in Example 4.1, all of the load combinations in ACI 9.2.1 must be investigated for this column except for those in Eqs. (9-5) and (9-7) that contain earthquake effects E.

A summary of the factored load combinations for beam BC is given in Table 4.5. In Eqs. (9-2) to (9-4) and (9-6), the "plus" sign preceding the load factor on W refers to sidesway to the right and the "minus" sign refers to sidesway to the left. For example, in Eq. (9-3), $1.2D + 1.6L_r + 0.8W$ corresponds to sidesway to the right and $1.2D + 1.6L_r − 0.8W$ corresponds to sidesway to the left.

In this example, taking one or more of the variable loads equal to zero in Eq. (9-2), (9-3), or (9-4) results in factored effects less than those shown in Table 4.5.

Unlike beam BC of Example 4.1, which of the factored load combinations that will produce the critical effects on column AB for combined axial compression and flexure is not readily obvious. The largest factored axial force and bending moment are obtained by Eq. (9-3) for rain loads R with sidesway to the left. However, Eq. (9-5) needs to be considered as well because the combined factored axial force and bending moment obtained from this combination occur in one of the critical areas of the interaction diagram for the column. The design of members subjected to combined axial load and bending is covered in Chap. 8.

In this example, the column remains in compression under all of the load combinations. It is possible that one or more of the load combinations—especially Eq. (9-6) or (9-7)—could yield a net factored tensile force on the column. This would have an important impact on the design of the column and that of the foundation that supports the column.

Equation Number	Load Combination	Location	Bending Moment M_u (ft kips)	Shear Force V_u (kips)
(9-1)	1.4D	Supports B and C	$1.4 \times (-32.0) = -44.8$	$1.4 \times 23.1 = 32.3$
		Midspan	$1.4 \times (74.9) = 104.9$	—
(9-2)	1.2D + 0.5R	Supports B and C	$[1.2 \times (-32.0)] + [0.5 \times (-14.1)] = -45.5$	$(1.2 \times 23.1) + (0.5 \times 10.2) = 32.8$
		Midspan	$(1.2 \times 74.9) + (0.5 \times 33.0) = 106.4$	—
(9-3)	1.2D + 1.6R + 0.8W	Support B	$[1.2 \times (-32.0)] + [1.6 \times (-14.1)] + (0.8 \times 10.0) = -53.0$	$(1.2 \times 23.1) + (1.6 \times 10.2) - (0.8 \times 1.1) = 43.2$
		Support C	$[1.2 \times (-32.0)] + [1.6 \times (-14.1)] + [0.8 \times (-10.0)] = -69.0$	$(1.2 \times 23.1) + (1.6 \times 10.2) + (0.8 \times 1.1) = 44.9$
		Midspan	$(1.2 \times 74.9) + (1.6 \times 33.0) = 142.7$	—
	1.2D + 1.6R − 0.8W	Support B	$[1.2 \times (-32.0)] + [1.6 \times (-14.1)] + [0.8 \times (-10.0)] = -69.0$	$(1.2 \times 23.1) + (1.6 \times 10.2) + (0.8 \times 1.1) = 44.9$
		Support C	$[1.2 \times (-32.0)] + [1.6 \times (-14.1)] + (0.8 \times 10.0) = -53.0$	$(1.2 \times 23.1) + (1.6 \times 10.2) - (0.8 \times 1.1) = 43.2$
		Midspan	$(1.2 \times 74.9) + (1.6 \times 33.0) = 142.7$	—
(9-4)	1.2D + 0.5R + 1.6W	Support B	$[1.2 \times (-32.0)] + [0.5 \times (-14.1)] + (1.6 \times 10.0) = -29.5$	$(1.2 \times 23.1) + (0.5 \times 10.2) - (1.6 \times 1.1) = 31.1$
		Support C	$[1.2 \times (-32.0)] + [0.5 \times (-14.1)] + [1.6 \times (-10.0)] = -61.5$	$(1.2 \times 23.1) + (0.5 \times 10.2) + (1.6 \times 1.1) = 34.6$
		Midspan	$(1.2 \times 74.9) + (0.5 \times 33.0) = 106.4$	—
	1.2D + 0.5R − 1.6W	Support B	$[1.2 \times (-32.0)] + [0.5 \times (-14.1)] + [1.6 \times (-10.0)] = -61.5$	$(1.2 \times 23.1) + (0.5 \times 10.2) + (1.6 \times 1.1) = 34.6$
		Support C	$[1.2 \times (-32.0)] + [0.5 \times (-14.1)] + (1.6 \times 10.0) = -29.5$	$(1.2 \times 23.1) + (0.5 \times 10.2) - (1.6 \times 1.1) = 31.1$
		Midspan	$(1.2 \times 74.9) + (0.5 \times 33.0) = 106.4$	—

TABLE 4.3   Summary of Load Combinations for Beam BC Given in Example 4.1 (*continued*)

Equation Number	Load Combination	Location	Bending Moment M_u (ft kips)	Shear Force V_u (kips)
(9-6)	0.9D + 1.6W	Support B	$[0.9 \times (-32.0)] + (1.6 \times 10.0) = -12.8$	$(0.9 \times 23.1) - (1.6 \times 1.1) = 19.0$
		Support C	$[0.9 \times (-32.0)] + [1.6 \times (-10.0)] = -44.8$	$(0.9 \times 23.1) + (1.6 \times 1.1) = 22.6$
		Midspan	$0.9 \times 74.9 = 67.4$	—
	0.9D − 1.6W	Support B	$[0.9 \times (-32.0)] + [1.6 \times (-10.0)] = -44.8$	$(0.9 \times 23.1) + (1.6 \times 1.1) = 22.6$
		Support C	$[0.9 \times (-32.0)] + (1.6 \times 10.0) = -12.8$	$(0.9 \times 23.1) - (1.6 \times 1.1) = 19.0$
		Midspan	$0.9 \times 74.9 = 67.4$	—

TABLE 4.3   Summary of Load Combinations for Beam BC Given in Example 4.1 (*continued*)

4.3 Design Strength

4.3.1 Overview

The design strength of a reinforced concrete member is equal to the nominal strength of the member, which is calculated in accordance with the provisions of the Code, multiplied by a strength reduction factor ϕ that is always less than 1 [see Eq. (4.2)]. The subscript n is used to denote nominal strength. For example, the notation for nominal flexural strength of a reinforced concrete member is M_n. Subsequent chapters of this book contain methods for determining the nominal strength. A discussion on the purpose of strength reduction factors follows.

ACI 9.4 places an upper limit of 80,000 psi for the yield strength of reinforcement that can be used in design calculations. A list of Code sections that have other limitations on reinforcement yield strength is given in ACI R9.4. It is shown in Section 4.4 that the deflection provisions of ACI 9.5 are directly related to f_y.

4.3.2 Strength Reduction Factors

Strength reduction factors are commonly referred to as resistance factors or ϕ-factors and play a key role in the determination of the design strength of a reinforced concrete member. The main purposes of these factors are as follows:

	Axial Force (kips)	Bending Moment (ft kips)	Shear Force (kips)
Dead load D	23.1	32.0	3.2
Roof live load L_r	3.1	4.3	0.4
Rain R	10.2	14.1	1.4
Snow S	3.9	5.4	0.5
Wind W	±1.1	±10.0	1.0

TABLE 4.4   Nominal Axial Forces, Bending Moments, and Shear Forces for Column AB Given in Example 4.2

Equation Number	Load Combination	Axial Force P_u (kips)	Bending Moment M_u (ft kips)	Shear Force V_u (kips)
(9-1)	$1.4D$	$1.4 \times 23.1 = 32.3$	$1.4 \times 32.0 = 44.8$	$1.4 \times 3.2 = 4.5$
(9-2)	$1.2D + 0.5L_r$	$(1.2 \times 23.1) + (0.5 \times 3.1) = 29.3$	$(1.2 \times 32.0) + (0.5 \times 4.3) = 40.6$	$(1.2 \times 3.2) + (0.5 \times 0.4) = 4.0$
	$1.2D + 0.5S$	$(1.2 \times 23.1) + (0.5 \times 3.9) = 29.7$	$(1.2 \times 32.0) + (0.5 \times 5.4) = 41.1$	$(1.2 \times 3.2) + (0.5 \times 0.5) = 4.1$
	$1.2D + 0.5R$	$(1.2 \times 23.1) + (0.5 \times 10.2) = 32.8$	$(1.2 \times 32.0) + (0.5 \times 14.1) = 45.5$	$(1.2 \times 3.2) + (0.5 \times 1.4) = 4.5$
(9-3)	$1.2D + 1.6L_r + 0.8W$	$(1.2 \times 23.1) + (1.6 \times 3.1) - (0.8 \times 1.1) = 31.8$	$(1.2 \times 32.0) + (1.6 \times 4.3) - (0.8 \times 10.0) = 37.3$	$(1.2 \times 3.2) + (1.6 \times 0.4) - (0.8 \times 1.0) = 3.7$
	$1.2D + 1.6S + 0.8W$	$(1.2 \times 23.1) + (1.6 \times 3.9) - (0.8 \times 1.1) = 33.1$	$(1.2 \times 32.0) + (1.6 \times 5.4) - (0.8 \times 10.0) = 39.0$	$(1.2 \times 3.2) + (1.6 \times 0.5) - (0.8 \times 1.0) = 3.8$
	$1.2D + 1.6R + 0.8W$	$(1.2 \times 23.1) + (1.6 \times 10.2) - (0.8 \times 1.1) = 43.2$	$(1.2 \times 32.0) + (1.6 \times 14.1) - (0.8 \times 10.0) = 53.0$	$(1.2 \times 3.2) + (1.6 \times 1.4) - (0.8 \times 1.0) = 5.3$
	$1.2D + 1.6L_r - 0.8W$	$(1.2 \times 23.1) + (1.6 \times 3.1) + (0.8 \times 1.1) = 33.6$	$(1.2 \times 32.0) + (1.6 \times 4.3) + (0.8 \times 10.0) = 53.3$	$(1.2 \times 3.2) + (1.6 \times 0.4) + (0.8 \times 1.0) = 5.3$
	$1.2D + 1.6S - 0.8W$	$(1.2 \times 23.1) + (1.6 \times 3.9) + (0.8 \times 1.1) = 34.8$	$(1.2 \times 32.0) + (1.6 \times 5.4) + (0.8 \times 10.0) = 55.0$	$(1.2 \times 3.2) + (1.6 \times 0.5) + (0.8 \times 1.0) = 5.4$
	$1.2D + 1.6R - 0.8W$	$(1.2 \times 23.1) + (1.6 \times 10.2) + (0.8 \times 1.1) = 44.9$	$(1.2 \times 32.0) + (1.6 \times 14.1) + (0.8 \times 10.0) = 69.0$	$(1.2 \times 3.2) + (1.6 \times 1.4) + (0.8 \times 1.0) = 6.9$
(9-4)	$1.2D + 0.5L_r + 1.6W$	$(1.2 \times 23.1) + (0.5 \times 3.1) - (1.6 \times 1.1) = 27.5$	$(1.2 \times 32.0) + (0.5 \times 4.3) - (1.6 \times 10.0) = 24.6$	$(1.2 \times 3.2) + (0.5 \times 0.4) - (1.6 \times 1.0) = 2.4$
	$1.2D + 0.5S + 1.6W$	$(1.2 \times 23.1) + (0.5 \times 3.9) - (1.6 \times 1.1) = 27.9$	$(1.2 \times 32.0) + (0.5 \times 5.4) - (1.6 \times 10.0) = 25.1$	$(1.2 \times 3.2) + (0.5 \times 0.5) - (1.6 \times 1.0) = 2.5$
	$1.2D + 0.5R + 1.6W$	$(1.2 \times 23.1) + (0.5 \times 10.2) - (1.6 \times 1.1) = 31.1$	$(1.2 \times 32.0) + (0.5 \times 14.1) - (1.6 \times 10.0) = 29.5$	$(1.2 \times 3.2) + (0.5 \times 1.4) - (1.6 \times 1.0) = 2.9$
	$1.2D + 0.5L_r - 1.6W$	$(1.2 \times 23.1) + (0.5 \times 3.1) + (1.6 \times 1.1) = 31.0$	$(1.2 \times 32.0) + (0.5 \times 4.3) + (1.6 \times 10.0) = 56.6$	$(1.2 \times 3.2) + (0.5 \times 0.4) + (1.6 \times 1.0) = 5.6$
	$1.2D + 0.5S - 1.6W$	$(1.2 \times 23.1) + (0.5 \times 3.9) + (1.6 \times 1.1) = 31.4$	$(1.2 \times 32.0) + (0.5 \times 5.4) + (1.6 \times 10.0) = 57.1$	$(1.2 \times 3.2) + (0.5 \times 0.5) + (1.6 \times 1.0) = 5.7$
	$1.2D + 0.5R - 1.6W$	$(1.2 \times 23.1) + (0.5 \times 10.2) + (1.6 \times 1.1) = 34.6$	$(1.2 \times 32.0) + (0.5 \times 14.1) + (1.6 \times 10.0) = 61.5$	$(1.2 \times 3.2) + (0.5 \times 1.4) + (1.6 \times 1.0) = 6.1$
(9-6)	$0.9D + 1.6W$	$(0.9 \times 23.1) - (1.6 \times 1.1) = 19.0$	$(0.9 \times 32.0) - (1.6 \times 10.0) = 12.8$	$(1.2 \times 3.2) - (1.6 \times 1.0) = 2.2$
	$0.9D - 1.6W$	$(0.9 \times 23.1) + (1.6 \times 1.1) = 22.6$	$(0.9 \times 32.0) + (1.6 \times 10.0) = 44.8$	$(1.2 \times 3.2) + (1.6 \times 1.0) = 5.4$

TABLE 4.5 Summary of Load Combinations for Column AB Given in Example 4.2

1. *To account for the understrength of a member due to variations in material strengths and dimension.* As was discussed in Chap. 2, the strength of concrete can vary because concrete is a composite material made of constituent materials whose properties vary. The strength of reinforcing steel can also vary but usually to a lesser degree than concrete.

   Member dimensions can differ from those specified in the construction documents because of construction and fabrication tolerances. The diameter of reinforcing bars can also fluctuate because of rolling and fabrication tolerances.

   Reinforcing bars in a concrete section can be placed at locations that are different from those specified in the construction documents. Tolerances on reinforcement placement are prescribed by the ACI.

2. *To allow for inaccuracies in the design equations.* As will be shown in subsequent chapters, a number of assumptions and simplifications are made in the design equations for nominal strength. These assumptions and simplifications introduce inaccuracies that must be accounted for when determining the design strength.

3. *To reflect the degree of ductility and required reliability of a member.* Reinforced concrete members that are more ductile, such as beams, are less sensitive to variations in concrete strength compared with members that are less ductile, such as columns.

   Spiral reinforcement confines the concrete in a column better than tied reinforcement. Thus, spirally reinforced columns are more ductile and have greater toughness than tied columns.

4. *To reflect the importance of a member.* The failure of a column in a structure is usually considered to be more detrimental than failure of a beam.

As an example of why strength reduction factors are used in design, consider a concrete mixture for a beam that is part of a cast-in-place concrete building. The engineer of record has specified a compressive strength of 4,000 psi at 28 days. At the time the beam is cast, specimens of the concrete mixture are collected in the field in cylinders and are subsequently tested in accordance with the provisions given in Chap. 5 of the Code. It is found that the test specimens yield an average strength of 3,750 psi. Assume that this average compressive strength satisfies the evaluation and acceptance provisions of ACI Chap. 5, which have been discussed in detail in Chap. 2 of this book. The strength factor that the engineer used in the design of the beam partly accounts for the lower compressive strength of the concrete that is actually used in the beam. Thus, redesign of the beam using the lower concrete compressive strength is not required.

ACI strength reduction factors are given in ACI 9.3.2 and are summarized in Table 4.6.

Tension-controlled and compression-controlled sections are defined in ACI 10.3.4 and 10.3.3, respectively, and are covered in more detail in Chap. 5. In very general terms, members with tension-controlled sections are usually beams, and members with compression-controlled sections are usually columns. Lower strength reduction factors are assigned to compression-controlled sections because they are less ductile and more sensitive to variations in concrete strength than tension-controlled sections. Also, columns tend to support areas much greater than those for beams, and as noted in Section 4.3, the consequences of column failure in a building or structure are generally

Tension-controlled sections		0.90
Compression-controlled sections	Members with spiral reinforcement conforming to ACI 10.9.3	0.75
	Other reinforced members	0.65
Shear and torsion		0.75
Bearing on concrete		0.65
Strut-and-tie models (Appendix A of the Code)		0.75

TABLE 4.6   ACI Strength Reduction Factors

more severe than those attributed to beam failure. The advantage of using spiral rein-forcement in a column is reflected in a strength reduction factor that is greater than that for other types of lateral reinforcement.

A linear transition in the strength reduction factor is permitted between the limits for tension-controlled and compression-controlled sections (ACI 9.3.2); this transition is illustrated in Fig. 4.2 for Grade 60 reinforcement ($f_y = 60,000$ psi). Equations for ϕ are provided in the figure as a function of the net tensile strain in the extreme layer of the longitudinal tension steel ε_t and the ratio of the depth of the neutral axis c to the distance from the extreme compression fiber to the centroid of the extreme layer of the longitudinal tension steel d_t. Also provided in the figure are the strain diagrams corre-sponding to compression-controlled and tension-controlled sections. These quantities are discussed in detail in the next chapter.

According to ACI 9.3.3, development lengths for reinforcement do not require a strength reduction factor. Strength reduction factors are also not required when de-termining lap splice lengths because such lengths are a function of the development length.

ACI 9.3.4 contains strength reduction factors for the design of members in structures assigned to higher levels of seismic risk.

4.4 Control of Deflections

4.4.1 Overview

Deflection control is part of the serviceability limit state that must be satisfied in the design of reinforced concrete members. The main material parameters that affect deflections are the modulus of elasticity E_c, the modulus of rupture f_r, shrinkage, and creep. More information on these parameters can be found in Chap. 2.

Excessive deflections of floor or roof members can result in damage to nonstructural components of a building, such as partitions, doors, or glass windows, to name a few. Such deflections can also result in the malfunction of sensitive equipment inside a building. Unsightly cracks can appear in structural members that have deflected in excess of the established limitations. Excessive deflection of roof members can cause ponding of water on the roof. Roof members will deflect under the weight of water, and if the deflection is excessive, additional water can accumulate on the roof, which can

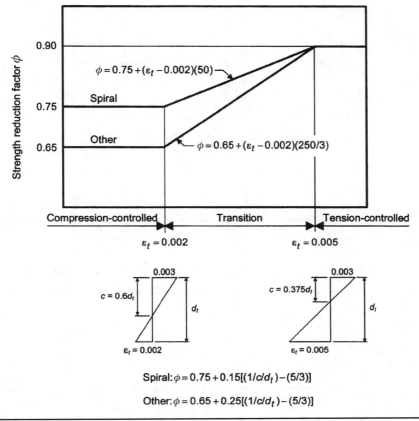

FIGURE 4.2 Variation of the strength reduction factor ϕ for Grade 60 reinforcement.

lead to even more deflection. In extreme cases, the progressive increase of deflection can result in the collapse of a roof. For these and other reasons, it is important to limit the deflection of reinforced concrete members.

ACI 9.5 provides two methods for controlling deflections due to service loads in one-way and two-way floor and roof systems: (1) minimum thickness limitations and (2) computed deflection limitations.

4.4.2 One-Way Construction

The minimum thickness limitations given in ACI Table 9.5(a), which appeared in the ACI Code at the time that the strength design method was introduced, are applicable to beams and one-way slabs that are not attached to partitions or other construction that is likely to be damaged by relatively large deflections. Deflection criteria are essentially satisfied for members that have a thickness equal to or greater than that given in the table. A summary of the minimum thicknesses for continuous members and cantilevered members is given in Fig. 4.3, where h is the overall thickness of the member.

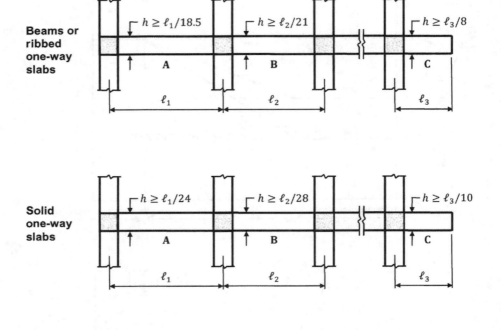

Beams or ribbed one-way slabs

$h \geq \ell_1/18.5$

$h \geq \ell_2/21$

$h \geq \ell_3/8$

A

B

C

ℓ_1 ℓ_2 ℓ_3

Solid one-way slabs

$h \geq \ell_1/24$

$h \geq \ell_2/28$

$h \geq \ell_3/10$

A

B

C

ℓ_1 ℓ_2 ℓ_3

A: One end continuous
B: Both ends continuous
C: Cantilever

FIGURE 4.3 The minimum thickness of beams and one-way slabs in accordance with ACI 9.5.2.

Values of minimum thickness shown in Fig. 4.3 are applicable to members with normal-weight concrete ($w_c = 145$ pcf) and Grade 60 reinforcement. For other conditions, the values in the figure must be modified as follows:

- For structural lightweight having w_c in the range 90 to 115 pcf, multiply the values by $(1.65 - 0.005w_c) \geq 1.09$.
- For f_y other than 60,000 psi, multiply the values by $(0.4 + f_y/100,000)$.

No correction factor is provided for concrete with w_c greater than 115 pcf because research has shown that the correction factor would be close to 1 where w_c is between 115 and 145 pcf.[8] The correction factor on yield stress of reinforcement produces conservative results for values of f_y between 40,000 and 80,000 psi.

For simply supported members, minimum thickness is $\ell/20$ and $\ell/16$ for one-way slabs and beams, respectively.

The minimum thickness limitations outlined earlier are an integral part of the design procedure for typical beams and one-way slabs: The thickness of a member will usually be determined first on the basis of strength requirements and then checked against the minimum thickness requirements for deflection.

In cases where members are supporting elements that are likely to be damaged by relatively large deflections, the provisions in ACI 9.5.2.2 through 9.5.2.6 must be used to determine deflections. Immediate and long-term deflections must both be calculated, and the magnitudes of the deflections must be less than or equal to the limiting values given in ACI Table 9.5(b). Methods for determining deflections of one-way reinforced concrete members are covered in Section 6.5.

4.4.3   Two-Way Construction

ACI 9.5.3 contains minimum thickness requirements for two-way construction. By definition, a two-way slab system has a ratio of long-to-short spans that is less than or equal to 2.

For two-way slab systems that do not have any interior beams, the minimum slab thicknesses are given in ACI Table 9.5(c) as functions of f_y and the clear span length in the long direction ℓ_n. For slabs without drop panels, the thickness must be at least 5 in, and for slabs with drop panels, the thickness must be at least 4 in. Drop panels are defined in ACI 13.2.5 and are covered in more detail in Chap. 7. It has been demonstrated through the years that slabs conforming to these minimum thickness requirements have performed adequately without any problems due to short-term or long-term deflections.

For two-way slab systems with interior beams, the minimum slab thickness is determined by Eqs. (9-12) and (9-13). In addition to f_y and ℓ_n, h is calculated as a function of the beam stiffness along the column lines. Comprehensive coverage of these equations is provided in Chap. 7.

The deflection limits of ACI Table 9.5(b) are also applicable to immediate and long-term calculated deflections for two-way slabs. Calculation of deflections for two-way slabs is complex, and more information on the methods used to determine such deflections can be found in Chap. 7.

References

1. Portland Cement Association (PCA). 2008. *Notes on ACI 318-08 Building Code Requirements for Structural Concrete*. PCA, Skokie, IL.
2. Ellingwood, B., Galambos, T. V., MacGregor, J. G., and Cornell, C. A. 1980. *Development of a Probability Based Load Criterion for American National Standard A58*, NBS Special Publication 577. National Bureau of Standards, U.S. Department of Commerce, Washington, DC.
3. Nowak, A. S., and Szerszen, M. M. 2001. *Reliability-based Calibration for Structural Concrete*, PCA R&D Serial No. 2558., Portland Cement Association (PCA), Skokie, IL.
4. Nowak, A. S., and Szerszen, M. M. 2004. *Reliability-based Calibration for Structural Concrete, Phase 2*, PCA R&D Serial No. 2674. Portland Cement Association (PCA), Skokie, IL.
5. Building Officials and Code Administration International (BOCA). *BOCA National Building Code*. BOCA, Country Club Hills, IL.
6. Southern Building Code Congress International (SBCCI). *Standard Building Code*. SBCCI, Birmingham, AL.
7. International Conference of Building Officials (ICBO). *Uniform Building Code*. ICBO, Whittier, CA.
8. American Concrete Institute (ACI), Committee 213. 2003. *Guide for Structural Lightweight Aggregate Concrete*, ACI 213R-03. ACI, Farmington Hills, MI.

Problems

4.1. A reinforced concrete beam is subjected to the following nominal bending moments:

Supports: $M_D = 75$ ft kips, $M_L = 20$ ft kips, and $M_W = \pm 55$ ft kips
Midspan: $M_D = 50$ ft kips and $M_L = 12$ ft kips

Determine the factored load combinations at the supports and at midspan.

4.2. A reinforced concrete column is subjected to the nominal effects given in Table 4.7.

	Axial Force (kips)	Bending Moment (ft kips)
Dead load D	80	20
Live load L	40	10
Roof live load L_r	15	0
Snow S	20	0
Rain R	25	0
Wind W	± 20	± 50

TABLE 4.7 Nominal Axial Forces and Bending Moments for Column AB Given in Problem 4.2

Determine the factored load combinations.

4.3. Given the five-story building and load data of Example 3.1, determine the factored load combinations for (a) column B2 in the fifth story and (b) column B2 in the first story.

4.4. An interior beam has a span of 24 ft in a continuous reinforced concrete frame. The beam is not supporting or attached to any type of construction likely to be damaged by large deflections. Material data are $w_c = 110$ pcf and $f_y = 60,000$ psi.
Determine the minimum thickness of the beam.

4.5. Given the floor system depicted in Fig. 3.21, determine the minimum thickness of the one-way slab assuming that it is not supporting or attached to any type of construction likely to be damaged by large deflections. Material data are $w_c = 145$ pcf and $f_y = 60,000$ psi.

4.6. Given the floor system depicted in Fig. 3.21, determine the minimum thickness of the beam assuming the following: (1) it is not supporting or attached to any type of construction likely to be damaged by large deflections, and (2) a single beam thickness must be provided for the entire floor system for economy. Material data are $w_c = 145$ pcf and $f_y = 60,000$ psi.

CHAPTER 5

General Principles of the Strength Design Method

5.1 Introduction

This chapter covers the fundamental principles and requirements of the strength design method. Presented are the design assumptions of the method and the basic techniques to determine the nominal strength of a reinforced concrete section subjected to flexure, axial load, or a combination of both.

Beginning in 1886, numerous theories on strength design of reinforced concrete have been published through the years. Reference 1 contains a summary of the significant aspects of these early theories, which were primarily based on results acquired from tests of reinforced concrete members and from analytical investigations.

As a reinforced concrete member approaches its ultimate strength, both the concrete and the reinforcing steel behave inelastically. This inelastic behavior must be captured in the design theory. It is evident from the information on material properties given in Chap. 2 that it is far easier to analytically express the inelastic behavior of reinforcing steel than that of concrete. As such, simplifying assumptions are made in the strength design method related to stress distribution in concrete (see Section 5.2).

The strength design method is based on the following two fundamental conditions:

1. *Static equilibrium.* The compressive and tensile forces acting on any cross-section of a member are in equilibrium.

2. *Compatibility of strains.* The strain in a reinforcing bar that is embedded in concrete is equal to the strain in concrete at that level.

The first condition must be satisfied at every cross-section of a member. It is shown later in this chapter that the basic equations of equilibrium are used in determining nominal strengths of reinforced concrete members.

The second condition implies that there is a perfect bond between the concrete and the reinforcing steel and that both the materials act together to resist the effects from external loads. Tests have shown that this condition is very close to being correct, especially where deformed reinforcing bars are utilized. In addition to the natural surface adhesion that exists between concrete and steel, bar deformations play an important role in limiting the amount of slip that occurs between the two materials (see Chap. 2).

5.2 Design Assumptions

The design assumptions used in the strength design method are outlined in ACI 10.2. They are applicable in the design of members subjected to flexure, axial loads, or a combination of both. The nominal strength of a reinforced concrete member is determined on the basis of these design assumptions.

Design Assumption No. 1:   The strains in the reinforcement and the concrete shall be assumed directly proportional to the distance from the neutral axis.

The first design assumption is the traditional assumption made in beam theory: Plane sections that are perpendicular to the axis of bending prior to bending remain plane after bending. This inherently implies that the concrete and the reinforcing steel act together to resist load effects (recall that this is the second of the conditions required in the strength design method; see Section 5.1).

Strictly speaking, this assumption is not correct for reinforced concrete members after cracking occurs because the strain on the tension side of the neutral axis varies significantly at any given level owing to the presence of cracks. However, many experimental tests have confirmed that the distribution of strain is essentially linear across a reinforced concrete cross-section, even near ultimate strength, when strains are measured across the same gage length on the compressive and tensile faces of a member.[2] The gage lengths that were used in the tests included several cracks on the tension face of the member.

For deep beams, which are defined in ACI 10.7, the strain is not linear, and a nonlinear distribution of strain must be utilized, or a strut-and-tie model as outlined in Appendix A of the Code may be used.

The strain distribution over the depth of a rectangular reinforced concrete section at ultimate strength is depicted in Fig. 5.1. For illustrative purposes, it is assumed that the strains are compressive above the neutral axis and are tensile below it. The strains in the concrete and the reinforcement are directly proportional to the distance from the neutral axis, which is located a distance c from the compression face of the section.

Because the strain distribution is linear, the strain in the concrete ε_c at the extreme compression fiber is directly proportional to the strains in the reinforcement. For example, the ratio of ε_c to

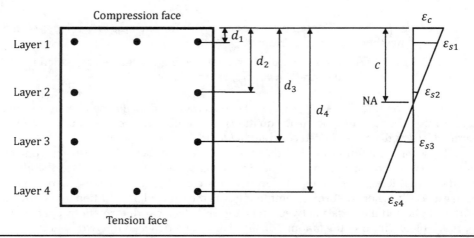

FIGURE 5.1   The assumed strain distribution in a reinforced concrete section.

the strain in the reinforcement farthest from the compression face ε_{s4} can be obtained by similar triangles:

$$\frac{\varepsilon_c}{\varepsilon_{s4}} = \frac{c}{d_4 - c} \tag{5.1}$$

Similar relationships can be established between ε_c and the other reinforcement strains and between the various reinforcement strains.

Note that the largest tensile strain occurs in the reinforcing steel farthest from the compression face. The concrete below the neutral axis is cracked at ultimate strength, and for all intents and purposes, it cannot resist any tensile strains (see design assumption no. 4). That is why no strain is shown on the tension face of the concrete in Fig. 5.1.

Design Assumption No. 2:   The maximum usable strain at the extreme concrete compression fiber is 0.0030.

The maximum compressive strain at crushing of concrete has been measured in many experimental tests of reinforced concrete members (beams and eccentrically loaded columns) and eccentrically loaded plain concrete prisms. Test data vary between 0.0030 and 0.0080 (see Ref. 2 for a summary of the test results). A maximum strain of $\varepsilon_c = 0.0030$ is a reasonably conservative value proposed for design (see the compressive stress–strain curves for concrete shown in Fig. 2.6).

Design Assumption No. 3:   The stress in the reinforcement f_s below its specified yield strength f_y is equal to the modulus of elasticity of the steel E_s times the steel strain ε_s. The stress in the reinforcement is equal to f_y for strains ε_s greater than or equal to f_y/E_s.

On the basis of the stress diagram of reinforcing steel (see Fig. 2.15), it is reasonable to assume that there is a linear relationship between stress and strain up to the yield strength f_y. As noted in Chap. 2, the modulus of elasticity can be taken as 29,000,000 psi for all grades of reinforcing steel (ACI 8.5.2).

The second part of the assumption implies that the effect of strain hardening of the steel above the yield point is neglected in strength computations. In other words, the stress in the reinforcement f_s is equal to f_y for any value of steel strain ε_s that is greater than the yield strain $\varepsilon_y = f_y/E_s$. The idealized stress–strain curve based on this assumption is illustrated in Fig. 5.2.

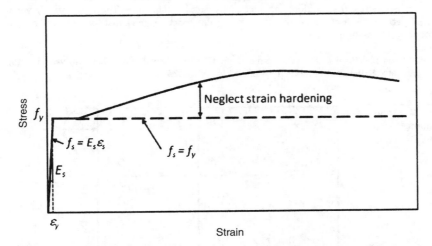

Figure 5.2   The idealized stress–strain curve of reinforcing steel used in the strength design method.

Design Assumption No. 4:   The tensile strength of concrete is neglected in the axial and flexural calculations of reinforced concrete.

It was discussed in Chap. 2 that the tensile strength of concrete is small compared with the tensile strength of reinforcing steel. Within the tension portion of a reinforced concrete cross-section, the tensile force in the cracked concrete is significantly less than the tensile force in the reinforcing steel. Thus, the tensile strength of concrete is conservatively taken as zero in the axial and flexural calculations of nominal strength.

The tensile resistance of concrete is used in other situations, most notably in serviceability calculations. For example, the modulus of rupture f_r, which is related to the tensile strength (see Chap. 2), is utilized in the determination of the immediate deflection of a reinforced concrete member.

Design Assumption No. 5:   The relationship between the concrete compressive stress distribution and the concrete strain shall be assumed to be rectangular, trapezoidal, parabolic, or of any other shape that results in prediction of strength in substantial agreement with the results of comprehensive tests.

Concrete behaves inelastically when subjected to a relatively high compressive stress, as is evident from the stress–strain curves in Fig. 2.6. Nonlinear behavior becomes pronounced after the stress reaches approximately 50% of the compressive strength f_c' (see Fig. 2.5).

Although a general nonlinear model for compressive stress distribution could be used when determining the nominal strength of a reinforced concrete member, such as the one illustrated in Fig. 5.3 for a flexural member, it is simpler to make use of a less complicated distribution as long as the simpler model yields results close to those from tests. Note that the shape of the stress distribution in the figure follows that of a stress–strain curve in compression where, as expected, zero stress occurs at the level of the neutral axis. The tension force T in the reinforcing steel must be equal to the resultant force C of the compressive stress in the concrete so that equilibrium is satisfied.

Numerous compressive stress distributions have been proposed through the years, and a summary of these can be found in Chap. 6 of Ref. 3. Additional information on the historical background of the distributions and a review of the tests that were performed to support the proposed stress distributions are given in Ref. 2.

Research has shown that models using rectangular, parabolic, trapezoidal, and other-shaped compressive stress distributions can adequately predict test results. The assumption given in ACI 10.2.7 permits the use of an equivalent rectangular concrete stress distribution, which is covered under design assumption no. 6.

Design Assumption No. 6:   The requirements of ACI 10.2.6 are satisfied by an equivalent rectangular concrete stress distribution, which is defined in ACI 10.2.7.

FIGURE 5.3 Stress conditions at nominal strength.

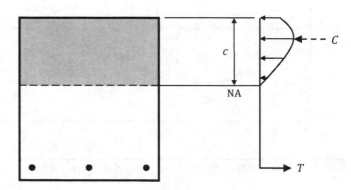

FIGURE 5.4 The equivalent rectangular concrete stress distribution.

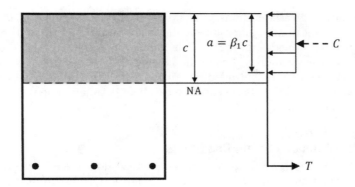

The Code permits the use of the equivalent rectangular concrete stress distribution defined in ACI 10.2.7, which is illustrated in Fig. 5.4. Although he was not the first to propose the use of a rectangular stress block, C. S. Whitney is best known in the United States for advocating it.[4]

A uniform stress equal to 85% of the concrete compressive strength f_c' is distributed over the depth a, which is equal to the factor β_1 times the depth to the neutral axis c. Although this assumed stress distribution does not represent the actual compressive stress distribution in the concrete at the ultimate state, it does provide basically the same results as those obtained from experimental investigations[2]; as noted previously, this is a requirement of the strength design method (see design assumption no. 5).

The need for the factor β_1 is due to the variation in shape of the stress–strain curves for different concrete strengths. It is evident that the stress–strain curves of higher-strength concretes are more linear and exhibit less inelastic behavior than those of lower-strength concretes (see Fig. 2.6). Up to compressive strengths of 4,000 psi, the ratio of the rectangular stress block depth a to the neutral axis depth c that best approximates the actual concrete stress distribution is equal to 0.85, that is, $\beta_1 = 0.85$. For compressive strengths greater than 4,000 psi, β_1 must be less than 0.85 in order to produce adequate results. ACI 10.2.7.3 requires that β_1 be reduced linearly at the rate of 0.05 for each 1,000 psi in excess of 4,000 psi for compressive strengths up to 8,000 psi; above 8,000 psi, $\beta_1 = 0.65$. The following equations define β_1:

- For 2,500 psi $\leq f_c' \leq$ 4,000 psi, $\beta_1 = 0.85$.
- For 4,000 psi $< f_c' \leq$ 8,000 psi, $\beta_1 = 1.05 - 0.00005 f_c'$.
- For $f_c' >$ 8,000 psi, $\beta_1 = 0.65$.

The lower limit of 0.65 was introduced in the 1976 supplement to the 1971 Code on the basis of the results of experiments that were performed on concrete specimens with compressive strengths exceeding 8,000 psi.[5,6]

5.3 General Principles and Requirements

5.3.1 Overview

The nominal strength of a reinforced concrete member is established using the fundamental conditions of equilibrium and strain compatibility (Section 5.1) and the assumptions presented in Section 5.2. Presented in this section are the basic principles utilized in the strength design method.

Design strength equations for cross-sections subjected to flexure or combined flexure and axial load were originally presented in Ref. 2 and were derived using essentially the same design assumptions as those summarized in Section 5.2. These equations form the basis of the nominal strength equations stipulated in the Code. The values of strength obtained from the design equations of Ref. 2 were compared with the test results of 364 beams whose strength was controlled by yielding of the reinforcing steel in tension (as opposed to strength controlled by crushing of the concrete in compression). A statistical analysis of the data revealed an excellent correlation between the analytical and experimental results.

5.3.2 Balanced Strain Conditions

A balanced strain condition exists at a cross-section of a reinforced concrete member when the strain in the tension reinforcement ε_s farthest from the compression face reaches the strain corresponding to yield (i.e., $\varepsilon_s = \varepsilon_y = f_y/E_s$) just as the strain in the extreme compression fiber of the concrete reaches its maximum value of 0.0030.

Consider the rectangular reinforced concrete cross-section with one layer of tension reinforcement depicted in Fig. 5.5. The balanced strain condition using design assumption no. 1 (linear strain distribution) is also shown in the figure. At the extreme compression fiber, the strain in the concrete is equal to the maximum value of 0.0030 (design assumption no. 2), and the strain in the tension reinforcement, which is located a distance d_t from the extreme compression fiber, is equal to ε_y (design assumption no. 3). By definition, d_t is the distance from the extreme compression fiber to the centroid of the longitudinal tension steel that is farthest from the extreme compression fiber. With only one layer of steel, the distance d_t is the same as that from the extreme compression fiber to the centroid of the longitudinal tension steel, which is designated as d. The significance of this definition will become evident shortly.

The ratio of the neutral axis depth c_b to the extreme depth d_t to produce a balanced strain condition in a section with tension reinforcement alone may be obtained by applying strain compatibility conditions. Referring to Fig. 5.5 and using similar triangles,

$$\frac{c_b}{\varepsilon_u} = \frac{d_t - c_b}{\varepsilon_y}.$$

(5.2)

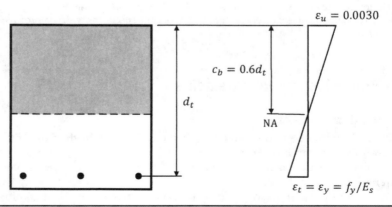

FIGURE 5.5 The balanced strain condition with Grade 60 reinforcement.

Substituting the definitions of $\varepsilon_u = 0.0030$ and $\varepsilon_y = f_y/E_s$ into Eq. (5.2) and rearranging the terms result in

$$\frac{c_b}{d_t} = \frac{\varepsilon_u}{\varepsilon_u + \varepsilon_y} = \frac{0.0030}{0.0030 + (f_y/E_s)} \tag{5.3}$$

For Grade 60 reinforcement, ACI 10.3.3 permits the yield strain ε_y of the reinforcement to be taken as 0.0020, as opposed to 0.00207 that would be obtained by dividing the yield stress (60 ksi) by the modulus of elasticity (29,000 ksi). Substituting 0.0020 for f_y/E_s in Eq. (5.3) results in the ratio $c_b/d_t = 0.6$. Note that this value applies to all sections with Grade 60 reinforcement and not just to rectangular sections.

5.3.3 Compression-controlled Sections

According to ACI 10.3.3, a section is compression-controlled if the net tensile strain in the extreme tension steel ε_t is less than or equal to the compression-controlled strain limit when the concrete in compression reaches its assumed strain limit ε_u of 0.0030. Note that ε_t is defined as the strain in the reinforcement after applicable strains due to creep, shrinkage, and temperature have been deducted from the total strain.

The compression-controlled strain limit is defined as the net tensile strain in the reinforcement at balanced conditions, which is equal to f_y/E_s (see the discussion on balanced strain conditions in the previous section). The strain distribution at the compression-controlled strain limit is illustrated in Fig. 4.2 for Grade 60 reinforcement where $\varepsilon_t = 0.0020$.

A brittle type of failure is generally expected in compression-controlled sections. This type of failure occurs suddenly with little or no warning. It is common for compression members (columns) to have compression-controlled sections, that is, sections where $\varepsilon_t \leq f_y/E_s$. However, compression members that are subjected to a relatively small axial compressive force and a relatively large bending moment, for example, may have sections that are not compression-controlled.

5.3.4 Tension-controlled Sections

A tension-controlled section is defined as a section where $\varepsilon_t \geq 0.0050$ when the concrete in compression reaches its assumed strain limit ε_u of 0.0030 (ACI 10.3.4). The strain distribution illustrated in Fig. 4.2 for the case in which $\varepsilon_t = 0.0050$ is also shown in Fig. 5.6.

The ratio of the neutral axis depth c_t to the extreme depth d_t at the tension-controlled limit may be obtained by applying strain compatibility conditions. Referring to Fig. 5.6 and using similar triangles,

$$\frac{c_t}{\varepsilon_u} = \frac{d_t - c_t}{\varepsilon_t} \tag{5.4}$$

Substituting $\varepsilon_u = 0.0030$ and $\varepsilon_t = 0.0050$ into Eq. (5.4) and rearranging the terms result in

$$\frac{c_t}{d_t} = \frac{\varepsilon_u}{\varepsilon_u + \varepsilon_t} = \frac{0.0030}{0.0030 + 0.0050} = 0.375 \tag{5.5}$$

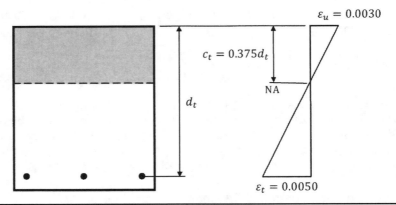

FIGURE 5.6   Strain distribution at the tension-controlled limit.

Flexural members, such as beams, usually have tension-controlled sections. Unlike compression-controlled sections, tension-controlled sections are ductile and generally exhibit significant deflections and cracking before failure.

Sections with ε_t between the compression-controlled strain limit and 0.0050 are said to be in a transition region between compression- and tension-controlled sections. A linear transition in the strength reduction factor occurs between the limits for compression- and tension-controlled sections (see Section 4.3). This transition is illustrated in Fig. 4.2 along with the applicable strength reduction factors for compression- and tension-controlled sections.

Although the preceding discussion has focused on a rectangular section with one layer of tension reinforcement, the basic methods are applicable to rectangular and nonrectangular sections with more than one layer of reinforcement. The effects of cross-section shape and multiple layers of reinforcement are automatically accounted for in the strain compatibility equations that are used to determine ε_t.

The concepts of compression-controlled, tension-controlled, and transition sections first appeared in 1992 as part of the Unified Design Method.[7] It has been demonstrated that this method produces results similar to those from previous strength design methods. A slightly modified version of the Unified Design Method appeared in Appendix B of the 1995 Code. Provisions of the method were moved into the main body of the 1999 Code, whereas those that were displaced from the main body were transferred to Appendix B.

5.4 Flexural Members

5.4.1 Overview

ACI 10.3.5 defines flexural members as members with a factored axial compressive force less than $0.1 f_c' A_g$, where A_g is the gross area of the concrete section. For purposes of discussion, it is assumed that flexural members have sections that are primarily subjected to bending moments.

FIGURE 5.7 The idealized moment–curvature diagram for a reinforced concrete beam with one layer of tension reinforcement.

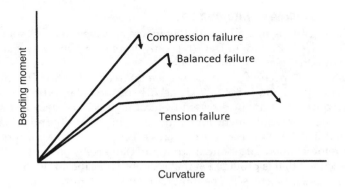

Presented next are the different types of flexural failure and the relationship between failure and the amount of reinforcement in a section. Also presented are the basic equations to determine the nominal strength of members or sections subjected to flexure.

5.4.2 Types of Flexural Failure

Consider a reinforced concrete beam with tension reinforcement only. Assume that the amount of tension reinforcement in the section is such that at failure, the reinforcing steel will yield in tension before the concrete crushes in compression. The relationship between bending moment and curvature of this beam is plotted to failure in Fig. 5.7. It is evident from the diagram that there is a long plastic region—that is, the beam exhibits a ductile response—up to failure. This is commonly referred to as *tension failure*, and the section is tension-controlled (see Section 5.3). In such cases, a member will typically exhibit large deformations and significant cracking prior to collapse, and it is anticipated that there will be ample warning prior to failure.

Now, assume that additional reinforcing steel is added to the section—with all of the other parameters remaining the same—so that at failure, the reinforcing steel yields at the same time the concrete crushes. This is commonly referred to as *balanced failure*. Referring to Fig. 5.5, balanced failure occurs when the strain in the extreme compression fiber of the concrete reaches the assumed crushing strain of 0.0030 at the same time that the strain in the reinforcing steel reaches the yield strain. The moment–curvature relationship for balanced failure is also plotted in Fig. 5.7. It is evident that no ductility is exhibited when such a failure occurs.

Finally, assume that even more reinforcing steel is added to the section so that the total amount is greater than that corresponding to balanced failure. In such cases, the concrete in the extreme compression fiber reaches the assumed crushing strain of 0.0030 prior to the reinforcing steel yielding. The moment–curvature diagram for such a member does not have the ductile postyielding response displayed by a member with an amount of reinforcement smaller than the balanced amount (see Fig. 5.7). This type of failure, which is called *compression failure*, occurs suddenly in a brittle manner without warning.

Tension failures are favored over compression failures in flexural members. The amount of reinforcement in a flexural member is limited by the Code to ensure that it is achieved in design.

5.4.3   Maximum Reinforcement

The Code requires that all flexural members and members with a factored axial compressive force less than $0.1 f'_c A_g$ (typically, beams and one-way slabs) have properties that ensure that tension failure occurs.

Instead of specifying a maximum reinforcement ratio, which was stipulated in editions of the Code prior to 2002 as 75% of the balanced reinforcement ratio, ACI 10.3.5 requires that nonprestressed flexural members be designed so that $\varepsilon_t \geq 0.0040$. In essence, this requirement limits the amount of tension reinforcement that can be provided at a section: For a given cross-section and material properties, the strain in the reinforcement at nominal strength is inversely proportional to the amount of reinforcement that is provided at that section. Thus, the strain decreases as the amount of reinforcement increases. The limitation on ε_t is slightly more conservative than that required previously (see ACI R10.3.5).

5.4.4   Minimum Reinforcement

ACI 10.5 contains minimum reinforcement requirements for beams, slabs, and footings. These requirements are generally applicable to members with cross-sections that are larger than that required for strength. For example, spandrel beams in a building may be deeper and have more strength than needed because they are incorporated into the architectural design of the facade.

When a small amount of tensile reinforcement is provided in such members, the strength of the reinforced concrete section based on a cracked section analysis becomes less than that of an unreinforced concrete section based on the modulus of rupture (see Chap. 2). Failure of reinforced sections with less than a minimum amount of reinforcement can occur suddenly.

To avoid this sudden type of failure, the Code prescribes a minimum amount of reinforcement $A_{s,min}$ that is to be provided at any positive or negative bending moment region of a flexural member where such tensile reinforcement is required [ACI Eq. (10-3)]:

$$A_{s,min} = \frac{3\sqrt{f'_c}b_w d}{f_y} \geq \frac{200\, b_w d}{f_y} \tag{5.6}$$

In this equation, the concrete compressive strength f'_c has the unit of pounds per square inch and b_w is defined as the web width of the member. The limit of $200 b_w d/f_y$ controls for concrete compressive strengths less than approximately 4,400 psi.

ACI 10.5.2 contains minimum reinforcement requirements for statically determinate members where the flange of the member is in tension. An example of such a member is given in Fig. 5.8. The flange of the cantilever beam, which has a width of b_f, is in tension due to the uniformly distributed load that is applied along the length of the member. According to ACI 10.5.2, $A_{s,min}$ is determined by ACI Eq. (10-3), where b_w in the equation is replaced by the larger of two times the actual web width of the member ($2b_w$) or the flange width b_f. It is especially important to provide a minimum amount of reinforcement in members such as cantilever beams where bending moments cannot be redistributed to adjoining framing members.

The aforementioned minimum areas of reinforcement need not be provided at any section that contains at least one-third of the required amount of reinforcement at that location (ACI 10.5.3). This exception is deemed sufficient for large members where the amount required by ACI 10.5.1 or 10.5.2 would be excessive.

FIGURE 5.8
A cantilever beam
with the flange in
tension.

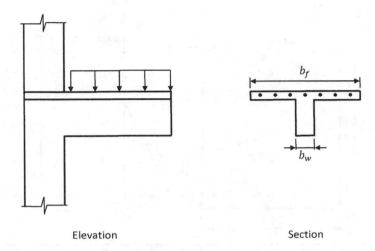

FIGURE 5.8
A cantilever beam
with the flange in
tension.

Elevation   Section

The minimum amount of reinforcement for slabs and footings of uniform cross-sections is equal to the minimum amount required by ACI 7.12.2.1 for shrinkage and temperature reinforcement (ACI 10.5.4):

- $A_{s,min} = 0.0020bh$ where Grade 40 or 50 deformed bars are used
- $A_{s,min} = 0.0018bh$ where Grade 60 deformed bars or welded wire reinforcement are used
- $A_{s,min} = (0.0018 \times 60,000)bh/f_y$ where reinforcement with $f_y > 60,000$ psi measured at a strain of 0.0035 is used

In these expressions, b is the width of the member and h is the overall thickness. A design width of 12 in is typically used in such cases, and reinforcement is specified in square inches per foot. The maximum spacing of this reinforcement is the smaller of three times the member thickness or 18 in.

5.4.5 Nominal Flexural Strength

Overview
The nominal flexural strength of a reinforced concrete member is determined using the two fundamental conditions given in Section 5.1—static equilibrium and compatibility of strains—and the design assumptions given in Section 5.2. Methods to determine the nominal flexural strength M_n are covered in this section.

Rectangular Sections
Single Layer of Tension Reinforcement   Consider the reinforced concrete beam with one layer of tension reinforcement depicted in Fig. 5.9. The strain distribution and equivalent rectangular stress distribution are also shown in the figure. Because there is only one layer of reinforcement in this beam, the distance d_t from the extreme compression fiber to the centroid of the extreme layer of longitudinal tension steel is equal to the distance d from the extreme compression fiber to the centroid of the longitudinal tension reinforcement.

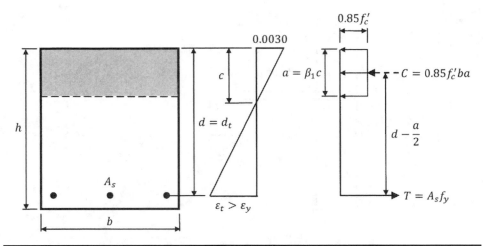

Figure 5.9 The strain and stress distributions in a rectangular beam with tension reinforcement.

The ultimate strain in the concrete is 0.0030, and the strain in the reinforcement is greater than the yield strain ε_y, assuming that the total area of reinforcement A_s is such that the reinforcing steel yields in tension before the concrete crushes in compression.

The resultant compressive force C in the concrete is equal to the compressive stress times the area over which the stress acts:

$$C = 0.85 f_c' ba \tag{5.7}$$

In Eq. (5.7), b is the width of the cross-section as shown in Fig. 5.9.

The tension force T in the reinforcement is equal to the total area of reinforcement A_s times the yield strength of the reinforcement f_y:

$$T = A_s f_y \tag{5.8}$$

In order for equilibrium to be satisfied, the sum of the forces and bending moments on the section must be equal to zero. From force equilibrium, $C = T$. The depth of the equivalent stress block a can be obtained by equating Eqs. (5.7) and (5.8) and solving for a:

$$a = \frac{A_s f_y}{0.85 f_c' b} \tag{5.9}$$

The nominal flexural strength of the section M_n is obtained from moment equilibrium. Moments can be summed about any point on the section. It is usually convenient to sum moments about either C or T. Summing moments about the point of application of the resultant force C yields the following expression for M_n:

$$M_n = A_s f_y \left(d - \frac{a}{2} \right) \tag{5.10}$$

Substituting Eq. (5.9) into Eq. (5.10) results in the following:

$$M_n = A_s f_y \left(d - \frac{0.59 A_s f_y}{b f_c'} \right) \tag{5.11}$$

Define the reinforcement ratio $\rho = A_s/bd$ and the reinforcement index $\omega = \rho f_y/f_c'$. Substituting these quantities into Eq. (5.11) results in the following nondimensional equation for M_n:

$$\frac{M_n}{bd^2 f_c'} = \omega \left(1 - 0.59\omega \right) \tag{5.12}$$

The amount of reinforcement in a section that is needed to resist factored load effects can be calculated by this equation (see Chap. 6).

The flowchart shown in Fig. 5.10 can be used to determine M_n for rectangular sections with one layer of tension reinforcement.

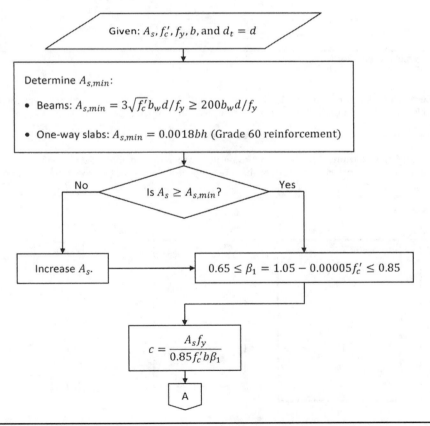

FIGURE 5.10 Nominal flexural strength—rectangular section with one layer of tension reinforcement. (*Continued*)

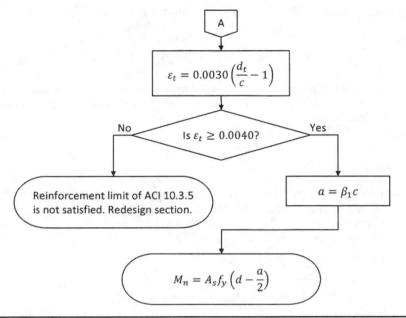

FIGURE 5.10 (Continued)

Example 5.1 Determine the nominal flexural strength M_n of the reinforced concrete beam depicted in Fig. 5.11. Assume $f'_c = 4{,}000$ psi and Grade 60 reinforcement ($f_y = 60{,}000$ psi).

Solution The flowchart shown in Fig. 5.10 is utilized to determine M_n.

Step 1: Determine $A_{s,min}$. The minimum amount of reinforcement is determined by ACI 10.5.1. Because the compressive strength of the concrete is less than 4,400 psi, the minimum amount is

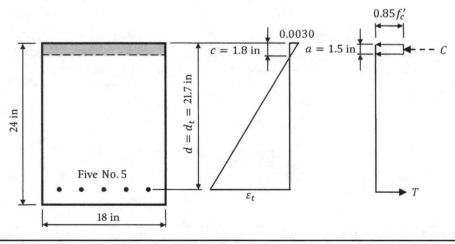

FIGURE 5.11 The reinforced concrete beam of Example 5.1.

determined by the lower limit given in that section:

$$A_{s,min} = \frac{200b_w d}{f_y} = \frac{200 \times 18 \times 21.7}{60{,}000} = 1.30\,\text{in}^2$$

$$A_{s,min} < A_s = 5 \times 0.31 = 1.55\,\text{in}^2$$

Step 2: Determine β_1. According to ACI 10.2.73, $\beta_1 = 0.85$ for $f_c' = 4{,}000$ psi (see Section 5.2).
Step 3: Determine neutral axis depth c.

$$c = \frac{A_s f_y}{0.85 f_c' b \beta_1} = \frac{1.55 \times 60{,}000}{0.85 \times 4{,}000 \times 18 \times 0.85} = 1.8\,\text{in}$$

Step 4: Determine ε_t. The strain ε_t is determined by similar triangles (see Fig. 5.11):

$$\frac{c}{0.0030} = \frac{d_t - c}{\varepsilon_t} \text{ or } \varepsilon_t = 0.0030\left(\frac{d_t}{c} - 1\right)$$

$$\varepsilon_t = 0.0030\left(\frac{21.7}{1.8} - 1\right) = 0.0332 > 0.0040$$

Because $\varepsilon_t > 0.0040$, the maximum reinforcement requirement of ACI 10.3.5 is satisfied. Also, the section is tension-controlled because $\varepsilon_t > 0.0050$ (ACI 10.3.4).
Step 5: Determine depth of the equivalent stress block a.

$$a = \beta_1 c = 0.85 \times 1.8 = 1.5\,\text{in}$$

Step 6: Determine the nominal flexural strength M_n. Equation (5.10) is used to determine M_n:

$$M_n = A_s f_y\left(d - \frac{a}{2}\right) = 1.55 \times 60{,}000\left(21.7 - \frac{1.5}{2}\right)/12{,}000 = 162.4\,\text{ft kips}$$

Comments
The reinforcement in this example is referred to as positive reinforcement because it is positioned near the bottom of the section to resist the effects from a positive bending moment. Accordingly, M_n in such cases is referred to as the positive nominal flexural strength. Positive moments typically occur within the span away from the supports for members subjected to only gravity loads that are supported at each end; however, a net positive bending moment could occur at the face of a support in such cases because of combined gravity and lateral effects.

Example 5.2 Determine the nominal flexural strength M_n of the one-way slab depicted in Fig. 5.12. Assume $f_c' = 3{,}000$ psi and Grade 60 reinforcement ($f_y = 60{,}000$ psi).

Solution The flowchart shown in Fig. 5.10 is utilized to determine M_n.

FIGURE 5.12 The one-way slab of Example 5.2.

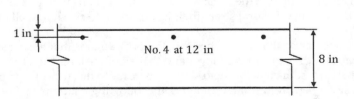

1 in

No. 4 at 12 in

8 in

Step 1: **Determine** $A_{s,min}$. When investigating one-way slab systems, it is customary to base the calculations on a 1-ft-wide design strip. Because the reinforcement is spaced at 12 in on center in this example, there is one No. 4 bar in the design strip.

For a one-way slab with Grade 60 reinforcement, the minimum area of steel is determined in accordance with ACI 7.12.2.1 (ACI 10.5.4):

$$A_{s,min} = 0.0018bh = 0.0018 \times 12 \times 8 = 0.17 \, in^2$$
$$A_{s,min} < A_s = 0.20 \, in^2$$

The reinforcement in this example is referred to as negative reinforcement because it is positioned near the top of the section to resist the effects from a negative bending moment. This negative bending moment produces tension at the top of the section and compression at the bottom of the section.

Step 2: **Determine** β_1. According to ACI 10.2.73, $\beta_1 = 0.85$ for $f_c' = 3,000$ psi (see Section 5.2).

Step 3: **Determine neutral axis depth** c.

$$c = \frac{A_s f_y}{0.85 f_c' b \beta_1} = \frac{0.20 \times 60,000}{0.85 \times 3,000 \times 12 \times 0.85} = 0.46 \, in$$

Step 4: **Determine** ε_t. The strain ε_t is determined by similar triangles:

$$\frac{c}{0.0030} = \frac{d_t - c}{\varepsilon_t} \text{ or } \varepsilon_t = 0.0030 \left(\frac{d_t}{c} - 1 \right)$$

Therefore,

$$\varepsilon_t = 0.0030 \left(\frac{7}{0.46} - 1 \right) = 0.0427 > 0.0040$$

Because $\varepsilon_t > 0.0040$, the maximum reinforcement requirement of ACI 10.3.5 is satisfied. Also, the section is tension-controlled because $\varepsilon_t > 0.0050$ (ACI 10.3.4).

Step 5: **Determine depth of the equivalent stress block** a.

$$a = \beta_1 c = 0.85 \times 0.46 = 0.39 \, in$$

Step 6: **Determine the nominal flexural strength** M_n. Equation (5.10) is used to determine M_n:

$$M_n = A_s f_y \left(d - \frac{a}{2} \right) = 0.20 \times 60,000 \left(7 - \frac{0.39}{2} \right) / 12,000 = 6.8 \, ft \, kips \, per \, foot \, width \, of \, slab$$

Multiple Layers of Tension Reinforcement   Under certain conditions, the required tension reinforcement cannot adequately fit within one layer in a section (spacing requirements for reinforcing steel is covered in Chap. 6). In such cases, the bars are provided in more than one layer, as shown in Fig. 5.13.

The nominal flexural strength is determined for sections with multiple layers of tension reinforcement in the same way as that for sections with one layer. When determining the nominal flexural strength, it is important to check that all of the reinforcement yields.

Considering the section shown in Fig. 5.13, assume that the bar size and the number of bars are the same in each layer and that all of the bars are located below the neutral axis. Also assume that the yield strain ε_y occurs at a distance of d_y from the extreme compression fiber. From similar triangles, the following relationship is established between

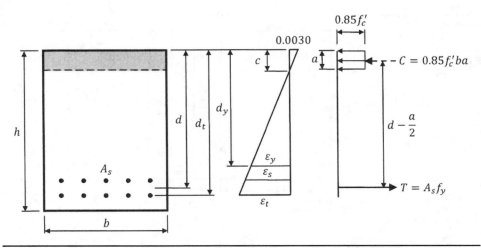

Figure 5.13 A reinforced concrete beam with multiple layers of tension reinforcement.

d_y and c:

$$\frac{c}{0.0030} = \frac{d_y - c}{\varepsilon_y} \tag{5.13}$$

Solving for d_y results in the following:

$$d_y = c \left(1 + \frac{\varepsilon_y}{0.0030}\right) \tag{5.14}$$

For Grade 60 reinforcement ($\varepsilon_y = 60/29{,}000 = 0.00207$), Eq. (5.14) reduces to

$$d_y = 1.7c \tag{5.15}$$

Reinforcement located a distance equal to or greater than d_y from the extreme compression fiber yields (i.e., $\varepsilon_s \geq \varepsilon_y$ and $f_s = f_y$). The nominal flexural strength is calculated by Eq. (5.10), assuming that the total area of reinforcement is concentrated at d. In situations where one or more layers of the reinforcement do not yield, the reinforcement at those levels is separated from the reinforcement in the layers that yield, and the actual stress in those bars ($f_s < f_y$) is used in the calculation of M_n.

Example 5.3   Determine the nominal flexural strength M_n of the beam depicted in Fig. 5.14. Assume $f_c' = 5{,}000$ psi and Grade 60 reinforcement ($f_y = 60{,}000$ psi).

Solution   The flowchart shown in Fig. 5.10 is utilized to determine M_n.

Step 1: Determine $A_{s,min}$. The minimum amount of reinforcement is determined by ACI 10.5.1. Because the compressive strength of the concrete is greater than 4,400 psi, the minimum amount is determined by ACI Eq. (10-3):

$$A_{s,min} = \frac{3\sqrt{f_c'}b_w d}{f_y} = \frac{3\sqrt{5{,}000} \times 12 \times 14.8}{60{,}000} = 0.63 \, \text{in}^2$$

$$A_{s,min} < A_s = 6 \times 0.44 = 2.64 \, \text{in}^2$$

FIGURE 5.14 The reinforced concrete beam of Example 5.3.

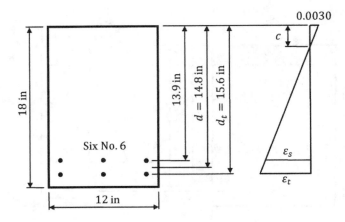

Step 2: Determine β_1.

$$\beta_1 = 1.05 - 0.00005 f_c' = 1.05 - (0.00005 \times 5{,}000) = 0.80 \text{ for } f_c'$$
$$= 5{,}000 \text{ psi (see ACI 10.2.7.3 and Section 5.2)}$$

Step 3: Determine the neutral axis depth c.

$$c = \frac{A_s f_y}{0.85 f_c' b \beta_1} = \frac{2.64 \times 60{,}000}{0.85 \times 5{,}000 \times 12 \times 0.80} = 3.9 \text{ in}$$

Step 4: Determine ε_t. The strain ε_t is determined by similar triangles (see Fig. 5.14):

$$\frac{c}{0.0030} = \frac{d_t - c}{\varepsilon_t} \text{ or } \varepsilon_t = 0.0030 \left(\frac{d_t}{c} - 1 \right)$$

Therefore,

$$\varepsilon_t = 0.0030 \left(\frac{15.6}{3.9} - 1 \right) = 0.0090 > 0.0040$$

Because $\varepsilon_t > 0.0040$, the maximum reinforcement requirement of ACI 10.3.5 is satisfied. Also, the section is tension-controlled because $\varepsilon_t > 0.0050$ (ACI 10.3.4).

Because Grade 60 reinforcement is specified in this example, use Eq. (5.15) to check the assumption that all of the reinforcement yields:

$$d_y = 1.7c = 1.7 \times 3.9 = 6.6 \text{ in}$$

Thus, the reinforcement layer located 13.9 in from the extreme compression fiber yields. Verify this by calculating the strain in the reinforcement at that layer:

$$\frac{\varepsilon_s}{13.9 - 3.9} = \frac{0.0030}{3.9} \text{ or } \varepsilon_s = 0.0077 > \varepsilon_y = 0.00207$$

Step 5: Determine the depth of the equivalent stress block a.

$$a = \beta_1 c = 0.80 \times 3.9 = 3.1 \text{ in}$$

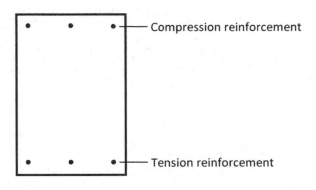

FIGURE 5.15 Doubly reinforced rectangular beam.

Compression reinforcement

Tension reinforcement

Step 6: Determine the nominal flexural strength M_n. Equation (5.10) is used to determine M_n:

$$M_n = A_s f_y \left(d - \frac{a}{2}\right) = 2.64 \times 60{,}000 \left(14.8 - \frac{3.1}{2}\right) / 12{,}000 = 174.9 \text{ ft kips}$$

Tension and Compression Reinforcement

Overview   The beam depicted in Fig. 5.15 contains reinforcement in both the tension and compression zones of the section, where the compression reinforcement A_s' is located a distance d' from the extreme compression fiber. Sections with both tension and compression reinforcement are commonly referred to as doubly reinforced sections.

Compression reinforcement can be added to a section to increase its design flexural strength ϕM_n (ACI 10.3.5.1). This additional strength can be achieved in situations where the dimensions of a beam are limited and the amount of tensile reinforcement that is required to resist the factored bending moments is greater than that permitted by ACI 10.3.5.

Reinforcement in the compression zone contributes to the total nominal flexural strength of a section, though the increase in M_n is usually relatively small. The presence of compression reinforcement in a section also results in larger values of ε_t, which essentially produces more ductile behavior. This has a direct impact on the magnitude of the strength reduction factor ϕ because ϕ is directly proportional to ε_t (see Section 4.3). For example, consider a rectangular beam where the strain ε_t is equal to 0.0045. The maximum reinforcement provisions of ACI 10.3.5 are satisfied because $\varepsilon_t > 0.0040$; however, the section is not tension-controlled because $\varepsilon_t < 0.0050$. Thus, the strength reduction factor is less than 0.90 (see Fig. 4.2). Adding a sufficient amount of reinforcement in the compression zone transforms the section from one that is in the transition region ($\phi < 0.90$) to one that is tension-controlled ($\phi = 0.90$). Chapter 6 contains additional information on why it is advantageous for flexural members to have tension-controlled sections.

Compression reinforcement is also added to help reduce long-term deflections. More information on this topic is also provided in Chap. 6.

Longitudinal reinforcement must be provided at both the top and the bottom of a reinforced concrete beam at certain locations regardless of whether it is needed or not for flexure. It is shown in Chap. 6 that transverse reinforcement, which is usually in the form of stirrups, is required in a beam to resist the effects from shear. Stirrups

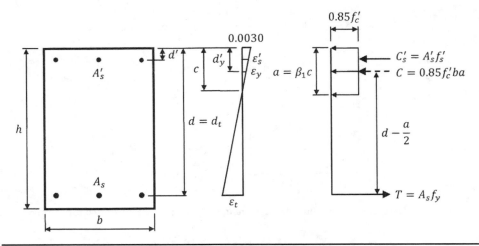

FIGURE 5.16 Strain and stress distributions in a doubly reinforced concrete beam.

must be anchored to the top and bottom longitudinal bars to properly develop them in tension (see ACI 12.13). Thus, longitudinal bars are needed wherever stirrups are required. Stirrups enclose the compression reinforcement and prevent it from buckling (see ACI 7.11).

Nominal flexural strength when compression reinforcement yields The strain and stress distributions in a doubly reinforced section are illustrated in Fig. 5.16. Similar to the case with multiple layers of tension steel, the nominal flexural strength of a doubly reinforced section depends on whether the compression reinforcement yields or not.

Assume that the yield strain ε_y occurs at a distance of d'_y from the extreme compression fiber. From similar triangles, the following relationship is established between d'_y and c:

$$\frac{c}{0.0030} = \frac{c - d'_y}{\varepsilon_y}$$

(5.16)

Solving for d'_y results in the following:

$$d'_y = c \left(1 - \frac{\varepsilon_y}{0.0030}\right)$$

(5.17)

For Grade 60 reinforcement ($\varepsilon_y = 60/29{,}000 = 0.00207$), Eq. (5.17) reduces to

$$d'_y = 0.31c$$

(5.18)

Compression reinforcement located a distance equal to or less than d'_y from the extreme compression fiber yields (i.e., $\varepsilon'_s \geq \varepsilon_y$ and $f'_s = f_y$).

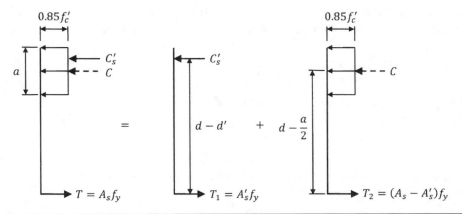

FIGURE 5.17   Force distribution in a doubly reinforced beam when the compression reinforcement yields.

When the compression steel yields, the depth of the equivalent stress block a can be obtained by satisfying force equilibrium:

$$T = C + C'_s \tag{5.19a}$$

$$A_s f_y = 0.85 f'_c ba + A'_s f_y \tag{5.19b}$$

$$a = \frac{(A_s - A'_s) f_y}{0.85 f'_c b} \tag{5.19c}$$

The total nominal flexural strength M_n is considered to be the sum of two parts. The first part M_{n1} is provided by the couple consisting of the force in the compression steel A'_s and the force in an equal area of tension steel (see Fig. 5.17):

$$M_{n1} = A'_s f_y (d - d') \tag{5.20}$$

The second part M_{n2} is provided by the couple consisting of the remaining tension steel $A_s - A'_s$ and the compression force in the concrete C:

$$M_{n2} = (A_s - A'_s) f_y \left(d - \frac{a}{2}\right) \tag{5.21}$$

Thus, the total nominal flexural strength of a doubly reinforced section where $f'_s = f_y$ is

$$M_n = (A_s - A'_s) f_y \left(d - \frac{a}{2}\right) + A'_s f_y (d - d') \tag{5.22}$$

Nominal flexural strength when compression reinforcement does not yield   When the compression reinforcement does not yield ($f'_s < f_y$), the depth of the stress block a cannot be determined by Eq. (5.19c) because the magnitude of f'_s is unknown. A relationship between f_s' and the neutral axis depth c can be obtained from strain compatibility. The strain in the compression reinforcement ε_s' is related to c as follows:

$$\frac{c}{0.0030} = \frac{c - d'}{\varepsilon_s'} \tag{5.23}$$

Substituting $\varepsilon_s' = f_s'/E_s$ into Eq. (5.23) and solving for f_s' results in

$$f_s' = 0.0030\, E_s \left(1 - \frac{d'}{c}\right) \tag{5.24}$$

The neutral axis depth c can be obtained by satisfying force equilibrium:

$$T = C + C_s' \tag{5.25a}$$

$$A_s f_y = 0.85 f_c' ba + A_s' f_s' \tag{5.25b}$$

Substituting $a = \beta_1 c$ and Eq. (5.24) into Eq. (5.25b) results in

$$A_s f_y = 0.85 f_c' b \beta_1 c + 0.003 A_s' E_s \left(1 - \frac{d'}{c}\right) \tag{5.26a}$$

$$a_1 c^2 + b_1 c - 87 A_s' d' = 0 \tag{5.26b}$$

$$c = \frac{-b_1 \pm \sqrt{b_1^2 + 348 a_1 A_s' d'}}{2 a_1} \tag{5.26c}$$

where $a_1 = 0.85 f_c' b \beta_1$
$b_1 = 87 A_s' - A_s f_y$
$E_s = 29{,}000$ ksi

Note that f_c' and f_y have the units of kips per square inch in the preceding equations.

Once the neutral axis depth c is determined by Eq. (5.26c), Eq. (5.24) can be used to calculate f_s'. The nominal flexural strength is obtained by satisfying moment equilibrium. Summing moments about the centroid of the tensile reinforcement results in the following equation for M_n where $f_s' < f_y$:

$$M_n = 0.85 f_c' ab \left(d - \frac{a}{2}\right) + A_s' f_s' (d - d') \tag{5.27}$$

Regardless of whether the compression steel yields or not, the strain ε_t must be equal to or greater than 0.0040 in order to satisfy the provisions of ACI 10.3.5.

The flowchart shown in Fig. 5.18 can be used to determine M_n for rectangular sections with tension and compression reinforcement.

It is important to note that the contribution of compression reinforcement can be conservatively neglected where it is not specifically required for strength. In such cases the nominal flexural strength can be computed using the equations developed previously for sections with only tension reinforcement.

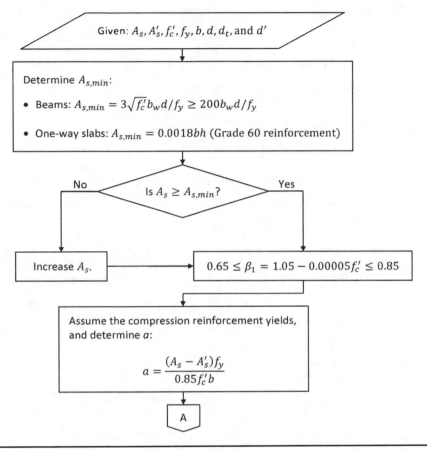

Figure 5.18 Nominal flexural strength—rectangular section with tension and compression reinforcement. (*Continued*)

Example 5.4 Determine the nominal flexural strength M_n of the beam depicted in Fig. 5.19. Assume $f'_c = 4{,}000$ psi and Grade 60 reinforcement ($f_y = 60{,}000$ psi).

Solution The flowchart shown in Fig. 5.18 is utilized to determine M_n.

Step 1: Determine $A_{s,min}$. The minimum amount of reinforcement is determined by ACI 10.5.1. Because the compressive strength of the concrete is less than 4,400 psi, the minimum amount is determined by the lower limit given in ACI 10.5.1:

$$A_{s,min} = \frac{200 b_w d}{f_y} = \frac{200 \times 12 \times 20.6}{60{,}000} = 0.82 \, \text{in}^2$$

$$A_{s,min} < A_s = 8 \times 0.60 = 4.80 \, \text{in}^2$$

Step 2: Determine β_1. According to ACI 10.2.73, $\beta_1 = 0.85$ for $f'_c = 4{,}000$ psi (see Section 5.2).

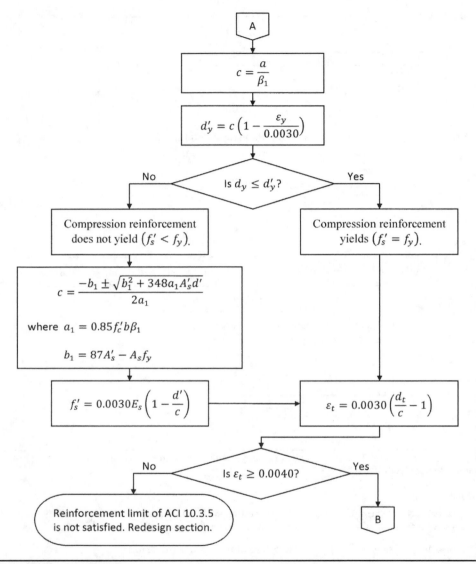

$$A$$

$$c = \frac{a}{\beta_1}$$

$$d'_y = c\left(1 - \frac{\varepsilon_y}{0.0030}\right)$$

Is $d_y \leq d'_y$?

No · Yes

Compression reinforcement does not yield $(f'_s < f_y)$.

Compression reinforcement yields $(f'_s = f_y)$.

$$c = \frac{-b_1 \pm \sqrt{b_1^2 + 348 a_1 A'_s d'}}{2a_1}$$

where $a_1 = 0.85 f'_c b \beta_1$

$$b_1 = 87 A'_s - A_s f_y$$

$$f'_s = 0.0030 E_s \left(1 - \frac{d'}{c}\right)$$

$$\varepsilon_t = 0.0030 \left(\frac{d_t}{c} - 1\right)$$

Is $\varepsilon_t \geq 0.0040$?

No · Yes

Reinforcement limit of ACI 10.3.5 is not satisfied. Redesign section.

$$B$$

FIGURE 5.18   (Continued)

Step 3: Determine the depth of the equivalent stress block a. Assuming that the compression steel yields, the depth of the equivalent stress block a is determined by Eq. (5.19c):

$$a = \frac{(A_s - A'_s)\, f_y}{0.85 f'_c b} = \frac{(4.80 - 0.40) \times 60}{0.85 \times 4 \times 12} = 6.5\,\text{in}$$

Step 4: Determine the neutral axis depth c.

$$c = \frac{a}{\beta_1} = \frac{6.5}{0.85} = 7.7\,\text{in}$$

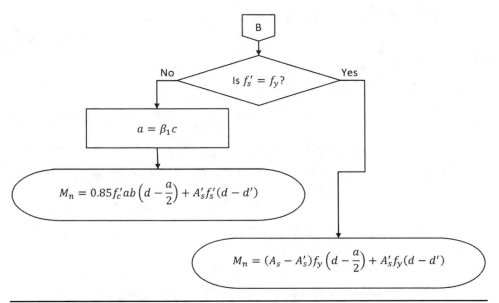

FIGURE 5.18   (Continued)

Step 5: Determine d'_y. Because Grade 60 reinforcement is specified in this example, use Eq. (5.18) to determine d'_y:

$$d'_y = 0.31c = 0.31 \times 7.7 = 2.4\,\text{in}$$

Thus, the compression reinforcement layer located 2.3 in from the extreme compression fiber yields because $d' < d'_y$. Verify this by determining the strain in the compression reinforcement:

$$\frac{\varepsilon'_s}{7.7 - 2.3} = \frac{0.0030}{7.7} \text{ or } \varepsilon'_s = 0.0021 > \varepsilon_y = 0.00207$$

FIGURE 5.19 The doubly reinforced concrete beam of Example 5.4.

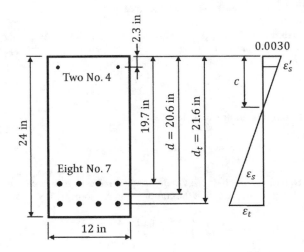

Step 6: Determine ε_t. The strain ε_t is determined by similar triangles (see Fig. 5.19):

$$\frac{c}{0.0030} = \frac{d_t - c}{\varepsilon_t} \text{ or } \varepsilon_t = 0.0030 \left(\frac{d_t}{c} - 1 \right)$$

Therefore,

$$\varepsilon_t = 0.0030 \left(\frac{21.6}{7.7} - 1 \right) = 0.0054 > 0.0040$$

Because $\varepsilon_t > 0.0040$, the maximum reinforcement requirement of ACI 10.3.5 is satisfied. Also, the section is tension-controlled because $\varepsilon_t > 0.0050$ (ACI 10.3.4).

Verify that the tension reinforcement located 19.7 in from the extreme compression fiber yields:

$$\frac{\varepsilon_s}{19.7 - 7.7} = \frac{0.0030}{7.7} \text{ or } \varepsilon_s = 0.0047 > \varepsilon_y = 0.00207$$

Step 7: Determine the nominal flexural strength M_n. Equation (5.22) is used to determine M_n:

$$M_n = \left(A_s - A_s' \right) f_y \left(d - \frac{a}{2} \right) + A_s' f_y \left(d - d' \right)$$
$$= \left[(4.80 - 0.40) \times 60{,}000 \times \left(20.6 - \frac{6.5}{2} \right) + 0.40 \times 60{,}000 \times (20.6 - 2.3) \right] / 12{,}000$$
$$= 381.7 + 36.6 = 418.3 \text{ ft kips}$$

Example 5.5   Determine the nominal flexural strength M_n of the beam shown in Fig. 5.19 with the following modification: use four No. 7 compression bars located at $d' = 2.4$ in instead of two No. 4 compression bars located at $d' = 2.3$ in.

Solution   The flowchart shown in Fig. 5.18 is utilized to determine M_n.

Step 1: Determine $A_{s,min}$.

$$A_{s,min} = 0.82 \text{ in}^2 < A_s = 8 \times 0.60 = 4.80 \text{ in}^2 \text{ (see Step 1 in Example 5.4)}$$

Step 2: Determine β_1.

$$\beta_1 = 0.85 \text{ for } f_c' = 4{,}000 \text{ psi (see Step 2 in Example 5.4)}.$$

Step 3: Determine the depth of equivalent stress block a. Assuming that the compression steel yields, the depth of the equivalent stress block a is determined by Eq. (5.19c):

$$a = \frac{\left(A_s - A_s' \right) f_y}{0.85 f_c' b} = \frac{(4.80 - 2.40) \times 60}{0.85 \times 4 \times 12} = 3.5 \text{ in}$$

Step 4: Determine neutral axis depth c.

$$c = \frac{a}{\beta_1} = \frac{3.5}{0.85} = 4.1 \text{ in}$$

Step 5: Determine d_y'. Because Grade 60 reinforcement is specified in Example 5.4, use Eq. (5.18) to check the assumption that the compression reinforcement yields:

$$d_y' = 0.31c = 0.31 \times 4.1 = 1.3 \text{ in}$$

Thus, the compression reinforcement layer located 2.4 in from the extreme compression fiber does not yield because $d' > d'_y$.

Step 6: Determine the revised neutral axis depth c. Equation (5.26c) is used to determine the revised neutral axis depth c:

$$c = \frac{-b_1 \pm \sqrt{b_1^2 + 348a_1 A'_s d'}}{2a_1}$$

$$a_1 = 0.85 f'_c b \beta_1 = 0.85 \times 4 \times 12 \times 0.85 = 34.7$$

$$b_1 = 87 A'_s - A_s f_y = (87 \times 2.40) - (4.80 \times 60) = -79.2$$

$$c = \frac{79.2 + \sqrt{(-79.2)^2 + (348 \times 34.7 \times 2.40 \times 2.4)}}{2 \times 34.7} = 5.1 \text{ in}$$

Step 7: Determine the stress in the compression reinforcement f'_s. Equation (5.24) is used to determine f'_s:

$$f'_s = 0.003 E_s \left(1 - \frac{d'}{c}\right) = 0.0030 \times 29{,}000 \left(1 - \frac{2.4}{5.1}\right) = 46.1 \text{ ksi}$$

Step 8: Determine ε_t. The strain ε_t is determined by similar triangles:

$$\frac{c}{0.0030} = \frac{d_t - c}{\varepsilon_t} \text{ or } \varepsilon_t = 0.0030 \left(\frac{d_t}{c} - 1\right)$$

Therefore,

$$\varepsilon_t = 0.0030 \left(\frac{21.6}{5.1} - 1\right) = 0.0097 > 0.0040$$

Because $\varepsilon_t > 0.0040$, the maximum reinforcement requirement of ACI 10.3.5 is satisfied. Also, the section is tension-controlled because $\varepsilon_t > 0.0050$ (ACI 10.3.4).

Verify that the tension reinforcement located 19.7 in from the extreme compression fiber yields:

$$\frac{\varepsilon_s}{19.7 - 5.1} = \frac{0.0030}{5.1} \text{ or } \varepsilon_s = 0.0086 > \varepsilon_y = 0.00207$$

Step 9: Determine the revised depth of the equivalent stress block a.

$$a = \beta_1 c = 0.85 \times 5.1 = 4.3 \text{ in}$$

Step 10: Determine the nominal flexural strength M_n. Equation (5.27) is used to determine M_n:

$$M_n = 0.85 f'_c a b \left(d - \frac{a}{2}\right) + A'_s f'_s (d - d')$$

$$= \left[0.85 \times 4{,}000 \times 4.3 \times 12 \times \left(20.6 - \frac{4.3}{2}\right) + 2.40 \times 46{,}100 \times (20.6 - 2.4)\right] / 12{,}000$$

$$= 269.7 + 167.8 = 437.5 \text{ ft kips}$$

Comments

Comparing the results from Examples 5.4 and 5.5, it is evident that the depth of the neutral axis c determined in step 6 of this example is less than that determined in step 4 of Example 5.4. A

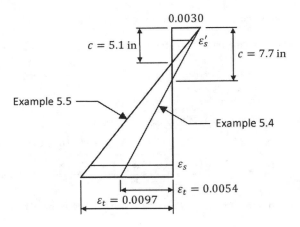

reduction in the neutral axis depth occurs whenever the compression steel does not yield ($\varepsilon'_s < \varepsilon_y$). It is also evident that increasing A'_s increases ε_t (approximately an 80% increase in these examples; see Fig. 5.20). In general, this can lead to larger values of the strength reduction factor ϕ. The sections in these examples are tension-controlled, so $\phi = 0.90$ in both cases; however, increasing A'_s in certain situations can transform a section in the transition zone to one that is tension-controlled.

Also, an increase in A'_s increases M_n. As expected, the increase is not linear: Approximately a 5% increase in M_n is realized when A'_s is increased by a factor of 6.

T-section and Inverted L-section with Tension Reinforcement

Overview   The discussions in the previous sections have focused on rectangular reinforced concrete beams without considering the slab that is supported by the beams. In typical cast-in-place concrete construction, the beams and slabs at a level are cast together with reinforcement that extends between the members; this forms a monolithic structure. Therefore, beams do not act alone but rather work together with the slab to resist the effects from applied loads.

A cast-in-place concrete floor/roof system is shown in Fig. 5.21. The interior beams are commonly referred to as T-beams, where the slab forms the flange of the T-beam and the concrete projection below the slab forms the web or stem of the T-beam. Similarly, the edge beam is referred to as an inverted L-beam.

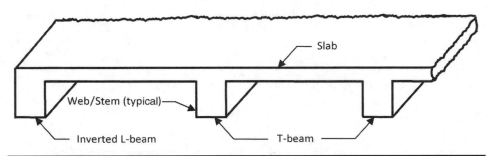

FIGURE 5.21   T-beam and inverted L-beam construction.

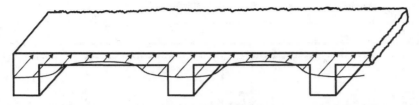

FIGURE 5.22   Distribution of compressive stress in a T-beam subjected to a positive moment.

Effective Flange Width   The beams depicted in Fig. 5.22 are subjected to a positive moment that produces compressive stresses at the top of the section above the neutral axis. As seen in the figure, the compressive stresses are greatest over the web and decrease between the webs. In lieu of using a variable compressive stress, the maximum uniform compressive stress is assumed to act over an effective slab (flange) width. The effective width is determined such that the compressive force obtained by multiplying the maximum compressive stress by the effective width is equal to the resultant force of the actual compressive stress distribution.

Effective flange widths depend on the geometry of the system (beam spacing, slab thickness, and span length) and are defined in ACI 8.12 for both T-beams and inverted L-beams as follows:

- For beams with slabs on both sides of the web (T-beams),

    Total effective flange width = the lesser of the following:

    Span length/4
    Web width + 16(slab thickness)
    Web width + one-half the clear distances to the next webs

- For beams with a slab on only one side of the web (inverted L-beams),

    Total effective flange width = the lesser of the following:

    Span length/4
    Web width + 6(slab thickness)
    Web width + one-half the clear distances to the next webs

The requirements of ACI 8.12 are summarized in Fig. 5.23 for the general case of varying web width and beam spacing. The nominal flexural strength of flanged sections is determined using this effective flange width.

Nominal Flexural Strength—Flange in Tension   The flange of a T-beam or an inverted L-beam will be in tension at locations of negative moment, which in a continuous system usually occur at the faces of a support. The strain and stress distributions for a T-beam in such a case are illustrated in Fig. 5.24, where it can be seen that the compression zone falls within the web of the member. An inverted L-beam would have similar distributions.

The nominal flexural strength is determined by the equations developed previously for rectangular sections with a single layer or multiple layers of tension reinforcement (i.e., two layers of reinforcement in the slab) where $b = b_w$. The flowchart shown in Fig. 5.10 can be used to determine M_n. If needed, the contribution of the compression

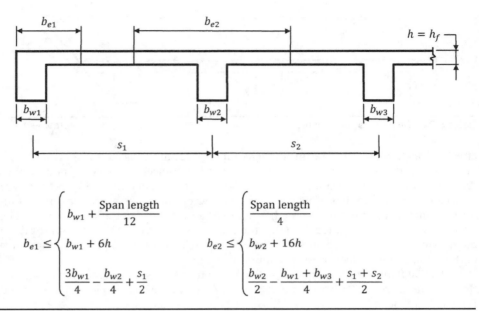

$$b_{e1} \le \begin{cases} b_{w1} + \dfrac{\text{Span length}}{12} \\[2mm] b_{w1} + 6h \\[2mm] \dfrac{3b_{w1}}{4} - \dfrac{b_{w2}}{4} + \dfrac{s_1}{2} \end{cases} \qquad b_{e2} \le \begin{cases} \dfrac{\text{Span length}}{4} \\[2mm] b_{w2} + 16h \\[2mm] \dfrac{b_{w2}}{2} - \dfrac{b_{w1} + b_{w3}}{4} + \dfrac{s_1 + s_2}{2} \end{cases}$$

Figure 5.23   Effective flange widths for a T-beam and an inverted L-beam.

reinforcement in the web can be included where the nominal flexural strength is determined by the equations developed previously for doubly reinforced sections (see the flowchart shown in Fig. 5.18).

Nominal Flexural Strength—Flange in Compression   At locations of positive moment, which usually occur away from the faces of a support in a continuous system, a portion of the flange or the entire flange of a T-beam or an inverted L-beam will be in compression. The determination of the nominal flexural strength depends on whether the depth of the stress block a is less than or greater than the thickness of the flange. Both cases are examined next.

   Depth of stress block less than or equal to flange thickness ($a \le h_f$)   When the depth of the compression zone a falls within the flange of a T-beam or an inverted L-beam, the compressive zone is rectangular with a width equal to the effective flange width b_e (see

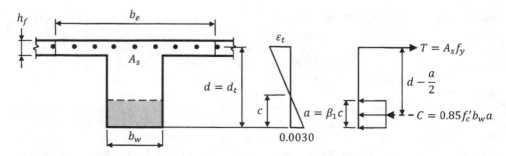

Figure 5.24   Strain and stress distributions in a T-beam with the flange in tension.

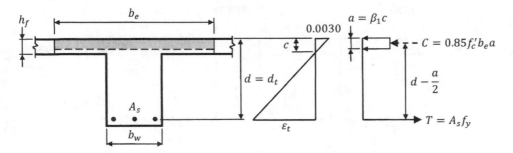

FIGURE 5.25   Strain and stress distributions in a T-beam with the flange in compression and $a \leq h_f$.

Fig. 5.25). The nominal flexural strength of the section is determined by the equations developed earlier for rectangular sections.

Depth of stress block greater than flange thickness ($a > h_f$)   When the depth of the stress block a falls within the web of the beam, the compressive zone is T or L shaped as opposed to rectangular (see Fig. 5.26). The resultant force C is equal to $0.85 f_c'$ times the area of the compressive zone and is located at its centroid. In such cases, it is convenient to divide the tensile reinforcement into two parts. The first part A_{sf} is defined as the area of steel that is required to balance the compressive force in the overhanging portions of the flange. This is depicted in Fig. 5.27. The following equation for A_{sf} is obtained from horizontal equilibrium, assuming that the tension reinforcement yields:

$$A_{sf} = \frac{0.85 f_c'(b_e - b_w)h_f}{f_y} \tag{5.28}$$

The nominal flexural resistance provided by the tensile reinforcement A_{sf} is obtained from moment equilibrium:

$$M_{n1} = A_{sf}f_y\left(d - \frac{h_f}{2}\right) \tag{5.29}$$

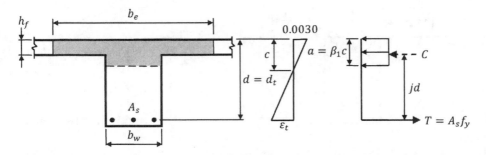

FIGURE 5.26   Strain and stress distributions in a T-beam with the flange in compression and $a > h_f$.

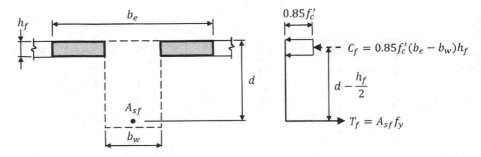

FIGURE 5.27   Stress distribution corresponding to overhanging flanges.

The remaining part of the tensile reinforcement $A_s - A_{sf}$ is balanced by the compression force in the rectangular part of the web (see Fig. 5.28). As usual, the depth of the stress block a is determined from horizontal equilibrium:

$$a = \frac{(A_s - A_{sf})f_y}{0.85 f'_c b_w} \tag{5.30}$$

The nominal flexural resistance provided by the tensile reinforcement $(A_s - A_{sf})$ is obtained from moment equilibrium:

$$M_{n2} = (A_s - A_{sf})f_y \left(d - \frac{a}{2}\right) \tag{5.31}$$

Thus, the total nominal flexural strength of the section M_n where $a > h_f$ is the addition of the two parts corresponding to the overhanging flanges and the web:

$$M_n = M_{n1} + M_{n2} = A_{sf}f_y \left(d - \frac{h_f}{2}\right) + (A_s - A_{sf})f_y \left(d - \frac{a}{2}\right) \tag{5.32}$$

The flowchart shown in Fig. 5.29 can be used to determine M_n for a T-beam or an inverted L-beam where the flange is in compression. Like in the case of rectangular

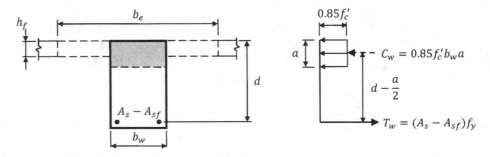

FIGURE 5.28   Stress distribution corresponding to the web.

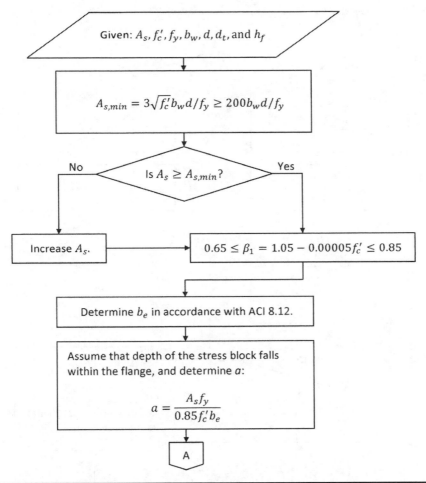

Given: $A_s, f'_c, f_y, b_w, d, d_t,$ and h_f

$$A_{s,min} = 3\sqrt{f'_c}\, b_w d / f_y \geq 200 b_w d / f_y$$

No | Is $A_s \geq A_{s,min}$? | Yes

Increase A_s.

$$0.65 \leq \beta_1 = 1.05 - 0.00005 f'_c \leq 0.85$$

Determine b_e in accordance with ACI 8.12.

Assume that depth of the stress block falls within the flange, and determine a:

$$a = \frac{A_s f_y}{0.85 f'_c b_e}$$

A

Figure 5.29   Nominal flexural strength—T-section or inverted L-section with the flange in compression. (*Continued*)

sections, the minimum and maximum reinforcement requirements of ACI 10.5 and 10.3.5, respectively, must be satisfied.

Example 5.6   Given the plan of the reinforced concrete floor system illustrated in Fig. 5.30, determine the nominal flexural strength M_n at the midspan of a typical wide-module joist in an end span. Assume $f'_c = 4{,}000$ psi and Grade 60 reinforcement ($f_y = 60{,}000$ psi).

Solution   The floor plan depicted in Fig. 5.30 is the same as the one depicted in Fig. 3.3 that was used in Examples 3.1 through 3.5. Figure 5.31 shows a section through the wide-module joist system. The flowchart shown in Fig. 5.29 is utilized to determine M_n.

Step 1: Determine $A_{s,min}$. The minimum amount of reinforcement is determined by ACI 10.5.1. Because the compressive strength of the concrete is less than 4,400 psi, the minimum amount is

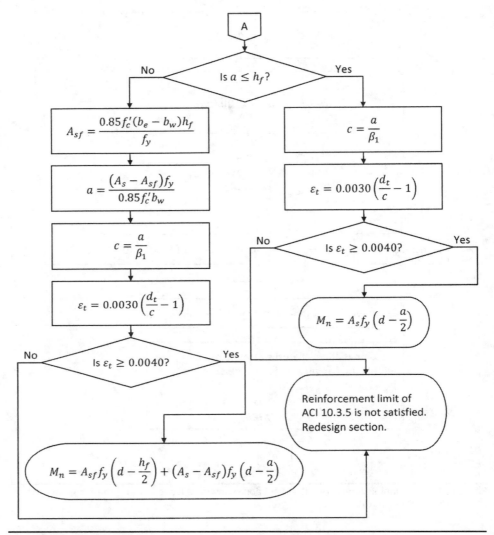

Figure 5.29 *(Continued)*

determined by the lower limit given in ACI 10.5.1:

$$A_{s,min} = \frac{200 b_w d}{f_y} = \frac{200 \times 7 \times 18.7}{60{,}000} = 0.44 \text{ in}^2$$

$$A_{s,min} < A_s = 2 \times 0.31 = 0.62 \text{ in}^2$$

Note that the web of a typical wide-module joist is tapered as shown in Fig. 5.31. The width of the web b_w used in calculations is generally taken as the least dimension at the bottom of the web, which in this case is 7 in. With a web taper of 1 to the horizontal and 12 to the vertical, the average web width is equal to $(9.7 + 7)/2 = 8.4$ in. The minimum area of steel corresponding

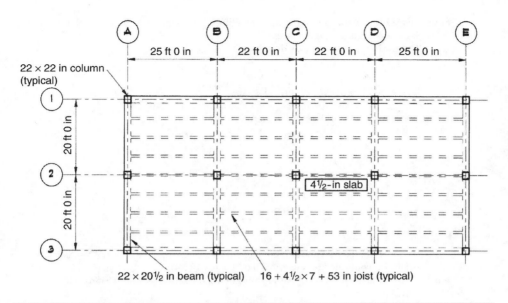

FIGURE 5.30 The plan of the floor system for Example 5.6.

to the average web width is $A_{s,min} = 200b_w d/f_y = 200 \times 8.4 \times 18.7/60,000 = 0.52$ in^2, which is less than that provided. The width of the equivalent stress block is also taken as 7 in.

Step 2: Determine β_1. According to ACI 10.2.73, $\beta_1 = 0.85$ for $f_c' = 4,000$ psi (see Section 5.2).

Step 3: Determine the effective flange width b_e. Figure 5.23 is used to determine b_e:

$$b_e = \begin{cases} \text{span length}/4 = (25 \times 12)/4 = 75 \text{ in} \\ b_w + 16h = 7 + (16 \times 4.5) = 79 \text{ in} \\ \text{spacing} = 60 \text{ in (governs)} \end{cases}$$

Step 4: Determine the depth of the equivalent stress block a. Assuming that the depth of the stress block falls within the flange, a is determined by Eq. (5.9) where $b = b_e$:

$$a = \frac{A_s f_y}{0.85 f_c' b_e} = \frac{0.62 \times 60}{0.85 \times 4 \times 60} = 0.18 \text{ in}$$

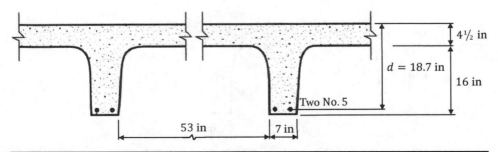

FIGURE 5.31 A section through the wide-module joist system of Example 5.6.

Because $a < h_f$, the assumption that the stress block falls within the flange is correct, and the section can be treated as a rectangular section.

Step 5: Determine the neutral axis depth c.

$$c = \frac{a}{\beta_1} = \frac{0.18}{0.85} = 0.21 \text{ in}$$

Step 6: Determine ε_t. The strain ε_t is determined by similar triangles:

$$\frac{c}{0.0030} = \frac{d_t - c}{\varepsilon_t} \text{ or } \varepsilon_t = 0.0030 \left(\frac{d_t}{c} - 1 \right)$$

Therefore,

$$\varepsilon_t = 0.0030 \left(\frac{18.7}{0.21} - 1 \right) = 0.2641 > 0.0040$$

Because $\varepsilon_t > 0.0040$, the maximum reinforcement requirement of ACI 10.3.5 is satisfied. Also, the section is tension-controlled because $\varepsilon_t > 0.0050$ (ACI 10.3.4).

Step 7: Determine the nominal flexural strength M_n. Equation (5.10) is used to determine M_n:

$$M_n = A_s f_y \left(d - \frac{a}{2} \right) = 0.62 \times 60,000 \times \left(18.7 - \frac{0.18}{2} \right) /12,000 = 57.7 \text{ ft kips}$$

Comments
It was found in step 6 that the section is tension-controlled, so the strength reduction factor $\phi = 0.9$ (see Section 4.3 and Fig. 4.2). Thus, the positive design strength of this wide-module joist is $\phi M_n = 0.9 \times 57.7 = 51.9$ ft kips. It was determined in Example 3.4 that the maximum factored positive moment $M_u = 42.2$ ft kips (see Table 3.4). Thus, the basic requirement for strength design is satisfied for this member at the positive moment section because $\phi M_n > M_u$ [see Eq. (4.2)].

Example 5.7   Given the floor plan illustrated in Fig. 5.30, determine the nominal flexural strength M_n at the face of the first interior support of a typical wide-module joist in an end span. Assume $f_c' = 4,000$ psi and Grade 60 reinforcement ($f_y = 60,000$ psi).

Solution   Figure 5.32 shows the section through the wide-module joist system. The spacing of the six No. 3 bars meets the requirements of ACI 10.6.6 for T-beam construction where the flange is in tension. Because the web of the wide-module joist is in compression, the flowchart shown in Fig. 5.10 is utilized to determine M_n for rectangular sections.

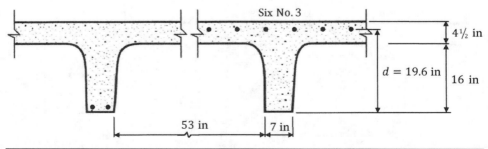

Figure 5.32   A section through the wide-module joist system of Example 5.7.

Step 1: Determine $A_{s,min}$. The minimum amount of reinforcement is determined by ACI 10.5.1. Because the compressive strength of the concrete is less than 4,400 psi, the minimum amount is determined by the lower limit given in ACI 10.5.1:

$$A_{s,min} = \frac{200 b_w d}{f_y} = \frac{200 \times 7 \times 19.6}{60,000} = 0.46 \, in^2$$

$$A_{s,min} < A_s = 6 \times 0.11 = 0.66 \, in^2$$

Step 2: Determine β_1. According to ACI 10.2.73, $\beta_1 = 0.85$ for $f'_c = 4,000$ psi (see Section 5.2).

Step 3: Determine the neutral axis depth c.

$$c = \frac{A_s f_y}{0.85 f'_c b_w \beta_1} = \frac{0.66 \times 60,000}{0.85 \times 4,000 \times 7 \times 0.85} = 2.0 \, in$$

Step 4: Determine ε_t. The strain ε_t is determined by similar triangles:

$$\frac{c}{0.0030} = \frac{d_t - c}{\varepsilon_t} \text{ or } \varepsilon_t = 0.0030 \left(\frac{d_t}{c} - 1 \right)$$

$$\varepsilon_t = 0.0030 \left(\frac{19.6}{2.0} - 1 \right) = 0.0264 > 0.0040$$

Because $\varepsilon_t > 0.0040$, the maximum reinforcement requirement of ACI 10.3.5 is satisfied. Also, the section is tension-controlled because $\varepsilon_t > 0.0050$ (ACI 10.3.4).

Step 5: Determine the depth of the equivalent stress block a.

$$a = \beta_1 c = 0.85 \times 2.0 = 1.7 \, in$$

Step 6: Determine the nominal flexural strength M_n. Equation (5.10) is used to determine M_n:

$$M_n = A_s f_y \left(d - \frac{a}{2} \right) = 0.66 \times 60,000 \left(19.6 - \frac{1.7}{2} \right) / 12,000 = 61.9 \, ft \, kips$$

Comments

It was found in step 4 that the section is tension-controlled, so the strength reduction factor $\phi = 0.9$ (see Section 4.3 and Fig. 4.2). Thus, the negative design strength of this wide-module joist is $\phi M_n = 0.9 \times 61.9 = 55.7$ ft kips. It was determined in Example 3.4 that the maximum factored negative moment $M_u = 51.7$ ft kips (see Table 3.4). Thus, the basic requirement for strength design is satisfied for this member at the negative moment section because $\phi M_n > M_u$ [see Eq. (4.2)].

Example 5.8   Determine the positive nominal flexural strength M_n of the T-beam in Fig. 5.33. Assume $f'_c = 3,000$ psi and Grade 60 reinforcement ($f_y = 60,000$ psi).

Solution

Step 1: Determine $A_{s,min}$. The minimum amount of reinforcement is determined by ACI 10.5.1. Because the compressive strength of the concrete is less than 4,400 psi, the minimum amount is determined by the lower limit given in ACI 10.5.1:

$$A_{s,min} = \frac{200 b_w d}{f_y} = \frac{200 \times 12 \times 14.5}{60,000} = 0.58 \, in^2$$

$$A_{s,min} < A_s = 6 \times 0.79 = 4.74 \, in^2$$

Step 2: Determine β_1. According to ACI 10.2.73, $\beta_1 = 0.85$ for $f'_c = 3,000$ psi (see Section 5.2).

FIGURE 5.33 The T-beam of Example 5.8.

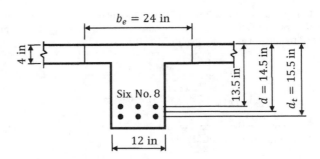

Step 3: Determine the effective flange width b_e. In Fig. 5.33, the effective flange width b_e is given as 24 in.

Step 4: Determine the depth of equivalent stress block a. Assuming that the depth of the stress block falls within the flange, a is determined by Eq. (5.9) where $b = b_e$:

$$a = \frac{A_s f_y}{0.85 f'_c b_e} = \frac{4.74 \times 60}{0.85 \times 3 \times 24} = 4.65 \text{ in}$$

Because $a > h_f$, the assumption that the stress block falls within the flange is not correct, and the section must be treated as a T-section.

Step 5: Determine A_{sf}. The area of steel A_{sf} that is required to balance the compressive force in the overhanging portion of the slab is determined by Eq. (5.28):

$$A_{sf} = \frac{0.85 f'_c (b_e - b_w) h_f}{f_y} = \frac{0.85 \times 3 \times (24 - 12) \times 4}{60} = 2.04 \text{ in}^2$$

Step 6: Determine the revised depth of the equivalent stress block a. The revised depth of the stress block is determined by Eq. (5.30):

$$a = \frac{(A_s - A_{sf}) f_y}{0.85 f'_c b_w} = \frac{(4.74 - 2.04) \times 60}{0.85 \times 3 \times 12} = 5.3 \text{ in}$$

Step 7: Determine the neutral axis depth c.

$$c = \frac{a}{\beta_1} = \frac{5.3}{0.85} = 6.2 \text{ in}$$

Step 8: Determine ε_t. The strain ε_t is determined by similar triangles:

$$\frac{c}{0.0030} = \frac{d_t - c}{\varepsilon_t} \quad \text{or} \quad \varepsilon_t = 0.0030 \left(\frac{d_t}{c} - 1 \right)$$

Therefore,

$$\varepsilon_t = 0.0030 \left(\frac{15.5}{6.2} - 1 \right) = 0.0045 > 0.0040$$

Because $\varepsilon_t > 0.0040$, the maximum reinforcement requirement of ACI 10.3.5 is satisfied. Also, the section is in the transition region because $0.0020 < \varepsilon_t < 0.0050$ (ACI 10.3.4).

Verify that the tension reinforcement located 13.5 in from the extreme compression fiber yields:

$$\frac{\varepsilon_s}{13.5 - 6.2} = \frac{0.0030}{6.2} \text{ or } \varepsilon_s = 0.0035 > \varepsilon_y = 0.00207$$

Step 9: Determine the nominal flexural strength M_n. Equation (5.32) is used to determine M_n:

$$M_n = M_{n1} + M_{n2} = A_{sf}f_y\left(d - \frac{h_f}{2}\right) + (A_s - A_{sf})f_y\left(d - \frac{a}{2}\right)$$

$$= \left[2.04 \times 60{,}000 \times \left(14.5 - \frac{4}{2}\right) + (4.74 - 2.04) \times 60{,}000 \times \left(14.5 - \frac{5.3}{2}\right)\right]/12{,}000$$

$$= 127.5 + 160.0 = 287.5 \text{ ft kips}$$

Comments
From Step 8, it was found that the section is in the transition region, so the strength reduction factor ϕ is determined in accordance with ACI 9.3.2 (see Section 4.3 and Fig. 4.2):

$$\phi = 0.65 + (\varepsilon_t - 0.002)\left(\frac{250}{3}\right) = 0.65 + (0.0045 - 0.002)\left(\frac{250}{3}\right) = 0.86$$

or

$$\phi = 0.65 + 0.25\left(\frac{1}{c/d_t} - \frac{5}{3}\right) = 0.65 + 0.25\left(\frac{1}{6.2/15.5} - \frac{5}{3}\right) = 0.86$$

Thus, the positive design strength of this beam is $\phi M_n = 0.86 \times 287.5 = 247.3$ ft kips.

5.5 Compression Members

5.5.1 Overview

The term *compression member* is used to refer to columns and other members, such as walls, that are subjected primarily to compressive forces. This section focuses on the determination of the nominal axial strength P_n for short, reinforced concrete columns that are subjected to essentially concentric axial loads. A *short column* is defined as one in which slenderness effects need not be considered. Slender columns are discussed in Chap. 8.

The nominal axial strength of a short column is related to the area of the column, the compressive strength of the concrete, the area and yield stress of the longitudinal reinforcement, and the type of transverse reinforcement. These quantities and their relationship to axial strength are discussed later.

5.5.2 Maximum Concentric Axial Load

Consider a reinforced concrete column subjected to a concentric axial load P. Assume that the longitudinal reinforcement is symmetrically distributed in the section and that lateral reinforcement that meets the size and spacing requirements of ACI 7.10 is provided. The type of lateral reinforcement is not relevant to the discussion at this time.

When subjected to P, the length of the column L decreases by an amount equal to the longitudinal strain ε times the original length L. For a concentric axial load, ε is uniform across the section. The strains in the concrete and the steel are equal because the concrete and the longitudinal steel are bonded together. For any given ε, it is possible to compute the stresses in the concrete and longitudinal steel using the stress–strain curves of the materials (see Chap. 2). The loads in the concrete and the longitudinal steel are equal to the stresses multiplied by the corresponding areas, and the total load that a short column can carry is the sum of the maximum loads carried by the concrete and the steel.

The maximum compressive axial load that can be resisted by the concrete P_c is equal to the following:

$$P_c = 0.85 f_c'(A_g - A_{st}) \tag{5.33}$$

In this equation, A_g is the gross area of the column and A_{st} is the total area of longitudinal reinforcement in the column; thus, $(A_g - A_{st})$ is equal to the area of the concrete. The factor 0.85 is based on the results of numerous tests.[8]

The maximum axial load that can be carried by the longitudinal reinforcement P_s is equal to the area times the yield strength of the reinforcement:

$$P_s = f_y A_{st} \tag{5.34}$$

Therefore, the maximum concentric axial load P_o that can be carried by a short column is equal to the summation of the maximum loads of the concrete and the steel:

$$P_o = 0.85 f_c'(A_g - A_{st}) + f_y A_{st} \tag{5.35}$$

Equation (5.35) forms the basis of the nominal axial strength, which is discussed next.

5.5.3 Nominal Axial Strength

In general, the maximum nominal axial strength $P_{n,max}$ is equal to a constant times the concentric axial load strength P_o. The constant depends on the type of transverse reinforcement utilized in the section and accounts for any accidental eccentricities—or, equivalently, any accidental bending moments—that may exist in a compression member and were not considered in the analysis. These eccentricities can arise from unbalanced moments in the beams framing into the column, misalignment of columns from floor to floor, or misalignment of the longitudinal reinforcement in the column, to name a few.

For members with spiral reinforcement, the constant is equal to 0.85. Therefore,

$$P_{n,max} = 0.85[0.85 f_c'(A_g - A_{st}) + f_y A_{st}] \tag{5.36}$$

The constant 0.85 produces nominal strength approximately equal to that from earlier Codes with the axial load applied at an eccentricity equal to 5% of the column depth.

Similarly, for members with tie reinforcement, the constant is equal to 0.80, and $P_{n,max}$ is

$$P_{n,max} = 0.80[0.85 f_c'(A_g - A_{st}) + f_y A_{st}] \tag{5.37}$$

In this case, the axial load is applied at an eccentricity equal to approximately 10% of the column depth.

Equations (10-1) and (10-2) in ACI 10.3.6 give the design axial strength $\phi P_n = \phi P_{n,max}$ for columns with spiral and tie lateral reinforcement, respectively. These equations are formed by multiplying the maximum nominal axial strength $P_{n,max}$ by the corresponding strength reduction factor ϕ. Thus,

- For members with spiral reinforcement

$$\phi P_{n,max} = 0.85\phi[0.85 f_c'(A_g - A_{st}) + f_y A_{st}] \tag{5.38}$$

- For members with tie reinforcement

$$\phi P_{n,max} = 0.80\phi[0.85 f_c'(A_g - A_{st}) + f_y A_{st}] \tag{5.39}$$

The strength reduction factor ϕ is equal to 0.75 for compression-controlled sections with spiral reinforcement and 0.65 for other reinforced members including tied reinforcement (see Section 4.3 and ACI 9.3.2.2). It was noted in Section 4.3 that the larger $\phi-$ factor for columns with spiral reinforcement reflects the more ductile behavior of such columns compared with columns with tied reinforcement.

5.5.4   Longitudinal Reinforcement Limits

ACI 10.9.1 prescribes the limits on the amount of longitudinal reinforcement for compression members, which are applicable to all such members regardless of the type of lateral reinforcement:

- Minimum $A_{st} = 0.01 A_g$
- Maximum $A_{st} = 0.08 A_g$

The lower limit is meant to provide resistance to bending, which may exist even though an analysis shows that it is not present, and to reduce the effects of creep and shrinkage of the concrete under sustained compressive stresses (see Chap. 2).

The upper limit is a practical maximum for longitudinal reinforcement in terms of economy and placement of the bars: For proper concrete placement and consolidation, the size and number of longitudinal bar sizes must be chosen to minimize reinforcement congestion, especially at beam–column joints. If column bars are lap spliced, the maximum area of longitudinal reinforcement should not exceed 4% of the gross column area at the location of the splice.

ACI 10.9.2 also contains requirements on the minimum number of longitudinal bars in compression members and the minimum volumetric reinforcement ratio for columns with spiral reinforcement. These and other requirements are covered in Chap. 8.

The flowchart shown in Fig. 5.34 can be used to determine $P_{n,max}$ for compression members.

Example 5.9   Determine the maximum nominal axial strength $P_{n,max}$ of the 24-in-diameter column shown in Fig. 5.35. Assume $f_c' = 6,000$ psi and Grade 60 reinforcement ($f_y = 60,000$ psi).

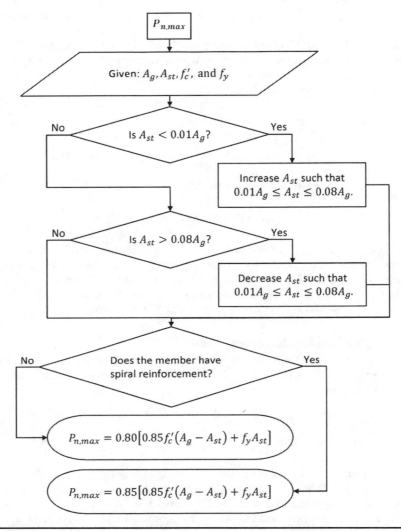

Figure 5.34 Maximum nominal axial strength for compression members.

Solution The flowchart shown in Fig. 5.34 is utilized to determine $P_{n,max}$ for this compression member.

Step 1: Check the minimum and maximum longitudinal reinforcement limits. The minimum and maximum amounts of longitudinal reinforcement permitted in a compression member are specified in ACI 10.9.1.

$$\text{Minimum: } A_{st} = 0.01 A_g = 0.01 \times \pi \times 24^2/4 = 4.52 \text{ in}^2$$
$$\text{Maximum: } A_{st} = 0.08 A_g = 0.08 \times \pi \times 24^2/4 = 36.2 \text{ in}^2$$

The provided area of longitudinal reinforcement $A_{st} = 8 \times 1.00 = 8.00 \text{ in}^2$ falls between the minimum and maximum limits.

FIGURE 5.35 The column of Example 5.9.

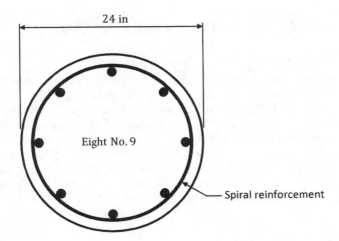

24 in

Eight No. 9

Spiral reinforcement

Step 2: Identify the type of lateral reinforcement. Spiral reinforcement is identified in Fig. 5.35.

Step 3: Determine the maximum nominal axial strength $P_{n,max}$. For compression members with spiral reinforcement conforming to ACI 7.10.4, $P_{n,max}$ is determined by Eq. (5.36):

$$P_{n,max} = 0.85[0.85 f_c'(A_g - A_{st}) + f_y A_{st}]$$
$$= 0.85[0.85 \times 6 \times (452.4 - 8.00) + (60 \times 8.00)] = 2{,}335 \text{ kips}$$

Comments

The design axial strength $\phi P_{n,max}$ is equal to the strength reduction factor ϕ, which is 0.75 for compression-controlled sections with spiral reinforcement [see Section 4.3 and ACI 9.3.2.2(a)], times $P_{n,max}$:

$$\phi P_{n,max} = 0.75 \times 2{,}335 = 1{,}751 \text{ kips}$$

5.6 Tension Members

Reinforced concrete tension members occur in certain specialty structures such as arches and trusses. As was discussed in Chap. 2, the tensile strength of concrete is relatively small compared with its compressive strength. As such, the tensile strength of concrete is neglected in the design of tension members, and it is assumed that the tension load is resisted solely by the longitudinal reinforcement. Therefore, the nominal tensile strength T_n of a symmetrical reinforced concrete tension member subjected to a concentric axial tension load is equal to the area times the yield strength of the longitudinal steel:

$$T_n = A_{st} f_y \tag{5.40}$$

The design strength ϕT_n is obtained by multiplying T_n in Eq. (5.40) by the strength reduction factor ϕ for tension, which is equal to 0.9 (see ACI 9.3).

5.7 Members Subjected to Flexure and Axial Load

5.7.1 Overview

Columns in reinforced concrete buildings are commonly subjected to the effects from more than just compressive loads. Consider columns AB and CD in the rigid frame depicted in Fig. 4.1. These members are subjected to axial compressive loads, bending moments, and shear forces from gravity loads and wind loads (see Table 4.4 in Example 4.2 for a summary of these effects).

Like all reinforced concrete members, the nominal strength of a section subjected to both flexure and axial load is determined using equilibrium, strain compatibility, and the design assumptions given in Section 5.2. Methods to determine the nominal strength of such members are presented later for the case of axial compression and flexure.

5.7.2 Nominal Strength

The general principles and assumptions of the strength design method are applied to the section depicted in Fig. 5.36, which is subjected to an axial compressive load and flexure. For purposes of discussion, assume that the longitudinal reinforcement in layer 1 is located closest to the extreme compression fiber of the section.

In accordance with design assumption no. 2 in Section 5.2, the maximum strain in the extreme compression fiber of the concrete $\varepsilon_c = 0.0030$. By assuming a value for the strain ε_{s3} in the reinforcement in layer 3, the depth to the neutral axis c can be determined from similar triangles (see Fig. 5.37). Because the reinforcement in layer 3 is farthest from the extreme compression of the section, $\varepsilon_{s3} = \varepsilon_t$ and

$$c = \frac{0.0030d_3}{\varepsilon_t + 0.0030} \tag{5.41}$$

Utilizing design assumption no. 6 of Section 5.2, the resultant compressive force C in the concrete is obtained by multiplying the stress $0.85f_c'$ by the area under the

FIGURE 5.36
A reinforced concrete section subjected to uniaxial compression and flexure.

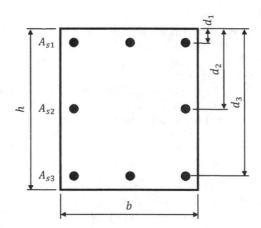

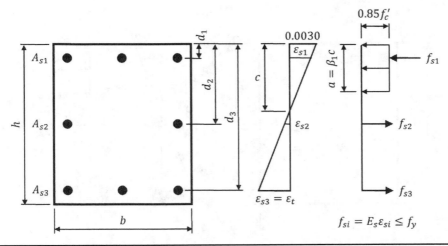

FIGURE 5.37 Strain and stress distributions in a section subjected to uniaxial compression and flexure.

equivalent rectangular stress block:

$$C = 0.85 f'_c a b \qquad (5.42)$$

In Eq. (5.42), the depth of the equivalent stress block $a = \beta_1 c$.

The strain ε_{si} in the reinforcement in layer i can also be determined from similar triangles:

$$\varepsilon_{si} = \frac{0.0030\,(c - d_i)}{c} \qquad (5.43)$$

For elastic–plastic reinforcement with the stress–strain curve defined by design assumption no. 3 given in Section 5.2, the stress in the reinforcement at each layer f_{si} is equal to the strain ε_{si} at that level determined by Eq. (5.43) times the modulus of elasticity of steel E_s. It is important to keep in mind that f_{si} must not exceed the yield stress f_y in tension or compression.

The magnitude of the force F_{si} in the reinforcement depends on whether the steel is in the equivalent compression zone or not:

- If d_i is greater than the depth of the equivalent stress block a,

$$F_{si} = f_{si} A_{si} \qquad (5.44)$$

   For the section depicted in Fig. 5.37, this equation would apply to layers 2 and 3.

- If d_i is less than the depth of the equivalent stress block a,

$$F_{si} = \left(f_{si} - 0.85 f'_c \right) A_{si} \qquad (5.45)$$

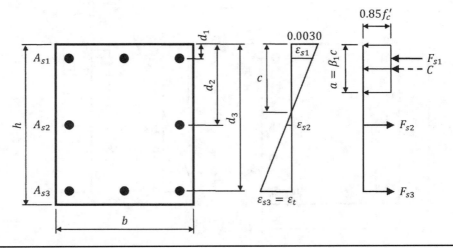

Figure 5.38   Nominal strength.

In this case, which is applicable to layer 1 shown in Fig. 5.37, the area of reinforcement in that layer has been included in the area ab used to compute the compressive force in the concrete C. Thus, $0.85f_c'$ must be subtracted from the steel stress f_{si} in that layer before computing the force F_{si}.

The nominal axial strength P_n for the assumed strain distribution is obtained by summing the axial forces on the section (see Fig. 5.38):

$$P_n = C + \sum F_{si} \tag{5.46}$$

Similarly, the nominal flexural strength M_n for the assumed strain distribution is determined by summing moments about the centroid of the column because this is the axis about which moments are computed in a conventional structural analysis (see Fig. 5.38):

$$M_n = 0.5C(h - a) + \sum F_{si}(0.5h - d_i) \tag{5.47}$$

The flowchart shown in Fig. 5.39 can be used to determine P_n and M_n as a function of the net tensile strain in the extreme layer of longitudinal tension steel ε_t.

As demonstrated earlier, the nominal strengths P_n and M_n are determined by Eqs. (5.46) and (5.47), respectively, for a particular assumed strain distribution. An *interaction diagram* for a column is a collection of P_n and M_n values that have been determined for a series of strain distributions (see Fig. 5.40). Nominal strengths represent a single point on the interaction diagram. The construction of interaction diagrams for columns is covered in Chap. 8.

Example 5.10   Determine the nominal strengths P_n and M_n corresponding to balanced failure for the rectangular column shown in Fig. 5.41. Assume that the extreme compression fiber occurs at the top of the section and that ties are utilized as the lateral reinforcement. Also assume $f_c' = 7{,}000$ psi and Grade 60 reinforcement ($f_y = 60{,}000$ psi).

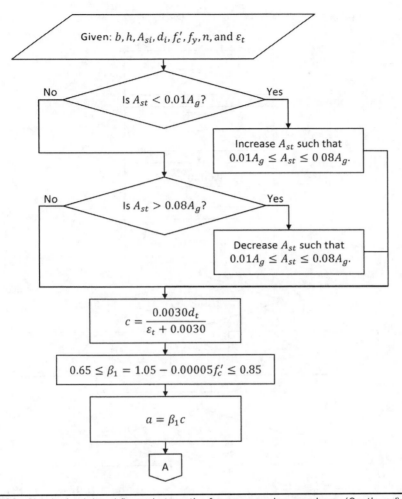

FIGURE 5.39 Nominal axial and flexural strengths for compression members. (*Continued*)

Solution The flowchart shown in Fig. 5.39 is utilized to determine P_n and M_n for this compression member.

Step 1: Check the minimum and maximum longitudinal reinforcement limits. The minimum and maximum amounts of longitudinal reinforcement permitted in a compression member are specified in ACI 10.9.1.

$$\text{Minimum: } A_{st} = 0.01 A_g = 0.01 \times 18 \times 24 = 4.32 \text{ in}^2$$
$$\text{Maximum: } A_{st} = 0.08 A_g = 0.08 \times 18 \times 24 = 34.6 \text{ in}^2$$

The provided area of longitudinal reinforcement $A_{st} = 10 \times 1.27 = 12.7 \text{ in}^2$ falls between the minimum and maximum limits.

Step 2: Determine the neutral axis depth c. Balanced failure occurs when crushing of the concrete and yielding of the reinforcing steel occur simultaneously (see Section 5.3). The balanced failure point also represents the change from compression failures for higher axial loads and tension

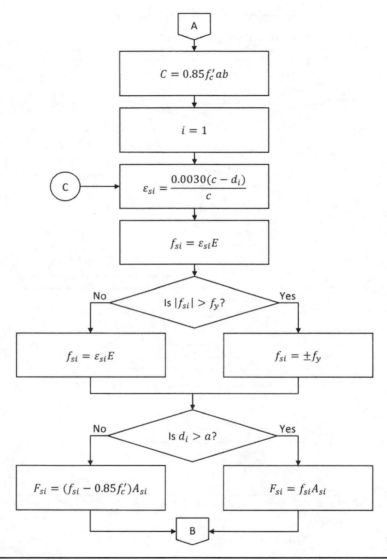

Figure 5.39   *(Continued)*

failures for lower axial loads for a given bending moment. ACI 10.3.3 permits the yield strain of the reinforcement to be taken as 0.0020 for Grade 60 reinforcement; thus, $\varepsilon_{s4} = \varepsilon_t = 0.0020$.

The neutral axis depth is determined by Eq. (5.41):

$$c = \frac{0.0030 d_t}{\varepsilon_t + 0.0030} = \frac{0.0030 \times 21.4}{0.0020 + 0.0030} = 12.8 \text{ in}$$

Step 3: Determine β_1.

$$\beta_1 = 1.05 - 0.00005 f_c' = 1.05 - (0.00005 \times 7{,}000) = 0.70 \text{ for } f_c'$$
$$= 7{,}000 \text{ psi (see ACI 10.2.7.3 and Section 5.2)}$$

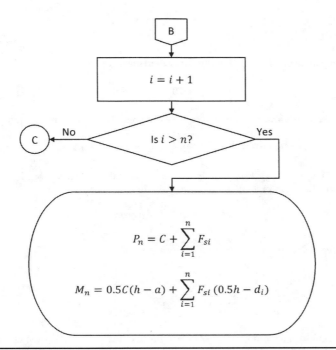

$$P_n = C + \sum_{i=1}^{n} F_{si}$$

$$M_n = 0.5C(h - a) + \sum_{i=1}^{n} F_{si}(0.5h - d_i)$$

FIGURE 5.39 (*Continued*)

FIGURE 5.40
A nominal strength
interaction diagram.

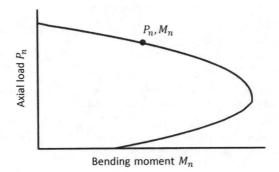

FIGURE 5.41 The
rectangular column
of Example 5.10.

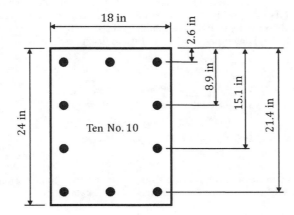

Step 4: Determine the depth of the equivalent stress block a.

$$a = \beta_1 c = 0.70 \times 12.8 = 9.0 \text{ in}$$

Step 5: Determine C. The concrete compression resultant force C is determined by Eq. (5.42):

$$C = 0.85 f'_c a b = 0.85 \times 7 \times 9.0 \times 18 = 963.9 \text{ kips}$$

Step 6: Determine ε_{si}. The strain in the reinforcement ε_{si} at the various layers is determined by Eq. (5.43) where compression strains are positive.

- Layer 1 ($d_1 = 2.6$ in): $\varepsilon_{s1} = \dfrac{0.0030\,(12.8 - 2.6)}{12.8} = 0.0024$

- Layer 2 ($d_2 = 8.9$ in): $\varepsilon_{s2} = \dfrac{0.0030\,(12.8 - 8.9)}{12.8} = 0.0009$

- Layer 3 ($d_3 = 15.1$ in): $\varepsilon_{s3} = \dfrac{0.0030\,(12.8 - 15.1)}{12.8} = -0.0005$

- Layer 4 ($d_4 = 21.4$ in): $\varepsilon_{s4} = \dfrac{0.0030\,(12.8 - 21.4)}{12.8} = -0.0020$ (checks)

It is evident that the top two layers of reinforcement are in compression and that the bottom two layers are in tension. Also, the layers of reinforcement closest to and farthest from the extreme compression fiber yield.

Step 7: Determine f_{si}. The stress in the reinforcement f_{si} at the various layers is determined by multiplying ε_{si} by the modulus of elasticity of the steel E_s.

- Layer 1: $f_{s1} = 0.0024 \times 29{,}000 = 69.6$ ksi > 60 ksi; use $f_{s1} = 60$ ksi
- Layer 2: $f_{s2} = 0.0009 \times 29{,}000 = 26.1$ ksi
- Layer 3: $f_{s3} = -0.0005 \times 29{,}000 = -14.5$ ksi
- Layer 4: $f_{s4} = -60$ ksi

Step 8: Determine F_{si}. The force in the reinforcement F_{si} at the various layers is determined by Eqs. (5.44) or (5.45), which depends on the location of the steel layer.

- Layer 1 ($d_1 = 2.6$ in $< a = 9.0$ in): $F_{s1} = [60 - (0.85 \times 7)] \times 3 \times 1.27 = 205.9$ kips
- Layer 2 ($d_2 = 8.9$ in $< a = 9.0$ in): $F_{s2} = [26.1 - (0.85 \times 7)] \times 2 \times 1.27 = 51.2$ kips
- Layer 3: $F_{s3} = -14.5 \times 2 \times 1.27 = -36.8$ kips
- Layer 4: $F_{s4} = -60 \times 3 \times 1.27 = -228.6$ kips

Note that the compression steel in the top two layers falls within the depth of the equivalent stress block; thus, Eq. (5.45) is used to determine the forces in the reinforcement in those layers.

Step 9: Determine P_n and M_n. The nominal axial strength P_n and the nominal flexural strength M_n of the section are determined by Eqs. (5.46) and (5.47), respectively:

$$P_n = C + \sum F_{si} = 963.9 + (205.9 + 51.2 - 36.8 - 228.6) = 955.6 \text{ kips}$$

$$
\begin{aligned}
M_n &= 0.5C(h - a) + \sum F_{si}(0.5h - d_i) \\
&= [0.5 \times 963.9 \times (24 - 9.0)] + [205.9(12 - 2.6) + 51.2(12 - 8.9) \\
&\quad + (-36.8)(12 - 15.1) + (-228.6)(12 - 21.4)] \\
&= 7{,}229.3 + 4{,}357.1 = 11{,}586.4 \text{ in kips} = 965.5 \text{ ft kips}
\end{aligned}
$$

Comments

This section is compression-controlled because ε_t is equal to the compression-controlled strain limit of 0.0020 (see ACI 10.3.3). Thus, in accordance with ACI 9.3.2.2, the strength reduction factor ϕ is equal to 0.65 for a compression-controlled section with lateral reinforcement consisting of ties (or, equivalently, a compression-controlled section without spiral reinforcement conforming to ACI 10.9.3). Therefore, the design axial strength ϕP_n and design flexural strength ϕM_n are

$$\phi P_n = 0.65 \times 955.6 = 621.1 \text{ kips}$$
$$\phi M_n = 0.65 \times 965.5 = 627.6 \text{ ft kips}$$

References

1. Hognestad, E. 1951. *A Study of Combined Bending and Axial Load in Reinforced Concrete Member*, Bulletin 399. University of Illinois Engineering Experiment Station, Urbana, IL, 128 pp.
2. Mattock, A. H., Kriz, L. B., and Hognestad, E. 1961. Rectangular concrete stress distribution in ultimate strength design." *ACI Journal Proceedings* 57(8):875-928.
3. Portland Cement Association (PCA). 2008. *Notes on ACI 318-08 Building Code Requirements for Structural Concrete*, 10th ed., EB708. PCA, Skokie, IL.
4. Whitney, C. S. 1940. Plastic theory of reinforced concrete design. *Proceedings of the American Society of Civil Engineers* 66(10):1749-1780.
5. Leslie, K. E., Rajagopalan, K. S., and Everard, N. J. 1976. Flexural behavior of high-strength concrete beams. *Journal of the American Concrete Institute* 73(9):517-521.
6. Karr, P. H., Hanson, N. W., and Capell, H. T. 1978. Stress–strain characteristics of high strength concrete. In: *Douglas McHenry International Symposium on Concrete and Concrete Structures*, SP-55. American Concrete Institute, Farmington Hills, MI, pp. 161-185.
7. Mast, R. F. 1992. Unified design provisions for reinforced and prestressed concrete flexural and compression members. *ACI Structural Journal* 89(2):185-199.
8. American Concrete Institute (ACI), Committee 105. 1930–1933. Reinforced concrete column investigation, *ACI Journal Proceedings* 26:601-612; 27:675-676; 28:157-178; 29:53-56; 30:78-90; 31:153-156.

Problems

5.1. Determine the nominal flexural strength M_n of a 32-in-wide and 20-in-deep beam reinforced with a single layer of 10 No. 10 bars (Grade 60). Assume $d = 17.4$ in and $f_c' = 5,000$ psi.

5.2. Determine the nominal flexural strength M_n of a 20-in-wide and 28-in-deep beam reinforced with a single layer of three No. 7 bars (Grade 60). Assume $d = 25.6$ in and $f_c' = 4,000$ psi.

5.3. Determine the nominal flexural strength M_n of a 28-in-wide and 18.5-in-deep beam reinforced with a single layer of 10 No. 8 bars (Grade 60). Assume $d = 16.0$ in and $f_c' = 4,000$ psi.

5.4. Determine the nominal flexural strength M_n of a 9-in-thick one-way slab reinforced with a single layer of No. 5 bars spaced 8 in on center (Grade 60). Assume $d = 7.9$ in and $f_c' = 4,000$ psi.

5.5. Determine the nominal flexural strength M_n of the beam depicted in Fig. 5.42. Assume that the extreme compression fiber is at the bottom of the section. Also assume $f_c' = 4,000$ psi and Grade 60 reinforcement.

5.6. Determine the nominal flexural strength M_n of the beam depicted in Fig. 5.43. Assume that the extreme compression fiber is at the bottom of the section. Also assume $f_c' = 4,000$ psi and Grade 60 reinforcement.

5.7. Determine the nominal flexural strength M_n of the beam depicted in Fig. 5.43. Assume that the extreme compression fiber is at the bottom of the section and that the two No. 3 bars are replaced with two No. 5 bars located 2.3 in from the bottom of the section. Also assume $f_c' = 4,000$ psi and Grade 60 reinforcement.

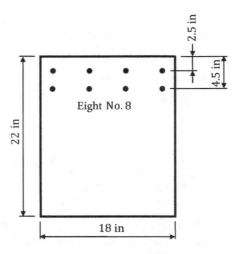

Figure 5.42 The reinforced concrete beam of Problem 5.5.

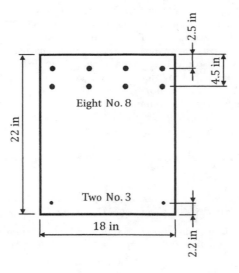

Figure 5.43 The reinforced concrete beam of Problem 5.6.

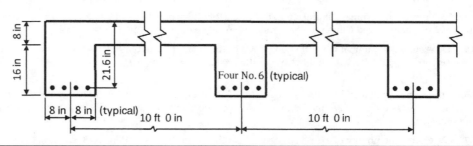

Figure 5.44 The reinforced concrete floor system of Problem 5.8.

5.8.   Given the reinforced concrete floor system in Fig. 5.44, determine the following: (1) the nominal flexural strength M_n at the midspan of an edge beam and (2) the nominal flexural strength M_n at the midspan of a typical interior beam. Assume $f'_c = 4,000$ psi and Grade 60 reinforcement. Also assume that the span is 25 ft.

FIGURE 5.45 The reinforced concrete beam of Problem 5.9.

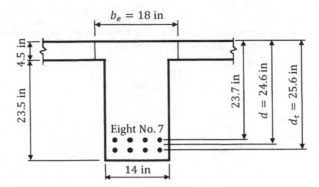

FIGURE 5.46 The compression member of Problem 5.10.

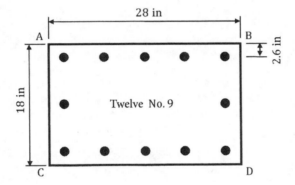

FIGURE 5.47 The compression member of Problem 5.12.

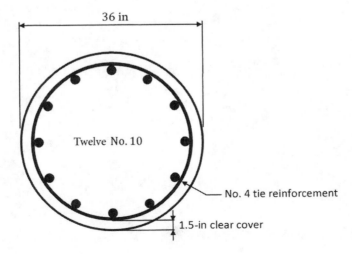

5.9.   Determine the nominal flexural strength M_n of the beam depicted in Fig. 5.45. Assume $f'_c = 4,000$ psi and Grade 60 reinforcement.

5.10.   Given the compression member depicted in Fig. 5.46, determine the following: (1) the maximum nominal axial strength $P_{n,max}$, (2) the nominal flexural strength M_n corresponding to zero axial load, (3) the nominal strengths P_n and M_n corresponding to $\varepsilon_t = 0$, (4) the nominal strengths P_n and M_n corresponding to balanced failure, (5) the nominal strengths P_n and M_n corresponding to $\varepsilon_t = -3\varepsilon_y$, and (6) the nominal tensile strength T_n. Assume $f'_c = 8,000$ psi and Grade 60 reinforcement. Also assume that the extreme compression fiber is on the AB side of the section.

5.11.   Repeat Problem 5.10 assuming the extreme compression fiber is on the AC side of the section.

5.12.   Repeat Problem 5.10 for the section depicted in Fig. 5.47.

CHAPTER 6

Beams and One-Way Slabs

6.1 Introduction

This chapter covers the design and detailing of beams and one-way slabs. The discussion is limited to beams that are not deep beams. A deep beam is defined in ACI 10.7 and 11.7 as a member that has a clear span equal to or less than four times the overall member depth or regions in a member that have concentrated loads within twice the member depth from the face of the support. Deep beams must be designed using a nonlinear strain distribution over the depth of the member or by the provisions of Code Appendix A, "Strut-and-Tie Models." The focus of this chapter is on members that can be designed using a linear distribution of strain in accordance with the general principles of the strength design method (see Chap. 5).

In general, the required cross-sectional dimensions and reinforcement for flexural members are determined using the general requirements for strength and serviceability given in Chap. 4 and the principles of the strength design method given in Chap. 5. Addressed in this chapter are the following criteria:

1. Flexure
2. Shear
3. Torsion
4. Deflection
5. Reinforcement details

Items 1 through 4 govern the size and amount of reinforcement in flexural members. Usually, the depth of a member is determined initially to satisfy deflection requirements. The cross-sectional dimensions of a member are chosen to ensure that strength requirements are fulfilled. Sizing members for economy, which is related to formwork and constructability, must also be considered.

The required amounts of flexural, shear, and torsional reinforcement are governed by strength and serviceability (including cracking). When choosing the size and spacing of the reinforcing bars, it is important to ensure that the reinforcement can be placed within a section without violating the requirements pertaining to the minimum and the maximum spacing. Additionally, the provided reinforcement must be fully developed so that it performs in accordance with the assumptions of the strength design method.

6.2 Design for Flexure

6.2.1 Overview

The cross-sectional dimensions of a flexural member and the required amount of flexural reinforcement at critical sections are determined using the strength and serviceability requirements of the Code. Typically, the member depth is determined first on the basis of the deflection requirements of ACI 9.5, which have been presented in Section 4.4 for one-way construction. The minimum thickness requirements given in ACI Table 9.5(a) are applicable to beams and one-way slabs that are not attached to partitions or other construction that is likely to be damaged by relatively large deflections. Section 6.5 contains methods to calculate deflections in any situation.

Once the depth of a member has been established, the width of the member and the required flexural reinforcement are determined using the basic requirement for flexural design strength:

$$\phi M_n \geq M_u \tag{6.1}$$

In this equation, ϕM_n is the design strength of the member at a particular section, which consists of the strength reduction factor ϕ (Section 4.3) and the nominal flexural strength M_n that is determined in accordance with the provisions and assumptions of the strength design method (Section 5.4). The required strength M_u is calculated at a section by combining the bending moments obtained from the analysis of the structure, using nominal loads in accordance with the load combinations of ACI Chap. 9 (Section 4.2).

The following sections provide the fundamental requirements and methods for sizing the cross-section and determining the required amount of flexural reinforcement for a reinforced concrete flexural member.

6.2.2 Sizing the Cross-Section

Establishing the dimensions of the cross-section is typically the initial step in the design of a reinforced concrete flexural member. The depth h is usually determined first on the basis of deflection requirements. This depth is sometimes modified for constructability, economy, or architectural reasons, to name a few. For a rectangular beam, the width b is subsequently determined on the basis of strength requirements, assuming that the section is tension-controlled, whereas for a one-way slab, the design width is commonly taken as 12 in. The following steps can be utilized to determine the dimensions of a reinforced concrete flexural member.

Step 1: Determine the depth of the member. As noted earlier, it is common to determine the depth of the member first on the basis of the serviceability requirements for deflection given in ACI 9.5 for one-way construction (i.e., members that bend in primarily one direction, such as beams and one-way slabs).

Consider the reinforced concrete beam depicted in Fig. 6.1. Although the following discussion focuses on beams, it applies equally to one-way slabs. If this member is not attached to partitions or other construction that is likely to be damaged by relatively large deflections, the minimum depth h that will satisfy the deflection requirements of

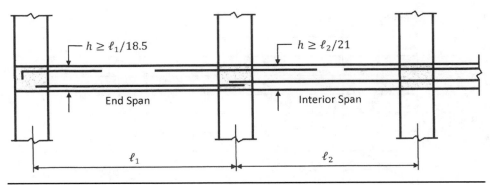

$h \geq \ell_1/18.5$

$h \geq \ell_2/21$

End Span

Interior Span

ℓ_1

ℓ_2

FIGURE 6.1 Minimum thickness requirements for continuous beams.

ACI 9.5.2 is determined by ACI Table 9.5(a). These requirements are summarized in Fig. 4.3. Alternatively, the member depth can be established on the basis of calculated deflections and the maximum permissible deflections given in ACI Table 9.5(b) (see Section 6.5).

It is important to consider economical formwork when choosing a member thickness from ACI Table 9.5(a). For the usual case of continuous construction and assuming normal-weight concrete and Grade 60 reinforcement, it is permitted to use a minimum thickness of $\ell_2/21$ for the interior spans and a minimum thickness of $\ell_1/18.5$ for the end spans. More than one beam depth along the same line of beams results in formwork that is not economical. Thus, the minimum depth of all of the beams should be determined on the basis of the span that yields the largest minimum h because this thickness will satisfy deflection criteria for all spans. In the case of equal end and interior spans or where $\ell_2 < \ell_1$ (see Fig. 6.1), the minimum h based on the end span governs, whereas in cases where $\ell_2 > 1.14\ell_1$, the minimum h based on the interior span governs. Recall that deflections need not be computed where a thickness equal to at least the minimum is provided.

The beam or one-way slab depth h that is actually constructed is specified in whole- or half-inch increments. For beams, whole-inch increments are usually used; however, this is not mandatory (in joist systems, half-inch increments for beam depths are common). An approximate value of d can be calculated as follows:

- Beams with one layer of reinforcement: $d = h - 2.5$ in
- One-way slabs: $d = h - 1.25$ in

The values of 2.5 and 1.25 in for beams and one-way slabs, respectively, are based on cover requirements and other reinforcement details, which are covered later in this chapter.

Step 2: Assume that the section is tension-controlled. The graph shown in Fig. 6.2 illustrates the effect of the strength reduction factor ϕ on the design strength ϕM_n for the case of 4,000 psi concrete and Grade 60 reinforcement. In particular, it shows what happens to the design strength when the limit for tension-controlled sections ($\phi = 0.9$) is passed. Similar curves can be generated for other material strengths. The reinforcement ratio corresponding to tension-controlled sections ($\varepsilon_t = 0.0050$) is ρ_t, and

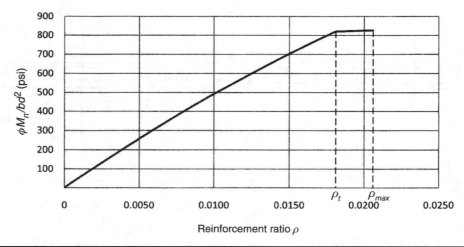

FIGURE 6.2   Design strength curve for 4,000 psi concrete and Grade 60 reinforcement.

the reinforcement ratio corresponding to the maximum permitted reinforcement ($\varepsilon_t = 0.0040$) is ρ_{max}.

The strength reduction factor ϕ for tension-controlled sections is equal to 0.9. If a section contains reinforcement greater than that corresponding to ρ_t, then $\phi < 0.9$ and the strength gain is minimal up to ρ_{max} (see Fig. 6.2). Any gain in strength with higher reinforcement ratios is canceled by the reduction in ϕ when net strains are less than 0.0050. Thus, for overall efficiency, flexural members should be designed as tension-controlled sections whenever possible.

Step 3: Determine the width of the member. The width of a beam is determined by setting $\phi M_n = M_u$ for an assumed reinforcement ratio ρ. A range for ρ is established as follows. Because a minimum amount of flexural reinforcement is required at any section, the assumed value of ρ must be greater than or equal to the minimum value prescribed in ACI 10.5 (see Section 5.4). Flexural members should be designed as tension-controlled sections whenever possible (see Step 2). As such, the assumed value of ρ should be less than or equal to ρ_t.

The reinforcement ratio ρ_t corresponding to tension-controlled sections with $\varepsilon_t = 0.0050$ can be derived using the basic principles and assumptions of the strength design method presented in Chap. 5. Figure 5.6 illustrates the strain condition in a rectangular, tension-controlled section with a single layer of tension reinforcement where $\varepsilon_t = 0.0050$. It is evident from the figure that the depth to the neutral axis $c_t = 0.375d_t$. Because $a_t = \beta_1 c_t$, $a_t = 0.375\beta_1 d_t$. Substituting a_t into Eq. (5.9) and solving for A_s result in the following:

$$A_s = \frac{0.319\beta_1 f_c' b d_t}{f_y} \tag{6.2}$$

Substituting $A_s = \rho_t b d_t$ into Eq. (6.2) and solving for ρ_t give

$$\rho_t = \frac{0.319\beta_1 f_c'}{f_y} \tag{6.3}$$

f_c' (psi)	ρ_t
3,000	0.0136
4,000	0.0181
5,000	0.0213

TABLE 6.1   Reinforcement Ratio ρ_t at Strain Limit of 0.0050 for Tension-controlled Sections, Assuming Grade 60 Reinforcement

Values of ρ_t are given in Table 6.1 as a function of the concrete compressive strength for Grade 60 reinforcement.

Once the assumed value of ρ has been established, it is used to determine the width of the beam b. The nominal flexural strength M_n is given by Eq. (5.11):

$$M_n = A_s f_y \left(d - \frac{0.59 A_s f_y}{b f_c'} \right) \tag{6.4}$$

Substituting $M_n = M_u/\phi$ into Eq. (6.4) and rearranging the terms result in the following:

$$\frac{M_u}{\phi b d^2} = \rho f_y \left(1 - \frac{0.59 \rho f_y}{f_c'} \right) = R_n \tag{6.5}$$

The strength reduction factor ϕ in Eq. (6.5) is equal to 0.9 because it is assumed that the section is tension-controlled. The term R_n is commonly referred to as the nominal strength coefficient of resistance.

Because d has been determined in Step 1, Eq. (6.5) can be solved for b:

$$b = \frac{M_u}{\phi R_n d^2} \tag{6.6}$$

It is important to reiterate that d has been calculated on the basis of deflection requirements. Thus, providing a beam width that is equal to greater than the value from this equation will result in cross-sectional dimensions that satisfy both the strength and deflection requirements of the Code.

A note on determining the beam width from Eq. (6.6) is in order at this time. Because beams are part of a continuous floor and/or roof system, the largest factored bending moment M_u along the spans should be used in this equation. In order to achieve economical formwork, the beam width determined from the maximum bending moment must be specified for all spans. Varying the amount of flexural reinforcement along the span lengths for different factored bending moments is by far more economical than varying the beam width (or depth).

As stated previously, the design width of a one-way slab is usually taken as 12 in.

The following guidelines should be followed when sizing members for economy:

- Use whole-inch increments for beam dimensions, if possible, and half-inch increments for slabs.
- Use beam widths in multiples of 2 or 3 in.

- Use constant beam size from span to span, and vary the reinforcement as required.
- Use wide, flat beams rather than narrow, deep beams.
- Use beam width equal to or greater than the column width.
- Use uniform width and depth of beams throughout the building, wherever possible.

Considerable cost savings can usually be realized by following these guidelines.

The flowchart shown in Fig. 6.3 can be used to determine h and b for a rectangular reinforced concrete beam.

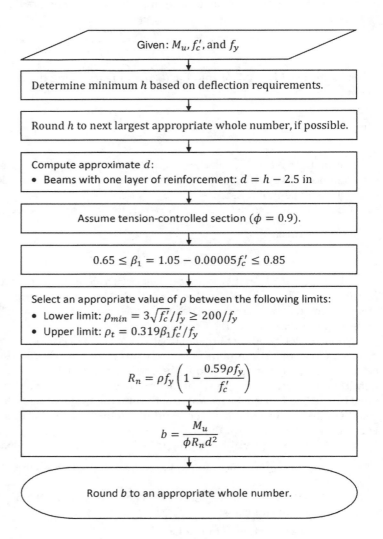

FIGURE 6.3 Determination of cross-section dimensions—rectangular beam.

Given: M_u, f_c', and f_y

Determine minimum h based on deflection requirements.

Round h to next largest appropriate whole number, if possible.

Compute approximate d:
- Beams with one layer of reinforcement: $d = h - 2.5$ in

Assume tension-controlled section ($\phi = 0.9$).

$0.65 \leq \beta_1 = 1.05 - 0.00005 f_c' \leq 0.85$

Select an appropriate value of ρ between the following limits:
- Lower limit: $\rho_{min} = 3\sqrt{f_c'}/f_y \geq 200/f_y$
- Upper limit: $\rho_t = 0.319\beta_1 f_c'/f_y$

$$R_n = \rho f_y \left(1 - \frac{0.59\rho f_y}{f_c'} \right)$$

$$b = \frac{M_u}{\phi R_n d^2}$$

Round b to an appropriate whole number.

Example 6.1 Check if the thickness of the one-way slab given in Example 3.5 satisfies the deflection requirements of the Code. Assume $f_c' = 4{,}000$ psi and Grade 60 reinforcement. Also assume that the one-way slabs are not attached to partitions or other construction that is likely to be damaged by relatively large deflections.

Solution The minimum thickness based on one end continuous governs [see ACI Table 9.5(a)]. Thus, for a 5-ft span length:

$$h = \frac{\ell}{24} = \frac{5 \times 12}{24} = 2.5 \text{ in} < \text{provided } h = 4.5 \text{ in}$$

Comments

A 4.5-in-thick one-way slab is used in this example to satisfy fire resistance requirements.

Example 6.2 Check if the dimensions of the wide-module joists given in Example 3.4 satisfy the deflection and strength requirements of the Code. Assume $f_c' = 4{,}000$ psi and Grade 60 reinforcement. Also assume that the wide-module joists are not attached to partitions or other construction that is likely to be damaged by relatively large deflections.

Solution The flowchart shown in Fig. 6.3 is utilized to check if the dimensions of the wide-module joists are adequate.

Step 1: Determine the minimum h based on deflection requirements. For economical formwork, the depth of the wide-module joists (which are beams) is determined on the basis of the exterior span because this will give the greatest minimum depth. Using ACI Table 9.5(a),

$$h = \frac{\ell}{18.5} = \frac{25 \times 12}{18.5} = 16.2 \text{ in}$$

Step 2: Round h to the next largest appropriate whole number, if possible. The available pan depths for wide-module joists that have a pan width of 53 in are 16, 20, and 24 in. A pan depth of 16 in is chosen in this example because it is the minimum depth that satisfies the deflection criteria. Thus, with a slab depth of 4.5 in (see Example 6.1), the overall depth of the wide-module joists is equal to $16 + 4.5 = 20.5$ in > 16.2 in.

Step 3: Compute approximate d. Assuming one layer of tension reinforcement, $d = 20.5 - 2.5 = 18.0$ in.

Step 4: Assume tension-controlled section. Sections that are tension-controlled are efficient for flexure. Thus, assume $\phi = 0.9$.

Step 5: Determine β_1. According to ACI 10.2.73, $\beta_1 = 0.85$ for $f_c' = 4{,}000$ psi.

Step 6: Select an appropriate value of ρ. For beams, the minimum value of the reinforcement ratio is given in ACI 10.5:

$$\rho_{min} = \frac{3\sqrt{f_c'}}{f_y} = \frac{3\sqrt{4{,}000}}{60{,}000} = 0.0032 < \frac{200}{f_y} = \frac{200}{60{,}000} = 0.0033 \text{ (governs)}$$

The upper limit must not be greater than ρ_t for tension-controlled sections:

$$\rho_t = \frac{0.319\beta_1 f_c'}{f_y} = \frac{0.319 \times 0.85 \times 4}{60} = 0.0181$$

As expected, this value matches that in Table 6.1 for 4,000 psi concrete and Grade 60 reinforcement.

In this example, a range of joist (beam) widths will be determined on the basis of the range of reinforcement ratios.

Step 7: Determine R_n. The nominal strength coefficient of resistance R_n is determined for the range of reinforcement ratios determined in step 6:

- For $\rho_{min} = 0.0033$,

$$R_n = \rho f_y \left(1 - \frac{0.59 \rho f_y}{f_c'}\right)$$

$$= 0.0033 \times 60 \left(1 - \frac{0.59 \times 0.0033 \times 60}{4}\right) = 0.192 \text{ ksi}$$

- For $\rho_t = 0.0181$,

$$R_n = 0.0181 \times 60 \left(1 - \frac{0.59 \times 0.0181 \times 60}{4}\right) = 0.912 \text{ ksi}$$

Step 8: Determine b. The width of the joists (beams) b is determined for the range of reinforcement ratios determined in step 6. For economical formwork, the largest factored moment in the spans is used to compute b. From Table 3.4, the maximum factored moment occurs at the exterior face of the first interior support in an end span: $M_u = 51.7$ ft kips. Thus,

- For $\rho_{min} = 0.0033$ ($R_n = 0.192$ ksi),

$$b = \frac{M_u}{\phi R_n d^2} = \frac{51.7 \times 12}{0.9 \times 0.192 \times 18^2} = 11.1 \text{ in}$$

- For $\rho_t = 0.0181$ ($R_n = 0.912$ ksi),

$$b = \frac{M_u}{\phi R_n d^2} = \frac{51.7 \times 12}{0.9 \times 0.912 \times 18^2} = 2.3 \text{ in}$$

In this example, a 7-in-wide joist is chosen. This is a common rib width for wide-module systems with 53-in-wide pan forms because the center-to-center joist spacing is equal to 5 ft.

Example 6.3   Check if the dimensions of the beam given in Example 3.3 satisfy the deflection and strength requirements of the Code. Assume $f_c' = 4,000$ psi and Grade 60 reinforcement. Also assume that the beams are not attached to partitions or other construction that is likely to be damaged by relatively large deflections.

Solution   The flowchart shown in Fig. 6.3 is utilized to check if the dimensions of the beam are adequate.

Step 1: Determine the minimum h based on deflection requirements. Both beams along column line B are exterior spans, and both have the same span length. Thus, the minimum depth of the beam h from ACI Table 9.5(a) is

$$h = \frac{\ell}{18.5} = \frac{20 \times 12}{18.5} = 13.0 \text{ in}$$

Step 2: Round h to the next largest appropriate whole number, if possible. The depth of the wide-module joists has been determined to be 20.5 in (see Example 6.2). Thus, for economical formwork, the depth of the beams should be specified as 20.5 in as well. This depth is greater than that required for deflection (see Step 1).

Step 3: Compute approximate d. Assuming one layer of tension reinforcement, $d = 20.5 - 2.5 = 18.0$ in.

Step 4: Assume tension-controlled section. Sections that are tension-controlled are efficient for flexure. Thus, assume $\phi = 0.9$.

Step 5: Determine β_1. According to ACI 10.2.73, $\beta_1 = 0.85$ for $f_c' = 4,000$ psi.

Step 6: Select an appropriate value of ρ. For beams, the minimum value of the reinforcement ratio is given in ACI 10.5:

$$\rho_{min} = \frac{3\sqrt{f_c'}}{f_y} = \frac{3\sqrt{4,000}}{60,000} = 0.0032 < \frac{200}{f_y} = \frac{200}{60,000} = 0.0033 \text{ (governs)}$$

The upper limit must not be greater than ρ_t for tension-controlled sections:

$$\rho_t = \frac{0.319\beta_1 f_c'}{f_y} = \frac{0.319 \times 0.85 \times 4}{60} = 0.0181$$

As expected, this value matches that in Table 6.1 for 4,000 psi concrete and Grade 60 reinforcement.

In this example, a range of beam widths will be determined on the basis of the range of reinforcement ratios.

Step 7: Determine R_n. The nominal strength coefficient of resistance R_n is determined for the range of reinforcement ratios determined in step 6:

- For $\rho_{min} = 0.0033$,

$$R_n = \rho f_y \left(1 - \frac{0.59\rho f_y}{f_c'}\right)$$

$$= 0.0033 \times 60 \left(1 - \frac{0.59 \times 0.0033 \times 60}{4}\right) = 0.192 \text{ ksi}$$

- For $\rho_t = 0.0181$,

$$R_n = 0.0181 \times 60 \left(1 - \frac{0.59 \times 0.0181 \times 60}{4}\right) = 0.912 \text{ ksi}$$

Step 8: Determine b. The width of the beam b is determined for the range of reinforcement ratios determined in step 6. For economical formwork, the largest factored moment in the two spans is used to compute b. From Table 3.3, the maximum factored moment occurs at the exterior face of the first interior support: $M_u = 198.1$ ft kips. Thus,

- For $\rho_{min} = 0.0033$ ($R_n = 0.192$ ksi),

$$b = \frac{M_u}{\phi R_n d^2} = \frac{198.1 \times 12}{0.9 \times 0.192 \times 18^2} = 42.5 \text{ in}$$

- For $\rho_t = 0.0181$ ($R_n = 0.912$ ksi),

$$b = \frac{M_u}{\phi R_n d^2} = \frac{198.1 \times 12}{0.9 \times 0.912 \times 18^2} = 8.9 \text{ in}$$

A wide range of beam widths that satisfy the requirements are available. In this example, a 22-in-wide beam is chosen to match the width of the column. If it is found later that there is congestion of reinforcement at the beam–column joint, the beam width can be made greater than the column width. As noted previously, a beam width that is narrower than the column width results in formwork that is not economical.

6.2.3   Determining Required Reinforcement

Once the cross-section dimensions of the flexural member have been determined, the required amount of flexural reinforcement can be calculated using $\phi M_n = M_u$. In continuous systems, both negative and positive reinforcement must be provided at the top and the bottom of a section, respectively, to resist the factored negative and positive bending moments determined from the applicable load combinations. Given later are methods to determine the required amount of flexural reinforcement for (1) rectangular sections with a single layer of tension reinforcement, (2) rectangular sections with multiple layers of tension reinforcement, (3) rectangular sections with tension and compression reinforcement, and (4) T-sections and inverted L-sections with tension reinforcement.

Rectangular Sections with a Single Layer of Tension Reinforcement

Equation (6.5) can be used to determine the required amount of flexural reinforcement A_s for a factored bending moment M_u by solving for the reinforcement ratio ρ:

$$\rho = \frac{0.85 f_c'}{f_y} \left[1 - \sqrt{1 - \frac{2R_n}{0.85 f_c'}} \right] \tag{6.7}$$

In this equation, $R_n = M_u/\phi b d^2$ and $\phi = 0.9$ for tension-controlled sections.

Once ρ is determined, the required area of flexural reinforcement is $A_s = \rho b d$. Reinforcing bars that provide an area of steel equal to or slightly greater than that required are chosen. It is important to check that the minimum and maximum reinforcement limits are satisfied. If it is found that the maximum limit is violated, the section must be redesigned. In cases where the section is found to be in the transition zone instead of tension-controlled, either the section and/or material properties can be revised, or compression reinforcement can be added.

In addition to satisfying strength requirements, the size and the number of flexural reinforcing bars must also satisfy the minimum and maximum spacing requirements of the Code. Additional information on these requirements is provided later.

The flowchart shown in Fig. 6.4 can be used to determine the required flexural reinforcement for a rectangular section with a single layer of tension reinforcement.

Rectangular Sections with Multiple Layers of Tension Reinforcement

Where the required tension reinforcement cannot adequately fit within one layer, the bars can be provided in more than one layer, as depicted in Fig. 5.13. A section with two layers of tension reinforcement can conservatively be designed using $d_t = d$. However, as discussed later, there is an advantage of using d_t and d separately.

The required A_s is determined in the same way as for sections with one layer of reinforcement, using d and d_t. The reinforcement ratio ρ based on d can be written in terms of the tensile force $T = A_s f_y$ as follows:

$$\rho = \frac{T}{b d f_y} \tag{6.8}$$

Similarly, the reinforcement ratio ρ_t corresponding to tension-controlled sections ($\varepsilon_t = 0.0050$) based on d_t is

$$\rho_t = \frac{T}{b d_t f_y} \tag{6.9}$$

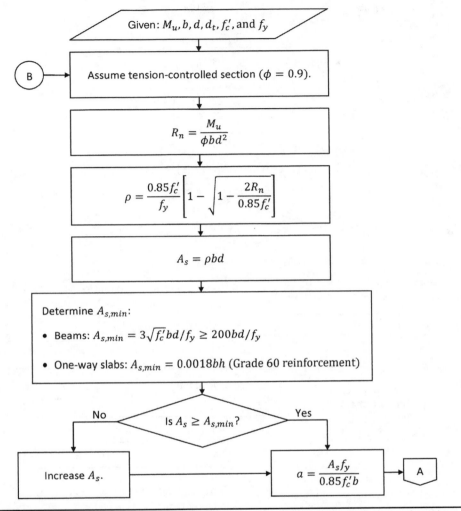

FIGURE 6.4 Required flexural reinforcement—rectangular section with a single layer of tension reinforcement. (*continued*)

Solving Eqs. (6.8) and (6.9) for T provides a relationship between ρ and ρ_t:

$$\rho = \rho_t \left(\frac{d_t}{d} \right) \tag{6.10}$$

It is important to determine whether the layer of reinforcement above the extreme steel layer yields or not (see Section 5.4). For design purposes, it is advantageous for this layer of reinforcement to yield. Grade 60 reinforcement located a distance equal to or greater than $d_y = 1.7c$ from the extreme compression fiber yields. For normally proportioned sections, this is usually the case.

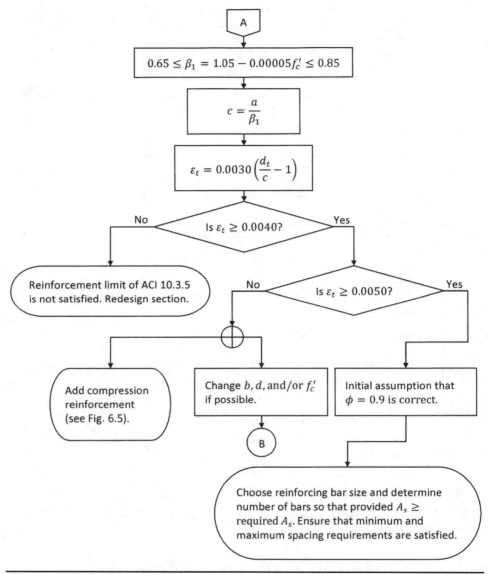

Figure 6.4 (Continued)

Rectangular Sections with Tension and Compression Reinforcement

For rectangular sections where it has been found that $\varepsilon_t < 0.0050$ (i.e., the section is not tension-controlled), compression reinforcement may be added to the section so that it becomes tension-controlled.

One of the design philosophies in this situation is to set $\varepsilon_t = 0.0050$, which means that the section is going to be designed at the tension-controlled net tensile strain limit. The need for compression reinforcement is determined by comparing the required nominal strength coefficient of resistance $R_n = M_u/\phi b d^2$ with the nominal strength coefficient of resistance R_{nt} corresponding to $\varepsilon_t = 0.0050$, which is obtained by the following equation [see Eq. (6.5)]:

$$R_{nt} = \rho_t f_y \left(1 - \frac{0.59 \rho_t f_y}{f_c'}\right) \tag{6.11}$$

In this equation, $\rho_t = 0.319 \beta_1 f_c'/f_y$ [see Eq. (6.3) and Table 6.1]. If $R_n > R_{nt}$, compression reinforcement is needed.

Once it has been established that compression reinforcement is needed, the next step is to determine the nominal flexural strength M_{nt} that corresponds to $\varepsilon_t = 0.0050$. From strength design, M_{nt} can be calculated as follows:

$$M_{nt} = A_s f_y \left(d - \frac{a}{2}\right)$$

$$= \rho f_y \left(1 - \frac{0.59 \rho f_y}{f_c'}\right) b d^2 \tag{6.12}$$

For a single layer of tension reinforcement, the reinforcement ratio to be used in Eq. (6.12) is $\rho = \rho_t = 0.319 \beta_1 f_c'/f_y$. In situations where compression reinforcement is required, it is common for two layers of tension reinforcement to be needed; therefore, the reinforcement ratio to be used in Eq. (6.12) is $\rho = \rho_t(d_t/d) = 0.319 \beta_1 f_c' d_t/d f_y$ [see Eq. (6.10)].

The required nominal flexural strength M_n' that needs to be resisted by the compression reinforcement is the difference between the required nominal flexural strength and the nominal flexural strength provided by the tension reinforcement:

$$M_n' = \frac{M_u}{\phi} - M_{nt} \tag{6.13}$$

where $\phi = 0.9$.

It was shown in Section 5.4 that compression reinforcement (Grade 60) that is located a distance equal to or less than $d_y' = 0.31c$ from the extreme compression fiber yields. In general, the stress in the compression reinforcement f_s' can be determined by the following equation [see Eq. (5.24)]:

$$f_s' = 0.0030 E_s \left(1 - \frac{d'}{c_t}\right) \leq f_y \tag{6.14}$$

In this equation, $c = c_t = 0.375 d_t$ because $\varepsilon_t = 0.0050$. The required area of compression reinforcement A_s' is determined from strength design (see Section 5.4):

$$A_s' = \frac{M_n'}{f_s'(d - d')} \tag{6.15}$$

Finally, the total required area of tension reinforcement A_s is the summation of the area of reinforcement corresponding to the tension-controlled net strain limit and A_s'

from Eq. (6.15) with $f_s' = f_y$:

$$A_s = \rho bd + \frac{M_n'}{f_y(d - d')} \qquad (6.16)$$

Like in the case of rectangular sections with tension reinforcement only, the size and the number of flexural reinforcing bars must satisfy the minimum and maximum spacing requirements in addition to satisfying strength requirements.

The flowchart shown in Fig. 6.5 can be used to determine the required flexural reinforcement for a rectangular section with tension and compression reinforcement.

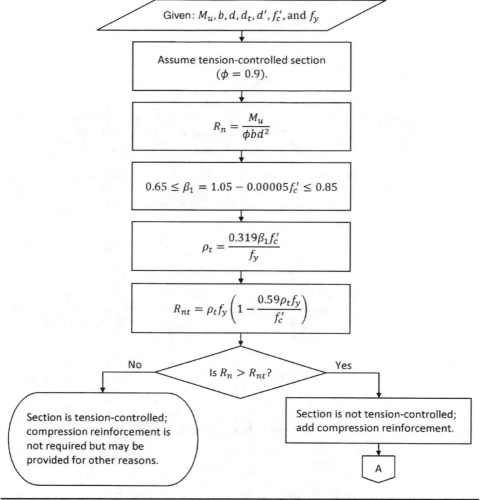

FIGURE 6.5 Determination of flexural reinforcement—rectangular section with tension and compression reinforcement. (*continued*)

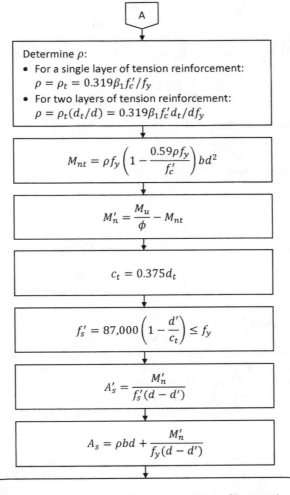

A

Determine ρ:
- For a single layer of tension reinforcement:
  $\rho = \rho_t = 0.319\beta_1 f_c'/f_y$
- For two layers of tension reinforcement:
  $\rho = \rho_t(d_t/d) = 0.319\beta_1 f_c' d_t/df_y$

$$M_{nt} = \rho f_y\left(1 - \frac{0.59\rho f_y}{f_c'}\right)bd^2$$

$$M_n' = \frac{M_u}{\phi} - M_{nt}$$

$$c_t = 0.375d_t$$

$$f_s' = 87{,}000\left(1 - \frac{d'}{c_t}\right) \leq f_y$$

$$A_s' = \frac{M_n'}{f_s'(d - d')}$$

$$A_s = \rho bd + \frac{M_n'}{f_y(d - d')}$$

Choose reinforcing bar size and determine number of bars so that provided $A_s \geq$ required A_s and provided $A_s' \geq$ required A_s'. Ensure that minimum and maximum spacing requirements are satisfied.

FIGURE 6.5 *(Continued)*

T-Sections and Inverted L-Section with Tension Reinforcement

A typical cast-in-place concrete floor/roof system is depicted in Fig. 5.21. It was discussed in Section 5.4 that because of the nature of concrete construction, the beams and the slab work together to resist the effects from applied loads. A portion of the slab supported by the beam web is considered effective in resisting the factored bending moment. Methods to determine the required area of flexural reinforcement where the flange is in tension and compression are presented next.

Flange in Tension   The flange of a T-beam or an inverted L-beam is in tension at locations of negative bending moment. Once the effective flange width b_e has been determined in accordance with ACI 8.12 (see Section 5.4 and Fig. 5.23), the required area of flexural reinforcement can be calculated using the methods presented previously for rectangular sections with tension reinforcement (see Fig. 6.4).

Special requirements for the distribution of the flexural reinforcement in the flange are given in ACI 10.6.6; these are discussed later in the chapter.

Flange in Compression   The flange of a T-beam or an inverted L-beam is in compression at locations of positive bending moment. Assuming that the section is tension-controlled with rectangular section behavior (i.e., the depth of the equivalent stress block $a \leq h_f$), a can be determined by the following equation:

$$a = \frac{a_1 d - \sqrt{(a_1 d)^2 - (2a_1 M_u/\phi)}}{a_1} \tag{6.17}$$

This equation is obtained by substituting Eq. (6.7) with $A_s = \rho b d$ into Eq. (5.9) where $a_1 = 0.85 f'_c b_e$.

If the value of a determined by Eq. (6.17) is equal to or less than the slab thickness h_f, the assumption that the section behaves as a rectangular section is correct, and A_s can be determined using the methods presented previously for rectangular sections with tension reinforcement (see Fig. 6.4).

If it is found that $a > h_f$, the assumption of rectangular section behavior is not correct, and the compressive zone is T or L shaped instead of rectangular. The next step is to calculate the area of reinforcement A_{sf} and the design strength ϕM_{n1} corresponding to the overhanging beam flange by Eqs. (5.26) and (5.27), respectively (see Fig. 5.27).

The required moment strength that needs to be carried by the beam web is determined by subtracting the design strength ϕM_{n1} from the total factored moment M_u:

$$M_{u2} = M_u - \phi M_{n1} \tag{6.18}$$

The required flexural reinforcement A_{sw} to develop M_{u2} can be determined by the following strength design equations:

$$R_{nw} = \frac{M_{u2}}{\phi b_w d^2} \tag{6.19}$$

$$\rho_w = \frac{0.85 f'_c}{f_y} \left[1 - \sqrt{1 - \frac{2R_{nw}}{0.85 f'_c}} \right] \tag{6.20}$$

$$A_{sw} = \rho_w b_w d \tag{6.21}$$

The total required flexural reinforcement is the sum of the steel areas from the overhanging flange and the web:

$$A_s = A_{sf} + A_{sw} \tag{6.22}$$

Once A_s is determined, the assumption that the section is tension-controlled needs to be checked; that is, verify that $\varepsilon_t \geq 0.0050$ or, equivalently, that $c = a_w/\beta_1 \leq 0.375d_t$, where $a_w = (A_s - A_{sf})f_y/0.85f_c'b_w$ [see Eq. (5.30)]. If it found that the assumption is correct, the size and the number of flexural reinforcing bars are chosen to satisfy the minimum and maximum spacing requirements. If the assumption is not correct, compression reinforcement may be added to the section to make it tension-controlled.

The flowchart shown in Fig. 6.6 can be used to determine the required flexural reinforcement for a T-section or an inverted L-section with tension reinforcement where the flange is in compression.

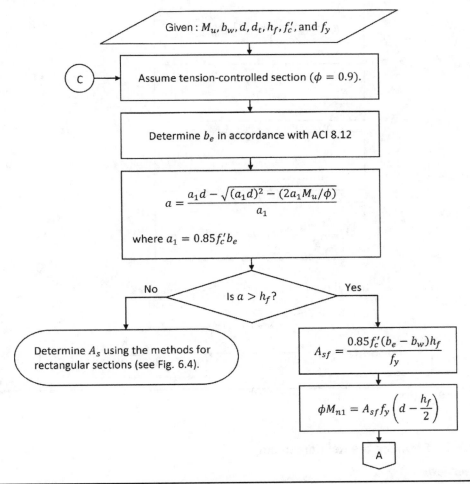

FIGURE 6.6 Determination of flexural reinforcement—T-section or inverted L-section with tension reinforcement and the flange in compression. (*continued*)

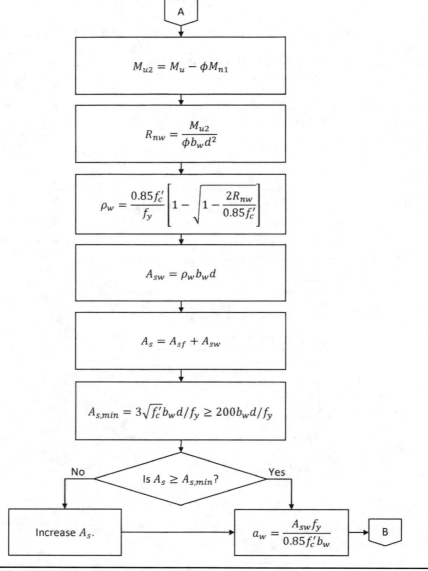

$$A$$

$$M_{u2} = M_u - \phi M_{n1}$$

$$R_{nw} = \frac{M_{u2}}{\phi b_w d^2}$$

$$\rho_w = \frac{0.85 f_c'}{f_y}\left[1 - \sqrt{1 - \frac{2R_{nw}}{0.85 f_c'}}\right]$$

$$A_{sw} = \rho_w b_w d$$

$$A_s = A_{sf} + A_{sw}$$

$$A_{s,min} = 3\sqrt{f_c'}\,b_w d/f_y \geq 200 b_w d/f_y$$

Is $A_s \geq A_{s,min}$?

No   Yes

Increase A_s.

$$a_w = \frac{A_{sw} f_y}{0.85 f_c' b_w}$$

$$B$$

Figure 6.6 (Continued)

6.2.4 Detailing the Reinforcement

Overview

Once the required area of steel has been determined using the methods presented earlier, the size and the number of reinforcing bars must be chosen to provide an area of steel that is equal to or greater than the amount that is required.

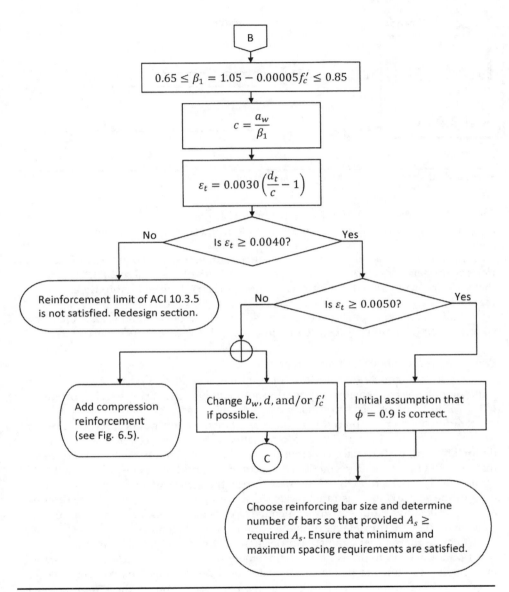

Figure 6.6   *(Continued)*

In general, the minimum and the maximum number of reinforcing bars in a layer of a given cross-section are a function of the cover and spacing requirements given in the Code. Minimum spacing between the longitudinal bars is required in order to adequately place the concrete. Concrete may not be able to flow in the voids between the bars if they are spaced too closely together, especially with concrete mixes with larger aggregates. Bars that are spaced too far apart could result in relatively large flexural crack widths. A maximum spacing between the longitudinal bars is required to limit

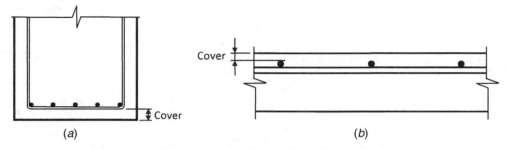

Figure 6.7 Concrete cover for (a) beams and (b) one-way slabs.

such crack widths. Finally, a minimum amount of concrete cover is required to protect the reinforcement from the effects of fire, weather, and corrosive environments, to name a few.

In addition to satisfying size and spacing requirements, the flexural reinforcing bars must be properly developed or anchored on both sides of a critical section. This ensures that the reinforcement will perform as intended in accordance with the strength design method.

Requirements for concrete cover, bar spacing, bar development and anchorage, splices of reinforcement, and structural integrity reinforcement are covered later.

Concrete Protection for Reinforcement

A discussion on concrete protection for reinforcement is needed prior to discussing requirements for bar spacing and bar development because it plays an important role in the formulation of those requirements.

Reinforcing bars are placed in a concrete member with a minimum concrete cover to protect it from weather, fire, and other effects. Minimum cover requirements for nonprestressed, cast-in-place concrete construction are given in ACI 7.7.1. For beams that have transverse reinforcement in the form of stirrups that enclose the main flexural reinforcing bars, concrete cover is measured from the surface of the concrete to the outer edge of the stirrups, as illustrated in Fig. 6.7a. For one-way slabs, which are designed and constructed without stirrups, concrete cover is measured from the surface of the concrete to the outermost layer of the flexural bars (see Fig. 6.7b).

Concrete that is cast against and permanently exposed to earth requires a clear cover of at least 3 in to protect the reinforcement from possible deleterious substances in the soil. For concrete that is exposed to earth or weather (i.e., concrete that has direct exposure to moisture changes and temperature changes), the minimum cover varies from 1.5 to 2 in, depending on the bar size. The minimum cover for members that are not exposed to ground or weather varies from 0.75 to 1.5 in; this minimum cover is typically utilized in members that are in enclosed, environmentally controlled structures.

Distribution of Flexural Reinforcement for Crack Control

Requirements for the distribution of flexural reinforcement in beams and one-way slabs are given in ACI 10.6. The intent of these requirements is to control flexural cracking. In general, a larger number of fine cracks are preferable to a few wide cracks mainly for reasons of durability and appearance. For example, wider cracks may not be acceptable in exposed concrete for architectural reasons and could give the erroneous impression that the structure is not safe.

Research has shown that crack width at service loads is directly proportional to the stress in the reinforcing bars.[1-3] The thickness of the concrete cover and the spacing of the bars are important variables that directly affect crack width. Control of cracking is improved when the reinforcing bars are well distributed in the flexural tension zones of a member. Using several smaller bars at a smaller spacing is more effective than using larger bars of equivalent area that are spaced farther apart.

Maximum Spacing of Reinforcing Bars in a Single Layer   A simple approach to address crack control in flexural members is given in ACI 10.6.4. The maximum center-to-center bar spacing s determined by ACI Eq. (10-4) is specifically meant to control cracking[4-6]:

$$s = 15 \left(\frac{40,000}{f_s} \right) - 2.5c_c \le 12 \left(\frac{40,000}{f_s} \right) \tag{6.23}$$

In this equation, f_s is the calculated stress in the flexural reinforcement closest to the tension face of the section, caused by the service loads. The stress f_s can be calculated by dividing the unfactored bending moment by the product of the steel area and the internal moment arm. This calculation requires knowledge of the working stress design method, which was prevalent prior to the strength design method. In lieu of that calculation, the Code permits $f_s = 2f_y/3$; this simplification is used throughout this discussion.

The term c_c is related to the clear cover of the reinforcement and is defined as the least distance from the surface of the reinforcement to the tension face of the member. For example, for a beam with No. 4 stirrups that is located inside of a building, $c_c = 2$ in [1.5-in concrete cover in accordance with ACI 7.7.1(c) plus the diameter of the No. 4 stirrup, which is 0.5 in; see Fig. 6.7]. Thus, according to Eq. (6.23), $s = 10$ in for Grade 60 flexural reinforcing bars. This means that the bars in this example must be spaced no greater than 10 in on center in order to satisfy crack control requirements. Note that s is independent of the size of the flexural bars.

In one-way slabs, c_c is usually equal to 0.75 in, so, $s = 12$ in for Grade 60 reinforcement. This value is generally less than the maximum spacing prescribed in ACI 10.5.4 for structural slabs of uniform thickness.

On the basis of the information given in Fig. 6.8, the following equation provides the minimum number of bars n_{min} required in a single layer to control cracking:

$$n_{min} = \frac{b_w - 2(c_c + 0.5d_b)}{s} + 1 \tag{6.24}$$

In this equation, s is limited to that obtained by Eq. (6.23). The values of n_{min} determined by Eq. (6.24) should be rounded up to the next whole number. Note that a minimum of two bars are required to anchor the stirrups in beams.

The minimum number of bars can be tabulated for various beam widths, as shown in Table 6.2. The information in this table is based on the following:

FIGURE 6.8
Maximum spacing
requirements of
flexural
reinforcement.

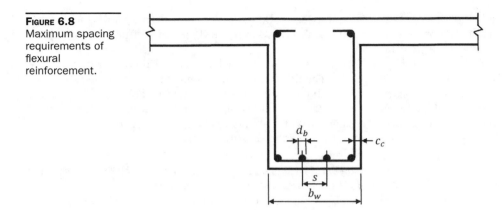

FIGURE 6.8
Maximum spacing
requirements of
flexural
reinforcement.

- Grade 60 reinforcement
- Least distance from the surface of the reinforcement to the tension face of the member $c_c = 2$ in
- Calculated stress in the flexural reinforcement closest to the tension face of the section, caused by the service loads $f_s = 40$ ksi

Given these assumptions, $s = 10$ in from Eq. (6.23).

Providing at least the number of flexural reinforcing bars in Table 6.2 for a given beam width automatically satisfies the crack control requirements of ACI 10.6.4.

Corrosive Environments   ACI 10.6.5 states that the crack control provisions of ACI 10.6.4 are not sufficient for structures that are subjected to very aggressive exposures (such as concrete exposed to moisture and external sources of chlorides) or structures that are designed to be watertight (such as tanks). Exposure tests have revealed that concrete quality, adequate concrete compaction, and ample cover to the reinforcing bars may

	Beam Width (in)												
Bar size	12	14	16	18	20	22	24	26	28	30	36	42	48
No. 4	2	2	3	3	3	3	3	4	4	4	5	5	6
No. 5	2	2	3	3	3	3	3	4	4	4	5	5	6
No. 6	2	2	3	3	3	3	3	4	4	4	5	5	6
No. 7	2	2	3	3	3	3	3	4	4	4	5	5	6
No. 8	2	2	3	3	3	3	3	4	4	4	5	5	6
No. 9	2	2	3	3	3	3	3	4	4	4	5	5	6
No. 10	2	2	3	3	3	3	3	4	4	4	5	5	6
No. 11	2	2	3	3	3	3	3	4	4	4	5	5	6

TABLE 6.2   Minimum Number of Reinforcing Bars Required in a Single Layer

FIGURE 6.9
Distribution of
tension
reinforcement in
a T-beam.

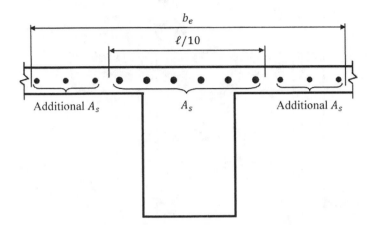

play a more important role in the prevention of corrosion of the reinforcing bars than crack width at the surface of the concrete. As such, special precautions should be taken when detailing the reinforcement.

Flanges of T-Beams and Inverted L-Beam in Tension   Requirements for the control of cracking in the flanges of T-beams and inverted L-beam that are in tension are given in ACI 10.6.6 and are summarized in Fig. 6.9 for the case of a T-beam.

The flexural reinforcement in the flange (slab) must be uniformly distributed over a width equal to the lesser of the effective width b_e defined in ACI 8.12 (see Fig. 5.23) or one-tenth of the span length. Additional reinforcement must be provided in the outer portions of the flange where b_e exceeds one-tenth of the span length to help ensure that these outer portions do not develop wide cracks. Although the Code does not specify the amount of reinforcement that needs to be provided in such cases, the temperature and shrinkage reinforcement prescribed in ACI 7.12 for slabs should be used as a minimum.

Beams with Depth Greater than 36 in   Research has shown that wide cracks can form on the side faces of relatively deep beams or joists between the flexural reinforcement and the neutral axis.[7,8] In some cases, these cracks have been found to be wider than those at the level of the flexural reinforcement.

To minimize these crack widths, ACI 10.6.7 requires that longitudinal skin reinforcement be provided on both faces of a member where $h > 36$ in. This reinforcement must be uniformly distributed over a distance of $h/2$ from the tension face. ACI Fig. R10.6.7 illustrates the skin reinforcement distribution for sections with positive and negative reinforcement.

The maximum spacing s of the skin reinforcement is calculated by Eq. (6.23) where c_c is the least distance from the surface of the skin reinforcement to the side face. The size of the skin reinforcement is not specified in ACI 10.6.7 because tests have shown that the spacing of the skin reinforcement is more important in crack control than the size.[8] However, ACI R10.6.7 recommends using Nos. 3 to 5 bars or welded wire reinforcement with a minimum area of 0.1 in^2 per foot of depth.

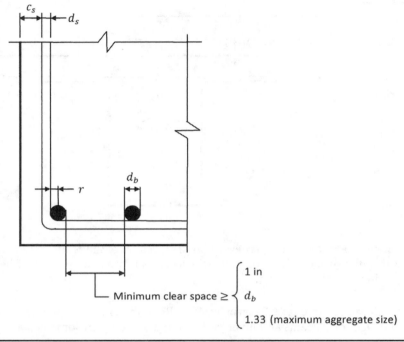

Figure 6.10   Spacing limits for reinforcing bars.

Minimum Spacing of Flexural Reinforcing Bars

Spacing limits for reinforcing bars are given in ACI 7.6 and 3.3.2. These limits have been established primarily so that concrete can flow readily into the spaces between adjoining bars and between bars and formwork. This helps to ensure that concrete fully surrounds the reinforcing bars without honeycombing and, thus, that a good bond is established between the concrete and the steel.

The spacing requirements are summarized in Fig. 6.10. The following equation provides the maximum number of bars n_{max} that can fit in a single layer on the basis of the spacing limits of ACI 7.6 and 3.3.2:

$$n_{max} = \frac{b_w - 2(c_s + d_s + r)}{(\text{clear space}) + d_b} + 1 \qquad (6.25)$$

where
$$
\begin{aligned}
c_s &= \text{clear cover to stirrups} \\
d_s &= \text{diameter of stirrups} \\
r &= \text{bend radius of stirrups (ACI 7.2.2)} \\
&= 2d_s \text{ for No. 5 stirrups and smaller} \\
&= \text{one-half the minimum bend diameters given in ACI Table 7.2 for} \\
&\quad \text{bars larger than No. 5} \\
\text{clear space} &= \text{minimum clear space defined in Fig. 6.10}
\end{aligned}
$$

The values of n_{max} determined by Eq. (6.25) should be rounded down to the next whole number.

Bar size	Beam Width (in)												
	12	14	16	18	20	22	24	26	28	30	36	42	48
No. 4	5	6	8	9	10	12	13	14	16	17	21	25	29
No. 5	5	6	7	8	10	11	12	13	15	16	19	23	27
No. 6	4	6	7	8	9	10	11	12	14	15	18	22	25
No. 7	4	5	6	7	8	9	10	11	12	13	17	20	23
No. 8	4	5	6	7	8	9	10	11	12	13	16	19	22
No. 9	3	4	5	6	7	8	8	9	10	11	14	17	19
No. 10	3	4	4	5	6	7	8	8	9	10	12	15	17
No. 11	3	3	4	5	5	6	7	8	8	9	11	13	15

TABLE 6.3   Maximum Number of Bars Permitted in a Single Layer

The maximum number of bars that can fit in a single layer can be tabulated for various beam widths, as shown in Table 6.3. The information in this table is based on the following:

- Grade 60 reinforcement.
- Clear cover to the stirrups $c_s = 1.5$ in.
- Maximum aggregate size of $3/4$ in.
- No. 3 stirrups are used for Nos. 4 to 6 bars, and No. 4 stirrups are used for No. 7 and larger bars.

Selecting the number of bars within the limits of Tables 6.2 and 6.3 provides automatic conformity with the Code requirements for cover and spacing given the assumptions noted earlier. Tables for other parameters can be generated by Eqs. (6.24) and (6.25).

Example 6.4   Determine the required positive and negative reinforcement for the one-way slab in Examples 3.5 and 6.1 (see Fig. 3.3). Assume $f_c' = 4,000$ psi and Grade 60 reinforcement.

Solution   The flowchart shown in Fig. 6.4 is utilized to determine A_s. The thickness of the slab is equal to 4.5 in from Example 6.1. A 12-in-wide design strip is utilized for this one-way slab.

Step 1: Assume tension-controlled section. Sections of flexural members should be tension-controlled whenever possible. Thus, assume that the strength reduction factor $\phi = 0.9$.

Step 2: Determine the nominal strength coefficient of resistance R_n. For a rectangular section, R_n is determined by Eq. (6.5), which is a function of the factored bending moment M_u. The negative factored bending moment was determined in Example 3.5 as 0.42 ft kips/ft. Also, because the thickness of the slab is 4.5 in, only one layer of reinforcement is able to fit in the section in the main direction of analysis. Therefore, the flexural reinforcement will be positioned at mid-depth, so that $d = h/2 = 4.5/2 = 2.25$ in. Thus,

$$R_n = \frac{M_u}{\phi b d^2} = \frac{0.42 \times 12,000}{0.9 \times 12 \times 2.25^2} = 92.2 \text{ psi}$$

Step 3: Determine the required reinforcement ratio ρ. The reinforcement ratio ρ is determined by Eq. (6.7):

$$\rho = \frac{0.85 f_c'}{f_y}\left[1 - \sqrt{1 - \frac{2R_n}{0.85 f_c'}}\right] = \frac{0.85 \times 4}{60}\left[1 - \sqrt{1 - \frac{2 \times 92.2}{0.85 \times 4{,}000}}\right] = 0.0016$$

Step 4: Determine the required area of tension reinforcement A_s. For a 1-ft-wide design strip, the required area of negative reinforcement is

$$A_s = \rho b d = 0.0016 \times 12 \times 2.25 = 0.04 \text{ in}^2/\text{ft}$$

Step 5: Determine the minimum required area of reinforcement $A_{s,min}$. For one-way slabs with Grade 60 reinforcement, $A_{s,min}$ is determined in accordance with ACI 10.5.4:

$$A_{s,min} = 0.0018bh = 0.0018 \times 12 \times 4.5 = 0.10 \text{ in}^2/\text{ft} > A_s$$

Use $A_s = 0.10 \text{ in}^2/\text{ft}$.

Step 6: Determine the depth of the equivalent rectangular stress block a.

$$a = \frac{A_s f_y}{0.85 f_c' b} = \frac{0.10 \times 60{,}000}{0.85 \times 4{,}000 \times 12} = 0.15 \text{ in}$$

Step 7: Determine β_1. According to ACI 10.2.73, $\beta_1 = 0.85$ for $f_c' = 4{,}000$ psi (see Section 5.2).

Step 8: Determine the neutral axis depth c.

$$c = \frac{a}{\beta_1} = \frac{0.15}{0.85} = 0.18 \text{ in}$$

Step 9: Determine ε_t.

$$\varepsilon_t = 0.0030\left(\frac{d_t}{c} - 1\right) = 0.0030\left(\frac{2.25}{0.18} - 1\right) = 0.0345 > 0.0040$$

Because $\varepsilon_t > 0.0040$, the maximum reinforcement requirement of ACI 10.3.5 is satisfied. Also, the section is tension-controlled because $\varepsilon_t > 0.0050$ (ACI 10.3.4), and the initial assumption that the section is tension-controlled is correct.

Step 10: Choose the size and spacing of the reinforcing bars. The required area of reinforcement is 0.10 in^2/ft. No. 3 bars spaced 12 in on center provide an area of reinforcement equal to 0.11 in^2/ft. Note that No. 4 bars spaced at 24 in on center also satisfy strength requirements (provided $A_s = 0.20 \times 12/24 = 0.10 \text{ in}^2/\text{ft}$).

Bar spacing is governed by the crack control requirements of ACI 10.6.4. Because the bars are located at the mid-depth of the slab, c_c is relatively large, and the maximum bar spacing is governed by the upper limit of Eq. (6.23):

$$s = 12\left(\frac{40{,}000}{f_s}\right) = 12\left(\frac{40{,}000}{2 \times 60{,}000/3}\right) = 12 \text{ in}$$

Check the maximum spacing requirements of ACI 10.5.4:

$$s = 3h = 3 \times 4.5 = 13.5 < 18.0 \text{ in}$$

Therefore, the maximum spacing of the bars is 12.0 in.
Use No. 3 bars at 12 in on center.

Comments

Although the No. 4 bars are adequate for strength, the 24-in spacing is greater than that re-quired for crack control. Specifying No. 4 bars at a 12-in spacing is adequate for both strength and crack control, but the provided reinforcement is twice as much as required, which is not econo-mical.

The positive factored bending moment is equal to 0.36 ft kips per foot width of slab (see Example 3.5). Because only one layer of reinforcement is being provided in the direction of analysis, use No. 3 bars at a spacing of 12 in throughout the span, as this amount of reinforcement is also adequate for the positive factored bending moment.

A minimum amount of temperature and shrinkage reinforcement must be provided in the slab perpendicular to the main flexural reinforcement (ACI 7.12.2.1). This amount is the same as that for minimum reinforcement with a maximum spacing equal to $5h$ = 22.5 in or 18 in (governs). Use No. 3 bars at a spacing of 12 in on center for temperature and shrinkage reinforcement (provided $A_s = 0.11$ in²/ft).

Example 6.5   Determine the required negative reinforcement at the exterior face of the first inte-rior support in the end span of a typical interior wide-module joist in Examples 3.4 and 6.2 (see Fig. 3.3). Assume f'_c = 4,000 psi, a ³/₄-in maximum aggregate, and Grade 60 reinforcement.

Solution   The height and the width of the wide-module joist are 20.5 and 7 in, respectively, which were determined in Example 6.2. For a negative bending moment, the reinforcement is located within the flange. Because the flange is in tension, the required area of flexural reinforcement can be determined using the methods for rectangular sections with tension reinforcement. Thus, the flowchart shown in Fig. 6.4 is utilized to determine A_s.

Step 1: Assume tension-controlled section. Sections of flexural members should be tension-controlled whenever possible. Thus, assume that the strength reduction factor $\phi = 0.9$.

Step 2: Determine the nominal strength coefficient of resistance R_n. For a rectangular section, R_n is determined by Eq. (6.5), which is a function of the factored bending moment M_u. The negative factored bending moment at the exterior face of the first interior support in the end span is equal to 51.7 ft kips (see Table 3.4). Also assume that $d = h - 2.5 = 20.5 - 2.5 = 18.0$ in.

$$R_n = \frac{M_u}{\phi b_w d^2} = \frac{51.7 \times 12,000}{0.9 \times 7 \times 18.0^2} = 303.9 \text{ psi}$$

Step 3: Determine the required reinforcement ratio ρ. The reinforcement ratio ρ is determined by Eq. (6.7):

$$\rho = \frac{0.85 f'_c}{f_y} \left[1 - \sqrt{1 - \frac{2R_n}{0.85 f'_c}} \right] = \frac{0.85 \times 4}{60} \left[1 - \sqrt{1 - \frac{2 \times 303.9}{0.85 \times 4,000}} \right] = 0.0053$$

Step 4: Determine the required area of tension reinforcement A_s.

$$A_s = \rho b d = 0.0053 \times 7 \times 18.0 = 0.67 \text{ in}^2$$

Step 5: Determine the minimum required area of reinforcement $A_{s,min}$. The minimum amount of reinforcement is determined by ACI 10.5.1. Because the compressive strength of the concrete is less than 4,400 psi, the minimum amount is determined by the lower limit given in that section:

$$A_{s,min} = \frac{200 b_w d}{f_y} = \frac{200 \times 7 \times 18.0}{60,000} = 0.42 \text{ in}^2 < 0.67 \text{ in}^2$$

Use $A_s = 0.67$ in².

Step 6: **Determine the depth of the equivalent rectangular stress block *a*.**

$$a = \frac{A_s f_y}{0.85 f'_c b} = \frac{0.67 \times 60,000}{0.85 \times 4,000 \times 7} = 1.7 \text{ in}$$

Step 7: **Determine β_1.** According to ACI 10.2.73, $\beta_1 = 0.85$ for $f'_c = 4,000$ psi (see Section 5.2).

Step 8: **Determine the neutral axis depth *c*.**

$$c = \frac{a}{\beta_1} = \frac{1.7}{0.85} = 2.0 \text{ in}$$

Step 9: **Determine ε_t.**

$$\varepsilon_t = 0.0030 \left(\frac{d_t}{c} - 1 \right) = 0.0030 \left(\frac{18.0}{2.0} - 1 \right) = 0.0240 > 0.0040$$

Because $\varepsilon_t > 0.0040$, the maximum reinforcement requirement of ACI 10.3.5 is satisfied. Also, the section is tension-controlled because $\varepsilon_t > 0.0050$ (ACI 10.3.4) and the initial assumption that the section is tension-controlled is correct.

Step 10: **Choose the size and spacing of the reinforcing bars.** The required area of reinforcement is 0.67 in². Because the flange of the wide-module joist is in tension, a portion of the reinforcement must be distributed over the lesser of the effective flange width b_e or one-tenth the span. Using Fig. 5.23, determine b_e:

$$b_e = \begin{cases} \text{span length}/4 = 25 \times 12/4 = 75.0 \text{ in} \\ b_w + 16h = 7 + (16 \times 4.5) = 79.0 \text{ in} \\ \text{joist spacing} = 60 \text{ in (governs)} \end{cases}$$

Also, span/10 = $25 \times 12/10 = 30.0$ in.

Because b_e exceeds one-tenth the span, provide the required reinforcement within the 30-in width and provide additional reinforcement in the 15-in-wide outer portions of the flange:

$$A_s = \frac{0.67}{30/12} = 0.27 \text{ in}^2/\text{ft}$$

Try No. 5 bars spaced at 12 in on center within the 2.5-ft-wide strip over the web (provided $A_s = 0.31$ in²/ft).

Bar spacing is governed by the crack control requirements of ACI 10.6.4 and the spacing limits of ACI 7.6 and 3.3.2. Assuming 1.5 in of cover to the flexural bars in accordance with ACI 7.7.1(c), $c_c = 1.5$ in. The maximum bar spacing is determined by Eq. (6.23):

$$s = 15 \left(\frac{40,000}{f_s} \right) - 2.5c_c = 15 \left(\frac{40,000}{2 \times 60,000/3} \right) - (2.5 \times 1.5) = 11.3 \text{ in (governs)}$$

$$\leq 12 \left(\frac{40,000}{f_s} \right) = 12 \left(\frac{40,000}{2 \times 60,000/3} \right) = 12.0 \text{ in}$$

Therefore the No. 5 bars at 12 in on center are not adequate for crack control. Try No. 5 spaced at 10 in on center.

Check the spacing limits of ACI 7.6 and 3.3.2.

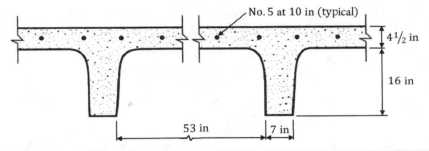

Figure 6.11 Details of the negative reinforcement in the wide-module joist given in Example 6.5.

From Fig. 6.10,

$$\text{Minimum clear space} = \begin{cases} 1.0\ \text{in} \\ d_b = 0.625\ \text{in} \\ 1.33\ (\text{maximum aggregate size}) = 1.33 \times 0.75 = 1.0\ \text{in} \end{cases}$$

Therefore, the minimum clear space is 1.0 in.
The provided clear space $= 10.0 - 0.625 = 9.4 > 1.0$ in.
Use No. 5 bars spaced 10 in on center. For simpler detailing, provide the same reinforcement in the outer portions of the flange.

Comments
No. 5 bars spaced at 11 in on center are adequate for both strength and spacing requirements. However, a 10-in spacing was chosen on the basis of the spacing of the wide-module joists (see Fig. 6.11).

Example 6.6 Determine the required positive reinforcement in the end span of a typical interior wide-module joist of Examples 3.4 and 6.2 (see Fig. 3.3). Assume $f_c' = 4,000$ psi, a $\frac{3}{4}$-in maximum aggregate, and Grade 60 reinforcement.

Solution The height and the width of the wide-module joist are 20.5 and 7 in, respectively, which were determined in Example 6.2. For a positive bending moment, the reinforcement is located within the web of the joist. Because the flange is in compression, the required area of flexural reinforcement is determined depending on whether the depth of the equivalent stress block a is less than or greater than the thickness of the flange. Thus, the flowchart shown in Fig. 6.6 is utilized to determine A_s.

Step 1: Assume tension-controlled section. Sections of flexural members should be tension-controlled whenever possible. Thus, assume that the strength reduction factor $\phi = 0.9$.

Step 2: Determine the effective flange width b_e. The effective flange width $b_e = 60$ in (see step 10 in Example 6.5).

Step 3: Determine the depth of the equivalent stress block a, assuming rectangular section behavior. The positive factored bending moment in the end span is equal to 42.2 ft kips (see Table 3.4). Assuming $d = 20.5 - 2.5 = 18.0$ in, a is determined from the following equation:

$$a = \frac{a_1 d - \sqrt{(a_1 d)^2 - (2a_1 M_u / \phi)}}{a_1}$$

$$= \frac{(204.0 \times 18.0) - \sqrt{(204.0 \times 18.0)^2 - (2 \times 204.0 \times 42.2 \times 12/0.9)}}{204.0} = 0.15\ \text{in}$$

where $a_1 = 0.85 f_c' b_e = 0.85 \times 4 \times 60 = 204.0$ kips/in.

Because $a < h_f = 4.5$ in, the section behaves as a rectangular section, and the flowchart shown in Fig. 6.4 can be used to determine A_s with $b = b_e$.

Step 4: Determine the nominal strength coefficient of resistance R_n.

$$R_n = \frac{M_u}{\phi b_e d^2} = \frac{42.2 \times 12,000}{0.9 \times 60 \times 18.0^2} = 28.9 \text{ psi}$$

Step 5: Determine the required reinforcement ratio ρ. The reinforcement ratio ρ is determined by Eq. (6.7):

$$\rho = \frac{0.85 f_c'}{f_y}\left[1 - \sqrt{1 - \frac{2R_n}{0.85 f_c'}}\right] = \frac{0.85 \times 4}{60}\left[1 - \sqrt{1 - \frac{2 \times 28.9}{0.85 \times 4,000}}\right] = 0.0005$$

Step 6: Determine the required area of tension reinforcement A_s.

$$A_s = \rho b d = 0.0005 \times 60 \times 18.0 = 0.54 \text{ in}^2$$

Step 7: Determine the minimum required area of reinforcement $A_{s,min}$. The minimum amount of reinforcement is determined by ACI 10.5.1. Because the compressive strength of the concrete is less than 4,400 psi, the minimum amount is determined by the lower limit given in that section:

$$A_{s,min} = \frac{200 b_w d}{f_y} = \frac{200 \times 7 \times 18.0}{60,000} = 0.42 \text{ in}^2 < 0.54 \text{ in}^2$$

Use $A_s = 0.54 \text{ in}^2$.

Step 8: Determine the depth of the equivalent rectangular stress block a.

$$a = \frac{A_s f_y}{0.85 f_c' b_e} = \frac{0.54 \times 60,000}{0.85 \times 4,000 \times 60} = 0.16 \text{ in}$$

Step 9: Determine β_1. According to ACI 10.2.73, $\beta_1 = 0.85$ for $f_c' = 4,000$ psi (see Section 5.2).

Step 10: Determine the neutral axis depth c.

$$c = \frac{a}{\beta_1} = \frac{0.16}{0.85} = 0.19 \text{ in}$$

Step 11: Determine ε_t.

$$\varepsilon_t = 0.0030\left(\frac{d_t}{c} - 1\right) = 0.0030\left(\frac{18.0}{0.19} - 1\right) = 0.2812 > 0.0040$$

Because $\varepsilon_t > 0.0040$, the maximum reinforcement requirement of ACI 10.3.5 is satisfied. Also, the section is tension-controlled because $\varepsilon_t > 0.0050$ (ACI 10.3.4), and the initial assumption that the section is tension-controlled is correct.

Step 12: Choose the size and spacing of the reinforcing bars. The required area of reinforcement is 0.54 in^2. Try two No. 5 bars (provided $A_s = 0.62$ in^2).

Bar spacing is governed by the crack control requirements of ACI 10.6.4 and the spacing limits of ACI 7.6 and 3.3.2. It is typically assumed that stirrups are not used in narrow members such as wide-module joists; therefore, $c_c = 1.5$ in. The maximum bar spacing is determined by Eq. (6.23):

$$s = 15\left(\frac{40,000}{f_s}\right) - 2.5c_c = 15\left(\frac{40,000}{2 \times 60,000/3}\right) - (2.5 \times 1.5) = 11.3 \text{ in (governs)}$$

$$\leq 12\left(\frac{40,000}{f_s}\right) = 12\left(\frac{40,000}{2 \times 60,000/3}\right) = 12.0 \text{ in}$$

FIGURE 6.12 Details of the positive reinforcement in the wide-module joist given in Example 6.6.

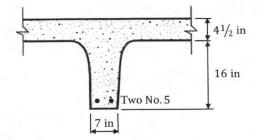

$4\frac{1}{2}$ in

16 in

Two No. 5

7 in

Check if two No. 5 bars can fit within the 7-in-wide section without violating the spacing limits of ACI 7.6 and 3.3.2.

From Fig. 6.10,

$$\text{Minimum clear space} = \begin{cases} 1.0 \text{ in} \\ d_b = 0.625 \text{ in} \\ 1.33 \text{ (maximum aggregate size)} = 1.33 \times 0.75 = 1.0 \text{ in} \end{cases}$$

Therefore, the minimum clear space is 1.0 in.

The provided clear space $= 7 - 2(1.5 + 0.625) = 2.75 > 1.0$ in.

Also, the provided bar spacing $2.75 + 0.625 = 3.375$ in is less than the maximum bar spacing for crack control, which is equal to 12 in.

Use two No. 5 bars.

A section of the joist near the midspan is shown in Fig. 6.12.

Comments

The actual effective depth d is equal to $20.5 - 1.5 - (0.625/2) = 18.7$ in. Using the actual d in the calculations will provide a design strength that is greater than that based on the assumed value of 18.0 in.

The positive design strength ϕM_n for the wide-module joist reinforced with two No. 5 bars is determined by Eq. (5.11):

$$\phi M_n = \phi A_s f_y \left(d - \frac{0.59 A_s f_y}{b_e f_c'} \right)$$

$$= 0.9 \times (2 \times 0.31) \times 60 \left(18.7 - \frac{0.59 \times 2 \times 0.31 \times 60}{60 \times 4} \right) / 12$$

$$= 51.9 \text{ ft kips} > M_u = 42.2 \text{ ft kips}$$

Example 6.7 Determine the required negative reinforcement at the exterior face of the first interior support for the beam along column line B of Examples 3.3 and 6.3 (see Fig. 3.3). Assume $f_c' = 4,000$ psi, a $3/4$-in maximum aggregate, and Grade 60 reinforcement.

Solution The height and the width of the beams are 20.5 and 22.0 in, respectively, which were determined in Example 6.3. For a negative bending moment, the reinforcement is located within the top portion of the beam. The flowchart shown in Fig. 6.4 is utilized to determine A_s.

Step 1: **Assume tension-controlled section.** Sections of flexural members should be tension-controlled whenever possible. Thus, assume that the strength reduction factor $\phi = 0.9$.

Step 2: **Determine the nominal strength coefficient of resistance R_n.** For a rectangular section, R_n is determined by Eq. (6.5), which is a function of the factored bending moment M_u. The negative factored bending moment at the exterior face of the first interior support in the end span is equal to 198.1 ft kips (see Table 3.3). Also assume that $d = h - 2.5 = 20.5 - 2.5 = 18.0$ in.

$$R_n = \frac{M_u}{\phi b_w d^2} = \frac{198.1 \times 12,000}{0.9 \times 22.0 \times 18.0^2} = 370.6 \text{ psi}$$

Step 3: Determine the required reinforcement ratio ρ. The reinforcement ratio ρ is determined by Eq. (6.7):

$$\rho = \frac{0.85 f_c'}{f_y} \left[1 - \sqrt{1 - \frac{2R_n}{0.85 f_c'}} \right] = \frac{0.85 \times 4}{60} \left[1 - \sqrt{1 - \frac{2 \times 370.6}{0.85 \times 4,000}} \right] = 0.0066$$

Step 4: Determine the required area of tension reinforcement A_s.

$$A_s = \rho b d = 0.0066 \times 22.0 \times 18.0 = 2.61 \text{ in}^2$$

Step 5: Determine the minimum required area of reinforcement $A_{s,min}$. The minimum amount of reinforcement is determined by ACI 10.5.1. Because the compressive strength of the concrete is less than 4,400 psi, the minimum amount is determined by the lower limit given in that section:

$$A_{s,min} = \frac{200 b_w d}{f_y} = \frac{200 \times 22.0 \times 18.0}{60,000} = 1.32 \text{ in}^2 < 2.61 \text{ in}^2$$

Use $A_s = 2.61$ in^2.

Step 6: Determine the depth of the equivalent rectangular stress block a.

$$a = \frac{A_s f_y}{0.85 f_c' b} = \frac{2.61 \times 60,000}{0.85 \times 4,000 \times 22.0} = 2.1 \text{ in}$$

Step 7: Determine β_1. According to ACI 10.2.73, $\beta_1 = 0.85$ for $f_c' = 4,000$ psi (see Section 5.2).

Step 8: Determine the neutral axis depth c.

$$c = \frac{a}{\beta_1} = \frac{2.1}{0.85} = 2.5 \text{ in}$$

Step 9: Determine ε_t.

$$\varepsilon_t = 0.0030 \left(\frac{d_t}{c} - 1 \right) = 0.0030 \left(\frac{18.0}{2.5} - 1 \right) = 0.0186 > 0.0040$$

Because $\varepsilon_t > 0.0040$, the maximum reinforcement requirement of ACI 10.3.5 is satisfied. Also, the section is tension-controlled because $\varepsilon_t > 0.0050$ (ACI 10.3.4), and the initial assumption that the section is tension-controlled is correct.

Step 10: Choose the size and spacing of the reinforcing bars. The required area of reinforcement is 2.61 in^2. Try six No. 6 bars (provided $A_s = 2.64$ in^2).

Bar spacing is governed by the crack control requirements of ACI 10.6.4 and the spacing limits of ACI 7.6 and 3.3.2. From Table 6.2, three No. 6 bars are required in a single layer for crack control for a 22-in-wide beam. From Table 6.3, 10 No. 6 bars can fit within a 22-in-wide beam without violating the spacing limits of ACI 7.6 and 3.3.2. Providing six No. 6 bars automatically satisfies the requirements of ACI 10.6.4, 7.6, and 3.3.2.

Use six No. 6 bars.

As an exercise, assume that a No. 3 stirrup ($d_s = 0.375$ in) is provided in the beam, and determine the minimum and maximum numbers of No. 6 bars that can be provided in the section.

The maximum bar spacing is determined by Eq. (6.23) where $c_c = 1.5 + 0.375 = 1.875$ in:

$$s = 15 \left(\frac{40,000}{f_s} \right) - 2.5c_c = 15 \left(\frac{40,000}{2 \times 60,000/3} \right) - (2.5 \times 1.875) = 10.3 \text{ in (governs)}$$

$$\leq 12 \left(\frac{40,000}{f_s} \right) = 12 \left(\frac{40,000}{2 \times 60,000/3} \right) = 12.0 \text{ in}$$

The minimum number of bars n_{min} required in a single layer to control cracking is determined by Eq. (6.24):

$$n_{min} = \frac{b_w - 2(c_c + 0.5d_b)}{s} + 1$$

$$= \frac{22 - 2[1.875 + (0.5 \times 0.75)]}{10.3} + 1 = 2.7$$

Thus, the minimum number of No. 6 bars required in the section is three.
From Fig. 6.10,

$$\text{Minimum clear space} = \begin{cases} 1.0 \text{ in} \\ d_b = 0.75 \text{ in} \\ 1.33 \text{ (maximum aggregate size)} = 1.33 \times 0.75 = 1.0 \text{ in} \end{cases}$$

Therefore, the minimum clear space is 1.0 in.
The maximum number of bars n_{max} that can fit in a single layer on the basis of the spacing limits of ACI 7.6 and 3.3.2 is determined by Eq. (6.25):

$$n_{max} = \frac{b_w - 2(c_s + d_s + r)}{(\text{clear space}) + d_b} + 1$$

$$= \frac{22 - 2[1.5 + 0.375 + (2 \times 0.375)]}{1.0 + 0.75} + 1 = 10.6$$

Thus, the maximum number of No. 6 bars permitted in the section is 10.
Figure 6.13 shows the reinforcement at the face of the support.

Example 6.8 Because of architectural constraints, a beam is limited to a width of 12 in and an overall depth of 24 in. Determine the required reinforcement for a positive factored bending moment of 370 ft kips. Assume $f_c' = 4,000$ psi, a $3/4$-in maximum aggregate, and Grade 60 reinforcement.

Solution Use the flowchart shown in Fig. 6.4 to determine the required flexural reinforcement.

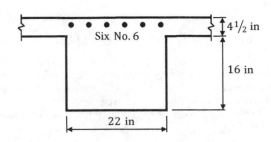

FIGURE 6.13 Details of the negative reinforcement in the beam given in Example 6.7.

Six No. 6

$4\frac{1}{2}$ in

16 in

22 in

Step 1: Assume tension-controlled section. Sections of flexural members should be tension-controlled whenever possible. Thus, assume that the strength reduction factor $\phi = 0.9$.

Step 2: Determine the nominal strength coefficient of resistance R_n. For a rectangular section, R_n is determined by Eq. (6.5), which is a function of the factored bending moment M_u. The positive factored bending moment is given as 370.0 ft kips. Because the beam is relatively narrow, assume that the tension reinforcement will be placed in two layers where $d_t = h - 2.5 = 21.5$ in and $d = h - 3.5 = 20.5$ in.

$$R_n = \frac{M_u}{\phi b_w d^2} = \frac{370.0 \times 12,000}{0.9 \times 12.0 \times 20.5^2} = 978.3 \text{ psi}$$

Step 3: Determine the required reinforcement ratio ρ. The reinforcement ratio ρ is determined by Eq. (6.7):

$$\rho = \frac{0.85 f'_c}{f_y}\left[1 - \sqrt{1 - \frac{2R_n}{0.85 f'_c}}\right] = \frac{0.85 \times 4}{60}\left[1 - \sqrt{1 - \frac{2 \times 978.3}{0.85 \times 4,000}}\right] = 0.0197$$

Step 4: Determine the required area of tension reinforcement A_s.

$$A_s = \rho bd = 0.0197 \times 12.0 \times 20.5 = 4.85 \text{ in}^2$$

Step 5: Determine the minimum required area of reinforcement $A_{s,min}$. The minimum amount of reinforcement is determined by ACI 10.5.1. Because the compressive strength of the concrete is less than 4,400 psi, the minimum amount is determined by the lower limit given in that section:

$$A_{s,min} = \frac{200 b_w d}{f_y} = \frac{200 \times 12.0 \times 20.5}{60,000} = 0.82 \text{ in}^2 < 4.85 \text{ in}^2$$

Use $A_s = 4.85 \text{ in}^2$.

Step 6: Determine the depth of the equivalent rectangular stress block a.

$$a = \frac{A_s f_y}{0.85 f'_c b} = \frac{4.85 \times 60,000}{0.85 \times 4,000 \times 12.0} = 7.1 \text{ in}$$

Step 7: Determine β_1. According to ACI 10.2.73, $\beta_1 = 0.85$ for $f'_c = 4,000$ psi (see Section 5.2).

Step 8: Determine the neutral axis depth c.

$$c = \frac{a}{\beta_1} = \frac{7.1}{0.85} = 8.4 \text{ in}$$

Step 9: Determine ε_t.

$$\varepsilon_t = 0.0030\left(\frac{d_t}{c} - 1\right) = 0.0030\left(\frac{21.5}{8.4} - 1\right) = 0.0047 > 0.0040$$

Because $\varepsilon_t > 0.0040$, the maximum reinforcement requirement of ACI 10.3.5 is satisfied. However, the section is not tension-controlled because $\varepsilon_t < 0.0050$ (ACI 10.3.4), and the initial assumption that the section is tension-controlled is not correct.

Thus, add compression reinforcement to change the section to tension-controlled. Use the flowchart shown in Fig. 6.5.

Note that Fig. 6.5 could also have been used to determine if the section was tension-controlled or not. In particular, calculate R_{nt} and compare it with R_n:

$$\rho_t = \frac{0.319\beta_1 f_c'}{f_y} = \frac{0.319 \times 0.85 \times 4}{60} = 0.0181$$

$$R_{nt} = \rho_t f_y \left(1 - \frac{0.59\rho_t f_y}{f_c'}\right)$$

$$= 0.0181 \times 60{,}000 \left(1 - \frac{0.59 \times 0.0181 \times 60}{4}\right)$$

$$= 912.0 \text{ psi} < R_n = 978.3 \text{ psi}$$

Because $R_n > R_{nt}$ the section is not tension-controlled.

Step 10: Determine the reinforcement ratio ρ. The reinforcement ratio ρ is determined from the following equation:

$$\rho = \rho_t \left(\frac{d_t}{d}\right) = 0.0181 \left(\frac{21.5}{20.5}\right) = 0.0190$$

Step 11: Determine M_{nt}. The nominal flexural strength M_{nt} that corresponds to $\varepsilon_t = 0.0050$ is determined by Eq. (6.12):

$$M_{nt} = \rho f_y \left(1 - \frac{0.59\rho f_y}{f_c'}\right) bd^2$$

$$= 0.0190 \times 60 \left(1 - \frac{0.59 \times 0.0190 \times 60}{4}\right) \times 12 \times 20.5^2/12 = 398.5 \text{ ft kips}$$

Step 12: Determine M_n'. The required nominal flexural strength M_n' that needs to be resisted by the compression reinforcement is determined by Eq. (6.13):

$$M_n' = \frac{M_u}{\phi} - M_{nt} = \frac{370}{0.9} - 398.5 = 12.6 \text{ ft kips}$$

Step 13: Determine c_t.

$$c_t = 0.375 d_t = 0.375 \times 21.5 = 8.1 \text{ in}$$

Step 14: Determine f_s'. Assuming that the distance d' from the extreme compression fiber to the centroid of the compression reinforcement is equal to $1.5 + 0.5 + 0.5 = 2.5$ in, the stress in the compression reinforcement is determined by Eq. (6.14):

$$f_s' = 0.0030 E_s \left(1 - \frac{d'}{c_t}\right) = 0.0030 \times 29{,}000 \left(1 - \frac{2.5}{8.1}\right) = 60.1 \text{ ksi} > f_y = 60 \text{ ksi}$$

Thus, the compression reinforcement yields ($f_s' = f_y = 60$ ksi).

Step 15: Determine A_s'. The required area of compression reinforcement A_s' is determined by Eq. (6.15):

$$A_s' = \frac{M_n'}{f_s'(d - d')} = \frac{12.6 \times 12}{60(20.5 - 2.5)} = 0.14 \text{ in}^2$$

FIGURE **6.14** Details of the tension and compression reinforcement in the beam given in Example 6.8.

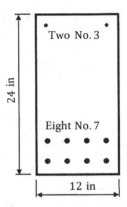

Step 16: Determine A_s. The total required area of tension reinforcement A_s is the summation of the area of reinforcement corresponding to the tension-controlled net strain limit and A_s' from Eq. (6.15) with $f_s' = f_y$ [see Eq. (6.16)]:

$$A_s = \rho b d + \frac{M_n'}{f_y(d - d')}$$
$$= (0.0190 \times 12 \times 20.5) + 0.14$$
$$= 4.67 + 0.14 = 4.81 \text{ in}^2$$

Step 17: Choose the size and spacing of reinforcing bars. The required area of tension reinforcement is 4.81 in^2. Try eight No. 7 bars in two layers (provided $A_s = 4.80 \text{ in}^2 \approx 4.81 \text{ in}^2$).

Bar spacing is governed by the crack control requirements of ACI 10.6.4 and the spacing limits of ACI 7.6 and 3.3.2. From Table 6.2, two No. 7 bars are required in a single layer for crack control for a 12-in-wide beam. From Table 6.3, four No. 7 bars can fit within a 12-in-wide beam without violating the spacing limits of ACI 7.6 and 3.3.2. Providing four No. 7 bars in a layer automatically satisfies the requirements of ACI 10.6.4, 7.6, and 3.3.2.

For the tension reinforcement, use eight No. 7 bars in two layers.

Similarly, it can be shown that two No. 3 bars (provided $A_s = 0.22 \text{ in}^2 > 0.14 \text{ in}^2$) are adequate for strength and spacing requirements.

For the compression reinforcement, use two No. 3 bars.

Figure 6.14 shows the reinforcement at the location of positive moment.

Development of Flexural Reinforcement

Overview   Flexural reinforcement must be properly developed or anchored in a concrete flexural member in order for the member to perform as intended in accordance with the strength design method. Prior to the 1971 Code, the concepts of flexural bond and anchorage bond were used to determine the required lengths of flexural reinforcing bars. In particular, to prevent bond failure or splitting, the calculated tension force in any bar at any section was required to be developed on each side of that section by proper embedment length, end anchorage, or hooks.

The development length concept was first introduced in the 1971 Code and is based on an average bond stress over the embedment length of the reinforcement.[9] Consider the reinforcing bar embedded near the bottom of the concrete beam illustrated in Fig. 6.15. The bond stress between the concrete and the steel varies along the length of the member, and its magnitude depends on the stress in the bar and the presence of cracks.

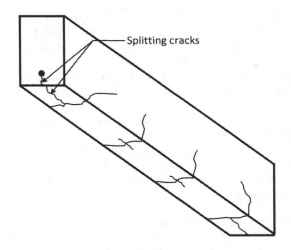

FIGURE 6.15
Splitting of concrete along a single bar of flexural reinforcement.

When subjected to high stresses due to external bending moments, the relatively thin concrete that is restraining the bar can split as shown. This splitting is primarily due to the ribs of the deformed bars bearing against the concrete. If the splitting extends to the end of a bar that is not properly anchored, complete bond failure occurs, and the beam can no longer resist the applied bending moment as originally designed. The bar must be long enough (or properly anchored) so that this type of failure does not occur.

For the more common situation where more than one reinforcing bar is present in the section, horizontal splitting along the plane of the bars can also occur (see Fig. 6.16). This type of splitting commonly begins at a diagonal crack due to flexure and/or shear. The required development length of the bars in this case would generally be greater than that for the single bar shown in Fig. 6.15.

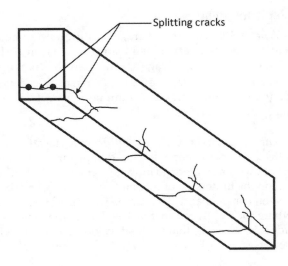

FIGURE 6.16
Splitting of concrete along multiple bars of flexural reinforcement.

The concept of development length can be simply stated as follows: Minimum lengths or extensions of reinforcement must be provided beyond the locations of peak stress in the reinforcement in order to fully develop the bars. These locations of peak stress are often referred to as critical sections, and they occur at points of maximum stress (maximum bending moment) and at locations where adjacent reinforcement is terminated. Development length or anchorage of reinforcement is required on both sides of a critical section.

In addition to embedment length, development or anchorage can be achieved with hooks, headed deformed bars, mechanical devices, or a combination thereof. The following discussion focuses on the development of flexural reinforcement where deformed bars or deformed wire are in tension. The Code provisions for the development length of reinforcing bars in tension must be examined prior to the general requirements for flexural reinforcement.

Development of Deformed Bars and Deformed Wire in Tension   ACI 12.2 contains provisions for the development of deformed bars and deformed wire in tension. The development length ℓ_d is determined using the provisions of ACI 12.2.2 or 12.2.3 along with the modification factors of ACI 12.2.4 and 12.2.5. Because the requirements of ACI 12.2.2 are based on the requirements of ACI 12.2.3, the latter requirements are covered first.

Method 1—ACI 12.2.3   The development length ℓ_d is given by ACI Eq. (12-1):

$$\ell_d = \left[\frac{3}{40} \frac{f_y}{\lambda \sqrt{f_c'}} \frac{\psi_t \psi_e \psi_s}{(c_b + K_{tr})/d_b} \right] d_b \geq 12 \text{ in} \tag{6.26}$$

The terms that make up this equation are examined next.

- *Modification factor for lightweight concrete* λ: The factor λ reflects the lower tensile strength of lightweight aggregate concrete. A lower tensile strength results in a reduction of splitting resistance. To account for this, the Code stipulates the following:

  $\lambda = 0.75$ for lightweight concrete.

  $\lambda = f_{ct}/6.7\sqrt{f_c'} \leq 1.0$, where the average splitting tensile strength of lightweight concrete f_{ct} has been determined by tests [Eq. (1) in ASTM C496/C496M]. Note that $6.7\sqrt{f_c'}$ is the average splitting tensile strength of normal-weight concrete. Also, $\sqrt{f_c'}$ must not exceed 100 psi in accordance with ACI 12.1.2; this requirement must be satisfied throughout ACI Chap. 12.

  $\lambda = 1.0$ for normal-weight concrete.

- *Reinforcement location factor* ψ_t: The reinforcement location factor ψ_t reflects the adverse effects that can occur to the top reinforcement in a member. During the placement of concrete, water and mortar migrate vertically through the member and collect on the underside of the reinforcing bars. It has been shown that the bond between concrete and steel can be weakened because of the presence of mortar where the depth of concrete cast below the bars exceeds 12 in.[10,11] To account for this, $\psi_t = 1.3$; that is, the development length is increased by 30%. In all other cases, $\psi_t = 1.0$.

- *Reinforcement coating factor ψ_e:* The reinforcement coating factor ψ_e accounts for the reduced bond strength between the concrete and the epoxy-coated reinforcing bars: The coating prevents adhesion and friction between the bar and the concrete.[12–14]

    In cases where the cover to the epoxy-coated bars is small ($<3d_b$) or where the clear spacing between the bars is small ($<6d_b$), splitting failure can occur, and the anchorage or bond strength is substantially reduced. Thus, in these situations, $\psi_e = 1.5$.

    Where the cover or clear spacing is greater than these limits, splitting failure is avoided and $\psi_e = 1.2$; this accounts for the reduced bond strength due to the epoxy coating.

    For uncoated and zinc-coated (galvanized) bars, $\psi_e = 1.0$.

    Note that the Code stipulates that the product of $\psi_t \psi_e$ need not be greater than 1.7. This limit takes into consideration that the bond of epoxy-coated bars is already reduced because of the loss of adhesion between the bars and the concrete.

- *Reinforcement size factor ψ_s:* The reinforcement size factor ψ_s reflects the more favorable performance of smaller-diameter reinforcement. For No. 6 and smaller deformed bars and wires, $\psi_s = 0.8$, whereas for No. 7 and larger bars, $\psi_s = 1.0$.

- *Spacing or cover dimension c_b:* The spacing or cover dimension c_b is defined as the smaller of (1) the distance from the center of the bar or wire being developed to the nearest concrete surface and (2) one-half the center-to-center spacing of the bars or wires being developed. These criteria are illustrated in Fig. 6.17. As discussed previously, small cover or clear spacing increases the likelihood that splitting failure can occur. Thus, an increase in the development length is warranted in such cases.

- *Transverse reinforcement index K_{tr}:* The transverse reinforcement index K_{tr} represents the role of confining reinforcement across potential splitting planes: Larger amounts of confining reinforcement reduce the potential for splitting failure, thus, reducing the overall required development length of the reinforcement. For beams, confining reinforcement is typically made up of stirrups that are required to resist shear forces (see Section 6.3).

FIGURE 6.17
Spacing or cover dimension c_b.

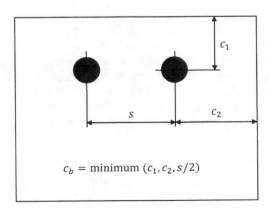

$$c_b = \text{minimum} \ (c_1, c_2, s/2)$$

The index K_{tr} is determined by ACI Eq. (12-2):

$$K_{tr} = \frac{40 A_{tr}}{sn} \qquad (6.27)$$

where A_{tr} = total cross-sectional area of all confining (transverse) reinforcement within a spacing of s that crosses the potential plane of splitting through the reinforcement being developed
  s = center-to-center spacing of the confining reinforcement
  n = number of bars or wires being developed across the plane of splitting

Because the presence of confining reinforcement has the potential to decrease development length, it is conservative to take $K_{tr} = 0$.

The Code prescribes that the confining term $(c_b + K_{tr})/d_b$ used in Eq. (12-1) must be equal to or less than 2.5. It has been shown that when this term is less than 2.5, splitting failures are likely to occur. A pullout failure of the reinforcement is more likely when this term is greater than 2.5, so an increase in the anchorage capacity due to an increase in cover or amount of confining reinforcement is not likely.

ACI 12.2.5 permits the development length ℓ_d to be reduced in cases where the provided flexural reinforcement is greater than that required from analysis. In such cases, the tensile stress in the bars being developed is less than the yield stress f_y. The reduction factor that can be applied to ℓ_d is equal to the required area of flexural reinforcement divided by the provided area of reinforcement. This reduction factor must not be used where development for f_y is required (ACI R12.2.5 provides some examples of this). This reduction factor is also not permitted in the design of members resisting the effects from seismic forces. Obviously, ignoring this reduction factor will lead to a conservative value of ℓ_d.

The flowchart shown in Fig. 6.18 can be used to determine the development length ℓ_d of deformed bars or wires in tension in accordance with ACI 12.2.3.

Method 2—ACI 12.2.2   The method given in ACI 12.2.2 to determine the development length ℓ_d is based on the requirements given in ACI 12.2.3 and the preselected values of the confining term $(c_b + K_{tr})/d_b$. Two cases are presented in ACI 12.2.2. In the first case, the spacing, cover, and confinement of the bars being developed meet one set of the conditions illustrated in Fig. 6.19. These conditions are presumed to occur frequently in practical construction cases. When these conditions are met, it is assumed that $(c_b + K_{tr})/d_b = 1.5$. Substituting this confining term into Eq. (6.26) results in the following:

$$\ell_d = \left(\frac{3}{40} \frac{f_y}{\lambda \sqrt{f_c'}} \frac{\psi_t \psi_e \psi_s}{1.5} \right) d_b = \left(\frac{f_y \psi_t \psi_e \psi_s}{20 \lambda \sqrt{f_c'}} \right) d_b \qquad (6.28)$$

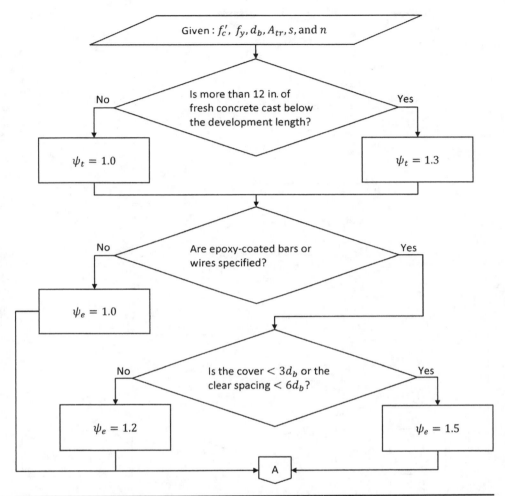

FIGURE 6.18 Development length of deformed bars or wires in tension—ACI 12.2.3. *(continued)*

For No. 7 and larger bars, $\psi_s = 1.0$, so Eq. (6.28) becomes

$$\ell_d = \left(\frac{f_y \psi_t \psi_e}{20\lambda\sqrt{f_c'}} \right) d_b \qquad (6.29)$$

Equation (6.29) is the equation given in the table under ACI 12.2.2 for No. 7 and larger bars. The equation in the table for No. 6 and smaller bars and deformed wires is obtained by multiplying Eq. (6.29) by $\psi_s = 0.8$, which results in a factor of $0.8/20 = 1/25$.

The second case is applicable where the conditions of the first case are not satisfied, and it is assumed that $(c_b + K_{tr})/d_b = 1.0$. Substituting this confining factor into Eq. (6.26) results in the equation given in the table under ACI 12.2.2 for "other cases" and

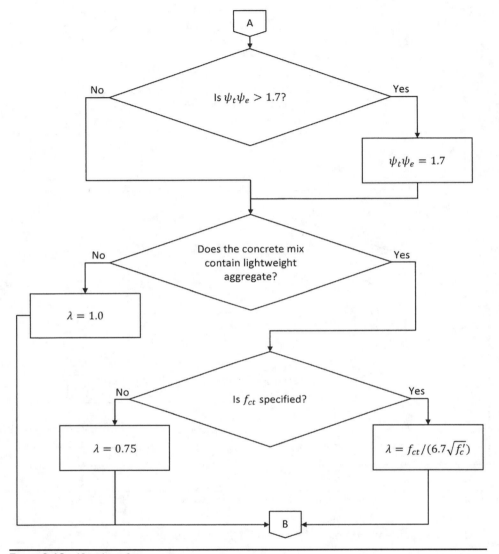

Figure 6.18 (Continued)

No. 7 and larger bars. Substitution of $\psi_s = 0.8$ into that equation results in the equation given in the table for No. 6 and smaller bars and deformed wires.

The flowchart shown in Fig. 6.20 can be used to determine the development length ℓ_d of deformed bars or wires in tension in accordance with ACI 12.2.2.

Development of Standard Hooks in Tension

Hooks are provided at the ends of reinforcing bars to provide additional anchorage where required development length cannot be attained with straight bars. Standard hooks are defined in ACI 7.1 and are illustrated in Fig. 6.21.

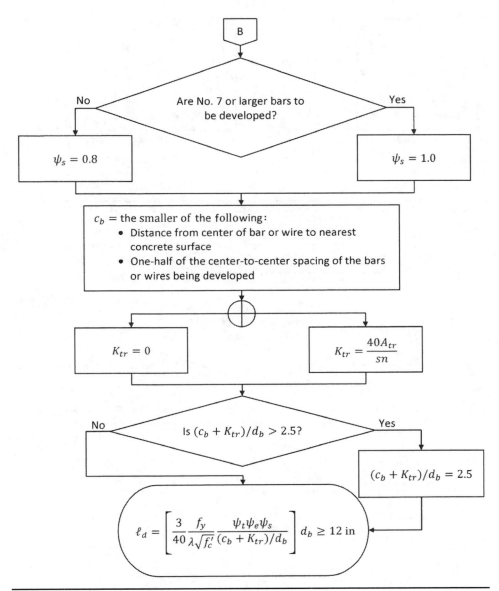

FIGURE 6.18   *(Continued)*

When subjected to tensile forces, high compressive stresses can occur on the inside of a hook. At failure, splitting of the concrete cover occurs in the plane of the hook because of these compressive stresses. Thus, the development of a hook in tension is directly proportional to the diameter of the bar because this governs the magnitude of the compressive stresses.

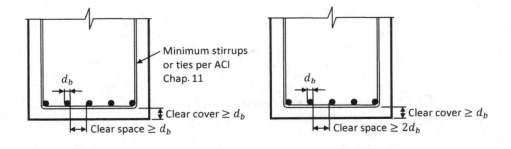

Condition 1   **Condition 2**

FIGURE **6.19** Spacing, cover, and confinement conditions of ACI 12.2.2.

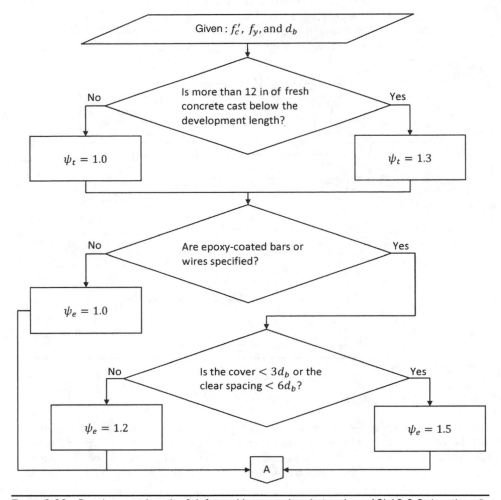

FIGURE **6.20** Development length of deformed bars or wires in tension—ACI 12.2.2. (*continued*)

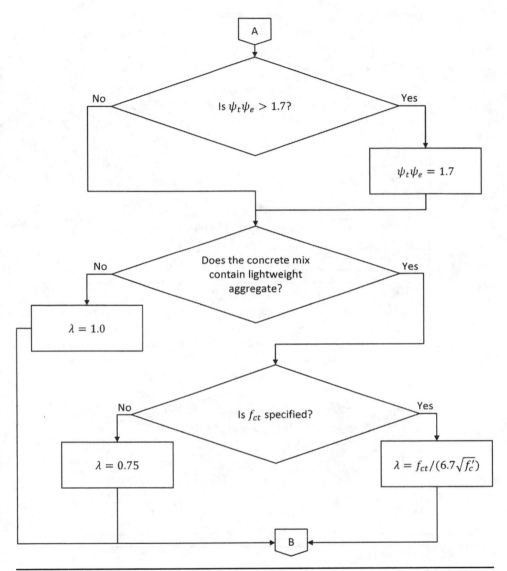

Figure 6.20   (Continued)

The development length ℓ_{dh} of a deformed reinforcing bar with a standard hook is given in ACI 12.5.2:

$$\ell_{dh} = \left(\frac{0.02\psi_e f_y}{\lambda\sqrt{f_c'}} \right) d_b \geq \text{larger of } 8d_b \text{ and } 6 \text{ in} \qquad (6.30)$$

In this equation, $\psi_e = 1.2$ for epoxy-coated bars and $\lambda = 0.75$ for lightweight concrete. In all other cases, ψ_e and λ are taken as 1.0.

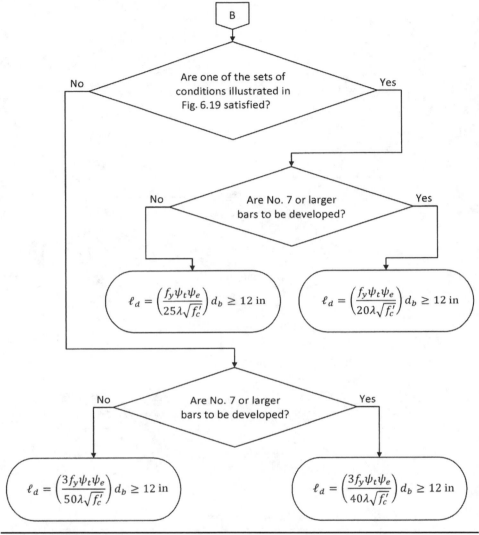

FIGURE 6.20 (*Continued*)

ACI Fig. R12.5 illustrates the details for development of bars with standard 90-degree and 180-degree hooks. The development length ℓ_{dh} is measured from the critical section to the outside end of the hook.

The values of ℓ_{dh} determined by Eq. (6.30) are permitted to be reduced by the factors given in ACI 12.5.3 wherever applicable. Reduction factors for increased cover, transverse ties or stirrups, and excess reinforcement are provided and may be applied where the conditions outlined in ACI 12.5.3 are met. ACI Fig. R12.5.3(a) and (b) illustrates the conditions for the use of the reduction factor given in ACI 12.5.3(b) for bars enclosed by ties or stirrups.

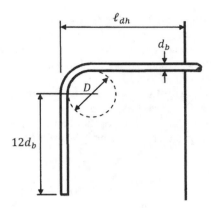

90-degree Hook

Bar No.	D
3–8	$6d_b$
9, 10, 11	$8d_b$
14, 18	$10d_b$

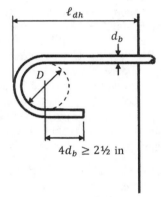

180-degree Hook

Hooked bars at discontinuous ends of members must be confined with ties or stir-rups where the side cover and the top or bottom cover are less than 2.5 in (ACI 12.5.4). The additional confinement provided by this transverse reinforcement is essential in order to prevent splitting failure. ACI Fig. R12.5.4 illustrates the details of the reinforce-ment. Note that the reduction factors given in ACI 12.5.3(b) and (c) are not applicable in this case.

Development of Headed and Mechanically Anchored Deformed Bars in Tension   ACI 12.6 contains provisions for the development of headed deformed bars and the development and anchorage of reinforcement using mechanical devices in concrete. These provisions are based on the anchorage requirements contained in Appendix D of the Code and the bearing requirements contained in ACI 10.14. A headed deformed reinforcing bar is illustrated in ACI Fig. R3.5.9.

The terms "development" and "anchorage" are distinguished in the provisions as follows: Development refers to cases where the force in a headed bar is transferred to the concrete through a combination of bearing force at the head and bond forces along the bar, whereas anchorage refers to cases where the force in a headed bar is transferred to the concrete through bearing force alone.

Headed deformed bars are permitted to be used only when the conditions of ACI 12.6.1 are satisfied. These conditions are based on tests that have been performed to establish the development length ℓ_{dt}. This development length, which is measured from the critical section to the bearing face of the head [see ACI Fig. R12.6(a)], is given in ACI 12.6.2:

$$\ell_{dt} = \left(\frac{0.016\psi_e f_y}{\sqrt{f_c'}}\right) d_b \geq \text{the larger of } 8d_b \text{ and 6 in} \tag{6.31}$$

In this equation, the compressive strength of the normal-weight concrete must be less than or equal to 6,000 psi. Also, ψ_e is equal to 1.2 for epoxy-coated bars and is equal to 1.0 in all other cases. Like in the case of straight bar development length, it is permitted to decrease ℓ_{dt} by the excess reinforcement factor.

Comparing Eq. (6.30) for hooked bars and Eq. (6.31) for headed bars, it is evident that the development length of headed bars is smaller than that for hooked bars. Staggering headed bars can avoid congestion at locations of termination, especially at beam–column joints. ACI Fig. R12.6(b) illustrates the termination of headed reinforcement at the far face of the confined core of a column.

ACI 12.6.4 permits the use of (1) headed deformed reinforcement that does not meet the requirements of ACI 3.5.9 or is not anchored in accordance with ACI 12.6.1 and 12.6.2 and (2) any other type of mechanical anchorage or device capable of developing the yield strength of the reinforcement, provided that test results that demonstrate that the system can adequately develop or anchor the bar are available.

Development of Positive and Negative Flexural Reinforcement   As noted at the beginning of this section, critical sections for development of flexural reinforcement occur at the following: (1) points of maximum stress, that is, at sections of maximum bending moment; and (2) locations where adjacent reinforcement is terminated. Development length or anchorage of reinforcement is required on both sides of a critical section.

In continuous beams and one-way slabs subjected to uniform loads, the maximum positive and negative bending moments typically occur near the midspan and at the faces of the supports, respectively. Positive and negative flexural reinforcing bars must be developed or anchored on both sides of these critical sections. The following discussion focuses on continuous members; additional requirements for simply supported members are given in ACI 12.10 and 12.11.

The required area of reinforcement at a critical section can be determined using the methods presented earlier in this chapter. Referring to Fig. 6.22, assume that the total required area of negative reinforcement for the maximum negative factored bending moment $(M_u^-)_A$ at critical section A is equal to A_s^- and that the total required area of positive reinforcement for the maximum positive factored bending moment $(M_u^+)_C$ at critical section C is equal to A_s^+. Also assume that the total number of negative reinforcing bars at section A is n and that the total number of positive reinforcing bars at section C is p.

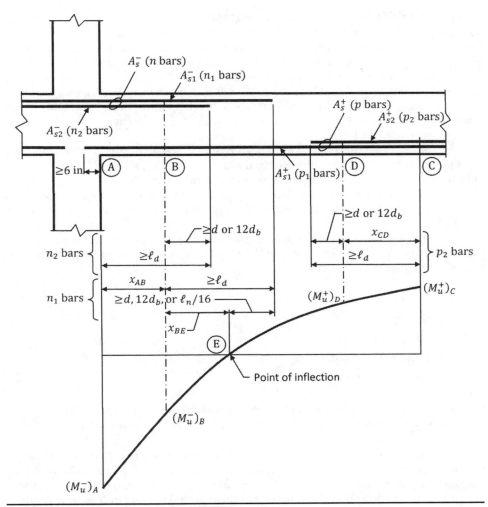

FIGURE 6.22 Development of flexural reinforcement.

It is evident that at sections away from the critical section, the required area of reinforcement decreases because the magnitude of the factored bending moment decreases. For cost savings, it is common for some of the reinforcing bars to be terminated (or cut off) at locations away from the critical sections. For example, reinforcing bars are no longer required past a point of inflection on the bending moment diagram. Also, a portion of the bars can be theoretically cut off prior to the point of inflection at a location where the continuing bars are adequate to supply the required design strength. Because a critical section occurs at a cutoff point, the bars must be properly developed at that location as well.

Negative reinforcement   Referring to Fig. 6.22, assume that a portion of the total negative reinforcement is cut off at section B. Thus, section B is a critical section. It is assumed that this reinforcement has an area equal to A_{s2}^- and consists of n_2 bars. The remaining portion of the negative reinforcement has an area equal to $A_{s1}^- = A_s^- - A_{s2}^-$ and consists of $n_1 = n - n_2$ bars that continue to the point of inflection. Requirements for the development of these two sets of reinforcing bars are discussed next.

The continuing reinforcement must be able to resist the negative factored bending moment $(M_u^-)_B$ at section B. Because section B is a critical section, n_1 bars must be adequately developed to the right of this section. In other words, these bars must extend a minimum distance of ℓ_d past section B as shown in Fig. 6.22 where ℓ_d is determined in accordance with ACI 12.2.

Also, at least one-third of the total negative reinforcement provided at a support must have an embedment length equal to the larger of d, $12d_b$, and $\ell_n/16$ past the point of inflection (ACI 12.12.3). This provision provides for possible shifting of the bending moment diagram at the point of inflection because the bending moment diagrams customarily used in design are approximate. Therefore, to satisfy the requirements of ACI 12.12.3, $n_1 \geq n/3$.

Given the requirements mentioned earlier, the minimum length of n_1 bars to the right of section A must be the larger of the lengths determined from items 1 and 2 of the following list (see Fig. 6.22):

1. $x_{AB} + \ell_d$
2. The larger of
    (a) $x_{AB} + x_{BE} + d$
    (b) $x_{AB} + x_{BE} + 12d_b$
    (c) $x_{AB} + x_{BE} + \ell_n/16$

where x_{AB} = distance from section A to the theoretical cutoff point at section B
$ x_{BE}$ = distance from section B to the point of inflection at section E

Because section A is a critical section (location of maximum negative factored bending moment), the bars that are cut off at section B must be developed to a distance that is equal to or greater than the tension development length ℓ_d beyond that section. Additionally, ACI 12.10.3 stipulates that these bars must extend beyond the point where they are no longer required by a distance equal to the larger of d and $12d_b$. The minimum length of n_2 bars to the right of section A must be the larger of the lengths determined from items 1 and 2 of the following list (see Fig. 6.22):

1. ℓ_d
2. The larger of
    (a) $x_{AB} + d$
    (b) $x_{AB} + 12d_b$

Reinforcement is not permitted to be terminated in a tension zone unless one of the conditions given in ACI 12.10.5.1, 12.10.5.2, or 12.10.5.3 is satisfied (ACI 12.10.5). Tests have shown that cutting off bars in a tension zone typically leads to reduced shear strength and loss of ductility. Also, flexural cracks tend to open early in such cases. If it is determined that the minimum length of n_2 bars based on development requirements is less than the length to the inflection point, which is equal to $x_{AB} + x_{BE}$ in Fig. 6.22, then there are essentially two options, which are to (1) increase the bar lengths past the point of inflection or (2) satisfy one or more of the following conditions:

(a) At the cutoff point, $V_u \leq 2\phi V_n/3$, where ϕV_n is the shear design strength of the section at that point (see Section 6.3 for determination of V_n).

(b) Stirrup area in excess of that required for shear and torsion is provided along each terminated bar or wire over a distance of $3d/4$ from the termination point where the excess stirrup area $\geq 60b_w s/f_{yt}$, stirrup spacing $s \leq d/(8\beta_b)$, and β_b = area of reinforcement cut off/total area of tension reinforcement at section.

(c) For No. 11 and smaller bars, continuing reinforcement provides at least double the area required for flexure at the cutoff point and $V_u \leq 3\phi V_n/4$.

The development of the negative bars to the left of section A depends on the location in the frame. At interior joints, like that depicted in Fig. 6.22, anchorage is achieved by continuing the negative reinforcement into the span that is to the left of the joint. At exterior columns, a standard hook is provided at the ends of the negative reinforcement, as depicted in ACI Fig. R12.12(a).

Positive reinforcement   Assume that a portion of the total positive reinforcement shown in Fig. 6.22 is cut off at section D. Thus, section D is a critical section. It is assumed that this reinforcement has an area equal to A_{s2}^+ and consists of p_2 bars. The remaining portion of the positive reinforcement that has an area equal to $A_{s1}^+ = A_s^+ - A_{s2}^+$ and consists of $p_1 = p - p_2$ bars must be able to resist the positive factored bending moment $(M_u^+)_D$ at section D. Requirements for the development of the positive reinforcement are discussed next.

The bars that are cut off at section D must be developed to a distance that is equal to or greater than the tension development length ℓ_d beyond the critical section at C (location of maximum factored positive bending moment), where ℓ_d is determined in accordance with ACI 12.2. Like in the case of the negative reinforcement, these bars must extend beyond the point where they are no longer required by a distance equal to the larger of d and $12d_b$ (ACI 12.10.3). Given that the distance between section C and the theoretical cutoff point located at section D is equal to x_{CD}, the minimum length of p_2 bars to the left of section C must be the larger of the lengths determined from items 1 and 2 of the following list (see Fig. 6.22):

1. ℓ_d
2. The larger of
    (a) $x_{CD} + d$
    (b) $x_{CD} + 12d_b$

The requirements of ACI 12.10.5 pertaining to reinforcement terminated in a tension zone are also applicable in this situation.

ACI 12.11.1 requires that at least one-fourth of the positive moment reinforcement in continuous members must extend along the same face of the member into the support.

FIGURE **6.23**
Maximum bar
size in
accordance with
ACI 12.11.3.

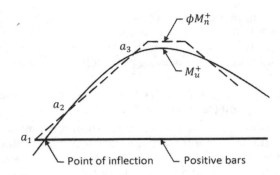

In the case of beams, this reinforcement must extend at least 6 in into the support. Therefore, $p_1 \geq p/4$.

If the bottom bars are used as compression reinforcement in conjunction with the top bars, the bottom bars must extend to the left of section A to a distance equal to the larger of (1) the tension development length ℓ_d determined in accordance with ACI 12.2.1 and (2) the compression development length determined in accordance with ACI 12.3. The following requirement must also be satisfied, if applicable.

For flexural members that are part of a seismic force–resisting system, such as moment frames, the positive reinforcement that is required by ACI 12.11.1 to extend into the supports must be anchored to develop the full yield strength of the reinforcement. During a seismic event, loads that are greater than those anticipated in design may be generated, and load reversals can occur at the supports. In such cases, the positive reinforcement at the bottom of the section must be capable of resisting the tension forces from the load reversals and, thus, must be properly anchored into the supports.

The diameter of the positive moment reinforcement is limited at points of inflection in accordance with the provisions of ACI 12.11.3. This requirement addresses the possibility that positive reinforcing bars located away from the critical section at the location of the maximum positive bending moment may not be developed.

Consider the required bending moment and design bending moment diagrams illustrated in Fig. 6.23. These diagrams are similar to those depicted in ACI Fig. R12.11.3(a) for a simply supported beam. Assume that ϕM_n^+ increases linearly from zero at the ends of the positive bars at point a_1 to a maximum value near the location of the maximum positive bending moment where $\phi M_n^+ \geq M_u^+$. It is evident from the figure that between points a_2 and a_3, $\phi M_n^+ < M_u^+$, which means that the stresses in the bars between these two points are larger than those that can be developed in the bars.

In order to circumvent this situation, the slope of the design bending moment diagram at the point of inflection (i.e., the location of zero bending moment) must be equal to or greater than the tangent of the required bending moment diagram:

$$\frac{d(\phi M_n^+)}{dx} = \frac{\phi M_n^+}{\ell_d} \geq \frac{d(M_u^+)}{dx} = V_u \qquad (6.32)$$

Thus, solving for ℓ_d,

$$\ell_d \leq \frac{\phi M_n^+}{V_u} \qquad (6.33)$$

Equation (6.33) forms the basis of ACI Eq. (12-5):

$$\ell_d \leq \frac{M_n}{V_u} + \ell_a \tag{6.34}$$

The nominal flexural strength M_n is used instead of the design flexural strength ϕM_n because it results in more conservative results. The quantity ℓ_a is equal to the larger of d and $12d_b$, which satisfies ACI 12.10.3. ACI Fig. R12.11.3(b) illustrates this case at points of inflection.

ACI Eq. (12-5) is not applicable in regions of negative bending moments because the shape of the bending moment diagram is concave downward as shown in Fig. 6.22. Thus, the critical location for bar development occurs only at the face of the support.

The preceding discussion on development lengths and bar cutoff points was developed for the most part on the basis of a single load case that produces a single bending moment diagram. In general, the diagrams for the maximum span bending moment and the maximum support bending moment should be used. These bending moment diagrams will have distinct inflection points in the span. A typical bending moment envelope is illustrated in ACI Fig. R12.10.2. This envelope is obtained, for example, by considering alternate span loading on a continuous member. Regardless of the bending moment diagram that is utilized, the basic principles for development length and cutoff points must be satisfied.

Splices of Reinforcement

Fabricators typically supply reinforcing bars with the following standard mill lengths: (1) 60 ft for No. 5 and larger bars and (2) 40 ft for No. 4 and smaller bars. Although it may be possible to acquire reinforcing bars that are longer than standard mill lengths, transporting longer bars may be problematic, because the standard length of a rail car is approximately 65 ft and the lengths of flatbed semitrailers range from 48 to 60 ft. One of the primary reasons that reinforcing bars are spliced together at the construction site has to do with transportation restrictions. Another reason has to do with handling and placing the reinforcement: It is usually more convenient to move and place shorter bars because they weigh less than longer bars.

Three types of reinforcement splices are commonly used:

- Lap splices
- Mechanical splices
- Welded splices

These types of splices are examined next for flexural reinforcement.

Lap Splices   Lap splices are frequently specified and are usually the most economical type of splice. In a lap splice, the bars are generally in contact over a specified length and are wired together. This is commonly referred to as a contact lap splice. The force in one bar is transferred to the surrounding concrete by bond, which subsequently transfers it to the adjoining bar. Splitting cracks can occur at the ends of a splice. In general, the following should be considered when specifying lap splices:

- Provide splices at locations away from maximum stress (maximum bending moment).
- Stagger the location of splices wherever possible.

The required lap splice length depends on the tension development length of the bars, the area of reinforcement provided over the length of the splice, and the percentage of reinforcement that is spliced at any one location. Because experimental data on lap splices using Nos. 14 and 18 bars are sparse, the use of tension lap splices for these bar sizes is prohibited (ACI 12.14.2.1).

Lap splices in tension are classified as Class A or Class B. The length of the lap splice is given as a multiple of the tension development length ℓ_d:

- Class A splice length $= 1.0\ell_d \geq 12$ in
- Class B splice length $= 1.3\ell_d \geq 12$ in

ACI 12.15.1 stipulates that ℓ_d be determined by the provisions of ACI 12.2; however, the 12-in minimum length specified in ACI 12.2.1 and the excess reinforcement modification factor of ACI 12.2.5 are not applicable (the splice length must be based on the full f_y because the provided area of reinforcement is accounted for in the definition of the splice classification). The effective clear spacing that is to be used in the calculation of ℓ_d is illustrated in Fig. 6.24 for a beam. The clear space that is to be used for staggered splices in one-way slabs is the minimum distance between adjacent splices [see ACI Fig. R12.15.1(b)].

The default splice classification for a lap splice is Class B. However, if both of the following two conditions are satisfied, a Class A splice is permitted:

1. The provided $A_s \geq 2$(required A_s) over entire splice length.
2. Less than or equal to $0.5A_s$ is spliced within the required lap length.

The first of these two conditions basically encourages splices to be located away from sections where the tensile stress in the bars is high. In situations without load reversals, negative reinforcing bars should be spliced near the midspan of a member, whereas positive reinforcing bars should be spliced over the supports.

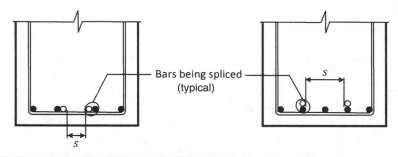

Bars being spliced
(typical)

FIGURE 6.24   Clear spacing of spliced bars in a beam.

Transverse reinforcement, such as stirrups, should be used over the length of a lap splice to improve its capacity. Tests have shown that transverse reinforcement delays or prevents the opening of splitting cracks that typically initiate at the ends of a splice. A decrease in splice length can also be realized by using a certain amount of transverse reinforcement (see ACI 12.2).

ACI 12.15.3 contains the requirements when bars of different size are lap spliced in tension. In such cases, the minimum lap splice length is the greater of the following: (1) tension development length ℓ_d of the larger bar and (2) tension lap splice length of the smaller bar.

Noncontact lap splices are permitted in flexural members. Unlike contact lap splices, the bars that are spliced in a noncontact lap splice are not in contact with each other. According to ACI 12.14.2.3, the bars must be spaced no farther than the smaller of (1) one-fifth the required lap splice length and (2) 6 in. A spacing any larger than the limiting value can result in an essentially unreinforced section of concrete between the bars. Contact lap splices are usually specified because the likelihood that the bars will displace during concrete placement is much smaller than that for noncontact splices.

Mechanical Splices   According to Ref. 15, a mechanical splice is defined as a "complete assembly of a coupler, a coupling sleeve, or an end-bearing sleeve, including any additional intervening material or other components required to accomplish the splicing of reinforcing bars." A variety of proprietary mechanical devices that can be used to splice flexural reinforcing bars are available. More information on the various systems can be found in Ref. 15.

Mechanical splices can be used in a number of situations and can be more cost-effective than lap splices under a number of conditions, including the following:

1. *When long lap splices are needed.* Long lap splices are commonly required when using No. 9 and larger bars and when using epoxy-coated bars.
2. *When lap splices cause reinforcement congestion.* This can occur at beam–column joints and other locations where bars are spliced in close proximity to each other.
3. *Where spacing of the flexural reinforcement is insufficient to permit lap splices.* This can happen in beams with relatively large reinforcement ratios and larger bar sizes.

Unlike lap splices, splitting failures are not a concern when utilizing mechanical splices because mechanical splices do not rely on the surrounding concrete to transfer the tensile force from one bar to the other. As such, the compressive strength of the concrete and the cover to the splice do not affect the strength of a mechanical splice.

ACI 12.14.3.2 requires that mechanical splices develop in tension or compression 125% of the specified yield strength of the bar. This is to ensure that some yielding occurs in the reinforcing bar adjacent to the mechanical splice prior to the failure of the splice. However, this requirement can be waived for No. 5 and smaller bars used in splices that meet the provisions of ACI 12.15.5 (ACI 12.14.3.5).

Welded Splices   The Code permits the use of welded splices for flexural reinforcement. The welding must conform to the provisions of Ref. 16, which cover aspects of welding reinforcing bars and criteria to qualify welding procedures. Because ASTM A615, A616, and A617 do not contain limits on the chemical elements that affect the weldability of the steel, these specifications must be supplemented to require a report of material properties necessary to conform to the requirements of Ref. 16 (ACI 3.5.2). ASTM A706 steel is intended for welding, and such supplements on material properties are not needed. Additional information on welded splices can be found in Ref. 17.

Like mechanical splices, a full welded splice, which is generally intended for No. 6 and larger bars, must be able to develop 125% of the specified yield strength of the bar. The exception given in ACI 12.14.3.5 is applicable in this case as well.

Structural Integrity Reinforcement

Although the probability of occurrence is generally low, it is possible that a structure can be subjected to extraordinary events during its lifetime. These events can arise from service or environmental conditions that are not considered explicitly in the design of ordinary buildings or structures. Examples of such events are explosions, vehicular impact, misuse by occupants, and tornadoes. The loads generated by these events are usually of short duration, but they can lead to damage or failure.

In an attempt to limit damage to relatively small areas, the overall integrity of a reinforced concrete structure can be substantially enhanced by relatively minor changes in reinforcement detailing. Thus, in addition to the requirements pertaining to detailing of flexural reinforcement that are given in Chap. 12 of the Code, the structural integrity requirements of ACI 7.13 must also be satisfied. The main purpose of these requirements is to improve the redundancy and ductility of reinforced concrete structures.

Table 6.4 contains a summary of the structural integrity requirements. A structure is essentially tied around its perimeter by requiring that a portion of the negative and positive reinforcements be continuous in perimeter or spandrel beams. The continuous reinforcement in these beams must be enclosed by closed stirrups or ties that satisfy the torsional detailing requirements of ACI 11.5.4.1. The transverse reinforcement must be anchored around the longitudinal bars using a 135-degree standard hook or a seismic hook, which is defined in ACI 2.2. A two-piece stirrup that satisfies the requirements of ACI 7.13.2.3 is depicted in ACI Fig. R7.13.2. A 90-degree hook is permitted where an adjoining slab or flange can prevent spalling, as shown in the figure. Note that pairs of U-stirrups that lap one another in accordance with ACI 12.13.5 are not permitted in perimeter beams because such stirrups usually cannot prevent themselves and the top reinforcement from tearing out of the concrete in the event that damage occurs to the side concrete cover.

Recommended Flexural Reinforcement Details

Recommended flexural reinforcement details for beams and one-way slabs are given in Figs. 6.25 and 6.26, respectively. The bar lengths in the figures are based on members subjected to uniformly distributed gravity loads. Adequate bar lengths must be determined by calculation for members subjected to the effects from other types of gravity loads and lateral loads. The bar lengths in these figures can also be used for members that have been designed using the approximate bending moment coefficients given in ACI 8.3.3 (see Section 3.3).

Additional information on reinforcement detailing can be found in Refs. 18 and 19.

Flexural Member	Requirements	ACI Section Number
Joists (defined in ACI 8.13.1 through 8.13.3)	At least one bottom bar shall be continuous or shall be spliced with a Class B tension splice or a mechanical or welded splice in accordance with ACI 12.14.3.	7.13.2.1
	At noncontinuous supports, the bottom bars shall be anchored to develop f_y at the face of the support, using a standard hook in accordance with ACI 12.5 or a headed deformed bar in accordance with ACI 12.6.	
Perimeter beams	At least one-sixth of the negative reinforcement required at the support, but not less than two bars, must be continuous over the span length and must pass through the region bounded by the longitudinal reinforcement of the column.	7.13.2.2(a)
	At least one-quarter of the positive reinforcement required at the midspan, but not less than two bars, must be continuous over the span length and must pass through the region bounded by the longitudinal reinforcement of the column.	7.13.2.2(b)
	At noncontinuous supports, the reinforcement shall be anchored to develop f_y at the face of the support, using a standard hook in accordance with ACI 12.5 or a headed deformed bar in accordance with ACI 12.6.	7.13.2.2
	The continuous negative and positive reinforcements required in ACI 7.13.2.2 shall be enclosed by transverse reinforcement in accordance with ACI 11.5.4.1. This transverse reinforcement shall be anchored in accordance with ACI 11.5.4.2 but need not be extended through the column.	7.13.2.3
	Where splices are used to satisfy ACI 7.13.2.4, the splices shall be Class B tension splices or mechanical or welded splices in accordance with ACI 12.14.3. The splice locations are as follows: • *Top reinforcement*: at or near the midspan • *Bottom reinforcement*: at or near a support	7.13.2.4
Beams other than perimeter beams	No additional requirements for longitudinal integrity reinforcement need to be satisfied where transverse reinforcement in accordance with ACI 7.13.2.3 is provided.	7.13.2.5
	Where transverse reinforcement in accordance with ACI 7.13.2.3 is not provided, the following requirements need to be fulfilled: • At least one-quarter of the positive reinforcement required at the midspan, but not less than two bars, must pass through the region bounded by the longitudinal reinforcement of the column. This reinforcement shall be continuous or shall be spliced over or near the support with a Class B tension splice or a mechanical or welded splice in accordance with ACI 12.14.3. • At noncontinuous supports, the reinforcement shall be anchored to develop f_y at the face of the support, using a standard hook in accordance with ACI 12.5 or a headed deformed bar in accordance with ACI 12.6.	

TABLE 6.4   Structural Integrity Requirements of ACI 7.13

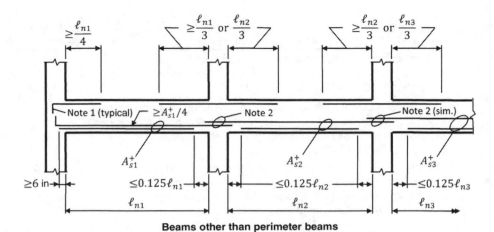

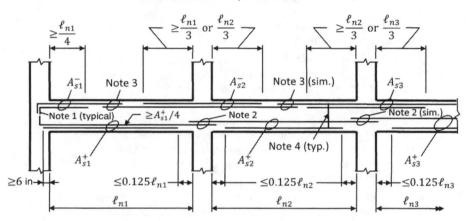

Beams other than perimeter beams

Perimeter beams

Notes:
1. Standard hook in accordance with ACI 12.5 or a headed deformed bar in accordance with ACI 12.6.
2. Greater than or equal to larger of $A_{s1}^+/4$ or $A_{s2}^+/4$ but not less than two bars continuous or spliced with Class B tension splices or mechanical or welded splices.
3. Greater than or equal to larger of $A_{s1}^-/6$ or $A_{s2}^-/6$ but not less than two bars continuous or spliced with Class B tension splices or mechanical or welded splices.
4. Closed transverse reinforcement in accordance with ACI 11.5.4.1.

FIGURE 6.25 Recommended flexural reinforcement details for beams.

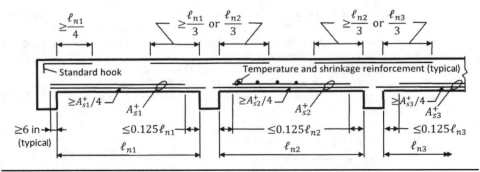

FIGURE 6.26 Recommended flexural reinforcement details for one-way slabs.

230

6.3 Design for Shear

6.3.1 Overview

In addition to flexure, reinforced concrete beams and one-way slabs must be designed for the effects of shear forces due to the weight of the member and any superimposed nominal loads. Typically, a reinforced concrete flexural member is designed for flexure prior to the design for shear. The dimensions of the cross-section and the amount of flexural reinforcement are determined using the strength design requirements given in the Code for flexure (see the previous sections). Included in those requirements are limits on the amount of flexural reinforcement, which ensure that the member behaves in a ductile manner.

Experiments have demonstrated that shear failure is brittle and usually occurs without any warning; thus, it is important to make certain that the shear strength of a member equals or exceeds the flexural strength at all sections so that a ductile failure is ensured.

In order to acquire an understanding on how to design for shear forces in a reinforced concrete beam, consider the rectangular beam depicted in Fig. 6.27. Assume that the beam is simply supported and subjected to a uniformly distributed load. Also assume that the beam is homogeneous, elastic, and uncracked. Two elements are shown in the figure: Element 1 is located above the neutral axis, and element 2 is located below the neutral axis. Both elements are subjected to flexural stresses f and shear stresses v due to the loads (Fig. 6.27a). As expected, the flexural stresses acting on element 1

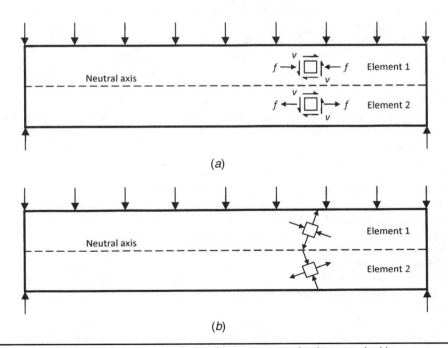

FIGURE 6.27 Flexural and shear stresses in a homogeneous, elastic, uncracked beam.

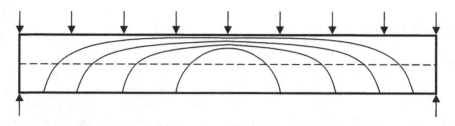

Figure 6.28 Principal compressive stress trajectories in a homogeneous, elastic, uncracked beam.

are compressive, whereas those on element 2 are tensile. Figure 6.27*b* shows the principal stresses acting on elements 1 and 2, which can be determined from the strength of materials using Mohr's circle.

The magnitudes of f and v change along the span because of changes in the magnitudes of the bending moment and shear forces, respectively. They also change vertically on the basis of the distance from the neutral axis. As such, the principal stresses change from one location to another. A plot of the principal compressive stress trajectories is shown in Fig. 6.28.

When the principal tensile stresses exceed the tensile strength of the concrete, the cracking pattern should resemble the lines depicted in Fig. 6.28 because the principal tensile stresses act perpendicular to the principal compressive stresses. In the center portion of the span where shear forces are small, the cracks are primarily vertical and are caused by the tension forces due to flexure. These cracks start at the bottom of the beam where the tension forces are the largest. Toward the ends of the member where shear forces are relatively large, the cracks are inclined and are due to flexure and shear. Such cracks are commonly referred to as *shear cracks* or *diagonal tension cracks*. Thus, in addition to flexural tension, *diagonal tension* due to combined flexure and shear must be considered in the design of a flexural member.

If there were no flexural reinforcement in the concrete beam shown in Figs. 6.27 and 6.28, a tension crack would form at the bottom of the beam at the location of maximum bending moment once the load was large enough so that the tensile stress exceeded the tensile strength of the concrete. This crack would immediately propagate to the top of the section, causing the beam to fail. In this case, shear forces have virtually no effect on the failure of the beam.

If longitudinal reinforcement is provided at the bottom of the beam, tension cracks would form as described earlier, and the reinforcement would provide the necessary tensile strength so that larger loads can be supported by the beam. If the reinforcement is detailed properly, the crack widths and lengths are relatively small.

Shear forces also increase with increasing loads. In regions where the bending moment is small and shear forces are large, *web-shear* cracks will form when the diagonal tension stress in the vicinity of the neutral axis exceeds the tensile strength of the concrete (see Fig. 6.29). These types of cracks are rare. It is more common for *flexure-shear* cracks to form at locations where both the bending moment and the shear force are large. Once the load that causes the formation of these diagonal cracks is reached, it is possible for the cracks to immediately propagate to the compression face of the member, thereby causing splitting failure. It is also possible for failure to occur shortly after the

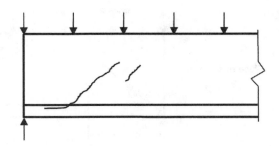

onset of diagonal cracking. In any case, it is evident from this discussion that flexural reinforcement alone is not sufficient to arrest this type of crack propagation; transverse reinforcement is needed to increase shear resistance.

Flexure-shear cracking cannot be predicted by determining the principal tensile stresses in an uncracked beam. Empirical equations based on experimental results have been derived to determine the nominal shear stress at which such cracking occurs.

Providing the proper amount of shear (transverse) reinforcement will enable a flexural member to develop its full bending moment capacity. Otherwise, its overall strength will be limited by its shear capacity, which is based on the tensile strength of concrete. The Code requirements for shear design are outlined in the following sections.

6.3.2 Shear Strength

Introduction

In general, the design for shear consists of calculating the maximum factored shear force V_u and requiring that it is equal to or less than the design shear strength ϕV_n:

$$V_u \leq \phi V_n \tag{6.35}$$

The factored shear force V_u is determined by combining the nominal shear forces determined from analysis in accordance with the load combinations given in ACI 9.2 (see Section 4.2). According to ACI 9.3.2.3, the strength reduction factor ϕ is equal to 0.75 for shear.

The nominal shear strength V_n is determined by ACI Eq. (11-2):

$$V_n = V_c + V_s \tag{6.36}$$

In this equation, V_c is the nominal shear strength provided by concrete and V_s is the nominal shear strength provided by shear reinforcement. These items are discussed in detail later.

Required Shear Strength

The factored shear force V_u is determined at a section using the load combinations given in ACI 9.2. More information on ACI load combinations can be found in Section 4.2. Recall from the discussion of analysis methods in Section 3.3 that an approximate shear force at the face of a support for gravity load effects can be determined using the coefficients given in ACI 8.3.3.

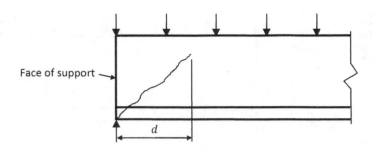

Face of support

d

The locations of the critical sections for shear are given in ACI 11.1.3. It is permitted to design sections located less than a distance d from the face of a support for V_u computed at the distance d, provided the following three criteria are satisfied:

1. Support reactions in the direction of the applied shear force introduce compression into the end regions of the member.

2. Loads are applied at or near the top of the member.

3. No concentrated loads occur between the face of the support and a distance d from the face of the support.

It has been shown from numerous experiments that the inclined shear crack that is closest to the support will extend upward from the face of the support, reaching the compression zone at a distance of approximately d from the face of the support (see Fig. 6.30).

Shown in Fig. 6.31 are the free-body diagrams of the beam sections above and below the location of the crack. The loads are applied to the top of the beam, and the stirrups across the crack have a force equal to the area of the stirrups A_v times the yield stress of the stirrups f_{yt}. It is evident from the upper free-body diagram that the loads applied to the beam between the face of the support and the section located a distance d from the face of the support are transferred directly to the support by compression in the web above the crack. Thus, the Code permits design for a maximum V_u at a distance d from the face of the support.

Figure 6.31 Free-
body diagrams
of the end
of the beam.

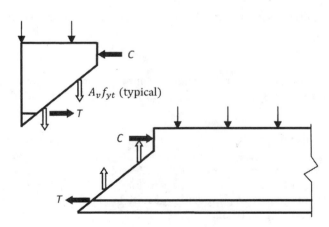

C

$A_v f_{yt}$ (typical)

T

C

T

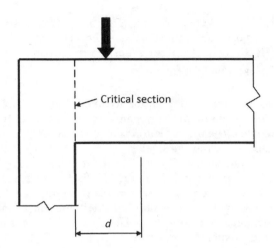

Figure 6.32 Critical section in members with concentrated forces near the support.

Stirrups are required across the potential crack plane between the face of the support and the distance d. If this member were loaded near the bottom of the beam instead of at the top, the transfer of forces would be different than that shown here, and the critical section for shear would be at the face of the support.

Generally, the critical location for shear will be at a distance d from the face of a support. Members supported by bearing at the bottom of the member and members in a continuous frame supporting uniformly distributed loads are typical conditions where this would be applicable [see ACI Fig. R11.1.3.1(c) and (d)].

Because there is a radical change in shear when a concentrated load occurs between the face of the support and a section located at a distance d from the face of the support, the critical section is taken at the face of the support (see Fig. 6.32).

The critical section for shear is also at the face of the support for members framing into a supporting member that is in tension [see ACI Fig. R11.1.3.1(e)].

Shear Strength Provided by Concrete

ACI 11.2 permits ACI Eq. (11-3) or (11-5) to be used to determine the shear strength provided by the concrete V_c for members subjected to flexure and shear:

$$V_c = 2\lambda\sqrt{f_c'}b_w d \tag{6.37}$$

$$V_c = \left(1.9\lambda\sqrt{f_c'} + 2{,}500\rho_w \frac{V_u d}{M_u}\right)b_w d \leq 3.5\lambda\sqrt{f_c'}b_w d \tag{6.38}$$

In both equations, f_c' has the units of pounds per square inch and λ is a modification factor that reflects the reduced mechanical properties of lightweight concrete (see ACI 8.6.1):

- $\lambda = 0.85$ for sand-lightweight concrete.
- $\lambda = 0.75$ for all-lightweight concrete.

- $\lambda = f_{ct}/6.7\sqrt{f_c'} \le 1.0$, where the average splitting tensile strength of lightweight concrete f_{ct} has been determined by tests [Eq. (1) in ASTM C496/C496M]. Note that $6.7\sqrt{f_c'}$ is the average splitting tensile strength of normal-weight concrete.
- $\lambda = 1.0$ for normal-weight concrete.

It is permitted to use linear interpolation to determine λ in cases where a concrete mixture contains normal-weight fine aggregate and a blend of lightweight and normal-weight coarse aggregates. The interpolation shall be between 0.85 and 1.0 on the basis of the volumetric fractions of the aggregates.

ACI 11.1.2 requires that values of $\sqrt{f_c'}$ be limited to 100 psi, except as allowed in ACI 11.1.2.1. This limitation is primarily due to the fact that there is a lack of test data and practical experience with concrete having compressive strengths greater than 10,000 psi. According to the exception given in Section 11.1.2.1, values of $\sqrt{f_c'}$ may be greater than 100 psi for reinforced concrete beams and joist construction that satisfy the minimum web reinforcement requirements given in ACI 11.4.6.3, 11.4.6.4, or 11.5.5.2.

Equation (6.38) is based on a large number of test results and conservatively predicts the nominal shear stress when flexure-shear cracks occur.[20] In this equation, the term $\lambda\sqrt{f_c'}$ is related to the tensile strength of the concrete (see Chap. 2) and ρ_w is the reinforcement ratio of the flexural reinforcement, that is, $\rho_w = A_s/b_w d$. Larger amounts of flexural reinforcement result in smaller and narrower tension cracks; thus, a larger area of uncracked concrete is available to resist shear, which translates into an increase in the shear at which diagonal cracks will form. Equation (6.38) also captures the influence that the ratio of the shear force to the bending moment has on the development of diagonal cracking that was described previously. The minimum value of M_u, which is equal to V_u times d, limits V_c near points of inflection (ACI 11.2.2.1).

It has been found that Eq. (6.38) overestimates the influence of f_c' on the shear strength of the concrete and underestimates the influence of ρ_w and $V_u d/M_u$.[20,21] Other research has indicated that member size has an influence on the shear strength of the concrete: The shear strength decreases as the overall depth of the member increases.[22]

Equation (6.37) is obtained by setting the second term in the parentheses in Eq. (6.38) equal to $0.1\sqrt{f_c'}$. This equation is essentially a lower bound to Eq. (6.38), and it is convenient to use in most designs.

The flowchart shown in Fig. 6.33 can be used to determine the design shear strength for concrete ϕV_c.

For joist construction that meets the size and spacing limitations of ACI 8.13.1 through 8.13.3, V_c is permitted to be increased by 10% more than that specified by Eq. (6.37) or (6.38). This increase is based on the relatively close spacing between the joists, which enables redistribution of loads to adjacent joists. It is also based on the satisfactory performance of joists designed by previous editions of the ACI Code which permitted higher shear strengths.

Shear Strength Provided by Shear Reinforcement
Types of Shear Reinforcement   ACI 11.4.1 permits the following types of shear reinforcement for flexural members:

1. Stirrups perpendicular to the axis of the member
2. Welded wire reinforcement with wires located perpendicular to the axis of the member

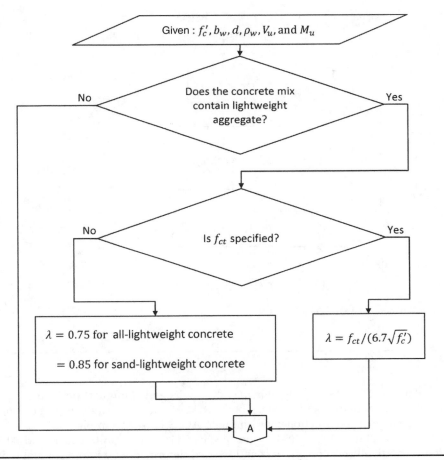

FIGURE 6.33   Design shear strength for concrete ϕV_c. *(continued)*

3. Spirals, circular ties, or hoops

4. Stirrups making an angle of 45 degrees or more with the longitudinal flexural reinforcement

5. Longitudinal reinforcement that is bent an angle of 30 degrees or more with respect to the longitudinal flexural reinforcement

Stirrups that are oriented perpendicular to the axis of the member and are anchored to the longitudinal flexural reinforcement are the most commonly used type of shear reinforcement in beams. However, in areas of moderate to high seismic risk, spirals or hoops must be used in accordance with ACI Chap. 21. Inclined stirrups and bent longitudinal bars are rarely used in practice and, thus, will not be covered in this book.

For the two-legged stirrups shown in Fig. 6.34, the total area of shear reinforcement A_v is equal to two times the area of the stirrup bar A_b. These are commonly referred to as U-stirrups because of their shape. In general, A_v is equal to the area of the stirrup bar times the number of legs that are provided. Figure 6.35 illustrates typical configurations of stirrups with multiple legs. The four-legged configuration is commonly used for

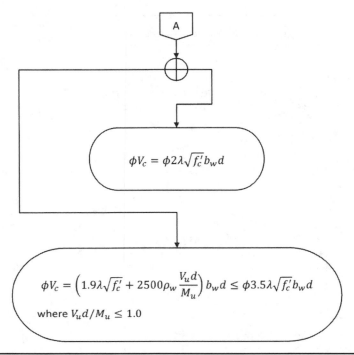

Figure 6.33   *(Continued)*

beams that are 24 to 48 in wide, whereas the six-legged configuration is utilized for beams that are more than 48 in wide. Note that ACI 12.13.2.5 permits the use of single-leg stirrups in joist construction conforming to ACI 8.11 (see Example 6.10).

ACI 11.4.4.2 limits the values of the specified yield strength to 60,000 and 80,000 psi for, respectively, reinforcing bars and welded deformed wire reinforcement that is

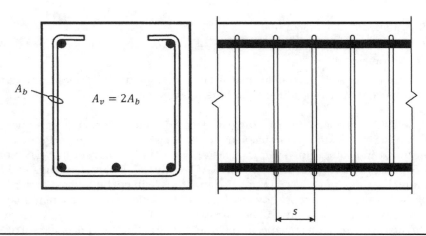

Figure 6.34   Two-legged U-stirrup.

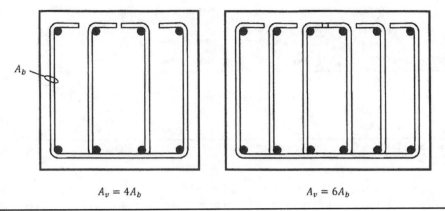

FIGURE 6.35 Multiple-legged stirrup configurations.

used as shear reinforcement. These limits are meant to control the width of diagonal cracks.

Development of Shear Reinforcement   Like in the case of flexural reinforcement, it is essential to properly develop and anchor shear reinforcement in order for it to be fully effective (i.e., in order for it to develop its full tensile force, which is equal to $A_v f_{yt}$).

Requirements for the development of shear (web) reinforcement are given in ACI 12.13 and are illustrated for the case of stirrups in Fig. 6.36. Note that stirrups are to be provided as close to the tension and compression faces of the member as cover requirements and other reinforcement in the section permits; this is stipulated because at or near ultimate load, cracks can extend over a large portion of the member. Each

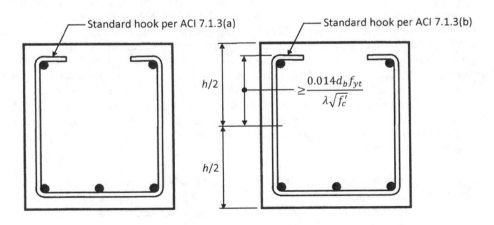

FIGURE 6.36 Anchorage details for U-stirrups.

Stirrup Size	Minimum Beam Height h (in.)			
	Concrete Compressive Strength f'_c (psi)			
	3,000	**4,000**	**5,000**	**6,000**
No. 6	26	23	21	20
No. 7	30	27	24	22
No. 8	34	30	27	25

TABLE 6.5   Minimum Beam Height to Accommodate No. 6, 7, or 8 Stirrups

bend in the continuous portion of the U-stirrup must enclose a longitudinal bar (ACI 12.13.3). The ends of the stirrups must be anchored around the longitudinal bars, using a standard hook defined in ACI 7.1.3(a) for No. 5 bars and smaller and in ACI 7.1.3(b) for Nos. 6 to 8 bars. In addition to a standard hook, a minimum embedment length equal to $0.014d_b f_{yt}/(\lambda\sqrt{f'_c})$ must be provided between the outside edge of the hook and the midheight of the member where No. 6, 7, or 8 stirrups are utilized (see ACI 12.13.2.2 and Fig. 6.36 of this book). This additional anchorage requirement takes into consideration the following: (1) It is not possible to bend a No. 6, 7, or 8 stirrup tightly around a longitudinal bar. (2) A large force can exist in the larger stirrup bars with $f_{yt} \geq 40,000$ psi.

The use of larger stirrup bars controls the beam height that must be provided in order to satisfy the development requirements of ACI 12.13.2.2. Table 6.5 contains minimum beam heights for various concrete compressive strengths, assuming normal-weight concrete, Grade 60 reinforcement, and a cover of 1.5 in to the stirrup hook. No. 6, 7, or 8 stirrups cannot be used in beams with heights less than those listed in the table.

Specific requirements for anchorage of welded plain wire reinforcement in the form of U-stirrups are given in ACI 12.13.2.3 and 12.13.2.4. ACI Fig. R12.13.2.3 illustrates the proper anchorage of such reinforcement in compression zones of beams.

ACI 12.13.5 contains provisions for closed stirrups that are formed from two U-stirrups. The legs of the stirrups must be lap spliced with a splice length equal to or greater than $1.3\ell_d$ but not less than 12 in where the tension development length ℓ_d is determined in accordance with ACI 12.2 (see Fig. 6.37). If the required lap length cannot fit within a member that has a height of at least 18 in, such stirrups can still be used, provided that the force in each leg is equal to or less than 9,000 lb. Thus, for Grade 60 reinforcement, only a No. 3 stirrup satisfies this requirement (force in stirrup leg = $0.11 \times 60,000 = 6,600$ lb).

Design of Shear Reinforcement   Shear reinforcement is required to augment the overall shear strength of a reinforced concrete flexural member. Prior to diagonal cracking, the stress in such reinforcement is essentially zero. After diagonal cracks develop, shear reinforcement restricts crack growth and penetration into the compression zone. The width of the cracks is also controlled.

The required shear reinforcement depends on the magnitude of the factored shear force V_u and the design shear strength of the concrete ϕV_c. The Code requirements are summarized in the following three cases.

Figure 6.37 Pairs of U-stirrups forming a closed stirrup.

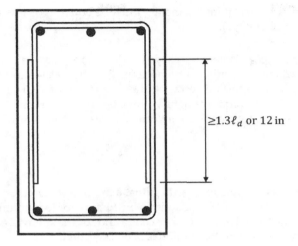

$\geq 1.3\ell_d$ or 12 in

Case 1: $V_u \leq \phi V_c/2$ Shear reinforcement is not requirement at any section where the factored shear force is less than 50% of the design shear strength of the concrete. For members subjected to uniform loads, these sections typically occur in the center segment of the span.

Case 2: $\phi V_c/2 < V_u \leq \phi V_c$ A minimum area of shear reinforcement is required where V_u is greater than 50% of ϕV_c (ACI 11.4.6). This requirement provides a minimum level of shear strength in otherwise-unreinforced portions of a member where the sudden formation of a diagonal crack because of an unexpected tensile force or overload could cause failure. The following members are exempt from this requirement:

- Footings and solid slabs
- Concrete joist construction defined by ACI 8.13
- Beams with $h \leq 10$ in
- Beams integral with slabs with (a) $h \leq 24$ in and (b) $h \leq$ the larger of $2.5h_f$ and $0.5b_w$
- Beams constructed from steel fiber–reinforced, normal-weight concrete with (a) $f'_c \leq 6{,}000$ psi, (b) $h \leq 24$ in, and (c) $V_u \leq \phi 2\sqrt{f'_c}b_w d$

The minimum area of shear reinforcement $A_{v,min}$ is determined by ACI Eq. (11-13):

$$A_{v,min} = \frac{0.75\sqrt{f'_c}b_w s}{f_{yt}} \geq \frac{50b_w s}{f_{yt}} \tag{6.39}$$

Tests have indicated that the minimum required area of shear reinforcement is dependent on the strength of the concrete.[23] The lower bound limit in Eq. (6.39) is applicable in cases where the concrete compressive strength is less than approximately 4,400 psi.

The spacing of shear reinforcement s must be equal to or less than the smaller of $d/2$ and 24 in (ACI 11.4.5.1).

Case 3: $V_u > \phi V_c$ Where V_u exceeds ϕV_c, more than the minimum amount of shear reinforcement is required at a section. Prior to calculating the required shear reinforcement, it is important to check the provisions of ACI 11.4.7.9: The nominal shear strength of shear reinforcement V_s must be equal to or less than $8\sqrt{f_c'}b_w d$. Using Eqs. (6.35) and (6.36), this requirement can be expressed as follows:

$$V_u - \phi V_c \le \phi 8\sqrt{f_c'}b_w d \tag{6.40}$$

This provision attempts to guard against excessive shear crack widths by limiting the maximum shear that can be transmitted by the stirrups to four times the design shear strength of the concrete. The size of the section and/or the strength of the concrete must be increased in cases where this limit has been exceeded.

Once the requirements of ACI 11.4.7.9 have been satisfied, the next step is to establish the segments along the span where shear reinforcement is required and where it is not required, using the information in cases 1 and 2. These segments are illustrated in Fig. 6.38.

In the segments where V_u exceeds ϕV_c, ACI Eq. (11-15) is used to determine the nominal shear strength provided by shear reinforcement:

$$V_s = \frac{A_v f_{yt} d}{s} \tag{6.41}$$

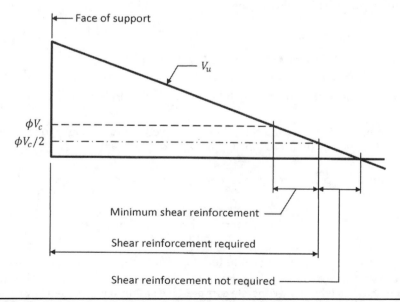

Figure 6.38 Segments along the span where shear reinforcement is and is not required.

Substituting Eqs. (6.35) and (6.36) into Eq. (6.41) results in an expression for the required area and spacing of shear reinforcement:

$$\frac{A_v}{s} = \frac{(V_u - \phi V_c)}{\phi f_{yt} d} \qquad (6.42)$$

It is evident from Eq. (6.42) that shear reinforcement is designed to carry the shear exceeding that which causes diagonal cracking.

Assuming a stirrup size and number of stirrup legs, Eq. (6.42) can be solved for the required stirrup spacing s:

$$s = \frac{\phi A_v f_{yt} d}{V_u - \phi V_c} \qquad (6.43)$$

The maximum spacing requirements are summarized as follows:

- Where $V_u - \phi V_c \leq \phi 4\sqrt{f_c'} b_w d$, maximum $s = d/2$ or 24 in (ACI 11.4.5.1).
- Where $V_u - \phi V_c > \phi 4\sqrt{f_c'} b_w d$, maximum $s = d/4$ or 12 in (ACI 11.4.5.3).

The purpose of the first of these two spacing requirements is to ensure that each 45-degree diagonal shear crack is intercepted by at least one stirrup. In situations where the shear force is relatively large, providing closer stirrup spacing leads to narrower inclined cracks; this is the reason why the maximum spacing in the second of the two requirements is one-half of that in the first requirement.

The required stirrup size and spacing is normally established at the critical section first. At sections away from the critical section, the spacing can be increased. For economy, stirrup spacing should be changed as few times as possible over the required length. If possible, no more than three different stirrup spacings should be specified along the span, with the first stirrup located 2 in from the face of the support. Also, larger stirrup sizes at a wider spacing are usually most cost-effective than smaller stirrup sizes at a closer spacing. The latter require disproportionately high costs for fabrication and placement.

The flowchart shown in Fig. 6.39 can be used to determine the required amount of shear reinforcement for members subjected to shear and flexure with stirrups perpendicular to the axis of the member.

Example 6.9 Determine the required shear strength of the one-way slab given in Examples 3.5, 6.1, and 6.4 (see Fig. 3.3), and compare it with the design shear strength. Assume normal-weight concrete with $f_c' = 4,000$ psi and Grade 60 reinforcement.

Solution It was determined in Example 3.5 that the maximum factored shear force $V_u = 0.6$ kips per foot width of slab. This shear force is located at the face of the first interior support and was determined using the approximate method of ACI 8.3.3. The critical section for shear occurs at a distance d from the face of the support; however, in this example, the shear strength will be conservatively checked at the face of the support.

The design shear strength of the slab is equal to the design shear strength of the concrete and is determined by ACI Eq. (11-3):

$$\phi V_c = \phi 2\lambda \sqrt{f_c'} b_w d = 0.75 \times 2 \times 1.0\sqrt{4,000} \times 12 \times 2.25/1,000$$
$$= 2.6 \text{ kips} > V_u = 0.6 \text{ kips}$$

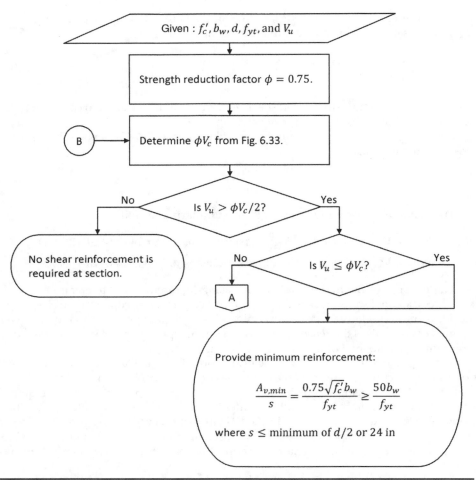

FIGURE 6.39 Required shear reinforcement for members subjected to shear and flexure. (*continued*)

Comments
If it had been determined that $V_u > \phi V_c$, then the design shear strength would have been increased by increasing the slab thickness and/or increasing the compressive strength of the concrete. It is evident from Eq. (6.37) that an increase in the slab thickness has a greater impact on shear strength than does the compressive strength. Rarely are stirrups used in slabs. It would be impossible to fabricate and develop stirrups for use in a thin slab like the one in this example.

Example 6.10 Determine the required shear reinforcement in the end span of a typical interior wide-module joist given in Examples 3.4, 6.2, 6.5, and 6.6 (see Fig. 3.3). Assume normal-weight concrete with $f'_c = 4,000$ psi and Grade 60 reinforcement.

Solution The flowchart shown in Fig. 6.39 will be used to determine the required shear reinforcement for this member subjected to shear and flexure.

   Step 1: Establish the strength reduction factor. In accordance with ACI 9.3.2.3, $\phi = 0.75$ for shear.

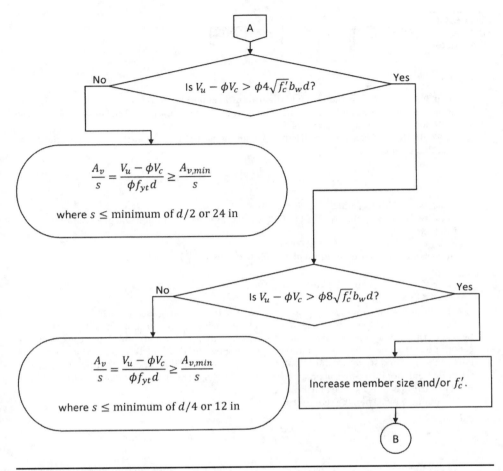

FIGURE 6.39 *(Continued)*

Step 2: Determine ϕV_c. The design shear strength ϕV_c is determined using the flowchart shown in Fig. 6.33:

(a) Determine the lightweight concrete modification factor λ.

Because normal-weight concrete is specified, $\lambda = 1.0$.

(b) Determine ϕV_c from ACI Eq. (11-3) or (11-5).

From Examples 6.5 and 6.6, $d = 18$ in. The width of the joist b_w at the bottom is equal to 7 in. Because the width increases at a slope of 12 to 1, an average value of b_w could be used to calculate ϕV_c. Conservatively use $b_w = 7$ in. Also, this system does not meet the size and spacing limitations for joist construction in accordance with ACI 8.13.1 through 8.13.3; thus, a 10% increase in V_c is not permitted (ACI 8.13.8), and the member must be designed as a beam (ACI 8.13.4).

Using ACI Eq. (11-3):

$$\phi V_c = \phi 2\lambda\sqrt{f_c'}b_wd = 0.75 \times 2 \times 1.0\sqrt{4,000} \times 7 \times 18/1,000 = 12.0 \text{ kips}$$

FIGURE 6.40
Single-leg
stirrup detail.

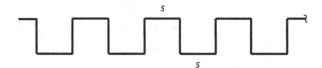

Step 3: Determine V_u and check if shear reinforcement is required. From Table 3.4 given in Example 3.4, the maximum factored shear force in the end span occurs at the exterior face of the first interior support and is equal to 14.7 kips.

The critical section occurs at a distance d from the face of the support, and the factored shear force at that location is equal to

$$V_u = 14.7 - [1.1 \times (18/12)] = 13.1 \text{ kips}$$

where $w_u = 1.1$ kips/ft from Example 3.4.

Because $V_u = 13.1$ kips $> \phi V_c/2 = 12.0/2 = 6.0$ kips, shear reinforcement is required at the critical section.

Also, $V_u = 13.1$ kips $> \phi V_c = 12.0$ kips.

Step 4: Determine the required shear reinforcement.

$$V_u - \phi V_c = 13.1 - 12.0 = 1.1 \text{ kips} < \phi 4\sqrt{f_c'}b_w d = 0.75 \times 4\sqrt{4,000} \times 7 \times 18/1,000 = 23.9 \text{ kips}$$

Therefore, the maximum stirrup spacing $= d/2 = 18/2 = 9$ in (governs) or 24 in.

Because of the sloping joist face and the narrow width of the joist, a single-leg stirrup will be used instead of a two-legged stirrup. A single-leg stirrup is a continuous bar located near the centerline of the joist, which is bent into the profile shown in Fig. 6.40. It is attached to the bottom bars of the joist and has a standard hook at the top. (Note that this type of shear reinforcement is permitted only for joist construction conforming to ACI 8.11; however, it is assumed that it can be used in narrow beams like the one in this example.)

Assuming a No. 3 bar, the required spacing is determined by Eq. (6.43):

$$s = \frac{\phi A_v f_{yt} d}{V_u - \phi V_c} = \frac{0.75 \times 0.11 \times 60 \times 18}{1.1} = 81.0 \text{ in} > 9.0 \text{ in (governs)}$$

Check the minimum shear reinforcement requirements. Because the compressive strength of the concrete is less than 4,400 psi, the lower bound limit in Eq. (6.39) governs:

$$A_{v,min} = \frac{50 b_w s}{f_{yt}} = \frac{50 \times 7 \times 9}{60,000} = 0.05 \text{ in}^2 < 0.11 \text{ in}^2$$

Determine the length over which stirrups are required. Stirrups are no longer required where $V_u \leq \phi V_c/2 = 6.0$ kips:

$$\text{Length} = \frac{V_u @ \text{support} - (\phi V_c/2)}{w_u} = \frac{(14.7 - 6.0)}{1.1} = 7.9 \text{ ft from the face of the support}$$

Use 12 No. 3 single-leg stirrups at a 9.0-in spacing, with the first stirrup located 2 in from the face of the support.

Reinforcement details for the wide-module joist are shown in Fig. 6.41. The lengths of the negative reinforcing bars were determined using Fig. 6.25 for beams other than perimeter beams.

Comments

Because the stirrups are spaced at the maximum permitted spacing, the same stirrup spacing is used over the entire length over which stirrups are required. The stirrups at the exterior end of the

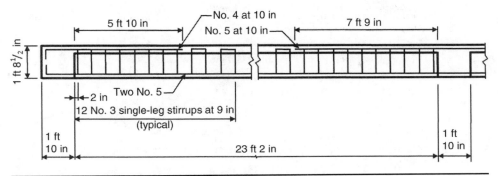

FIGURE 6.41 Reinforcement details for wide-module joist.

wide-module joist are detailed the same as those at the interior end, even though the factored shear forces are less (see Table 3.4).

Example 6.11 Determine the required shear reinforcement for the beam along column line B given in Examples 3.3, 6.3, and 6.7 (see Fig. 3.3). Assume normal-weight concrete with $f'_c = 4{,}000$ psi and Grade 60 reinforcement.

Solution The flowchart shown in Fig. 6.39 will be used to determine the required shear reinforcement for this member subjected to shear and flexure.

   Step 1: Establish the strength reduction factor. In accordance with ACI 9.3.2.3, $\phi = 0.75$ for shear.
   Step 2: Determine ϕV_c. The design shear strength ϕV_c is determined using the flowchart shown in Fig. 6.33.

(a) Determine the lightweight concrete modification factor λ.

   Because normal-weight concrete is specified, $\lambda = 1.0$.

(b) Determine ϕV_c from ACI Eq. (11-3) or (11-5).

   From Examples 6.3 and 6.7, $d = 18$ in and $b_w = 22$ in.
   Using ACI Eq. (11-3),

$$\phi V_c = \phi 2\lambda \sqrt{f'_c} b_w d = 0.75 \times 2 \times 1.0 \sqrt{4{,}000} \times 22 \times 18/1{,}000 = 37.6 \text{ kips}$$

Alternatively, calculate ϕV_c by ACI Eq. (11-5):

$$\phi V_c = \phi \left(1.9\lambda \sqrt{f'_c} + 2{,}500 \rho_w \frac{V_u d}{M_u} \right) b_w d \leq \phi 3.5\lambda \sqrt{f'_c} b_w d$$

For a factored positive bending moment of 127.3 ft kips (see Table 3.3), it can be determined that four No. 6 bars are adequate. Thus,

$$\rho_w = \frac{A_s}{b_w d} = \frac{4 \times 0.44}{22 \times 18} = 0.0044$$

At the critical section, which is located a distance equal to $d = 18$ in from the face of the exterior face of the first interior support, V_u and M_u are (see Table 3.3)

$$V_u = 56.4 - [5.4 \times (18/12)] = 48.3 \text{ kips}$$
$$M_u = 198.1 + [5.4 \times (18/12)^2/2] - [56.4 \times (18/12)] = 119.6 \text{ ft kips}$$
$$V_u d/M_u = 48.3 \times (18/12)/119.6 = 0.61 < 1.0$$

Thus,

$$\phi V_c = 0.75[(1.9 \times 1.0\sqrt{4,000}) + (2,500 \times 0.0044 \times 0.61)] \times 22 \times 18/1,000 = 37.7 \text{ kips}$$
$$< 0.75 \times 3.5 \times 1.0\sqrt{4,000} \times 22 \times 18/1,000 = 65.7 \text{ kips}$$

The values of ϕV_c obtained from ACI Eqs. (11-3) and (11-5) are essentially the same. Use $\phi V_c = 37.6$ kips.

Step 3: Determine V_u and check if shear reinforcement is required. The value of V_u at the critical section was determined in Step 2 to be 48.3 kips.

Because $V_u = 48.3$ kips $> \phi V_c/2 = 37.6/2 = 18.8$ kips, shear reinforcement is required at the critical section.

Also, $V_u = 48.3$ kips $> \phi V_c = 37.6$ kips.

Step 4: Determine the required shear reinforcement.

$$V_u - \phi V_c = 48.3 - 37.6 = 10.7 \text{ kips} < \phi 4\sqrt{f_c'}b_w d = 0.75 \times 4\sqrt{4,000} \times 22 \times 18/1,000 = 75.1 \text{ kips}$$

Therefore, the maximum stirrup spacing $= d/2 = 18/2 = 9$ in (governs) or 24 in.

Assuming No. 3 U-stirrups with two legs, the required spacing is determined by Eq. (6.43):

$$s = \frac{\phi A_v f_{yt} d}{V_u - \phi V_c} = \frac{0.75 \times 2 \times 0.11 \times 60 \times 18}{10.7} = 16.7 \text{ in} > 9.0 \text{ in (governs)}$$

Check the minimum shear reinforcement requirements. Because the compressive strength of the concrete is less than 4,400 psi, the lower bound limit in Eq. (6.39) governs:

$$A_{v,min} = \frac{50 b_w s}{f_{yt}} = \frac{50 \times 22 \times 9}{60,000} = 0.17 \text{ in}^2 < 0.22 \text{ in}^2$$

Determine the length over which stirrups are required. Stirrups are no longer required where $V_u \le \phi V_c/2 = 18.8$ kips:

$$\text{Length} = \frac{V_u \text{ at support} - (\phi V_c/2)}{w_u} = \frac{(56.4 - 18.8)}{5.4} = 7.0 \text{ ft from the face of the support}$$

Use 11 No. 3 U-stirrups with two legs at a 9.0-in spacing, with the first stirrup located 2 in from the face of the support.

Reinforcement details for the beam are shown in Fig. 6.42. The lengths of the negative reinforcing bars were determined using Fig. 6.25 for beams that are not at the perimeter.

Comments

Because the stirrups are spaced at the maximum permitted spacing, the same stirrup spacing is used over the entire length over which stirrups are required. The stirrups at the exterior end of the beam are detailed the same as those at the interior end even though the factored shear forces are less (see Table 3.3).

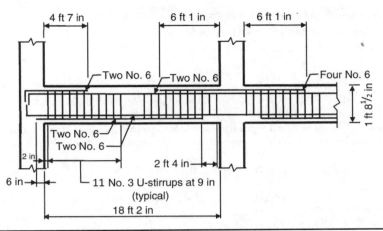

FIGURE 6.42   Reinforcement details for beam on column line B.

Example 6.12   Determine the required shear reinforcement for the reinforced concrete beam subjected to the factored loads in Fig. 6.43. Assume $b_w = 16$ in, $h = 22$ in, and $d = 19.5$ in. Also assume sand-lightweight concrete with $f_c' = 5,000$ psi and Grade 60 reinforcement.

Solution   The flowchart shown in Fig. 6.39 will be used to determine the required shear reinforcement for this member subjected to shear and flexure.

Step 1: **Establish the strength reduction factor.** In accordance with ACI 9.3.2.3, $\phi = 0.75$ for shear.

Step 2: **Determine ϕV_c.** The design shear strength ϕV_c is determined using the flowchart shown in Fig. 6.33:

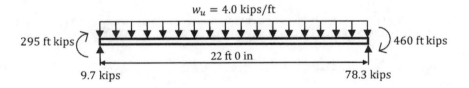

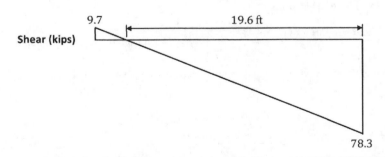

FIGURE 6.43   Beam given in Example 6.12.

(a) Determine the lightweight concrete modification factor λ.

Because sand-lightweight concrete is specified, $\lambda = 0.85$.

(b) Determine ϕV_c from ACI Eq. (11-3) or (11-5).

Using ACI Eq. (11-3),

$$\phi V_c = \phi 2\lambda \sqrt{f_c'} b_w d = 0.75 \times 2 \times 0.85\sqrt{5,000} \times 16 \times 19.5/1,000 = 28.1 \text{ kips}$$

Step 3: Determine V_u and check if shear reinforcement is required. The critical section occurs at a distance d from the face of the support, and the factored shear force at that location is equal to

$$V_u = 78.3 - [4.0 \times (19.5/12)] = 71.8 \text{ kips}$$

Because $V_u = 71.8$ kips $> \phi V_c/2 = 28.1/2 = 14.0$ kips, shear reinforcement is required at the critical section.

Also, $V_u = 71.8$ kips $> \phi V_c = 28.1$ kips.

Step 4: Determine the required shear reinforcement.

$$V_u - \phi V_c = 71.8 - 28.1 = 43.7 \text{ kips} < \phi 4\sqrt{f_c'} b_w d = 0.75 \times 4\sqrt{5,000} \times 16 \times 19.5/1,000 = 66.2 \text{ kips}$$

Therefore, the maximum stirrup spacing $= d/2 = 19.5/2 = 9.8$ in (governs) or 24 in.

Assuming No. 3 U-stirrups with two legs, the required spacing is determined by Eq. (6.43):

$$s = \frac{\phi A_v f_{yt} d}{V_u - \phi V_c} = \frac{0.75 \times 2 \times 0.11 \times 60 \times 19.5}{43.7} = 4.4 \text{ in}$$

Because the required spacing of the No. 3 U-stirrups is small, recalculate the spacing using No. 4 U-stirrups:

$$s = \frac{\phi A_v f_{yt} d}{V_u - \phi V_c} = \frac{0.75 \times 2 \times 0.20 \times 60 \times 19.5}{43.7} = 8.0 \text{ in}$$

Check the minimum shear reinforcement requirements. Because the compressive strength of the concrete is greater than 4,400 psi, ACI Eq. (11-13) governs:

$$A_{v,min} = \frac{0.75\sqrt{f_c'} b_w s}{f_{yt}} = \frac{0.75\sqrt{5,000} \times 16 \times 8}{60,000} = 0.11 \text{ in}^2 < 0.40 \text{ in}^2$$

Determine the length over which stirrups are required. Stirrups are no longer required where $V_u \leq \phi V_c/2 = 14.0$ kips (see Fig. 6.43):

$$\text{Length} = \frac{V_u @ \text{ support} - (\phi V_c/2)}{w_u} = \frac{(78.3 - 14.0)}{4.0} = 16.1 \text{ ft from the face of the support}$$

Determine the distance x where the maximum stirrup spacing of 9.0 in may be used:

$$s = \frac{\phi A_v f_{yt} d}{V_u - \phi V_c}$$

$$9 = \frac{0.75 \times 2 \times 0.20 \times 60 \times 19.5}{V_u - 28.1}$$

$$V_u = 67.1 \text{ kips}$$

From the face of the support,

$$x = \frac{78.3 - 67.1}{4.0} = 2.8 \text{ ft}$$

Use five No. 4 U-stirrups with two legs at an 8.0-in spacing, with the first stirrup located 2 in from the face of the support. Use No. 4 U-stirrups spaced at a 9.0-in spacing for the remainder of the span.

Comments
Even though the factored shear force at the left support is less than $\phi V_c/2$, shear reinforcement is provided over the entire span length.

6.4 Design for Torsion

6.4.1 Overview

Reinforced concrete beams must be designed for the effects from torsional loads wherever applicable. Spandrel beams and beams supporting transverse spans that differ significantly in length are two examples where torsion is likely to play a major role in the design of the member.

Both transverse reinforcement and longitudinal reinforcement are needed to resist the effects from torsional loads. The former type is added to that required for shear, and the latter type is added to that required for flexure. Because of the way torsional cracks propagate in a member, closed stirrups are required. Background information on the Code design requirements are provided later.

Design for torsion in the ACI Code is based on a thin-walled tube, space truss analogy. A beam is idealized as a tube where the center portion of a solid beam is conservatively neglected (see Fig. 6.44). Prior to cracking, torsion is resisted through a constant shear flow q acting around the centerline of the tube, as depicted in Fig. 6.44. The following equation is obtained from equilibrium of the external torque T and the internal stresses:

$$T = 2A_o q = 2A_o \tau t \tag{6.44}$$

FIGURE 6.44
Thin-walled tube.

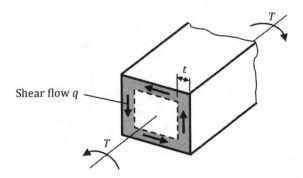

Shear flow q

Figure 6.45 Area enclosed by shear flow path.

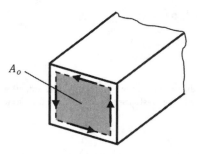

In this equation, A_o is the gross area enclosed by the shear flow path (see Fig. 6.45); τ is the shear stress at any point along the perimeter of the tube; and t is the thickness of the wall where τ is being computed.

Equation (6.44) can be rearranged as follows:

$$q = \tau t = \frac{T}{2A_o} \tag{6.45}$$

Diagonal cracks form around a beam when it is subjected to a torsional moment in excess of that which causes cracking. After cracking, the tube is idealized as a space truss (see Fig. 6.46). The resultant of the shear flow in the wall tubes induces forces in the truss members. The truss diagonals, which are inclined at an angle θ, are in compression and consist of concrete "compression struts." The truss members that are in tension consist of the longitudinal reinforcement and closed stirrups and are called "tension ties." Therefore, once a reinforced concrete beam has cracked in torsion, its torsional resistance is provided primarily by closed stirrups and longitudinal bars. Any concrete outside of the closed stirrups is essentially ineffective in resisting torsion.

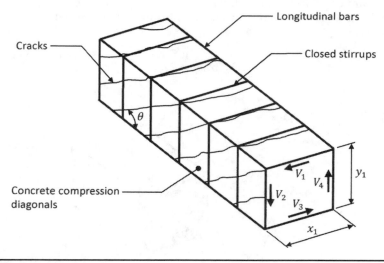

Figure 6.46 Space truss analogy.

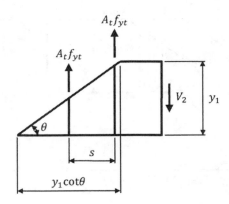

FIGURE 6.47
Free-body
diagram—vertical
equilibrium.

A free-body diagram of a portion of the truss depicted in Fig. 6.46 is shown in Fig. 6.47. The shear flow q creates the shear forces V_1 to V_4 around the walls of the tube. The shear force V_i on any face of the tube is equal to the shear flow q times the height or width of the corresponding wall. For example, the shear force V_2 is equal to q times the height of the wall y_1. Assuming that the stirrups are designed to yield when the maximum torque is reached and that the height of the wall y_1 is equal to the center-to-center vertical length of the closed stirrups, the following equation can be obtained from vertical equilibrium:

$$V_2 = \frac{A_t f_{yt}}{s} y_1 \cot \theta \tag{6.46}$$

In this equation, A_t is the area of one leg of the closed stirrup and s is the center-to-center spacing of the stirrups. Also note that $y_1 \cot \theta$ is the horizontal projection of the inclined surface.

As noted previously, the shear flow is constant over the height of the wall; thus, using Eq. (6.45),

$$V_2 = q y_1 = \frac{T}{2A_o} y_1 \tag{6.47}$$

Equating Eqs. (6.46) and (6.47) results in the following:

$$T = \frac{2A_o A_t f_{yt}}{s} y_1 \tag{6.48}$$

It is shown later that this equation forms the basis of ACI Eq. (11-21) that is used to determine the nominal torsional moment strength T_n of a member or the required amount of torsional reinforcement A_t/s.

The vertical shear force V_i can be resolved into diagonal compressive stresses and axial tension forces as shown in the free-body diagram of the beam in Fig. 6.48. The diagonal compressive component D_i is resisted by the concrete and is equal to $V_i/\sin\theta$, and the axial tension force N_i is resisted by the longitudinal reinforcement and is equal to $V_i \cot\theta$. Because q is constant along the walls, N_i is centered at the midheight of the wall. The top and bottom chords (i.e., the top and bottom longitudinal reinforcement)

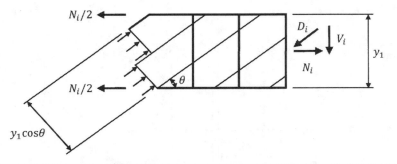

Figure 6.48 Free-body diagram—horizontal equilibrium.

resist a tension force of $N_i/2$. Assuming that the longitudinal reinforcement A_ℓ yields when the maximum torque is reached, horizontal equilibrium results in the following:

$$A_\ell f_y = \sum N_i = \sum V_i \cot\theta = q \cot\theta \sum y_i = \frac{T}{2A_o} \cot\theta \sum y_i \qquad (6.49)$$

In this equation, $\sum y_i = 2(x_1 + y_1)$, which is equal to the perimeter of the centerline of the closed stirrups p_h. Rearranging Eq. (6.49) and substituting p_h for $\sum y_i$ results in the following:

$$A_\ell = \frac{T p_h \cot\theta}{2A_o f_y} \qquad (6.50)$$

This equation forms the basis of ACI Eq. (11-22) that is used to determine the amount of longitudinal reinforcement that is required to resist torsional effects.

The transverse reinforcement and the longitudinal reinforcement are added to the reinforcement required for shear and flexure, respectively (ACI 11.5.3.8).

6.4.2 Threshold Torsion

Threshold torsion is defined as the torsional moment below which torsion effects can be neglected. According to the Code, torsion can be neglected in a section where the factored torsional moment T_u is less than one-fourth of the cracking torque T_{cr}. For the case of pure torsion, T_{cr} is derived using an equivalent thin-walled tube that has a wall thickness t equal to the following prior to cracking:

$$t = \frac{3A_{cp}}{4p_{cp}} \qquad (6.51)$$

In this equation, A_{cp} is the area that is enclosed by the outside perimeter of the concrete cross-section and p_{cp} is the outside perimeter of the concrete cross-section. The area enclosed by the wall centerline A_o is

$$A_o = \frac{2A_{cp}}{3} \qquad (6.52)$$

For spandrel beams and other members that are cast monolithically with a slab, a portion of the slab may be able to contribute to the torsional resistance of the section. The overhanging flange defined in ACI 13.2.4 and illustrated in ACI Fig. R13.2.4 is permitted to be used in the calculation of A_{cp} and p_{cp} (ACI 11.5.1). However, overhanging flanges shall be neglected in cases where A_{cp}^2/p_{cp} calculated for a beam with flanges is less than A_{cp}^2/p_{cp} calculated for the beam without flanges.

In a beam subjected to pure torsion, the principal tensile stress is equal to the torsional stress:

$$\tau = \frac{T}{2A_o t} \tag{6.53}$$

Cracking is assumed to occur when the principal tensile stress is equal to or greater than $4\lambda\sqrt{f_c'}$. Substituting $\tau = 4\lambda\sqrt{f_c'}$ into Eq. (6.53) results in

$$4\lambda\sqrt{f_c'} = \frac{T}{2A_o t} \tag{6.54}$$

An equation for the cracking torque T_{cr} is obtained by substituting Eqs. (6.51) and (6.52) into Eq. (6.54):

$$T_{cr} = 4\lambda\sqrt{f_c'}\left(\frac{A_{cp}^2}{p_{cp}}\right) \tag{6.55}$$

Therefore, the threshold torsional moment is equal to $T_{cr}/4$; that is, torsion can be neglected when the following equation is satisfied:

$$T_u < \phi\lambda\sqrt{f_c'}\left(\frac{A_{cp}^2}{p_{cp}}\right) \tag{6.56}$$

The strength reduction factor for torsion is equal to 0.75 in accordance with ACI 9.3.2.3. Like in the case for shear, ACI 11.1.2 requires that values of $\sqrt{f_c'}$ be limited to 100 psi for members subjected to torsion.

6.4.3 Calculation of Factored Torsional Moment

Equilibrium Torsion
Once a beam cracks because of torsion, the torsional stiffness of the member decreases. It has been shown that the reduction in torsional stiffness after cracking is much larger than the reduction in flexural stiffness after cracking.[24] If a torsional moment T_u at a section is greater than the threshold torsion and if T_u cannot be reduced by redistribution of internal forces in the structure, a member must be designed for T_u determined from analysis (ACI 11.5.2.1). This type of torsion is referred to as *equilibrium torsion* because the torsional moment is required for the structure to be in equilibrium. ACI Fig. R11.5.2.1 illustrates a typical condition where torsional moment redistribution is not possible; there are no adjoining members in the structure that can assist in redistribution of the torsional moment.

Compatibility Torsion

In indeterminate structures, redistribution of internal forces can occur. This type of torsion is referred to as *compatibility torsion*. In cases like the one illustrated in ACI Fig. 11.5.2.2 for typical cast-in-place construction, a large twist occurs at the onset of torsional cracking (i.e., when $T_u = T_{cr}$); this results in a large redistribution of forces in the structure.[25,26] As such, the member can be designed for the reduced cracking torque T_{cr} at the critical section instead of the torsional moment obtained from analysis:

$$T_u = \phi 4 \lambda \sqrt{f_c'} \left(\frac{A_{cp}^2}{p_{cp}} \right) \tag{6.57}$$

Adjoining members must be designed for the redistributed bending moments and shear forces due to the application of the compatibility torsional moment (ACI 11.5.2.2). In other words, a beam that frames into the edge beam shown in ACI Fig. 11.5.2.2 must be designed for a concentrated bending moment at its end equal to the compatibility torsional moment given by Eq. (6.57), in addition to the bending moments and shear forces corresponding to gravity and other loads.

In cases where the factored torsional moment T_u obtained from analysis is greater than the threshold torsional moment and is less than the compatibility torsional moment defined in Eq. (6.57), the section should be designed to resist the factored torsional moment from analysis.

The flowchart shown in Fig. 6.49 can be used to determine the factored torsional moment T_u. Torsional section properties for edge and interior beams are given in Fig. 6.50.

Critical Section

Like in the case of shear, the critical section for torsion is located at a distance d from the face of the support (ACI 11.5.2.4). In other words, sections located less than a distance d from the face of a support are permitted to be designed for T_u computed at a distance d. Note that T_u is determined by either equilibrium torsion or compatibility torsion as discussed previously.

The critical section for torsion occurs at the face of the support if a concentrated torque occurs between the face of the support and a distance d from it. This commonly occurs where a beam frames into the side of girder near the support of the girder.

6.4.4   Torsional Moment Strength

Adequacy of Cross-section

Prior to determining the required transverse and longitudinal reinforcements for torsion, the adequacy of the cross-section must be checked in accordance with ACI 11.5.3.1. Cross-sectional dimensions are limited to help reduce unsightly cracking and to prevent crushing of the inclined concrete compression struts due to shear and torsion. ACI Eq. (11-18) must be satisfied for solid sections:

$$\sqrt{\left(\frac{V_u}{b_w d} \right)^2 + \left(\frac{T_u p_h}{1.7 A_{oh}^2} \right)^2} \leq \phi \left(\frac{V_c}{b_w d} + 8\sqrt{f_c'} \right) \tag{6.58}$$

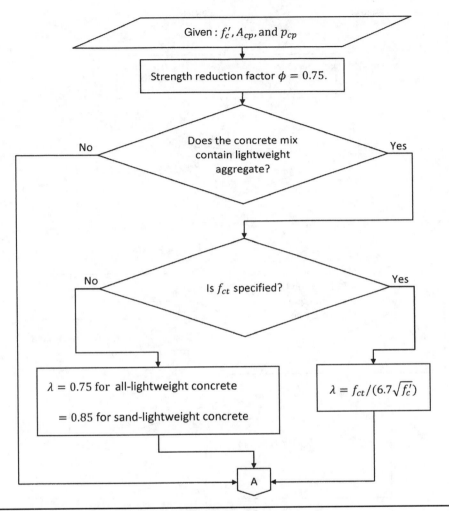

FIGURE 6.49 Factored torsional moment T_u. (*continued*)

The terms on the left-hand side of Eq. (6.58) are the stresses due to shear and torsion. This equation essentially sets an upper limit on the maximum stresses that a section can resist when subjected to shear and torsion. This limit is analogous to the one prescribed in ACI 11.4.7.9 for shear alone (see Section 6.3).

In solid sections, stresses due to shear are resisted by the full width of a member, whereas those due to torsion are resisted by a thin-walled tube [see ACI Fig. R11.5.3.1(b)]. That is why the stresses in Eq. (6.58) are combined using the square root of the sum of the squares rather than by direct addition.

The parameters A_{oh} and p_h are defined as the area enclosed by the centerline of the outermost closed transverse torsional reinforcement and the perimeter of the centerline of the outermost closed transverse torsional reinforcement, respectively. Equations for

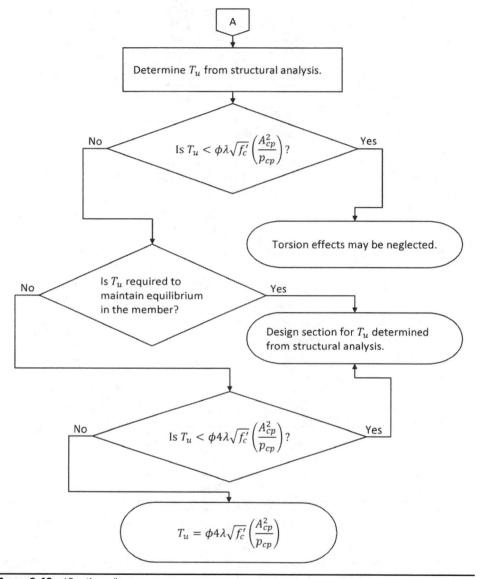

A

Determine T_u from structural analysis.

No   Is $T_u < \phi\lambda\sqrt{f_c'}\left(\dfrac{A_{cp}^2}{p_{cp}}\right)$?   Yes

Torsion effects may be neglected.

No   Is T_u required to maintain equilibrium in the member?   Yes

Design section for T_u determined from structural analysis.

No   Is $T_u < \phi 4\lambda\sqrt{f_c'}\left(\dfrac{A_{cp}^2}{p_{cp}}\right)$?   Yes

$$T_u = \phi 4\lambda\sqrt{f_c'}\left(\frac{A_{cp}^2}{p_{cp}}\right)$$

FIGURE 6.49   *(Continued)*

these parameters are provided in Fig. 6.50 for edge and interior beams. On the basis of these definitions, it is evident that the concrete cover surrounding the transverse reinforcement is ignored in the analysis.

Dimensions of a cross-section must be modified if Eq. (6.58) is not satisfied. Increasing the cross-sectional dimensions of a section will typically have a greater impact than increasing the compressive strength of the concrete.

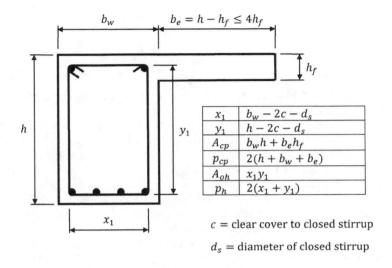

$$b_e = h - h_f \leq 4h_f$$

x_1	$b_w - 2c - d_s$
y_1	$h - 2c - d_s$
A_{cp}	$b_w h + b_e h_f$
p_{cp}	$2(h + b_w + b_e)$
A_{oh}	$x_1 y_1$
p_h	$2(x_1 + y_1)$

c = clear cover to closed stirrup

d_s = diameter of closed stirrup

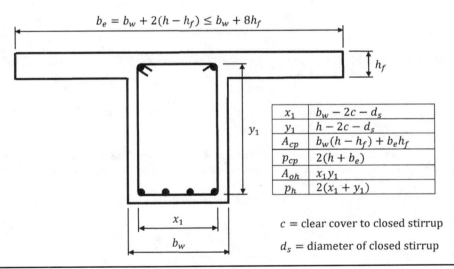

$$b_e = b_w + 2(h - h_f) \leq b_w + 8h_f$$

x_1	$b_w - 2c - d_s$
y_1	$h - 2c - d_s$
A_{cp}	$b_w(h - h_f) + b_e h_f$
p_{cp}	$2(h + b_e)$
A_{oh}	$x_1 y_1$
p_h	$2(x_1 + y_1)$

c = clear cover to closed stirrup

d_s = diameter of closed stirrup

FIGURE 6.50 Torsional section properties.

Design Torsional Strength and Required Torsional Reinforcement

In sections where the factored torsional moment T_u exceeds the threshold torsion $\phi \lambda \sqrt{f_c'}(A_{cp}^2/p_{cp})$, ACI Eq. (11-20) must be satisfied:

$$\phi T_n \geq T_u \tag{6.59}$$

After a section has cracked, it assumed that T_u is resisted by transverse reinforcement and longitudinal reinforcement only; that is, the concrete contribution to the total nominal torsional moment strength is assumed to be zero. The nominal torsional moment strength T_n was derived previously from equilibrium and is given by ACI

Eq. (11-21) [see Eq. (6.48)]:

$$T_n = \frac{2 A_o A_t f_{yt}}{s} \cot \theta \tag{6.60}$$

It is assumed that the nominal shear strength provided by the concrete V_c is not affected by the presence of torsion.

As noted earlier, the concrete outside of the stirrups is essentially ineffective once torsional cracking develops. Thus, the gross area A_o enclosed by the shear flow path around the tube after cracking must be determined. Reference 27 provides a rigorous theoretical method to determine A_o. Alternatively, A_o can simply be taken as $A_o = 0.85 A_{oh}$ (ACI 11.5.3.6).

The angle of the concrete compression diagonals θ can also be determined by analysis.[27] ACI 11.5.3.6 sets a range for θ between 30 and 60 degrees and permits it to be 45 degrees for nonprestressed members.

Substituting Eq. (6.59) into Eq. (6.60) results in the following equation that can be used to determine the required transverse torsional reinforcement:

$$\frac{A_t}{s} = \frac{T_u}{2 \phi \cot \theta A_o f_{yt}} \tag{6.61}$$

The additional longitudinal reinforcement that is required for torsion is determined by ACI Eq. (11-22) [see Eq. (6.50)]:

$$A_\ell = \frac{A_t}{s} p_h \left(\frac{f_{yt}}{f_y} \right) \cot^2 \theta \tag{6.62}$$

In this equation, the term A_t/s is computed by Eq. (6.61), but the modifications to A_t/s given in ACI 11.5.5.2 and 11.5.5.3 related to minimum reinforcement are not applicable when calculating A_ℓ by Eq. (6.62).

The minimum area of longitudinal reinforcement for torsion is determined by ACI Eq. (11-24):

$$A_{\ell,min} = \frac{5 \sqrt{f_c'} A_{cp}}{f_y} - \frac{A_t}{s} p_h \left(\frac{f_{yt}}{f_y} \right) \tag{6.63}$$

In this equation, A_t/s that is determined by Eq. (6.61) must be taken equal to or greater than $25 b_w / f_{yt}$.

Compressive stresses in the flexural compression zone of a member can offset a part of the longitudinal tensile stresses due to torsion. Consequently, the required area of the longitudinal torsional reinforcement can be reduced at the top of a continuous beam near the midspan and at the bottom near the supports. ACI 11.5.3.9 permits the area of the longitudinal torsional reinforcement to be reduced by an amount equal to $M_u/(0.9 f_y d)$, where M_u is the factored bending moment that occurs at the section simultaneously with the factored torsional moment T_u. Note that the reduced area of longitudinal steel must be greater than the minimum values specified in ACI 11.5.5.3 and 11.5.6.2.

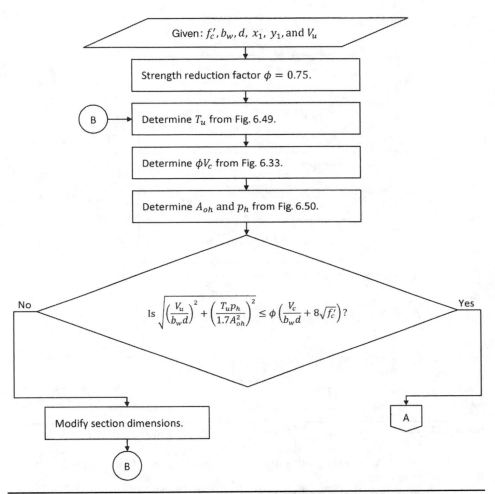

FIGURE 6.51 Transverse and longitudinal reinforcements for torsion. (*continued*)

ACI 11.5.3.4 limits the values of the yield strength of the longitudinal and transverse torsional reinforcements to 60,000 psi. The purpose of this requirement is to control the widths of diagonal cracks.

The flowchart shown in Fig. 6.51 can be used to determine the required transverse and longitudinal reinforcements for torsion.

6.4.5 Details of Torsional Reinforcement

The longitudinal bars must be enclosed by one or more of the following:

- Closed stirrups or closed ties perpendicular to the axis of the member
- A closed cage of welded wire reinforcement with transverse wires perpendicular to the axis of the member
- Spiral reinforcement

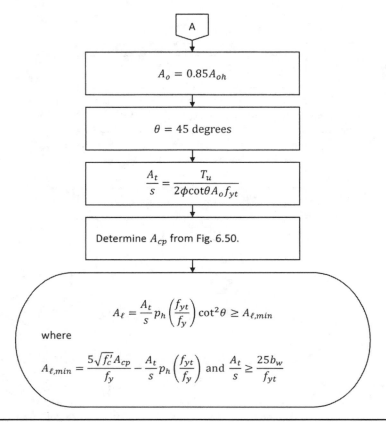

$$A$$

$$A_o = 0.85A_{oh}$$

$$\theta = 45 \text{ degrees}$$

$$\frac{A_t}{s} = \frac{T_u}{2\phi\cot\theta A_o f_{yt}}$$

Determine A_{cp} from Fig. 6.50.

$$A_\ell = \frac{A_t}{s} p_h \left(\frac{f_{yt}}{f_y}\right) \cot^2\theta \geq A_{\ell,min}$$

where

$$A_{\ell,min} = \frac{5\sqrt{f_c'}A_{cp}}{f_y} - \frac{A_t}{s}p_h\left(\frac{f_{yt}}{f_y}\right) \text{ and } \frac{A_t}{s} \geq \frac{25b_w}{f_{yt}}$$

FIGURE 6.51   *(Continued)*

Stirrups must be closed because the inclined cracking due to torsion can occur on all faces of a member.

Tests have shown that the corners of beams subjected to torsion can spall off because of the inclined compressive stresses in the concrete diagonals as the maximum torsional moment is reached. The diagonal forces bear against the longitudinal corner reinforcement, and the component perpendicular to the longitudinal reinforcement is transferred to the transverse reinforcement. Thus, the transverse reinforcement must be properly anchored so that it does not fail. Spalling is essentially prevented in beams with a slab on one or both sides of the web; in such cases, ACI 11.5.4.2 permits the transverse reinforcement to be anchored by standard 90-degree hooks like in the case for shear anchorage (see Fig. 6.52). Note that the 90-degree hooks on the closed stirrups for the beam shown in Fig. 6.52 are located on the side that is adjacent to the slab, which restrains spalling.

For beams where spalling cannot be restrained, the transverse reinforcement must be anchored by a 135-degree hook or a seismic hook [see ACI 11.5.4.2(a) and Fig. 6.53 of this book]. Tests have shown that closed stirrups anchored by 90-degree hooks fail when there is no restraint.[28] Lapped U-stirrups (Fig. 6.37) have also been found to be inadequate for resisting torsion due to a loss of bond when the concrete cover spalls.

FIGURE 6.52
Transverse
reinforcement
detail for
torsion—spalling
restrained.

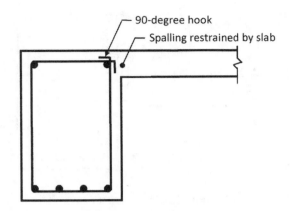

According to ACI 11.5.6.1, the spacing of transverse torsional reinforcement is limited to the smaller of $p_h/8$ and 12 in. This limitation helps to ensure the following: (1) the ultimate torsional strength of the member is developed; (2) excessive loss of torsional stiffness after cracking is prevented; and (3) crack widths are controlled.

Longitudinal torsional reinforcement must also be developed at both ends for tension. Proper anchorage is especially important at the ends of a beam that is subjected to high torsional moments.

The following detailing requirements must be satisfied for the longitudinal reinforcement for torsion (ACI 11.5.6):

1. The longitudinal reinforcement must be distributed around the perimeter of the closed stirrups at a maximum spacing of 12 in (see Fig. 6.54). As was shown earlier, the longitudinal tensile forces due to torsion act along the centroidal axis of the section. Thus, the additional longitudinal reinforcement for torsion should approximately coincide with the centroid of the section. This is accomplished by requiring the longitudinal torsional reinforcement to be distributed around the perimeter of the closed stirrups.

2. At least one longitudinal bar is required in each corner of the stirrups. These bars provide anchorage for the stirrups and have also been found to be very effective in developing torsional strength and in controlling cracks.

FIGURE 6.53
Transverse
reinforcement
detail for
torsion—spalling
not restrained.

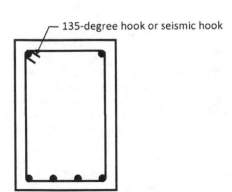

FIGURE **6.54**
Detailing
requirements for
longitudinal
torsional
reinforcement.

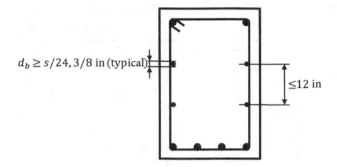

$d_b \geq s/24, 3/8$ in (typical)

≤ 12 in

3. The diameter of the longitudinal bars must be equal to or greater than $s/24$ but not less than three-eighths of an inch, the latter of which corresponds to a No. 3 bar. This is meant to prevent buckling of the longitudinal reinforcement due to the transverse component of the stress in the diagonal concrete compression struts.

Torsional reinforcement must be provided for a distance that is equal to or greater than $b_t + d$ beyond the point that it is theoretically required, where b_t is the width of that part of the cross-section that contains the closed stirrups resisting torsion (ACI 11.5.6.3). This distance is larger than that used for shear reinforcement and flexural reinforcement because torsional diagonal tension cracks develop in a helical form around a member.

6.4.6 Design for Combined Torsion, Shear, and Bending Moment

In members subjected to torsion, shear, and bending moments, the amounts of transverse reinforcement and longitudinal reinforcement required to resist all actions are determined using superposition (ACI 11.5.3.8).

The total required amount of transverse reinforcement per stirrup leg is equal to that required for shear plus that required for torsion:

$$\frac{A_v}{2s} + \frac{A_t}{s} \geq \text{the greater of} \begin{cases} \dfrac{0.375\sqrt{f_c'}b_w}{f_{yt}} \\ \\ \dfrac{25b_w}{f_{yt}} \end{cases} \tag{6.64}$$

Only the two legs of the stirrups that are adjacent to the sides of the beam are effective for torsion; this is consistent with the methodology that was presented earlier, where only a thin-walled tube resists torsion. Therefore, A_v in Eq. (6.64) is equal to the area of two legs of a closed stirrup. One-half of A_v is used in the equation because the summation is based on one stirrup leg. The area of transverse torsional reinforcement A_t was derived previously on the basis of one leg of a closed stirrup.

The minimum values of combined transverse reinforcement on the right-hand side of Eq. (6.64) are consistent with those presented in Section 6.2 for shear.

Longitudinal reinforcement for torsion is added to that required for flexure and is distributed around the perimeter of the beam in accordance with ACI 11.5.6.3. The minimum amount of longitudinal torsional reinforcement is given in Eq. (6.63). An area of steel equal to $A_\ell/4$ is added to each face of the beam.

The following must also be considered when designing and detailing the reinforcement:

- The most restrictive requirements for reinforcement spacing, cutoff points, and placement for torsion, shear, and flexure must be satisfied.

- Negative and positive flexural reinforcement may be cut off using the provisions of ACI 12.10 through 12.12 (see Section 6.2). However, when determining theoretical cutoff points, the area of longitudinal torsional reinforcement ($A_\ell/4$) must be subtracted from the total area of longitudinal steel provided at that face; the design flexural strength of the member at that section must be determined on the basis of the area of reinforcement required for flexure only.

- The structural integrity requirements of ACI 7.13 must also be satisfied when detailing the reinforcement.

The following example illustrates the design of a beam subjected to torsion, shear, and bending moments.

Example 6.13   Determine the required reinforcement for beam CD in the reinforced concrete floor system shown in Fig. 6.55. Assume normal-weight concrete with $f'_c = 4,000$ psi and Grade 60 reinforcement. Additional data are as follows:

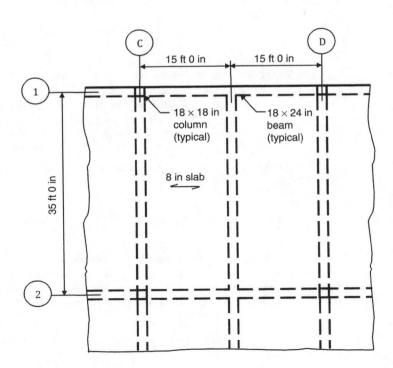

FIGURE 6.55 Partial floor plan of the floor system given in Example 6.13.

Superimposed dead load = 30 psf

Live load = 100 psf

Story height = 12 ft

Solution

Step 1: Determine T_u and check if torsional effects need to be considered. The flowchart shown in Fig. 6.49 is used to determine T_u.

(a) Strength reduction factor $\phi = 0.75$ for torsion and shear.

(b) Because normal-weight concrete is specified, $\lambda = 1.0$.

(c) The maximum factored bending moment at the exterior end of the 18 × 24 in beam that frames into the side of beam CD is determined from analysis and is equal to 243.2 ft kips. From equilibrium, this bending moment is transferred to beam CD as a torsional moment that acts at the midspan of the beam. Therefore, the maximum factored torsional moment T_u at the face of beam CD is equal to 243.2/2 = 121.6 ft kips. This is also T_u at the critical section because the torsional moment is a constant from the face of the support to the midspan of the beam.

(d) Torsion can be neglected when the factored torsional moment from analysis is less than the threshold torsion determined by Eq. (6.56):

$$T_u < \phi\lambda\sqrt{f_c'}\left(\frac{A_{cp}^2}{p_{cp}}\right)$$

Because the beam and slab are cast monolithically, A_{cp} and p_{cp} for beam CD can include a portion of the adjoining slab (ACI 11.5.1.1). In accordance with ACI 13.2.4, the overhanging flange width b_e is the smaller of the following (see Fig. 6.50):

$$h - h_f = 24 - 8 = 16 \text{ in (governs)}$$
$$4h_f = 4 \times 8 = 32 \text{ in}$$

Thus, the torsional properties of the section are as follows:

$$A_{cp} = b_w h + b_e h_f = (18 \times 24) + (16 \times 8) = 560 \text{ in}^2$$
$$p_{cp} = 2(h + b_w + b_e) = 2 \times (24 + 18 + 16) = 116 \text{ in}$$
$$A_{cp}^2/p_{cp} = 560^2/116 = 2,704 \text{ in}^3$$

(This is greater than A_{cp}^2/p_{cp} for the beam without flanges, which is equal to $(18 \times 24)^2/2(24 + 18) = 2,222 \text{ in}^3$.)

$$T_u = 121.6 \text{ ft kips} > 0.75 \times 1.0\sqrt{4,000}\left(\frac{560^2}{116}\right)/12,000 = 10.7 \text{ ft kips}$$

Therefore, torsional effects must be considered in beam CD.

(e) Because beam CD is part of an indeterminate system in which redistribution of internal forces can occur following torsional cracking, the maximum factored torsional moment at the critical section need not exceed the compatibility torsional moment determined by Eq. (6.57):

$$T_u = \phi 4\lambda\sqrt{f_c'}\left(\frac{A_{cp}^2}{p_{cp}}\right) = 4 \times 10.7 = 42.8 < 121.6 \text{ ft kips}$$

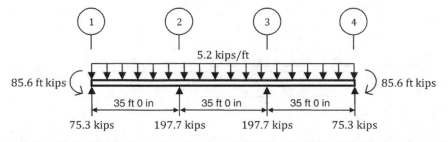

FIGURE 6.56 Free-body diagram of three-span beam.

Therefore, design beam CD for a torsional moment of 42.8 ft kips at the critical section.

Note that ACI 11.5.2.2 requires that the concentrated torsional moment of 2 × 42.8 =85.6 ft kips at the center of the span must be used in determining the redistribution of bending moments and shear forces in the beam that frames into beam CD. The reactions that are obtained after redistribution are transferred to beam CD (see Step 2).

Step 2: Determine the shear forces and bending moments in the beam that frames into beam CD. The total factored gravity load is as follows:

$$w_D = \left[\left(\frac{8}{12} \times 15\right) + \left(\frac{18 \times 16}{144}\right)\right] \times 0.150 + (0.030 \times 15) = 2.3 \text{ kips/ft}$$

$$w_L = 0.100 \times 15 = 1.5 \text{ kips/ft}$$

$$w_u = (1.2 \times 2.3) + (1.6 \times 1.5) = 5.2 \text{ kips/ft}$$

The free-body diagram of this three-span beam is shown in Fig. 6.56. Note that the bending moments at lines 1 and 4 must be equal to 85.6 ft kips after redistribution. The 75.3-kip reaction at line 1 must be transferred to beam CD.

Step 3: Determine the shear forces, bending moments, and torsional moments in beam CD. The total uniformly distributed load on beam CD is

$$w_D = \left(\frac{18 \times 24}{144}\right) \times 0.150 = 0.5 \text{ kips/ft}$$

$$w_u = 1.2 \times 0.5 = 0.6 \text{ kips/ft}$$

A two-dimensional analysis of the frame along column line 1 was performed assuming the far ends of the columns to be fixed (see Section 3.3 and Fig. 6.57).

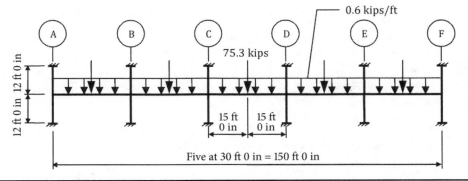

FIGURE 6.57 Two-dimensional analysis model for the frame along column line 1.

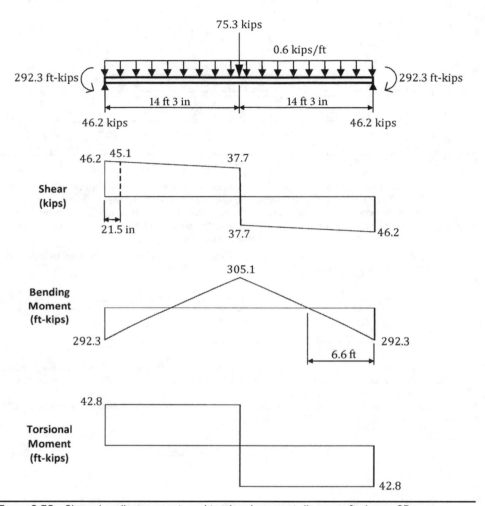

Figure 6.58 Shear, bending moment, and torsional moment diagrams for beam CD.

The shear diagram, bending moment diagram, and torsional moment diagram for beam CD are shown in Fig. 6.58.

Step 4: Determine the transverse and longitudinal reinforcements required for torsion. The flowchart shown in Fig. 6.51 is used to determine A_t/s and A_ℓ.

(a) As determined in step 1, reduction factor $\phi = 0.75$ for torsion and shear.

(b) The torsional moment T_u was determined in step 1 to be equal to 42.8 ft kips at the critical section (see Fig. 6.58).

(c) Determine ϕV_c from Fig. 6.33.

Because normal-weight concrete is specified, $\lambda = 1.0$.

Assume $d = 24 - 2.5 = 21.5$ in.

$$\phi V_c = \phi 2\lambda \sqrt{f_c'} b_w d = 0.75 \times 2 \times 1.0\sqrt{4,000} \times 18 \times 21.5/1,000 = 36.7 \text{ kips}$$

(d) Determine A_{oh} and p_h from Fig. 6.50.

Assuming 1.5-in clear cover to No. 4 closed stirrups in the beam web only,

$$x_1 = 18 - (2 \times 1.5) - 0.5 = 14.5 \text{ in}$$
$$y_1 = 24 - (2 \times 1.5) - 0.5 = 20.5 \text{ in}$$
$$A_{oh} = x_1 y_1 = 14.5 \times 20.5 = 297.3 \text{ in}^2$$
$$p_h = 2(x_1 + y_1) = 2 \times (14.5 + 20.5) = 70.0 \text{ in}$$

(e) Check the adequacy of cross-sectional dimensions.

For solid section, check Eq. (6.58):

$$\sqrt{\left(\frac{V_u}{b_w d}\right)^2 + \left(\frac{T_u p_h}{1.7 A_{oh}^2}\right)^2} \leq \phi \left(\frac{V_c}{b_w d} + 8\sqrt{f_c'}\right)$$

From Fig. 6.58, the factored shear force and torsional moment at the critical section, which is located 21.5 in from the face of the support, are

$$V_u = 45.1 \text{ kips}$$
$$T_u = 42.8 \text{ ft kips}$$

Also, $V_c = 36.7/0.75 = 48.9$ kips.

$$\sqrt{\left(\frac{V_u}{b_w d}\right)^2 + \left(\frac{T_u p_h}{1.7 A_{oh}^2}\right)^2} = \sqrt{\left(\frac{45.1 \times 1,000}{18 \times 21.5}\right)^2 + \left(\frac{42.8 \times 12,000 \times 70.0}{1.7 \times 297.3^2}\right)^2} = 266 \text{ psi}$$

$$\phi \left(\frac{V_c}{b_w d} + 8\sqrt{f_c'}\right) = 0.75 \left(\frac{48.9 \times 1,000}{18 \times 21.5} + 8\sqrt{4,000}\right) = 474 \text{ psi} > 266 \text{ psi}$$

Thus, the cross-sectional dimensions are adequate.

(f) Determine A_o:

$$A_o = 0.85 A_{oh} = 0.85 \times 297.3 = 252.7 \text{ in}^2$$

(g) Determine θ.

In accordance with ACI 11.5.3.6(a), θ may be taken equal to 45 degrees.

(h) Determine A_t/s.

Equation (6.61) is used to determine A_t/s:

$$\frac{A_t}{s} = \frac{T_u}{2\phi \cot\theta A_o f_{yt}} = \frac{42.8 \times 12,000}{2 \times 0.75 \times \cot 45 \times 252.7 \times 60,000} = 0.0226 \text{ in}^2/\text{in}$$

(i) Determine A_{cp} from Fig. 6.50.

From step 1, $A_{cp} = 560 \text{ in}^2$.

(j) Determine A_ℓ.

Equation (6.62) is used to determine A_ℓ:

$$A_\ell = \frac{A_t}{s} p_h \left(\frac{f_{yt}}{f_y}\right) \cot^2\theta = 0.0226 \times 70.0 \times \left(\frac{60}{60}\right) \times 1.0 = 1.58 \text{ in}^2$$

Check $A_{\ell,min}$ using Eq. (6.63):

$$A_{\ell,min} = \frac{5\sqrt{f_c'}A_{cp}}{f_y} - \frac{A_t}{s}p_h\left(\frac{f_{yt}}{f_y}\right)$$

In this equation, A_t/s is the larger of 0.0226 in²/in from Eq. (6.61) (governs) and $25b_w/f_{yt} = 25 \times 18/60,000 = 0.0075$ in²/in.
Therefore,

$$A_{\ell,min} = \frac{5 \times \sqrt{4,000} \times 560}{60,000} - 0.0226 \times 70.0 \times \left(\frac{60}{60}\right) = 1.37 \text{ in}^2 < 1.58 \text{ in}^2$$

Use $A_\ell = 1.58$ in².

Step 5: Determine the transverse reinforcement required for shear. The flowchart shown in Fig. 6.39 is used to determine A_v/s.

(a) As determined in Step 1, reduction factor $\phi = 0.75$ for torsion and shear.

(b) Determine ϕV_c from Fig. 6.33.

From Step 4, $\phi V_c = 36.7$ kips.

(c) Determine V_u and check if shear reinforcement is required.

At the critical section, $V_u = 45.1$ kips $> \phi V_c/2 = 36.7/2 = 18.4$ kips (see Fig. 6.58); therefore, shear reinforcement is required at the critical section.

Also, because $V_u = 45.1$ kips $> \phi V_c = 36.7$ kips, more than minimum shear reinforcement is required at the critical section.

(d) Determine the required shear reinforcement:

$$V_u - \phi V_c = 45.1 - 36.7 = 8.4 \text{ kips} < \phi 4\sqrt{f_c'}b_w d = 0.75 \times 4\sqrt{4,000} \times 18 \times 21.5/1,000 = 73.4 \text{ kips}$$

The required shear reinforcement is determined by Eq. (6.42):

$$\frac{A_v}{s} = \frac{(V_u - \phi V_c)}{\phi f_{yt} d} = \frac{8.4}{0.75 \times 60 \times 21.5} = 0.0087 \text{ in}^2/\text{in}$$

Step 6: Determine the total transverse reinforcement. The total required amount of transverse reinforcement per stirrup leg is determined by Eq. (6.64):

$$\frac{A_v}{2s} + \frac{A_t}{s} = \frac{0.0087}{2} + 0.0226 = 0.0270 \text{ in}^2/\text{in} > \begin{cases} \dfrac{0.375\sqrt{f_c'}b_w}{f_{yt}} = \dfrac{0.375\sqrt{4,000} \times 18}{60,000} = 0.0071 \text{ in}^2/\text{in} \\[3mm] \dfrac{25b_w}{f_{yt}} = \dfrac{25 \times 18}{60,000} = 0.0075 \text{in}^2/\text{in} \end{cases}$$

$$\text{Maximum spacing } s = \begin{cases} \dfrac{p_h}{8} = \dfrac{70.0}{8} = 8.8 \text{ in (governs)} \\[3mm] 12 \text{ in} \\[3mm] \dfrac{d}{2} = \dfrac{21.5}{2} = 10.8 \text{ in} \end{cases}$$

Location	M_u(ft kips)	A_s(in^2)
Face of support	292.3	3.26
Midspan	305.1	3.42

TABLE 6.6   Required Longitudinal Reinforcement for Flexure

For a No. 4 closed stirrup, the required spacing at the critical section is

$$s = \frac{A_b}{\frac{A_v}{2s} + \frac{A_t}{s}} = \frac{0.20}{0.0270} = 7.4 \text{ in} < 8.8 \text{ in}$$

Provide No. 4 closed stirrups spaced at 7.0 in on center at the critical section. From the torsional moment diagram in Fig. 6.58, it is evident that transverse torsional reinforcement is required over the entire span length. Because the maximum permitted spacing of 8.0 in is close to the provided spacing of 7.0 in at the critical section, use No. 4 closed stirrups over the entire span length for simpler detailing.

Step 7: Determine the longitudinal reinforcement required for flexure. The flowchart shown in Fig. 6.4 can be used to determine both the negative and positive flexural reinforcement. A summary of the required longitudinal reinforcement for flexure is given in Table 6.6.

Step 8: Determine the total longitudinal reinforcement. In accordance with ACI 11.5.6.2, the longitudinal torsional reinforcement must be distributed around the perimeter of the section with a maximum spacing of 12 in and must be combined with that required for flexure. Assign approximately one-fourth of this reinforcement to each face ($1.58/4 = 0.40$ in^2).

Use two No. 4 bars on each side face (area $= 2 \times 0.20 = 0.40$ in^2; bar diameter $= 0.50$ in $> s/24 = 7/24 = 0.29$ in and 0.375 in). The spacing of these bars on the 24-in-deep sides of the beam is less than 12 in, which satisfies the spacing requirement of ACI 11.5.6.2.

The remaining longitudinal reinforcement for torsion is distributed equally between the top and the bottom of the section: $0.5[1.58 - (2 \times 0.40)] = 0.39$ in^2.

- *Face of support*: Total top steel required $= 3.26 + 0.39 = 3.65$ in^2. For 4,000 psi concrete, minimum $A_s = 200 b_w d / f_y = 1.29$ in$^2 < 3.65$ in^2. Use four No. 9 bars ($A_s = 4.00$ in$^2 > 3.65$ in^2). These bars also satisfy the Code requirements for cover and spacing (see Tables 6.2 and 6.3).

- *Midspan*: Total bottom steel required $= 3.42 + 0.39 = 3.81$ in^2. Use four No. 9 bars ($A_s = 4.00$ in$^2 > 3.81$ in^2).

Step 9: Detail the reinforcement. According to the structural integrity requirements of ACI 7.13.2, at least one-sixth of the negative reinforcement (but not less than two bars) and at least one-quarter of the positive reinforcement (but not less than two bars) must be continuous over the span length (see Table 6.4). Thus provide two No. 9 bars at both the top and the bottom of the section, which are either continuous or are spliced using Class B tension splices or mechanical or welded splices. The splice locations of the top and bottom bars should be at or near the midspan and at or near the supports, respectively.

Two of the four No. 9 top bars can be theoretically cut off at the location where the factored bending moment is equal to the design flexural strength of the section based on a total area of steel equal to the area of two No. 9 bars minus the area of steel required for torsion: $2.00 - 0.39 = 1.61$ in^2. Thus, with $A_s = 1.61$ in^2 and using Fig. 5.10, $\phi M_n = 150.1$ ft kips. The distance x from the face of the support to the location where $M_u = 150.1$ ft kips is obtained by summing moments about the section at this location (see Fig. 6.58):

$$\frac{0.6x^2}{2} - 46.2x + 292.3 = 150.1$$

Solution of this equation gives $x = 3.1$ ft from the face of the support. The two No. 9 bars must extend a distance of $d = 21.5$ in (governs) or $12d_b = 12 \times 1.128 = 13.5$ in beyond the distance x (see Fig. 6.22). Thus, from the face of the support, the total bar length must be at least equal to $3.1 + (21.5/12) = 4.9$ ft.

The bars must also extend a full development length beyond the face of the support (see Fig. 6.22). The development length of the No. 9 bars is determined by Eq. (6.26):

$$\ell_d = \left[\frac{3}{40} \frac{f_y}{\lambda \sqrt{f_c'}} \frac{\psi_t \psi_e \psi_s}{(c_b + K_{tr})/d_b} \right] d_b \geq 12 \text{ in}$$

where $\lambda = 1.0$ for normal-weight concrete
$\psi_t = 1.3$ for top bars
$\psi_e = 1.0$ for uncoated reinforcement
$\psi_s = 1.0$ for No. 9 bars

$$c_b = 1.5 + 0.5 + \frac{1.128}{2} = 2.6 \text{ in}$$

$$= \frac{18 - 2(1.5 + 0.5) - 1.128}{2 \times 3} = 2.2 \text{ in (governs)}$$

$$K_{tr} = 0 \text{ (conservative)}$$

$$\frac{c_b + K_{tr}}{d_b} = \frac{2.2 + 0}{1.128} = 2.0 < 2.5$$

Therefore,

$$\ell_d = \left(\frac{3}{40} \frac{60,000}{1.0\sqrt{4,000}} \frac{1.3 \times 1.0 \times 1.0}{2.0} \right) \times 1.128 = 52.2 \text{ in} = 4.4 \text{ ft}$$

Thus, the total length of the No. 9 bars must be at least 4.9 ft beyond the face of the support.

According to ACI 12.10.5, flexural reinforcement shall not be terminated in a tension zone unless one or more of the conditions in that section are satisfied. The point of inflection is located approximately 6.6 ft from the face of the support (see Fig. 6.58). Thus, the No. 9 bars cannot be terminated at 4.9 ft. Check if the condition given in ACI 12.10.5.1 is satisfied; that is, check if the factored shear force V_u at the cutoff point is equal to or less than $2\phi V_n/3$.

With No. 4 closed stirrups spaced at 7.0 in on center, ϕV_n is determined by Eqs. (6.36), (6.37), and (6.41):

$$\phi V_n = \phi(V_c + V_n) = 0.75 \left(48.9 + \frac{0.4 \times 60 \times 21.5}{7.0} \right) = 92.0 \text{ kips}$$

At 4.9 ft from the face of the support, $V_u = 46.2 - (0.6 \times 4.9) = 43.3$ kips, which is less than $2 \times 92.0/3 = 61.3$ kips. Therefore, the two No. 9 bars can be terminated at 4.9 ft from the face of the support.

It is assumed in this example that the four No. 9 positive bars are continuous with Class B splices over the columns.

Figure 6.59 shows the reinforcement details for the beam.

Comments
Because there is a constant torsional moment along the entire span length, the transverse reinforcement and the two No. 4 side bars must be provided over the entire span. A reduction in the longitudinal torsional reinforcement in the flexural compression zones in accordance with ACI 11.5.3.9 has not been considered in this example.

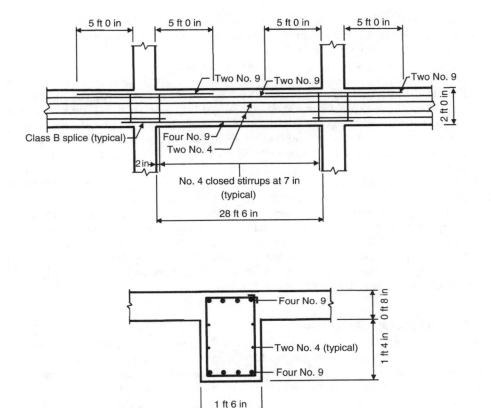

FIGURE 6.59 Reinforcement details for beam CD.

6.4.7   Alternative Design for Torsion

ACI 11.5.7 permits the use of torsion design procedures other than those presented in the Code for solid members with an aspect ratio h/b_t of 3 or greater, where b_t is the width of that part of a cross-section containing the closed stirrups. Such procedures must produce results that are in substantial agreement with results from comprehensive tests. Also, the detailing requirements of ACI 11.5.4 and the spacing of torsional reinforcement requirements of ACI 11.5.6 must be satisfied regardless of the procedure that is used.

References 29 through 31 contain examples of torsion design procedures that meet the requirements of ACI 11.5.7.

6.5   Deflections

6.5.1   Overview

It was discussed in Section 4.4 that two methods for controlling deflections are provided in ACI 9.5: (1) minimum thickness limitations and (2) computed deflection limitations.

Figure 4.3 contains a summary of the minimum thickness limitations for beams and one-way slabs that are not attached to partitions or other construction that is likely to be damaged by relatively large deflections.

The provisions given in ACI 9.5.2.2 through 9.5.2.6 may be used to determine the deflection of any reinforced concrete beam or one-way slab regardless of whether the minimum thickness requirements of ACI 9.5.2.1 are met or not. However, these provisions must be used where members are supporting elements that are likely to be damaged by relatively large deflections. Both immediate and long-term deflections of a member must be calculated, and these deflections must be equal to or less than the limiting deflections given in ACI Table 9.5(b) in order for this serviceability requirement to be satisfied.

Procedures for determining the deflection of beams and one-way slabs are presented later. Additional information on deflections can be found in Ref. 32. Although there are numerous methods available to determine deflections in reinforced concrete structures, it is important to note that these methods can only estimate deflections within an accuracy range of 20% to 40%. This is primarily due to the variability in the properties of the constituent materials of concrete (see Chap. 2) and to tolerances in construction. It is important for the designer to be aware of this range of accuracy, especially in the design of deflection-sensitive members.

6.5.2 Immediate Deflections

When a one-way reinforced concrete flexural member is loaded such that the maximum bending moment at service loads produces a tensile stress less than the modulus of rupture f_r of the concrete, the section is uncracked, and the immediate deflection can be calculated using methods (such as the moment-area method) or formulas for elastic deflections. In such cases, the gross moment of inertia I_g of the section (neglecting reinforcement) can be used in the calculations.

Once the tensile stress due to the applied loads equals or exceeds f_r, tension cracks occur, and the stiffness of the member decreases. Cracks occur when the bending moment exceeds the cracking moment M_{cr}, which is determined by ACI Eq. (9-9):

$$M_r = \frac{f_r I_g}{y_t} \qquad (6.65)$$

The modulus of rupture f_r is defined in ACI Eq. (9-10) as $7.5\lambda\sqrt{f'_c}$ and y_t is the distance from the centroidal axis of the gross section (neglecting reinforcement) to the tension face of the member.

At cracked sections, the moment of inertia can no longer be calculated using gross section properties; instead, a cracked moment of inertia I_{cr} must be used. The moment of inertia I_{cr} of a cracked rectangular beam with tension reinforcement can be computed using the information given in Fig. 6.60. A cracked section is obtained by transforming the reinforcing steel into an equivalent area of concrete. This is achieved by multiplying the area of steel A_s by the modular ratio n, which is the ratio of the modulus of elasticity of the reinforcing steel E_s to that of the concrete E_c. The cracked concrete below the neutral axis is ignored in the analysis, and the distance from the extreme compression face to the neutral axis is defined as a constant k times the effective depth d.

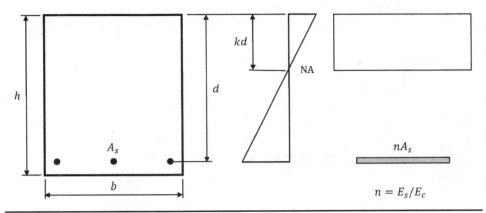

Figure 6.60   Cracked transformed section of a rectangular beam with tension reinforcement.

Taking the moment of areas about the neutral axis results in the following:

$$b \times kd \times \frac{kd}{2} = nA_s \times (d - kd) \tag{6.66}$$

Define $a_1 = b/nA_s$ and solve Eq. (6.66) for kd:

$$kd = \frac{\sqrt{2a_1 d + 1} - 1}{a_1} \tag{6.67}$$

Thus, the cracked moment of inertia I_{cr} is

$$I_{cr} = \frac{b(kd)^3}{3} + nA_s(d - kd)^2 \tag{6.68}$$

Similar equations can be derived for rectangular sections with tension and compression reinforcement and for flanged sections. A summary of these equations can be found in Fig. 6.61.

On the basis of the earlier discussion, it is evident that two different values of the moment of inertia would be required for calculating short-term or immediate deflections. In order to eliminate the need for two distinct moments of inertia, the Code permits the use of an effective moment of inertia I_e.

An idealized bilinear relationship between bending moment and immediate deflection is depicted in Fig. 6.62. When the applied moment M_a is less than the cracking moment M_{cr}, the section is uncracked, and the gross moment of inertia of the section I_g would be used to calculate the deflection. Once M_a equals or exceeds M_{cr}, cracking occurs, and the reduced moment of inertia of the cracked section I_{cr} is needed. In essence, I_e provides a transition between the upper bound I_g and the lower bound I_{cr} as a function of the applied bending moment.

Gross section	Cracked transformed section	Cracked moment of inertia I_{cr}
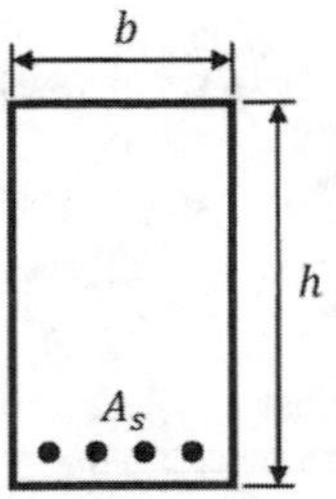	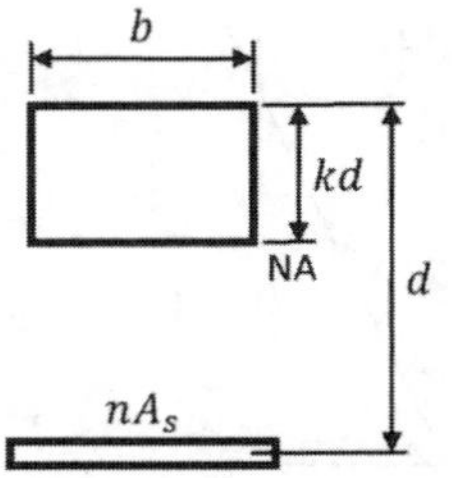	$I_{cr} = \dfrac{b(kd)^3}{3} + nA_s(d-kd)^2$ where $kd = \dfrac{\sqrt{2da_1+1}-1}{a_1}$
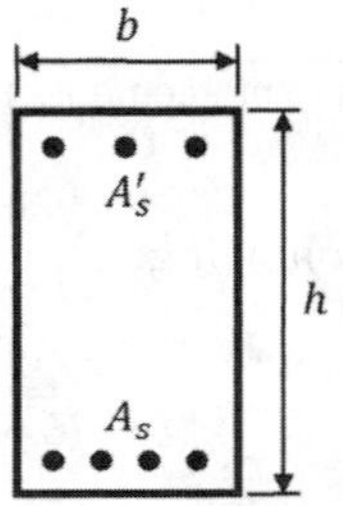	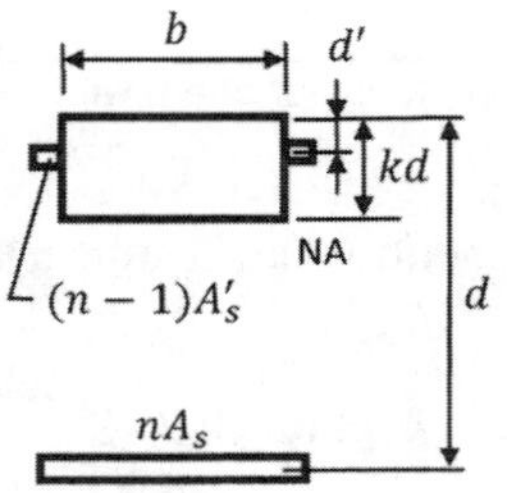	$I_{cr} = \dfrac{b(kd)^3}{3} + nA_s(d-kd)^2$ $\quad +(n-1)A_s'(kd-d')^2$ where $kd = \dfrac{\sqrt{2da_1+[1+(a_2d')/d]+(1+a_2)^2}}{a_1}$ $\qquad -\dfrac{(1+a_2)}{a_1}$

$n = E_s/E_c$
$I_g = bh^3/12$
$a_1 = b/nA_s$
$a_2 = (n-1)A_s'/nA_s$

(a)

Figure 6.61   Cracked section properties: (a) rectangular sections and (b) flanged sections. (*continued*)

ACI Eq. (9-8) is used to determine I_e for simply supported and cantilevered members:

$$I_e = \left(\frac{M_{cr}}{M_a}\right)^3 I_g + \left[1 - \left(\frac{M_{cr}}{M_a}\right)^3\right] I_{cr} \le I_g \tag{6.69}$$

In this equation, M_a is the maximum bending moment due to applicable service loads. In most cases, I_e will be less than I_g. However, in certain heavily reinforced flanged sections, I_e may be larger than I_g, which is not permitted.

For continuous members, ACI 9.5.2.4 permits I_e to be taken as the average of the values calculated by Eq. (9-8) at the critical negative and positive locations. It is also permitted to calculate I_e at the midspan for prismatic members because it has been shown that the midspan rigidity has the dominant effect on deflections.[33,34] The following equation from the 1989 edition of the Code can be used to determine I_e in continuous

Gross section	Cracked transformed section	Cracked moment of inertia I_{cr}
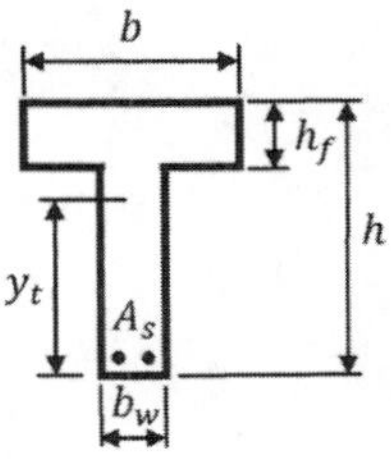	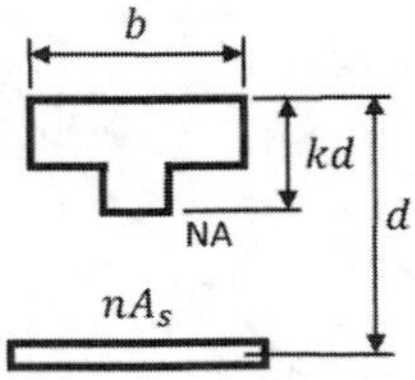	$I_{cr} = \dfrac{(b - b_w)h_f^3}{12} + \dfrac{b_w(kd)^3}{3}$ $+ (b - b_w)h_f\left(kd - \dfrac{h_f}{2}\right)^2$ $+ nA_s(d - kd)^2$ where $kd = \dfrac{\sqrt{a_3(2d + h_f a_4) + (1 + a_4)^2}}{a_3}$ $- \dfrac{(1 + a_4)}{a_3}$
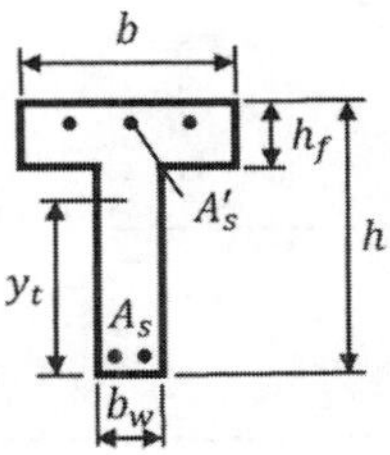	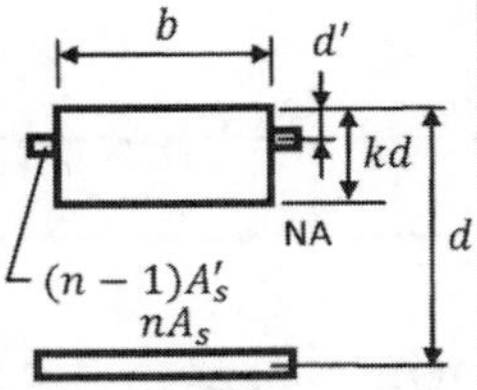	$I_{cr} = \dfrac{(b - b_w)h_f^3}{12} + \dfrac{b_w(kd)^3}{3}$ $+ (b - b_w)h_f\left(kd - \dfrac{h_f}{2}\right)^2$ $+ nA_s(d - kd)^2$ $+ (n - 1)A_s'(kd - d')^2$ where $kd = \dfrac{\sqrt{a_3(2d + h_f a_4 + 2a_2 d') + (1 + a_2 + a_4)^2}}{a_3}$ $- \dfrac{(1 + a_2 + a_4)}{a_3}$

$y_t = \left[(b - b_w)(h - 0.5h_f)h_f + 0.5b_w h^2\right]/\left[(b - b_w)h_f + b_w h\right]$

$I_g = \dfrac{(b - b_w)h_f^3}{12} + (b - b_w)h_f(h - 0.5h_f - y_t)^2 + \dfrac{b_w h^3}{12} + b_w h(y_t - 0.5h)^2$

$a_3 = b_w/nA_s$

$a_4 = h_f(b - b_w)/nA_s$

(b)

FIGURE 6.61 (Continued)

members, which is based on the average values at the supports ($I_{e(1)}$ and $I_{e(2)}$) and at the midspan ($I_{e(m)}$):

$$I_e = 0.50I_{e(m)} + 0.25(I_{e(1)} + I_{e(2)}) \tag{6.70}$$

In lieu of Eq. (6.70), the 1983 Code Commentary to Section 9.5.2.4 suggested the following weighted averages for continuous members:

For members continuous on both ends, $I_e = 0.70I_{e(m)} + 0.15(I_{e(1)} + I_{e(2)})$ (6.71)

For members continuous on one end, $I_e = 0.85I_{e(m)} + 0.15I_{e(1)}$ (6.72)

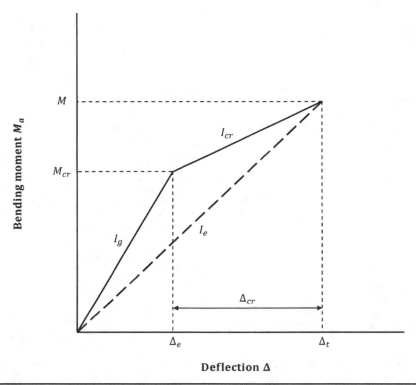

FIGURE 6.62 Bilinear bending moment–deflection relationship.

Note that in Eq. (6.72), $I_{e(1)}$ is the effective moment of inertia at the internal support; the effective moment of inertia at the external support is not required in the calculation of I_e for a member that is continuous on one end.

Also given in the 1983 Code Commentary to Section 9.5.2.4 is the following elastic equation that can be used to determine the immediate deflection Δ_i at the tips of cantilevers and at the midspan of simply supported and continuous members:

$$\Delta_i = \frac{5KM_a\ell^2}{48E_cI_e} \tag{6.73}$$

In this equation, M_a is the support bending moment for cantilevers and the midspan bending moment for simply supported and continuous beams. Values of the deflection coefficient K are given in Table 6.7 for members subjected to uniformly distributed loads with different span conditions. Values of K for other types of loading can be found in Ref. 35.

6.5.3 Long-term Deflections

Section 2.2 of this book describes the time-dependent effects of creep and shrinkage due to sustained loads. In one-way flexural members, creep and shrinkage cause

Span Condition	K
Cantilever	2.0
Simple	1.0
Continuous	$1.2 - 0.2(M_o/M_a)$, where $M_o = w\ell^2/8$

TABLE 6.7   Deflection Coefficient K

long-term deflections that can be two to three times greater than the immediate deflections. Not accounting for long-term deflection in reinforced concrete members can result in a significant underestimation of total deflection.

The Code method for determining long-term deflections is given in ACI 9.5.2.5. Long-term deflections due to creep and shrinkage Δ_{cs} are obtained by multiplying the immediate deflection Δ_i due to sustained loads by the factor λ_Δ given in ACI Eq. (9-11), which was developed on the basis of deflection data for rectangular, flanged, and box beams[36]:

$$\lambda_\Delta = \frac{\xi}{1 + 50\rho'} \tag{6.74}$$

In this equation, ξ is the time-dependent factor for sustained loads, which is given in ACI 9.5.2.5 and ACI Fig. R9.5.2.5 (see Table 6.8). The influence that compression reinforcement ($\rho' = A'_s/bd$) has on long-term deflections is also accounted for in this equation, where ρ' is determined at the midspan of simply supported and continuous spans and at the support of cantilevers.

Thus, the long-term deflection due to creep and shrinkage Δ_{cs} can be determined from the following equation:

$$\Delta_{cs} = \lambda_\Delta \Delta_i = \frac{\xi}{1 + 50\rho'} \Delta_i \tag{6.75}$$

It is important to reiterate that only the dead load and any portion of the live load that is sustained need to be considered in the calculation of long-term deflections.

Reference 32 contains a method to determine long-term deflections, which considers the effects of creep and shrinkage separately. This method is usually utilized in cases

Sustained Load Duration	ξ
5 years or more	2.0
12 months	1.4
6 months	1.2
3 months	1.0

TABLE 6.8   Time-dependent Factor for Sustained Loads

where part of the live load is considered as a sustained load. Other methods available in the literature are also given in that reference.

6.5.4 Maximum Permissible Computed Deflections

ACI Table 9.5(b) contains the maximum permissible deflections that are applicable when deflections are computed using the methods presented earlier. This table is the result of an effort to simplify an extensive set of limitations that would be required to cover all possible types of construction and loading conditions.

As expected, the deflection limitations are more stringent for members that support or are attached to nonstructural elements that are likely to be damaged by large deflections. The limitations given in ACI Table 9.59(b) may not be adequate in preventing instability (such as ponding of water on a roof) or adverse effects on other structural members; in such cases, the deflections must be limited (or, equivalently, member stiffness must be increased) to prevent the occurrence of such detrimental events.

Example 6.14 Determine the immediate and long-term deflections for the floor beam on line C between lines 1 and 2 in the reinforced concrete floor system shown in Fig. 6.55, given the design data in Example 6.13. Assume that the beam is not supporting or attached to nonstructural elements likely to be damaged by large deflections. Also assume that 30% of the live load is sustained and that the beam is reinforced with five No 9 bars at the midspan and five No. 10 bars at the interior support.

Solution

Step 1: Determine the service loads and bending moments.

$$w_D = \left[\left(\frac{8}{12} \times 150\right) + 30\right] \times 15 + \left(\frac{18 \times 16}{144} \times 150\right) = 2{,}250\,\text{plf}$$

$$w_L = 100 \times 15 = 1{,}500\,\text{plf}$$

The approximate moment coefficients given in ACI 8.3.3 are used to determine the service bending moments (see Fig. 3.17). The bending moments at the external support are not required for the calculation of the effective moment of inertia I_e for a member that is continuous on one end [see Eqs. (6.69) and (6.72)].

- At the midspan,

$$M_D^+ = \frac{w_D \ell_n^2}{14} = \frac{2.25 \times 33.5^2}{14} = 180.4\,\text{ft kips}$$

$$M_L^+ = \frac{w_L \ell_n^2}{14} = \frac{1.50 \times 33.5^2}{14} = 120.2\,\text{ft kips}$$

Sustained bending moment $M_{sus}^+ = M_D^+ + 0.3 M_L^+ = 180.4 + (0.3 \times 120.2) = 216.5\,\text{ft kips}$

- At the exterior face of the first interior support,

$$M_D^- = \frac{w_D \ell_n^2}{10} = \frac{2.25 \times 33.5^2}{10} = 252.5\,\text{ft kips}$$

$$M_L^- = \frac{w_L \ell_n^2}{10} = \frac{1.50 \times 33.5^2}{10} = 168.3\,\text{ft kips}$$

Sustained bending moment $M_{sus}^- = M_D^- + 0.3 M_L^- = 252.5 + (0.3 \times 168.3) = 303.0\,\text{ft kips}$

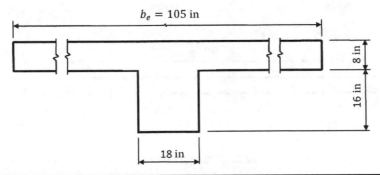

FIGURE 6.63   Gross section of beam at positive moment section.

Step 2: Determine the material properties of the concrete and the steel.

For normal-weight concrete, $f_r = 7.5\lambda\sqrt{f_c'} = 7.5 \times 1.0\sqrt{4,000} = 474\,\text{psi}$

From ACI 8.5.1, $E_c = 57,000\sqrt{f_c'} = 57,000\sqrt{4,000} = 3,605,000\,\text{psi}$

From ACI 8.5.2, $E_s = 29,000,000\,\text{psi}$

The modular ratio $n = E_s/E_c = 8.0$

Step 3: Determine the gross and cracked moments of inertia. Determine the effective width of the flange b_e from Fig. 5.23:

$$b_e = \begin{cases} \text{span length}/4 = 35 \times 12/4 = 105.0 \text{ in (governs)} \\ b_w + 16h_f = 18 + (16 \times 8.0) = 146.0 \text{ in} \\ \text{beam spacing} = 15 \times 12 = 180 \text{ in} \end{cases}$$

- *Positive moment section*: The gross section at the positive moment section at the midspan is depicted in Fig. 6.63.

  The gross section properties are (see Fig. 6.61)

$$y_t = \frac{(b_e - b_w)(h - 0.5h_f)h_f + 0.5b_w h^2}{(b_e - b_w)h_f + b_w h}$$

$$= \frac{(105 - 18) \times [24 - (0.5 \times 8)] \times 8 + (0.5 \times 18 \times 24^2)}{[(105 - 18) \times 8] + (18 \times 24)} = 16.9 \text{ in}$$

$$I_g = \frac{1}{12}(b_e - b_w)h_f^3 + (b_e - b_w)h_f(h - 0.5h_f - y_t)^2 + \frac{1}{12}b_w h^3 + b_w h(y_t - 0.5h)^2$$

$$= \frac{1}{12}(105 - 18)(8)^3 + (105 - 18)(8)(24 - 4 - 16.9)^2 + \frac{1}{12}(18)(24)^3 + (18)(24)(16.9 - 12)^2$$

$$= 41,509 \text{ in}^4$$

Assuming a rectangular compression area, the cracked section properties are the following (see Figs. 6.61 and 6.64):

$$a_1 = \frac{b_e}{nA_s} = \frac{105}{8 \times (5 \times 1.0)} = 2.6/\text{in}$$

$$kd = \frac{\sqrt{2da_1 + 1} - 1}{a_1} = \frac{\sqrt{(2 \times 21.5 \times 2.6) + 1} - 1}{2.6} = 3.7 \text{ in} < h_f = 8.0 \text{ in}$$

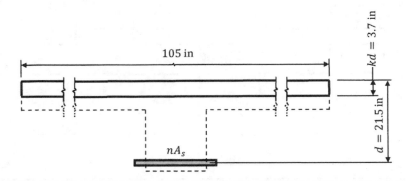

Figure 6.64   Cracked transformed section of beam at positive moment section.

Therefore, the assumption of a rectangular compression area is correct.

$$I_{cr} = \frac{b_e(kd)^3}{3} + nA_s(d - kd)^2$$

$$= \frac{105 \times 3.7^3}{3} + 8 \times 5 \times (21.5 - 3.7)^2 = 14{,}447 \text{ in}^4$$

- *Negative moment section*: The negative moment section is the 18 × 24 in web. Thus,

$$I_g = \frac{1}{12}b_w h^3 = \frac{1}{12} \times 18 \times 24^3 = 20{,}736 \text{ in}^4$$

$$a_1 = \frac{b_w}{nA_s} = \frac{18}{8 \times (5 \times 1.27)} = 0.35/\text{in}$$

$$kd = \frac{\sqrt{2da_1 + 1} - 1}{a_1} = \frac{\sqrt{(2 \times 21.5 \times 0.35) + 1} - 1}{0.35} = 8.6 \text{ in}$$

$$I_{cr} = \frac{b_w(kd)^3}{3} + nA_s(d - kd)^2$$

$$= \frac{18 \times 8.6^3}{3} + 8 \times 6.35 \times (21.5 - 8.6)^2 = 12{,}270 \text{ in}^4$$

Step 4: Determine the effective moments of inertia.

- *Positive moment section*: The cracking moment is determined by Eq. (6.65):

$$M_{cr} = \frac{f_r I_g}{y_t} = \frac{474 \times 41{,}509}{16.9 \times 12{,}000} = 97.0 \text{ ft kips}$$

The effective moment of inertia I_e is determined by Eq. (6.69) for dead, sustained, and dead plus live load cases:

$$\frac{M_{cr}}{M_D^+} = \frac{97.0}{180.4} = 0.54$$

$$(I_e)_D^+ = \left(\frac{M_{cr}}{M_D^+}\right)^3 I_g + \left[1 - \left(\frac{M_{cr}}{M_D^+}\right)^3\right] I_{cr}$$

$$= [(0.54)^3 \times 41{,}509] + [1 - (0.54)^3] \times 14{,}447 = 18{,}708 \text{ in}^4 < I_g$$

$$\frac{M_{cr}}{M_{sus}^+} = \frac{97.0}{216.5} = 0.45$$

$$(I_e)_{sus}^+ = \left(\frac{M_{cr}}{M_{sus}^+}\right)^3 I_g + \left[1 - \left(\frac{M_{cr}}{M_{sus}^+}\right)^3\right] I_{cr}$$

$$= [(0.45)^3 \times 41{,}509] + [1 - (0.45)^3] \times 14{,}447 = 16{,}913 \text{ in}^4 < I_g$$

$$\frac{M_{cr}}{M_{D+L}^+} = \frac{97.0}{300.6} = 0.32$$

$$(I_e)_{D+L}^+ = \left(\frac{M_{cr}}{M_{D+L}^+}\right)^3 I_g + \left[1 - \left(\frac{M_{cr}}{M_{D+L}^+}\right)^3\right] I_{cr}$$

$$= [(0.32)^3 \times 41{,}509] + [1 - (0.32)^3] \times 14{,}447 = 15{,}334 \text{ in}^4 < I_g$$

- *Negative moment section*:

$$M_{cr} = \frac{f_r I_g}{y_t} = \frac{474 \times 20{,}736}{12 \times 12{,}000} = 68.3 \text{ ft kips}$$

$$\frac{M_{cr}}{M_D^-} = \frac{68.3}{252.5} = 0.27$$

$$(I_e)_D^- = \left(\frac{M_{cr}}{M_D^-}\right)^3 I_g + \left[1 - \left(\frac{M_{cr}}{M_D^-}\right)^3\right] I_{cr}$$

$$= [(0.27)^3 \times 20{,}736] + [1 - (0.27)^3] \times 12{,}270 = 12{,}437 \text{ in}^4 < I_g$$

$$\frac{M_{cr}}{M_{sus}^-} = \frac{68.3}{303.0} = 0.23$$

$$(I_e)_{sus}^- = \left(\frac{M_{cr}}{M_{sus}^-}\right)^3 I_g + \left[1 - \left(\frac{M_{cr}}{M_{sus}^-}\right)^3\right] I_{cr}$$

$$= [(0.23)^3 \times 20{,}736] + [1 - (0.23)^3] \times 12{,}270 = 12{,}373 \text{ in}^4 < I_g$$

$$\frac{M_{cr}}{M_{D+L}^-} = \frac{68.3}{420.8} = 0.16$$

$$(I_e)_{D+L}^- = \left(\frac{M_{cr}}{M_{D+L}^-}\right)^3 I_g + \left[1 - \left(\frac{M_{cr}}{M_{D+L}^-}\right)^3\right] I_{cr}$$

$$= [(0.16)^3 \times 20{,}736] + [1 - (0.16)^3] \times 12{,}270 = 12{,}305 \text{ in}^4 < I_g$$

- *Average effective moments of inertia*: The average effective moments of inertia are obtained from Eq. (6.72) for one end continuous:

$$(I_e)_D = 0.85(I_e)_D^+ + 0.15(I_e)_D^- = (0.85 \times 18{,}708) + (0.15 \times 12{,}437) = 17{,}767 \text{ in}^4$$
$$(I_e)_{sus} = 0.85(I_e)_{sus}^+ + 0.15(I_e)_{sus}^- = (0.85 \times 16{,}913) + (0.15 \times 12{,}373) = 16{,}232 \text{ in}^4$$
$$(I_e)_{D+L} = 0.85(I_e)_{D+L}^+ + 0.15(I_e)_{D+L}^- = (0.85 \times 15{,}334) + (0.15 \times 12{,}305) = 14{,}880 \text{ in}^4$$

Step 5: Determine the immediate deflections and check the maximum permissible deflection for live load. The immediate deflections Δ_i are determined using Eq. (6.73) and Table 6.7:

$$\Delta_i = \frac{5KM_a \ell^2}{48E_c I_e}$$

For continuous members,

$$K = 1.2 - 0.2(M_o/M_a)$$

where $\quad M_o = w\ell^2/8$
$\qquad M_a = w\ell^2/14$

Thus, $K = 1.2 - 0.2(14/8) = 0.85$.

$$(\Delta_i)_D = \frac{5KM_D^+\ell^2}{48E_c(I_e)_D} = \frac{5 \times 0.85 \times 180.4 \times 12,000 \times (33.5 \times 12)^2}{48 \times 3,605,000 \times 17,767} = 0.48 \text{ in}$$

$$(\Delta_i)_{sus} = \frac{5KM_{sus}^+\ell^2}{48E_c(I_e)_{sus}} = \frac{5 \times 0.85 \times 216.5 \times 12,000 \times (33.5 \times 12)^2}{48 \times 3,605,000 \times 16,232} = 0.64 \text{ in}$$

$$(\Delta_i)_{D+L} = \frac{5KM_{D+L}^+\ell^2}{48E_c(I_e)_{D+L}} = \frac{5 \times 0.85 \times 300.6 \times 12,000 \times (33.5 \times 12)^2}{48 \times 3,605,000 \times 14,880} = 0.96 \text{ in}$$

$$(\Delta_i)_L = (\Delta_i)_{D+L} - (\Delta_i)_D = 0.96 - 0.48 = 0.48 \text{ in}$$

For a floor member that is not supporting or attached to nonstructural elements likely to be damaged by large deflections, the maximum permissible immediate live load deflection from ACI Table 9.5(b) is $\ell/360 = 33.5 \times 12/360 = 1.1 \text{ in} > 0.48 \text{ in}$.

Step 6: Determine the long-term deflections and check the maximum permissible deflection. Determine the factor λ_Δ by Eq. (6.74) and Table 6.8, assuming a 5-year duration of loading:

$$\lambda_\Delta = \frac{\xi}{1 + 50\rho'} = \frac{2.0}{1 + 0} = 2.0$$

The long-term deflection due to creep and shrinkage Δ_{cs} is determined by Eq. (6.75), using the immediate deflection for the sustained loads:

$$\Delta_{cs} = \lambda_\Delta(\Delta_i)_{sus} = 2.0 \times 0.64 = 1.3 \text{ in}$$

The sum of the long-term deflection due to the sustained loads and the immediate deflection due to any additional live load is $1.3 + 0.48 = 1.8 \text{ in}$.

For a floor member that is not supporting or attached to nonstructural elements likely to be damaged by large deflections, the maximum permissible deflection from ACI Table 9.5(b) is $\ell/240 = 33.5 \times 12/240 = 1.7 \text{ in} < 1.8 \text{ in}$.

Comments
Generally, the most efficient way to decrease deflections is to increase the depth of a member. However, this may not always be possible because of architectural or other considerations.

In this case, because the long-term deflection due to the sustained loads plus the immediate deflection due to any additional live load are slightly larger than the permissible value, increasing the width of the beam and the compressive strength of the concrete are two viable options for decreasing deflections. Also, using compressive reinforcement would decrease the factor λ_Δ and the long-term deflection Δ_{cs}.

References

1. Gergely, P., and Lutz, L. A. 1968. Maximum crack width in reinforced concrete flexural members. In: *Causes, Mechanism, and Control of Cracking in Concrete*, SP-20. American Concrete Institute (ACI), Farmington Hill, MI, pp. 87-117.

2. Karr, P. H. 1966. High strength bars as concrete reinforcement, part 8: Similitude in flexural cracking of t-beam flanges. *Journal of the PCA Research and Development Laboratories* 8(2): 2-12.

3. Base, G. D., Reed, J. B., Beeby, A. W., and Taylor, H. P. J. 1966. *An Investigation of the Crack Control Characteristics of Various Types of Bar in Reinforced Concrete Beams*, Research Report No. 18. Cement and Concrete Association, London, p. 44.

4. Beeby, A. W. 1979. The prediction of crack widths in hardened concrete. *The Structural Engineer* 57A(1):9-17.

5. Frosch, R. J. 1999. Another look at cracking and crack control in reinforced concrete. *ACI Structural Journal* 96(3):437-442.

6. ACI Committee 318. 1999. Closure to public comments on ACI 318-99. *Concrete International* 21(5):318-1-318-50.

7. Frantz, G. C., and Breen, J. E. 1980. Cracking on the side faces of large reinforced concrete beams. *ACI Journal, Proceedings* 77(5):307-313.

8. Frosch, R. J. 2002. Modeling and control of side face beam cracking. *ACI Structural Journal* 99(3):376-385.

9. American Concrete Institute (ACI), Committee 408. 1966. Bond stress—The state of the art. *ACI Journal, Proceedings* 63(11):1161-1188.

10. Jirsa, J. O., and Breen, J. E. 1981. *Influence of Casting Position and Shear on Development and Splice Length—Design Recommendations*, Research Report 242-3F. Center for Transportation Research, Bureau of Engineering Research, University of Texas at Austin, TX.

11. Jeanty, P. R., Mitchell, D., and Mirza, M. S. 1988. Investigation of "top bar" effects in beams. *ACI Structural Journal* 85(3):251-257.

12. Treece, R. A., and Jirsa, J. O. 1989. Bond strength of epoxy-coated reinforcing bars. *ACI Materials Journal* 86(2):167-174.

13. Johnston, D. W., and Zia. P. 1982. *Bond Characteristics of Epoxy-coated Reinforcing Bars*, Report No. FHWA/NC/82-002. Department of Civil Engineering, North Carolina State University, Raleigh, NC.

14. Mathey, R. G., and Clifton, J. R. 1976. Bond of coated reinforcing bars in concrete. *Journal of the Structural Division, American Society of Civil Engineers* 102(ST1):215-228.

15. American Concrete Institute (ACI), Committee 439. 2007. *Types of Mechanical Splices for Reinforcing Bars*, ACI 439.3. ACI, Farmington Hills, MI.

16. American Welding Society (AWS). 2005. *Structural Welding Code—Reinforcing Steel*, AWS D1.4/D1.4M. AWS, Miami, FL.

17. Concrete Reinforcing Steel Institute (CRSI). 2008. *Reinforcement Anchorage and Splices*, 5th ed. CRSI, Schaumburg, IL.

18. American Concrete Institute (ACI). 2004. *ACI Detailing Manual*, SP-66. ACI, Farmington Hills, MI.

19. Concrete Reinforcing Steel Institute (CRSI). 2002. *CRSI Design Handbook*, 10th ed. CRSI, Schaumburg, IL.

20. American Concrete Institute/American Society of Civil Engineers (ACI-ASCE) Committee 326. 1962. Shear and diagonal tension. *ACI Journal, Proceedings* 59(1-3):1-30, 277-344, and 352-396.

21. Kani, G. N. J. 1966. Basic facts concerning shear failure. *ACI Journal, Proceedings* 63(6):675-692.

22. Kani, G. N. J. 1967. How safe are our large reinforced concrete beams? *ACI Journal, Proceedings* 64(3):128-141.

23. Roller, J. J., and Russell, H. G. 1990. Shear strength of high-strength concrete beams with web reinforcement. *ACI Structural Journal* 87(2):191-198.

24. Lampert, P. 1973. Postcracking stiffness of reinforced concrete beams in torsion and bending. In: *Analysis of Structural Systems for Torsion*, SP-35. American Concrete Institute, Farmington Hills, MI, pp. 385-433.

25. Collins, M. P., and Lampert, P. 1973. Redistribution of moments at cracking—the key to simpler torsion design? In: *Analysis of Structural Systems for Torsion*, SP-35. American Concrete Institute, Farmington Hills, MI, pp. 343-383.

26. Hsu, T. T. C., and Burton, K. T. 1974. Design of reinforced concrete spandrel beams. *Proceedings of the American Society of Civil Engineers* 100(ST1):209-229.

27. Hsu, T. T. C. 1990. Shear flow zone in torsion of reinforced concrete. *Journal of Structural Engineering* 116(11):3206-3226.

28. Mitchell, D., and Collins, M. P. 1976. Detailing for torsion. *ACI Journal, Proceedings* 73(8):506-511.

29. Zia, P., and McGee, W. D. 1974. Torsion design of prestressed concrete. *PCI Journal* 19(2):46-65.

30. Zia, P., and Hsu, T. T. C. 2004. Design for torsion and shear in prestressed concrete flexural members. *PCI Journal* 49(3):34-42.

31. Collins, M. P., and Mitchell, D. 1980. Shear and torsion design of prestressed and non-prestressed concrete beams. *PCI Journal* 25(4):32-100.

32. American Concrete Institute (ACI), Committee 435. 1995. *Control of Deflection in Concrete Structures*, ACI 435. ACI, Farmington Hills, MI.

33. American Concrete Institute (ACI). 1974. *Deflections of Concrete Structures*, SP-43. ACI, Farmington Hills, MI.

34. American Concrete Institute (ACI), Committee 435. 1973 (Reapproved 1989). *Deflections of Continuous Concrete Beams*, ACI 435.5. ACI, Farmington Hills, MI.

35. Branson, D. E. 1977. *Deformation of Concrete Structures*. McGraw-Hill, New York.

36. Branson, D. E. 1971. Compression steel effect on long-time deflections. *ACI Journal, Proceedings* 68(8):555-559.

Problems

6.1. Determine the slab thickness to satisfy the deflection criteria of ACI 9.5 for a six-span continuous one-way slab supported on beams. The center-to-center span lengths are 18 ft 0 in for the end spans and 20 ft 0 in for all interior spans. The widths of the edge beams and interior beams are 20 and 24 in, respectively. Assume normal-weight concrete with $f_c' = 4,000$ psi.

6.2. Given the one-way slab system described in Problem 6.1, determine the required negative and positive reinforcements in an end span and in an interior span, assuming a 9.5-in-thick slab, a superimposed service dead load of 20 psf, a service live load of 50 psf, and Grade 60 reinforcement.

6.3. Determine the required flexural reinforcement for a 20-in-wide and 28-in-deep cantilever beam subjected to a factored bending moment equal to 140 ft kips. Assume normal-weight concrete with $f_c' = 4,000$ psi, a $3/4$-in maximum aggregate size, and Grade 60 reinforcement. Draw a section of the beam showing the reinforcement.

6.4. Determine the required positive flexural reinforcement of a 32-in-wide and 20-in-deep beam subjected to a positive factored bending moment equal to 700 ft kips. Assume normal-weight concrete with $f_c' = 5,000$ psi, a 1-in maximum aggregate size, and Grade 60 reinforcement.

6.5. Given the information provided in Problem 6.4, determine the required flexural reinforcement for a positive factored bending moment equal to 825 ft kips and a tension-controlled section.

6.6. Determine the beam thickness to satisfy the deflection criteria of ACI 9.5 for a three-span continuous beam supported on columns. The center-to-center span lengths are 25 ft 0 in for one of the end spans, 32 ft 6 in for the interior span, and 29 ft 7 in for the other end span. The edge columns are 18 × 18 in, and the interior columns are 22 × 22 in. The beams are spaced 12 ft 0 in on center, and they are not supporting or attached to partitions or other construction likely to be damaged by large deflections. The slab thickness is 7 in. Assume lightweight concrete with a density of 110 pcf, $f_c' = 4,000$ psi, and Grade 60 reinforcement.

6.7. Given the information provided in Problem 6.6, determine the required positive and negative flexural reinforcement for the interior beam. Assume that the beam is 26 in wide and 20 in deep. Also assume a superimposed dead load of 20 psf and a live load of 100 psf.

6.8. Given the information provided in Problems 6.6 and 6.7, determine the required shear reinforcement for the interior beam.

6.9. Determine the required spacing of shear reinforcement over the span of a 28-in-wide and 18-in-deep beam with a clear span equal to 30 ft 2 in. The beam supports a total factored uniformly distributed load of 5.0 kips/ft. Assume No. 4 U-stirrups with two legs (Grade 60) and normal-weight concrete with $f'_c = 4{,}000$ psi.

6.10. A 36-in-wide and 24-in-deep beam is reinforced with 10 No. 8 top bars at a support. It has been determined that 6 of the 10 No. 8 bars can be theoretically cut off at a distance of 3.5 ft from the face of the support. Determine the total required length of the six No. 8 bars from the face of the support. Assume normal-weight concrete with $f'_c = 4{,}000$ psi, Grade 60 reinforcement, and a clear cover to the No. 8 bars of 1.875 in.

6.11. Given the information provided in Example 6.13, determine the positive and negative flexural reinforcement in the beam on line 2 between C and D for the given gravity loads and a nominal wind bending moment at each end of the beam equal to 50 ft kips. Also determine the theoretical cutoff points of the negative reinforcement.

6.12. Determine the required spacing of the closed stirrups at the critical section of a 20-in-wide and 32-in-deep beam subjected to a factored shear force $V_u = 100$ kips and factored torsional moment $T_u = 75$ ft kips. Assume normal-weight concrete with $f'_c = 4{,}000$ psi and No. 4 closed stirrups (Grade 60) with a clear cover of 1.5 in.

6.13. Given the information provided in Problem 6.12, determine the required longitudinal reinforcement for torsion A_ℓ, assuming $A_t/s = 0.021$ in^2/in.

6.14. Given the information provided in Example 6.13, determine the reinforcement for combined torsion, shear, and bending moment for beam CD, assuming that there are two beams framing into beam CD instead of one. The beams are spaced at 10 ft 0 in on center.

6.15. Determine immediate and long-term deflections of a 36-in-wide and 22-in-deep rectangular beam reinforced with 12 No. 8 bars at the interior supports and 9 No. 8 bars at the midspan. The clear span of the beam is 28 ft 0 in, and the beam is continuous at both ends. The beams are spaced 32 ft 6 in on center. The total dead load, including the weight of the beam, is 130 psf, and the live load is 100 psf (20% of the live load is sustained). Assume normal-weight concrete with $f'_c = 4{,}000$ psi. Also assume that the beam is not supporting or attached to nonstructural elements likely to be damaged by large deflections.

CHAPTER 7

Two-Way Slabs

7.1 Introduction

A two-way slab is defined on the basis of the ratio of its panel dimensions. Where the ratio of the long to the short side of a slab panel is two or more, load transfer is predominantly by bending in the short direction, and the panel essentially acts as a one-way slab. In such cases, the slab can be designed and detailed using the methods and procedures outlined in Chap. 6. As the ratio of the sides of a slab panel approaches unity (i.e., as a panel approaches a square shape), significant load is transferred by bending in both orthogonal directions, and the panel must be treated as a two-way slab rather than a one-way slab. Thus, by definition, a two-way slab system has a long-to-short panel (span) ratio that is two or less. Design and detailing requirements for two-way slab systems are covered in this chapter.

Descriptions of the different types of two-way slab systems are given, and information is provided on when these systems are economical for various span and loading conditions.

The minimum slab thickness requirements of the Code, based on serviceability and shear strength, are presented for slabs with and without interior beams. A brief discussion on the calculation of immediate and long-term deflections in two-way systems is also included.

Analysis methods for two-way slabs are covered, including approximate methods (the Direct Design Method and the Equivalent Frame Method) that can be used to analyze typical concrete framing systems. These approximate methods greatly simplify the determination of the positive and negative bending moments at the critical sections in the slab system.

Design requirements are presented for flexure, serviceability, and shear. Included are methods and procedures on how to determine a minimum slab thickness that satisfies all three of these requirements. As in the case of beams and one-way slabs, the required amount of flexural reinforcement in two-way slabs is determined using the strength design method presented in Chap. 5. The size and spacing of the reinforcement must be chosen so that the provisions related to strength and serviceability are not violated. Flexural reinforcement must be fully developed. Furthermore, reinforcement details are presented for typical two-way systems.

7.2 Two-Way Slab Systems

7.2.1 Beam-Supported Slab

A solid slab supported on beams on all four sides is depicted in Fig. 7.1. This system, which was the original slab system in reinforced concrete, can accommodate a wide range of span and loading conditions. However, it is not as economical as other two-way systems with similar span and loading conditions, due to formwork costs and costs associated with deeper overall floor thickness. As such, column-line beams are not used as often as they once were, except in cases where the demands for lateral force resistance are relatively large. For example, because flat-plate systems are not permitted to be the primary seismic force–resisting system in areas of high seismicity, moment-resisting frames with column-line beams must be used as the seismic force–resisting system.

The minimum thickness of the slab h that satisfies serviceability requirements depends on the longer of the two spans and the average flexural stiffness of the beams on the perimeter of the panel (see ACI 9.5.3.3). Distribution of bending moment and shear force between the slab and beams also depends on the relative stiffness of the beams. Two-way (punching) shear, which is discussed later, is usually not a concern for this type of two-way system. The largest required slab and beam thickness from all of the panels should be used over the entire floor or roof area whenever possible for economy in formwork.

The beams in this system are designed and detailed using the methods presented in Chap. 6 on the basis of the portion of load that is assigned to them in accordance with the provisions in Chap. 13 of the Code.

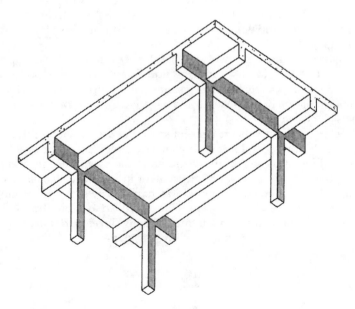

Figure 7.1
Beam-supported slab.

FIGURE 7.2 Flat-plate system.

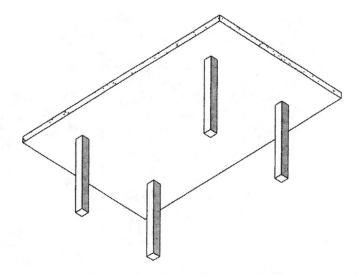

7.2.2 Flat Plate

A flat-plate floor system is a two-way concrete slab supported directly on columns with reinforcement in two orthogonal directions (Fig. 7.2). This system, which is popular in residential buildings (e.g., hotels and apartments), has the advantages of simple construction and formwork and a flat ceiling, the latter of which reduces ceiling finishing costs because the architectural finish can be applied directly to the underside of the slab. Even more significant are the cost savings associated with the low story heights made possible by the shallow floor system. Smaller vertical runs of cladding, partition walls, mechanical systems, plumbing, elevators, and a number of other items of construction translate into large cost savings, especially for medium- and high-rise buildings. Moreover, where the total height of a building is restricted, using a flat plate can result in more stories accommodated within the set height.

Flat plates are typically economical for span lengths between 15 and 25 ft when subjected to moderate live loads. The thickness h of a flat plate will usually be controlled by the deflection requirements of ACI 9.5.3.2 for relatively short spans and live loads of 50 psf or less. Flexural reinforcement at the critical sections will be approximately the minimum amount specified in ACI 13.3 in such cases. Therefore, utilizing a slab thickness greater than the minimum required for serviceability is not economical because a thicker slab requires more concrete without an accompanying reduction in reinforcement. Because the minimum slab thickness requirements of ACI 9.5.3.2 are independent of the concrete compressive strength, a 4,000 psi concrete mixture is usually the most economical; using a concrete strength that is greater than 4,000 psi increases cost without a reduction in slab thickness.

Two-way (or punching) shear plays an important role in determining the thickness of a flat plate, especially where the spans are relatively long and/or the live load is 100 psf or greater. In order to satisfy the shear strength requirements of ACI 11.12, the required thickness is usually found to be greater than that required for serviceability. Shear stresses at edge columns and corner columns are particularly critical because

FIGURE 7.3 Flat-slab system.

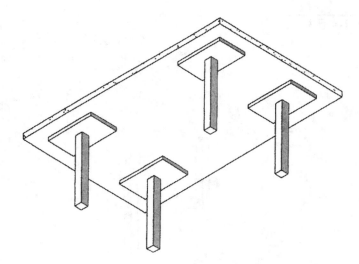

relatively large unbalanced moments can occur at those locations. Providing spandrel beams significantly increases shear strength at perimeter columns, but as noted previously, there is additional material and forming costs associated with such members, and they may not fit into the architectural scheme.

Headed shear stud reinforcement provides an economical means of resisting shear stresses and helps to alleviate congestion at slab–column joints (see ACI 11.11.5). More information on this type of shear reinforcement is given in Section 7.6.

For a live load of 50 psf or less, flat plates are economically viable for spans between 15 and 25 ft. The economical range for live loads of 100 psf is 15 to 20 ft. A flat-plate floor subjected to a 100 psf live load is only approximately 8% more expensive than one subjected to a 50 psf live load, primarily due to the minimum thickness requirements for deflection, which typically control for smaller live loads.

7.2.3 Flat Slab

A flat-slab floor system is similar to a flat-plate floor system, with the exception that the slab is thickened around the columns as shown in Fig. 7.3. The thickened portions of the slab are called drop panels, and they must conform to the dimensional requirements of ACI 13.2.5, which are illustrated in Fig. 7.4. In the figure, ℓ_A and ℓ_B are the

FIGURE 7.4
Minimum drop panel dimensions.

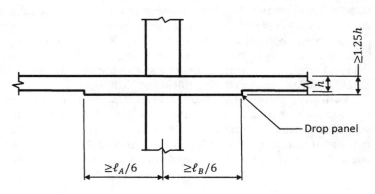

FIGURE 7.5 Shear cap.

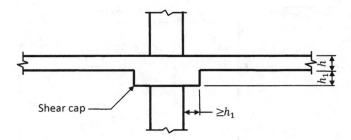

Shear cap — $\geq h_1$

center-to-center span lengths in the directions shown; similar dimensional requirements must be satisfied in the orthogonal direction as well.

The main purpose of the drop panels is to increase the shear strength around the columns. Additionally, properly proportioned drop panels result in a reduction in the required amount of negative reinforcement and in the overall slab thickness [see ACI 9.5.3.2 and ACI Table 9.5(c)].

Shear caps are thickened concrete elements that extend horizontally below the slab a minimum distance from the edge of the column equal to the thickness of the projection below the slab soffit (see Fig. 7.5). These elements are similar to drop panels, but they are provided exclusively to increase shear strength (see ACI 13.2.6).

The specified thickness of a drop panel or shear cap is controlled by formwork considerations. Using depths other than the standard depths indicated in Fig. 7.6, which are dictated by lumber dimensions, will unnecessarily increase formwork costs. It is common to initially check shear strength requirements, using a total drop panel or shear cap thickness that is 2.25 in greater than the required slab thickness. If this thickness is not adequate, then a 4.25-in thickness is checked. This process continues using the thicknesses in Fig. 7.6 until all required criteria are satisfied.

Another way to reduce shear around columns is to flare the top of a column, creating column capitals (see Fig. 7.7). For purposes of design, a column capital is part of the

FIGURE 7.6 Drop panel and shear cap formwork details.

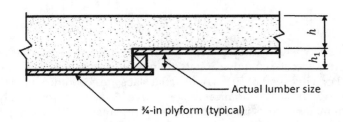

Actual lumber size

¾-in plyform (typical)

Nominal lumber size	Actual lumber size (in)	Drop panel or shear cap thickness h_1 (in)
2X	$1^1/_2$	$2^1/_4$
4X	$3^1/_2$	$4^1/_4$
6X	$5^1/_2$	$6^1/_4$
8X	$7^1/_2$	$8^1/_4$

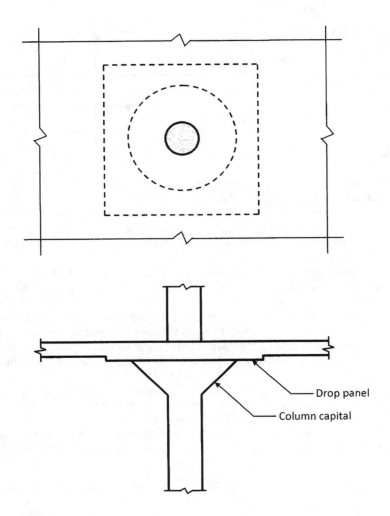

FIGURE 7.7 Column capitals and drop panels.

Drop panel

Column capital

column, whereas a drop panel or shear cap is part of the slab. Because of relatively large formwork costs, column capitals are not commonly used any more.

For a live load of 50 psf or less, flat slabs are economically viable for spans between 25 and 30 ft. The economical range for live loads of approximately 100 psf is 20 to 25 ft. Total material costs increase by only approximately 4% when going from a 50 psf live load to a 100 psf live load because the material quantities are usually controlled by deflections.

7.2.4 Two-Way Joist

Two-way joist construction, which is commonly referred to as a waffle slab system, consists of rows of concrete joists at right angles to each other with solid heads at the columns, which are needed for shear strength (see Fig. 7.8). The joists are formed by using standard square "dome" forms that are 30, 41, and 52 in wide, resulting in 3-, 4-, and 5-ft modules, respectively. Depending on the dome width, the depth of the dome

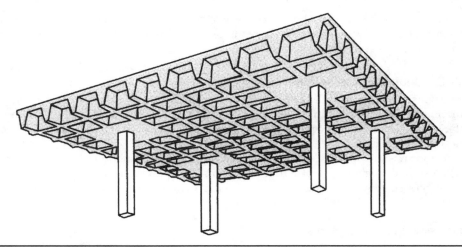

FIGURE 7.8   Two-way joist (waffle) system.

varies from 8 to 24 in. Waffle slabs are economically viable for long spans (40 to 50 ft) with heavy loads and are used in office buildings, warehouses, libraries, museums, and industrial buildings.

For design purposes, waffle slabs are considered as flat slabs with the solid heads acting as drop panels. Thus, the minimum thickness requirements of ACI 9.5.3.2 must be satisfied. This is accomplished by transforming the actual cross-section of the floor system into an equivalent section of uniform thickness. In other words, a slab thickness that provides the same moment of inertia as the two-way joist section is determined.

Waffle slab construction allows a considerable reduction in dead load compared with conventional flat-slab construction because the slab thickness can be minimized owing to the short span between the joists. Thus, this system is particularly advantageous where long spans and/or heavy loads are desired without the use of deepened drop panels or support beams. The geometric shape formed by the joist ribs is generally considered to be architecturally desirable and is often left exposed.

Like beam-supported slabs, waffle slabs are not specified as often as they once were. This is primarily due to the cost attributed to the formwork. Increasing the live load from 50 to 100 psf results in an approximately 7% increase in the overall cost of this system.

7.3 Minimum Thickness Requirements

7.3.1 Overview

The first step in the design of a two-way slab system is to determine a preliminary slab thickness. A minimum slab thickness must be provided to control deflections and to provide adequate shear strength.

Serviceability requirements for two-way construction with nonprestressed reinforcement are given in ACI 9.5.3. The governing provisions depend on whether beams are present or not in the slab system.

Both one- and two-way shear must be investigated at the critical sections around the supports. Two-way shear requirements play an important role in the selection of slab thickness in systems without beams.

7.3.2   Control of Deflections

Provisions for minimum thickness of two-way slab systems based on deflection requirements are contained in ACI 9.5.3 and are summarized later for slabs (1) without interior beams spanning between the supports and (2) with beams spanning between the supports.

The provisions in the Code greatly simplify the determination of minimum slab thickness for deflection control in routine designs. Complex deflection calculations need not be performed if the overall thickness is equal to or greater than that determined by ACI 9.5.3.

Slabs without Interior Beams

The information presented in Fig. 7.9 is a summary of the minimum slab thickness requirements of ACI Table 9.5(c) for slabs without interior beams that utilize Grade 60 reinforcement. Similar to beams and one-way slabs, the minimum thickness of a two-way slab without drop panels (flat plates) and with drop panels (flat slabs) is a function of the clear span length ℓ_n. In slabs without beams, ℓ_n is the length of the clear span in the long direction measured face-to-face of supports. In all other cases, ℓ_n is measured face-to-face of beams or other supports.

It is evident that flat slabs with drop panels that meet the minimum size requirements illustrated in Fig. 7.4 are permitted to have an overall thickness that is 10% less than that required for flat plates. In cases where a drop panel is provided with overall dimensions greater than the minimum specified (e.g., the thickness is greater than the minimum in order to increase shear strength), a corresponding decrease in the minimum slab thickness is not permitted unless deflections are computed.

A decrease of 10% in the minimum thickness is also permitted in exterior panels with relatively stiff edge beams. In particular, this reduction is allowed in panels where the stiffness ratio α_f is equal to or greater than 0.8. Methods on how to determine α_f are given later.

ACI 9.5.3.2(a) requires that a 5-in minimum slab thickness be provided for slabs without drop panels and a 4-in minimum thickness be provided for slabs with drop panels. It has been demonstrated that slabs conforming to the limitations of ACI 9.5.3.2 have not exhibited any deflection problems for both short- and long-term loads.

The greatest minimum slab thickness that is determined from all panels is used for the entire floor or roof system. Like the case for beams and one-way slabs, the most economical solution is obtained by varying the reinforcement and not the slab thickness.

Slabs with Interior Beams

Provisions for minimum slab thickness of two-way systems with beams spanning between the supports on all sides of a panel are given in ACI 9.5.3.3. These provisions are based on the average value of the relative beam stiffness α_{fm}. A discussion on the computation of this stiffness ratio follows.

Stiffness Ratios α_f and α_{fm}   The term α_f is the ratio of the flexural stiffness of a beam section to the flexural stiffness of a width of slab bounded laterally by centerlines of

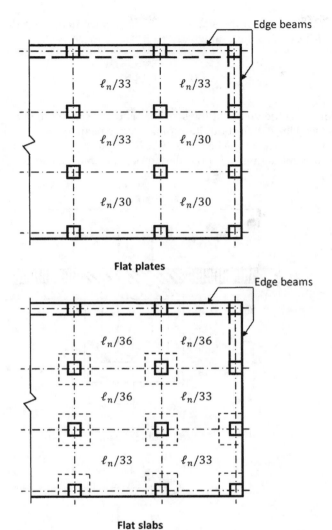

FIGURE 7.9 Minimum slab thickness for two-way slabs without interior beams (Grade 60 reinforcement).

any adjacent panels on each side of the beam [see ACI Eq. (13-3)]:

$$\alpha_f = \frac{E_{cb}I_b}{E_{cs}I_s} \tag{7.1}$$

In this equation, E_{cb} and E_{cs} are the moduli of elasticity of the concrete for the beam and slab, respectively, which are determined in accordance with ACI 8.5.1 (also see Section 2.2.2). In most monolithic cast-in-place structures, the same concrete is used for the beams and slabs, so, $E_{cb} = E_{cs}$.

The moment of inertia of the slab section I_s is determined using the full width of the slab that is tributary to the beam:

$$I_s = \frac{1}{12}\ell_2 h_f^3 \tag{7.2}$$

In this equation, ℓ_2 is the width of the slab that is tributary to the beam and h_f is the slab thickness (see Fig. 7.10 for both interior and edge beams).

The moment of inertia of the beam section I_b is determined using the web portion of the section in combination with the flange portion that has an effective width b_e,

FIGURE 7.10
Effective beam and slab sections for computation of stiffness ratio for interior and edge beams.

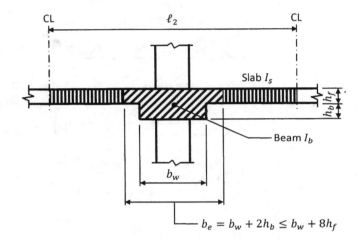

Interior beam

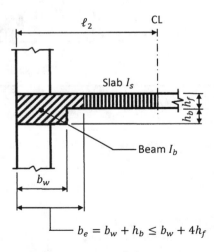

Edge beam

which is defined in ACI 13.2.4. Figure 7.10 contains interior and edge beams where the beam web projects below the slab; in general, the greater of the projections above or below the slab is used to calculate b_e. Beams that project above the slab are commonly referred to as upturned beams and are usually located at the perimeter of a structure.

The following equation can be used to determine I_b for both interior and edge beams:

$$I_b = \frac{1}{12} b_w h_b^3 + b_w h_b \left(y_b - \frac{h_b}{2} \right)^2 + \frac{1}{12} b_e h_f^3 + b_e h_f \left(h_b + \frac{h_f}{2} - y_b \right)^2 \qquad (7.3)$$

The term y_b is the distance from the bottom of the beam section to the centroid of the combined section:

$$y_b = \frac{b_e h_f [(h_b + (h_f/2))] + (b_w h_b^2/2)}{b_e h_f + b_w h_b} \qquad (7.4)$$

The stiffness ratio α_{fm} is defined as the average value of the stiffness ratios α_f for all beams on the edges of a panel. Thus, for a panel with four beams, $\alpha_{fm} = (\alpha_{f1} + \alpha_{f2} + \alpha_{f3} + \alpha_{f4})/4$.

Minimum Slab Thickness Once α_{fm} has been established, the minimum slab thickness $h = h_f$ is determined by ACI 9.5.3.3. A summary of these requirements is given in Table 7.1.

In ACI Eqs. (9-12) and (9-13), ℓ_n is the length of the clear span in the long direction measured face-to-face of supporting beams and β is the ratio of the long to the short clear span dimensions of the panel.

It is evident from Table 7.1 that panels with an average relative stiffness ratio α_{fm} less than 0.2 must have a minimum thickness that is equal to or greater than that required for two-way systems without interior beams. In other words, the beams in such cases are not stiff enough to warrant a decrease in minimum thickness.

ACI Eqs. (9-12) and (9-13) must be modified where beams at discontinuous panel edges do not meet certain minimum stiffness requirements. In particular, the minimum thickness values obtained by these equations must be increased by at least 10% where an edge beam has a stiffness ratio α_f less than 0.80. This requirement exemplifies the positive effect that stiff edge beams have in controlling overall deflections.

α_{fm}	ACI Equation Number	Minimum h
$\alpha_{fm} \leq 0.2$	—	Use the provisions of ACI 9.5.3.2.
$0.2 < \alpha_{fm} \leq 2.0$	9–12	$\dfrac{\ell_n[0.8 + (f_y/200{,}000)]}{36 + 5\beta(\alpha_{fm} - 0.2)} \geq 5.0$ in
$\alpha_{fm} > 2.0$	9–13	$\dfrac{\ell_n[0.8 + (f_y/200{,}000)]}{36 + 9\beta} \geq 3.5$ in

TABLE 7.1 Minimum Slab Thickness for Two-way Slabs with Beams Spanning Between the Supports

Calculated Deflections

ACI 9.5.3.4 permits the use of a two-way slab thickness less than that determined by the appropriate provisions of ACI 9.5.3.1 to 9.5.3.3 if it can be shown that calculated immediate and long-term deflections are equal to or less than the limiting values given in ACI Table 9.5(b). In general, the size and shape of the panel and the support conditions must be taken into account in the model.

The calculation of deflections for two-way slab systems is complex even when the calculations are based on linearly elastic behavior. A number of approximate procedures that make the calculations more tenable have been developed, and a summary of these procedures can be found in Refs. 1 and 2. Regardless of the approximate procedure that is employed, the results must be in reasonable agreement with those from comprehensive tests.

In the calculation of immediate deflections, the modulus of elasticity E_c defined in ACI 8.5.1 and the effective moment of inertia I_e defined in ACI 9.5.2.3 for one-way construction may be used for two-way construction as well (see Section 6.5).[3] ACI 9.5.3.4 permits the long-term multiplier given in ACI 9.5.2.5 for one-way construction to be used in the calculation of long-term deflections in two-way systems. Data on long-term deflections in such systems are limited, so a more sophisticated procedure to determine such deflections is not justified.

It is important for the designer to be aware of the range of accuracy of estimated deflections in two-way construction. Approximate methods can provide only approximate deflections, so a conservative approach should be followed in the design of deflection-sensitive systems.

7.3.3 Shear Strength Requirements

When establishing a slab thickness in the preliminary design stage, it is important to check the shear strength requirements, especially for slab systems without beams. Both one- and two-way shear must be investigated at the critical sections around the supports. More often than not, slab thickness will be controlled by two-way shear requirements, so it is important to establish a preliminary slab thickness based on shear strength at the onset of the design procedure. One- and two-way shear requirements for two-way slab systems are given in Section 7.6.

Because two-way shear requirements of ACI 11.11 are related to flexural requirements (namely, the assumed distribution of shear stress around the critical section of a column includes the effects of unbalanced bending moments at a support), a slab thickness that satisfies both sets of requirements cannot be obtained in the preliminary design stage unless some simplifying assumptions are made. Figure 7.11 can be used to obtain a preliminary slab thickness for flat plates where the slab thickness is typically controlled by two-way shear requirements. The information in the figure is based on the following assumptions:

- Square edge column of size c_1 bending perpendicular to the slab edge with a three-sided critical section
- Column supporting a tributary area A
- Square bays

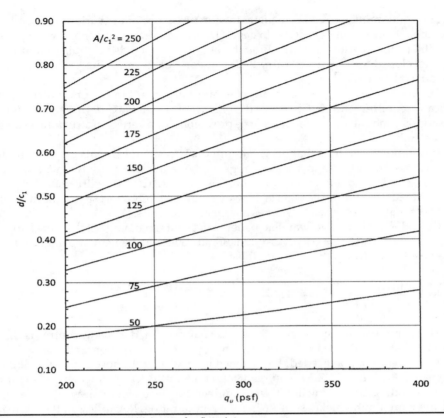

FIGURE 7.11   Preliminary slab thickness for flat plates.

- Gravity load moment transferred between the slab and the edge column in accordance with the Direct Design Method requirement of ACI 13.6.3.6
- 4,000 psi normal-weight concrete

The term q_u is the total factored load, which must include an estimate of the slab weight. The ratio d/c_1 is determined from Fig. 7.11 as a function of q_u and the area ratio A/c_1^2. A preliminary slab thickness h can be obtained by adding 1.25 in to d acquired from the figure. The purpose of this design aid is to help decrease the number of iterations that are needed to establish a viable slab thickness based on shear strength requirements; it is not meant to replace shear strength calculations. Determining a preliminary slab thickness based on the conditions at an edge column was chosen because the shear requirements are usually the most critical at that location.

7.3.4   Fire Resistance Requirements
Even though concrete floor systems offer inherent fire resistance, the thickness of the slab must be chosen in the preliminary design stage to satisfy fire resistance requirements.

State and municipal codes regulate the fire resistance of the various elements and assemblies of a building structure. In general, structural frames, floor and roof systems, and other load-bearing elements must be able to withstand the strains imposed by fully developed fires and must carry their tributary loads, including their own weight, without collapsing.

Fire resistance rating requirements generally vary from 1 to 4 hours, with buildings typically requiring 2 hours. Chap. 7 of the *International Building Code* (IBC) contains both prescriptive and calculated fire resistance provisions for a variety of structural elements and assemblies.[4]

It is important to note that the minimum thickness requirements given in the ACI Code do not consider fire resistance. In general, the thickness of a two-way slab that satisfies strength and serviceability requirements of the ACI Code will usually be greater than that required for fire resistance. However, for relatively short spans, the minimum thickness may be governed by fire resistance requirements.

Consider a square, two-way slab panel with beams spanning between the supports where $\alpha_{fm} = 2.5$ and $\ell_n = 11$ ft. Assuming Grade 60 reinforcement, the minimum slab thickness from ACI Eq. (9-13) is (see Table 7.1)

$$h = \frac{\ell_n[0.8 + (f_y/200{,}000)]}{36 + 9\beta} = \frac{(11 \times 12)[0.8 + (60{,}000/200{,}000)]}{36 + (9 \times 1.0)} = 3.2 \text{ in} < 3.5 \text{ in}$$

Therefore, the minimum slab thickness required for deflection control is 3.5 in. Assume that the IBC has been adopted by the local jurisdiction and that a 2-hour fire resistance rating is needed. For a concrete mixture with normal-weight siliceous aggregate, IBC Table 721.2.2.1 requires at least a 4.6-in-thick slab to achieve a 2-hour fire resistance rating. Specifying lightweight aggregate instead of normal-weight aggregate requires a minimum thickness of 3.6 in for a 2-hour rating. Regardless of the aggregate type, the minimum slab thickness is governed in this case by fire resistance requirements instead of deflection requirements.

It is good practice to check the requirements of the local building code governing a project early in design to ensure that minimum fire resistance requirements are satisfied.

Example 7.1   Determine the minimum slab thickness that satisfies the deflection criteria of ACI 9.5.3 for the flat-plate system depicted in Fig. 7.12. Assume Grade 60 reinforcement.

Solution   For slabs without interior beams spanning between the supports, the provisions of ACI 9.5.3.2 must be used to determine the minimum slab thickness.

The minimum slab thickness is determined using the information given in ACI Table 9.5(c) for slab systems without drop panels and edge beams and with Grade 60 reinforcement (see Fig. 7.9):

- Exterior panel: longest $\ell_n = (23.5 \times 12) - 20 = 262$ in

$$h_{min} = \frac{\ell_n}{30} = \frac{262}{30} = 8.7 \text{ in}$$

- Interior panel: longest $\ell_n = (23.5 \times 12) - 24 = 258$ in

$$h_{min} = \frac{\ell_n}{33} = \frac{258}{33} = 7.8 \text{ in}$$

Five at 23 ft 6 in

Five at 21 ft 8 in

N

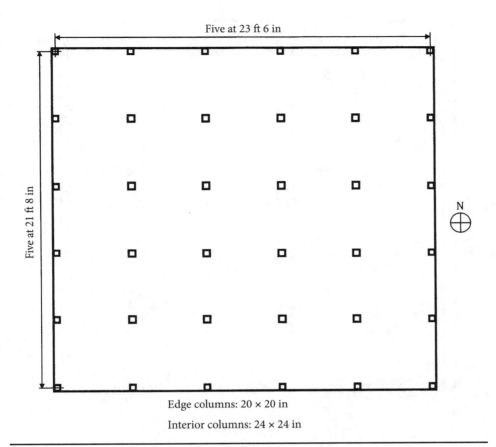

Edge columns: 20 × 20 in

Interior columns: 24 × 24 in

FIGURE 7.12 The flat-plate system of Example 7.1.

Thus, the minimum slab thickness for an exterior panel governs.

Use a 9-in-thick slab. This thickness is greater than the minimum thickness of 5 in prescribed in ACI 9.5.3.2(a) for slabs without drop panels.

Example 7.2 Determine the minimum slab thickness that satisfies the deflection criteria of ACI 9.5.3 for the flat-slab system depicted in Fig. 7.13. Assume Grade 60 reinforcement.

Solution For slabs without interior beams spanning between the supports, the provisions of ACI 9.5.3.2 must be used to determine the minimum slab thickness.

Check if the plan dimensions of the drop panel meet the minimum requirements of ACI 13.2.5(b):

$$\text{Minimum extension from centerline of support} = \frac{\ell}{6} = \frac{23.5}{6} = 3.9 \text{ ft}$$
$$\text{Minimum width of drop panel} = 2 \times 3.9 = 7.8 \text{ ft}$$

Note that the largest center-to-center distance between columns was used to determine the minimum drop panel extension.

The requirements of ACI 13.2.5(b) are satisfied for the plan dimensions of the drop panels because 8 × 8 ft drop panels are provided. Once a slab thickness has been determined, the thickness

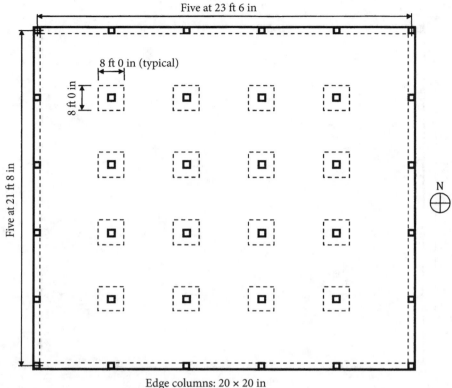

Five at 23 ft 6 in

8 ft 0 in (typical)

8 ft 0 in

Five at 21 ft 8 in

N

Edge columns: 20 × 20 in

Interior columns: 24 × 24 in

Edge beams: 20 × 24 in

FIGURE 7.13   The flat-slab system of Example 7.2.

of the drop panel below the slab must be equal to at least one-quarter of the slab thickness in order to satisfy the requirements of ACI 13.2.5(a).

The minimum slab thickness is determined using the information given in ACI Table 9.5(c) for slab systems with drop panels and edge beams and with Grade 60 reinforcement (see Fig. 7.9). The minimum thickness values in the table for exterior panels with edge beams are based on stiffness ratio $\alpha_f \geq 0.8$. Because the slab thickness is not known at this stage, α_f cannot be determined. In lieu of assuming a slab thickness, a minimum slab thickness will be determined assuming $\alpha_f \geq 0.8$; this assumption will be checked after the slab thickness has been calculated.

• Exterior panel: longest $\ell_n = (23.5 \times 12) - 20 = 262$ in

$$h_{min} = \frac{\ell_n}{36} = \frac{262}{36} = 7.3 \text{ in}$$

• Interior panel: longest $\ell_n = (23.5 \times 12) - 24 = 258$ in

$$h_{min} = \frac{\ell_n}{36} = \frac{258}{36} = 7.2 \text{ in}$$

Thus, the minimum slab thickness for an exterior panel governs.

Try a 7.5-in slab thickness, and calculate α_f for an edge beam, using Eq. (7.1):

$$\alpha_f = \frac{E_{cb}I_b}{E_{cs}I_s}$$

Effective width $b_e = \begin{cases} b_w + h_b = 20 + (24 - 7.5) = 36.5 \text{ in (governs)} \\ b_w + 4h_f = 20 + (4 \times 7.5) = 50.0 \text{ in} \end{cases}$ (see Fig. 7.10)

Using Eq. (7.4),

$$y_b = \frac{b_e h_f [h_b + (h_f/2)] + (b_w h_b^2/2)}{b_e h_f + b_w h_b} = \frac{(36.5 \times 7.5)(16.5 + 3.75) + (0.5 \times 20 \times 16.5^2)}{(36.5 \times 7.5) + (20 \times 16.5)} = 13.7 \text{ in}$$

Using Eq. (7.3),

$$\begin{aligned} I_b &= \frac{1}{12}b_w h_b^3 + b_w h_b \left(y_b - \frac{h_b}{2}\right)^2 + \frac{1}{12}b_e h_f^3 + b_e h_f \left(h_b + \frac{h_f}{2} - y_b\right)^2 \\ &= \left(\frac{1}{12} \times 20 \times 16.5^3\right) + (20 \times 16.5)\left(13.7 - \frac{16.5}{2}\right)^2 + \left(\frac{1}{12} \times 36.5 \times 7.5^3\right) \\ &\quad + (36.5 \times 7.5)\left(16.5 + \frac{7.5}{2} - 13.7\right)^2 \\ &= 30{,}317 \text{ in}^4 \end{aligned}$$

Also, from Eq. (7.2),

$$I_s = \frac{1}{12}\ell_2 h_f^3 = \frac{1}{12} \times \left(\frac{23.5 \times 12}{2} + \frac{20}{2}\right) \times 7.5^3 = 5{,}309 \text{ in}^4$$

Note that I_s was calculated using the largest ℓ_2; this results in the maximum I_s, which is required in order to obtain the minimum α_f.

Therefore,

$$\alpha_f = \frac{E_{cb}I_b}{E_{cs}I_s} = \frac{30{,}317}{5{,}309} = 5.7 > 0.8$$

Thus, the assumption that $\alpha_f \geq 0.8$ is correct.

Use a 7.5-in-thick slab. This thickness is greater than the minimum thickness of 4 in prescribed in ACI 9.5.3.2(b) for slabs with drop panels.

The minimum thickness of the projection of the drop panel below the slab must be $0.25 \times 7.5 = 1.9$ in (see Fig. 7.4). From Fig. 7.6, use a 2.25-in drop panel projection. The overall thickness of the drop panel and the slab adjacent to the drop panel need to be checked for two-way shear.

Example 7.3   Determine the minimum slab thickness that satisfies the deflection criteria of ACI 9.5.3 for the beam-supported slab system depicted in Fig. 7.14. Assume Grade 60 reinforcement.

Solution   For slabs with interior beams spanning between the supports, the provisions of ACI 9.5.3.3 must be used to determine the minimum slab thickness.

The minimum slab thickness is determined on the basis of the stiffness ratio α_{fm}. Because the slab thickness is not known at this stage, α_f and α_{fm} cannot be determined. In lieu of assuming a slab thickness, a minimum slab thickness will be determined assuming $\alpha_{fm} > 2.0$; this assumption will be checked after the slab thickness has been calculated.

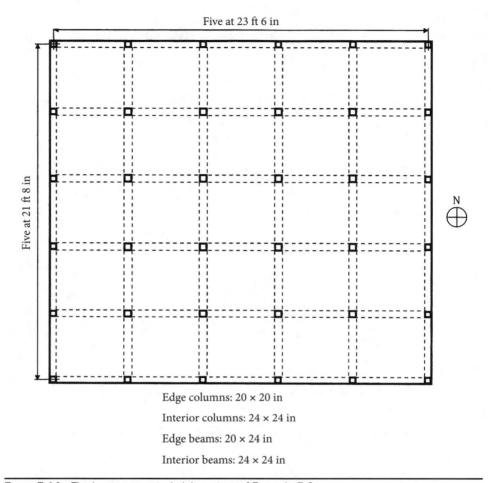

Five at 23 ft 6 in

Five at 21 ft 8 in

N

Edge columns: 20 × 20 in

Interior columns: 24 × 24 in

Edge beams: 20 × 24 in

Interior beams: 24 × 24 in

FIGURE 7.14 The beam-supported slab system of Example 7.3.

For an edge bay (see Table 7.1),

$$h_{min} = \frac{\ell_n[0.8 + (f_y/200{,}000)]}{36 + 9\beta} = \frac{[(23.5 \times 12) - 20]\,[0.8 + (60{,}000/200{,}000)]}{36 + [9 \times (262/238)]} = 6.3\,\text{in} > 3.5\,\text{in}$$

In this equation, the longest clear span was used for ℓ_n (this occurs between two edge columns). This results in a maximum value of h_{min}. The same value of h_{min} is obtained using the clear spans in a corner bay.

Try a 6.5-in slab thickness, and calculate α_f for the beams and α_{fm} for the panels.

North-South Edge Beams

$$\text{Effective width } b_e = \begin{cases} b_w + h_b = 20 + (24 - 6.5) = 37.5\,\text{in (governs)} \\ b_w + 4h_f = 20 + (4 \times 6.5) = 46.0\,\text{in} \end{cases} \quad \text{(see Fig. 7.10)}$$

Using Eq. (7.4),

$$y_b = \frac{b_e h_f [h_b + (h_f/2)] + (b_w h_b^2/2)}{b_e h_f + b_w h_b} = \frac{(37.5 \times 6.5)(17.5 + 3.25) + (0.5 \times 20 \times 17.5^2)}{(37.5 \times 6.5) + (20 \times 17.5)} = 13.7 \text{ in}$$

Using Eq. (7.3),

$$\begin{aligned}
I_b &= \frac{1}{12} b_w h_b^3 + b_w h_b \left(y_b - \frac{h_b}{2} \right)^2 + \frac{1}{12} b_e h_f^3 + b_e h_f \left(h_b + \frac{h_f}{2} - y_b \right)^2 \\
&= \left(\frac{1}{12} \times 20 \times 17.5^3 \right) + (20 \times 17.5) \left(13.7 - \frac{17.5}{2} \right)^2 + \left(\frac{1}{12} \times 37.5 \times 6.5^3 \right) \\
&\quad + (37.5 \times 6.5) \left(17.5 + \frac{6.5}{2} - 13.7 \right)^2 \\
&= 30{,}481 \text{ in}^4
\end{aligned}$$

Also, from Eq. (7.2),

$$I_s = \frac{1}{12} \ell_2 h_f^3 = \frac{1}{12} \times \left(\frac{23.5 \times 12}{2} + \frac{20}{2} \right) \times 6.5^3 = 3{,}456 \text{ in}^4$$

Therefore,

$$\alpha_f = \frac{E_{cb} I_b}{E_{cs} I_s} = \frac{30{,}481}{3{,}456} = 8.8$$

East-West Edge Beams
Because the beam size in the east-west direction is the same as that in the north-south direction, $I_b = 30{,}481 \text{ in}^4$.

$$I_s = \frac{1}{12} \ell_2 h_f^3 = \frac{1}{12} \times \left(\frac{21.67 \times 12}{2} + \frac{20}{2} \right) \times 6.5^3 = 3{,}204 \text{ in}^4$$

$$\alpha_f = \frac{E_{cb} I_b}{E_{cs} I_s} = \frac{30{,}481}{3{,}204} = 9.5$$

North-South Interior Beams

$$\text{Effective width } b_e = \begin{cases} b_w + 2h_b = 24 + [2 \times (24 - 6.5)] = 59.0 \text{ in (governs)} \\ b_w + 8h_f = 24 + (8 \times 6.5) = 76.0 \text{ in} \end{cases} \quad \text{(see Fig. 7.10)}$$

Using Eq. (7.4),

$$y_b = \frac{b_e h_f [h_b + (h_f/2)] + (b_w h_b^2/2)}{b_e h_f + b_w h_b} = \frac{(59.0 \times 6.5)(17.5 + 3.25) + (0.5 \times 24 \times 17.5^2)}{(59.0 \times 6.5) + (24 \times 17.5)} = 14.5 \text{ in}$$

Using Eq. (7.3),

$$\begin{aligned}
I_b &= \frac{1}{12} b_w h_b^3 + b_w h_b \left(y_b - \frac{h_b}{2} \right)^2 + \frac{1}{12} b_e h_f^3 + b_e h_f \left(h_b + \frac{h_f}{2} - y_b \right)^2 \\
&= \left(\frac{1}{12} \times 24 \times 17.5^3 \right) + (24 \times 17.5) \left(14.5 - \frac{17.5}{2} \right)^2 + \left(\frac{1}{12} \times 59.0 \times 6.5^3 \right) \\
&\quad + (59.0 \times 6.5) \left(17.5 + \frac{6.5}{2} - 14.5 \right)^2 \\
&= 40{,}936 \text{ in}^4
\end{aligned}$$

Also, from Eq. (7.2),

$$I_s = \frac{1}{12}\ell_2 h_f^3 = \frac{1}{12} \times (23.5 \times 12) \times 6.5^3 = 6{,}454\,\text{in}^4$$

Therefore,

$$\alpha_f = \frac{E_{cb}I_b}{E_{cs}I_s} = \frac{40{,}936}{6{,}454} = 6.3$$

East-West Interior Beams

Because the beam size in the east-west direction is the same as that in the north-south direction, $I_b = 40{,}936\,\text{in}^4$.

$$I_s = \frac{1}{12}\ell_2 h_f^3 = \frac{1}{12} \times (21.67 \times 12) \times 6.5^3 = 5{,}951\,\text{in}^4$$

$$\alpha_f = \frac{E_{cb}I_b}{E_{cs}I_s} = \frac{40{,}936}{5{,}951} = 6.9$$

For a corner panel, $\alpha_{fm} = (8.8 + 9.5 + 6.3 + 6.9)/4 = 7.9$.
For an edge panel on the east or west face, $\alpha_{fm} = (8.8 + 6.9 + 6.3 + 6.9)/4 = 7.2$.
For an edge panel on the north or south face: $\alpha_{fm} = (6.3 + 9.5 + 6.3 + 6.9)/4 = 7.3$
For an interior panel: $\alpha_{fm} = (6.3 + 6.9 + 6.3 + 6.9)/4 = 6.6$
Because $\alpha_{fm} > 2.0$ for all panels, the initial assumption is correct.
Use a 6.5-in-thick slab. This thickness is greater than the minimum thickness of 3.5 in prescribed in ACI 9.5.3.3(c) for slabs with beams spanning between the supports and $\alpha_{fm} > 2.0$.

7.4 Analysis Methods

7.4.1 Overview

Methods of analysis for reinforced concrete two-way slabs are given in ACI 13.5. Any method based on the fundamental principles of structural mechanics that satisfy equilibrium and geometric compatibility is permitted, provided the results are in reasonable agreement with test data.

Two methods of analysis for two-way slab systems under gravity loads are addressed in ACI Chap. 13. The approximate Direct Design Method gives reasonably conservative bending moment values for slab systems that meet the specified limitations. This method is much simpler to use than the more complex Equivalent Frame Method, which gives more exact results.

Before the Direct Design Method and the Equivalent Frame Method are covered, the fundamental assumptions that are made when analyzing two-way slab systems are examined, including a discussion on the variation of bending moments in two-way slabs. This discussion gives additional insight into the requirements of the analysis methods.

Approximate methods of analysis for slab systems subjected to the effects of lateral loads are also covered, as are the requirements pertaining to openings.

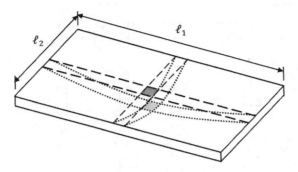

FIGURE 7.15
Deflection of a
two-way slab on
simple edge
supports.

Simple supports on all edges

7.4.2 Gravity Loads

Variation of Bending Moments

When a two-way slab is subjected to uniform gravity loads, it bends in two directions as illustrated in Fig. 7.15 for the case of simple edge supports. This behavior is different from that of a one-way slab, which bends primarily in one direction.

Strips of slab located at the center of the panel in each direction are also shown in Fig. 7.15. In order to satisfy compatibility requirements, the deflections of the strips must be the same at the point of intersection at the center of the span. Equating the deflections of both simply supported strips results in the following:

$$\frac{5q_1\ell_1^4}{384EI} = \frac{5q_2\ell_2^4}{384EI} \tag{7.5}$$

In this equation, q_1 and q_2 are the portions of the total load q carried in the long and short directions, respectively. Because EI is a constant, Eq. (7.5) can be rewritten as follows:

$$q_2 = \frac{\ell_1^4}{\ell_2^4}q_1 \tag{7.6}$$

Because ℓ_1 is greater than ℓ_2, the larger portion of the total load q is carried in the short direction.

When other than the center slab strips are considered in each direction, it becomes evident that these strips must bend and twist in order to maintain equilibrium. Thus, in general, the total load q is carried in both directions by bending moments and by twisting moments.

The largest bending moment in the slab occurs where the curvature is the sharpest, which is at the midspan of the strip in the short direction (see Fig. 7.15); other strips in that direction have smaller bending moments. A similar situation occurs in the perpendicular direction. Therefore, bending moments vary across both the width and the length of a two-way slab.

To illustrate the variation of bending moments in other than edge-supported slabs, consider the flat plate supported by the four columns shown in Fig. 7.16. Assume that the slab is subjected to a uniformly distributed gravity load q. The following discussion,

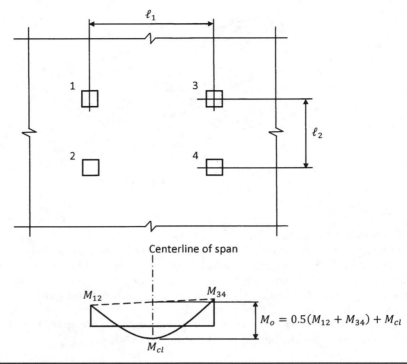

FIGURE 7.16 Bending moment diagram in the direction of ℓ_1.

which is applicable to any column-supported two-way slab system, focuses on the variation in bending moments in the direction parallel to ℓ_1 and perpendicular to ℓ_2. A similar discussion is applicable in the perpendicular direction.

Figure 7.16 shows the bending moment diagram along span ℓ_1. In this direction, the slab may be considered a beam that has a width equal to ℓ_2 where the load per width of span is equal to $q\ell_2$. If strips are considered in the short and long directions, it can be shown from statics that 100% of the applied gravity load must be carried in each direction. This requirement is essential in the determination of bending moments in a column-supported two-way slab system.

In any span of a continuous system, the sum of the midspan positive moment and the average of the negative moments at the supports is equal to the midspan positive moment of a simply supported beam of the same length and applied load. This bending moment is represented by M_o in Fig. 7.16 and can be expressed by the following equation:

$$M_o = \frac{q\ell_2\ell_1^2}{8} = \frac{M_{12} + M_{34}}{2} + M_{cl} \tag{7.7}$$

Approximate methods, such as the Direct Design Method, provide ways to determine the negative bending moments at the supports and the positive bending moments in the span as a function of M_o.

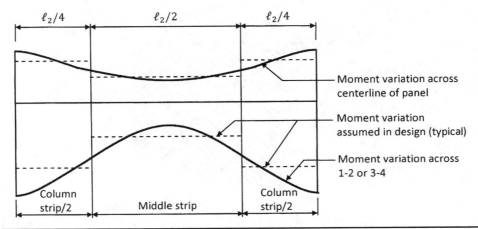

Moment variation across centerline of panel

Moment variation assumed in design (typical)

Moment variation across 1-2 or 3-4

$\ell_2/4$ $\ell_2/2$ $\ell_2/4$

Column strip/2   Middle strip   Column strip/2

Figure 7.17 Variation of bending moments across width of critical section in column-supported two-way slab.

The variation in bending moment across the width of the critical sections at 1-2 (or 3-4) and the critical section at the centerline of the span are depicted in Fig. 7.17. The exact variation of the bending moments depends on the magnitude of the applied load, the presence of beams on the column lines, and the presence of drop panels or column capitals. It is evident from the figure that the larger bending moments are concentrated along the column lines.

Also shown in Fig. 7.17 are column strips and middle strips. In order to simplify the design, it is assumed that the bending moments are constant within these strips. Column strips and middle strips are defined in ACI 13.2 and depend on the span lengths ℓ_1 and ℓ_2, where ℓ_1 is the center-to-center distance between supports in the direction of analysis (i.e., in the direction bending moments are being computed) and ℓ_2 is the center-to-center span length perpendicular to ℓ_1. A column strip is a design strip with a width on each side of a column centerline equal to 25% of ℓ_1 or ℓ_2, whichever is less (see Fig. 7.18). Note that column-line beams, if present, are included in the column strips. Middle strips are bounded by two column strips.

Approximate methods of analysis, such as the Direct Design Method, provide distribution factors for determining the bending moments in the column strips and middle strips at each critical location along the span.

Direct Design Method

Limitations   The Direct Design Method of ACI 13.6 is an approximate method that can be utilized to determine the bending moments in two-way slab systems. Before discussing the details of the method, it is important to know the limitations under which this method can be used.

The limitations contained in ACI 13.6.1 are as follows (see Fig. 7.19):

1. Three continuous spans must be present in each direction.

2. Slab panels must be rectangular with a ratio of the longer to the shorter span, centerline-to-centerline of supports, equal to or less than 2.

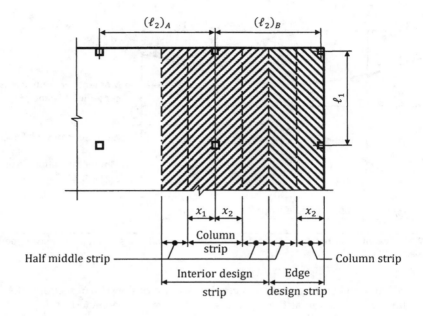

$$x_1 = \text{the lesser of } \ell_1/4 \text{ or } (\ell_2)_A/4$$

$$x_2 = \text{the lesser of } \ell_1/4 \text{ or } (\ell_2)_B/4$$

FIGURE 7.18   Column strips and middle strips.

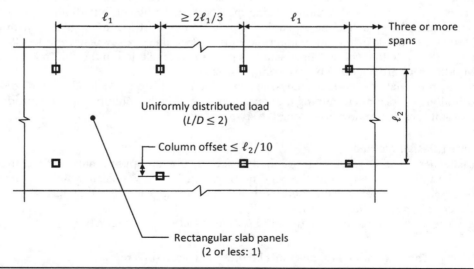

FIGURE 7.19   Limitations of the Direct Design Method.

3. Successive span lengths, centerline-to-centerline of supports, in each direction must not differ by more than one-third of the longer span.

4. Columns must not be offset more than 10% of the span in the direction of offset from either axis between the centerlines of successive columns.

5. Loads applied to the slab must be uniformly distributed gravity loads where the ratio of the unfactored live load to the unfactored dead load is equal to or less than 2.

6. For panels with column-line beams on all sides, ACI Eq. (13-2) must be satisfied.

7. Redistribution of bending moments in accordance with ACI 8.4 is not permitted.

It is important to note that the Direct Design Method is based on tests where only uniformly distributed gravity loads were considered.[5] A frame analysis is required where lateral forces, such as wind or seismic forces, act on a structure. Combining the results from the Direct Design Method with those from the frame analysis is permitted (ACI 13.5.1.3). The rationale behind other limitations of this method can be found in ACI R13.6.1.

ACI 13.6.1.8 permits the use of the Direct Design Method even when the limitations are not satisfied, provided it can be shown by analysis that a particular limitation does not apply to the structure.

Analysis Procedure   In cases where all of the applicable limitations outlined in ACI 13.6.1 are satisfied, the Direct Design Method may be used to determine the bending moments in the slab. The three-step analysis procedure is summarized next.

Step 1: Determine the total factored static moment M_o in each span. The total factored static moment M_o is determined by ACI Eq. (13-4), which is similar to Eq. (7.7) derived earlier:

$$M_o = \frac{q_u \ell_2 \ell_n^2}{8} \tag{7.8}$$

In this equation, q_u is the total factored gravity loads acting on the slab; ℓ_n is the clear span in the direction of analysis; and ℓ_2 is the centerline-to-centerline span length perpendicular to ℓ_n. Where the transverse spans of panels on either side of the centerline of supports are not the same (see Fig. 7.18), ℓ_2 is set equal to the average of these transverse span lengths (i.e., $\ell_2 = [(\ell_2)_A + (\ell_2)_B]/2$). Also, where the span adjacent and parallel to an edge is considered, the value of ℓ_2 that is to be used in Eq. (7.8) is equal to the distance from the edge of the slab to the panel centerline. Thus, for the edge design strip depicted in Fig. 7.18, ℓ_2 is equal to $(\ell_2)_B/2$ plus one-half of the column dimension parallel to $(\ell_2)_B$.

Requirements on how to determine the clear span length ℓ_n are given in ACI 13.6.2.5 and are illustrated in Fig. 7.20. In general, ℓ_n is to extend face-to-face of columns, capitals, brackets, or walls. In cases where the supporting member does not have a rectangular cross-section or if the sides of the rectangle are not parallel to the spans, such members are to be treated as a square support that has the same area as that of the actual support. ACI 13.6.2.5 also requires that ℓ_n be equal to or greater than 65% of the span length ℓ_1.

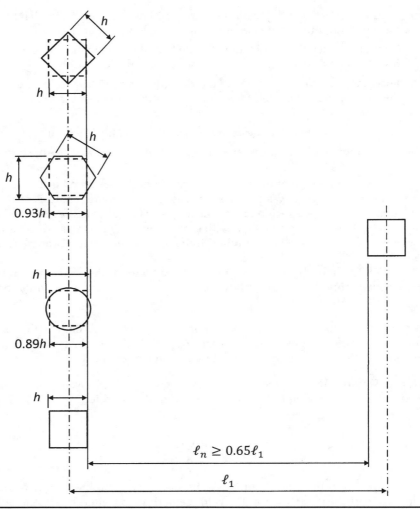

Figure 7.20   Definition of clear span.

Step 2: Distribute M_o into negative and positive bending moments in each span. Once M_o has been calculated, it is divided into negative and positive moments within each span in accordance with the distribution factors given in ACI 13.6.3. The resulting bending moments are the total bending moments in the design strip in the direction of analysis. The negative factored bending moments are located at the face of a support (ACI 13.6.3).

According to ACI 13.6.3.2, the total negative factored bending moment at the face of a support in an interior span is equal to 65% of M_o. The positive factored bending moment is equal to 35% of M_o.

A summary of the bending moment coefficients of ACI 13.6.3.3 for an end span is given in Table 7.2. These coefficients are based on the equivalent column stiffness expressions derived in Refs. 6 through 8. An unrestrained edge would correspond to

	Exterior Edge Unrestrained	Slab with Beams Between All Supports	Slab without Beams Between All Supports		Exterior Edge Fully Restrained
			Without Edge Beam	With Edge Beam	
Exterior negative	0	0.16	0.26	0.30	0.65
Positive	0.63	0.57	0.52	0.50	0.35
Interior negative	0.75	0.70	0.70	0.70	0.65

TABLE 7.2 Bending Moment Coefficients for an End Span

a slab that is simply supported on a masonry or concrete wall. A fully restrained edge would include a slab that is integrally constructed with a concrete wall that has a flexural stiffness much greater than that of the slab so that little rotation occurs at the slab–wall connection. Beam-supported slabs are slabs with beams between all supports, and flat plates and flat slabs are slabs without beams between all supports (see Section 7.2). The coefficients given in Table 7.2 for these systems yield upper-bound values for positive and interior negative bending moments. As such, exterior negative bending moments are close to lower-bound values.

The total bending moments in the design strip are obtained by multiplying M_o by the bending moment coefficients given in Table 7.2.

An important requirement related to shear stresses that develop at edge columns is given in ACI 13.6.3.6: The gravity load bending moment that is to be transferred between the slab and an edge column bending perpendicular to the edge must be 30% of M_o when the Direct Design Method is utilized. This bending moment is the unbalanced moment at the edge column and contributes to the total shear stress at that location. More details on shear requirements are given in Section 7.6.

Step 3: Distribute the total negative and positive bending moments in the design strip to the columns strips and middle strips. ACI 13.6.4 to 13.6.6 provide the percentages of the negative and positive bending moments at the critical sections that are to be assigned to the column strips, beams (if any), and middle strips, respectively. In general, the percentages depend on the relative beam-to-slab stiffness ratio α_{f1} in the direction of analysis, the torsional stiffness parameter β_t, and the panel width-to-length ratio ℓ_2/ℓ_1. These percentages are based on studies of linearly elastic slabs with different beam stiffness.[9]

Column strip—negative factored bending moments at interior supports The percentages for interior negative factored bending moments in a column strip are given in ACI 13.6.4.1 as a function of $\alpha_{f1}\ell_2/\ell_1$ and ℓ_2/ℓ_1, where α_{f1} is determined by Section 7.3. In lieu of using the table in that section, the percentage can be calculated by the following equation[2]:

$$75 + 30\left(\frac{\alpha_{f1}\ell_2}{\ell_1}\right)\left(1 - \frac{\ell_2}{\ell_1}\right) \tag{7.9}$$

Column strip—negative factored bending moments at exterior supports Percentages for exterior negative factored bending moments in a column strip, which are given in ACI

13.6.4.2, depend not only on the aforementioned parameters but also on the torsional stiffness parameter β_t of the edge beam, which is defined in ACI Eq. (13-5):

$$\beta_t = \frac{E_{cb}C}{2E_{cs}I_s} \tag{7.10}$$

In this equation, the shear modulus of the concrete has been taken as 50% of the modulus of elasticity of the concrete for the beam E_{cb}.

The cross-sectional constant C is determined by dividing the beam section into its component rectangles, each having a smaller dimension x and a larger dimension y, and by summing the contributions of each rectangle:

$$C = \sum \left(1 - 0.63\frac{x}{y}\right)\frac{x^3 y}{3} \tag{7.11}$$

The subdivision can be done in such as way as to maximize C. Equations for the calculation of C for an edge beam are given in Fig. 7.21. The larger of C_A and C_B is to be used in Eq. (7.10).

In lieu of the table given in ACI 13.6.4.2, the percentage of negative factored bending moment at the exterior support in a column strip can be calculated by the

Figure 7.21
Calculation of cross-sectional constant C.

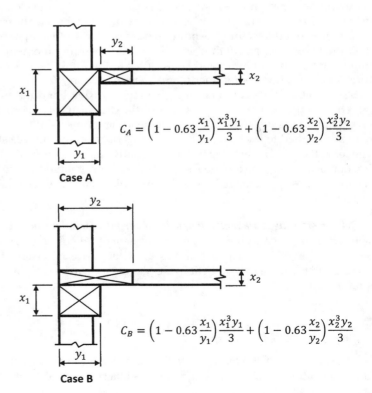

$$C_A = \left(1 - 0.63\frac{x_1}{y_1}\right)\frac{x_1^3 y_1}{3} + \left(1 - 0.63\frac{x_2}{y_2}\right)\frac{x_2^3 y_2}{3}$$

Case A

$$C_B = \left(1 - 0.63\frac{x_1}{y_1}\right)\frac{x_1^3 y_1}{3} + \left(1 - 0.63\frac{x_2}{y_2}\right)\frac{x_2^3 y_2}{3}$$

Case B

following equation[2]:

$$100 - 10\beta_t + 12\beta_t \left(\frac{\alpha_{f1}\ell_2}{\ell_1}\right)\left(1 - \frac{\ell_2}{\ell_1}\right) \tag{7.12}$$

In cases where β_t is greater than 2.5, use $\beta_t = 2.5$ in Eq. (7.12).

It is evident from the previous discussion that all of the exterior negative factored bending moment is assigned to the column strip unless an edge beam that has a relatively large torsional stiffness compared with the flexural stiffness of the slab is provided (i.e., where $\beta_t > 2.5$).

For purposes of analysis, walls along column lines in the direction of analysis can be regarded as stiff beams with $\alpha_{f1}\ell_2/\ell_1 > 1.0$. If an exterior support consists of a masonry wall that is perpendicular to the direction of analysis, it is conservative to assume that the wall has no torsional resistance (i.e., assume $\beta_t = 0$). However, if the wall is concrete instead of masonry and if it is monolithic with the slab, significant torsional resistance occurs at that location and β_t can be taken as 2.5.

ACI 13.6.4.3 contains an exception to the percentages given earlier, when proportioning factored bending moments in column strips where the supports, consisting of columns or walls, have a large width perpendicular to the direction of analysis. When the transverse width of a column or wall extends for a distance equal to or greater than 75% of the width of the design strip ℓ_2, negative factored bending moments are to be uniformly distributed across ℓ_2.

Column strip—positive factored bending moments   The percentage of positive factored bending moments in a column strip is obtained from the table given in ACI 13.6.4.4 or from the following equation[2]:

$$60 + 30\left(\frac{\alpha_{f1}\ell_2}{\ell_1}\right)\left(1.5 - \frac{\ell_2}{\ell_1}\right) \tag{7.13}$$

For cases where $\alpha_{f1}\ell_2/\ell_1$ is greater than 1.0, use $\alpha_{f1}\ell_2/\ell_1 = 1.0$ in Eqs. (7.9), (7.12), and (7.13).

Factored bending moments in beams   In slabs with column-line beams, the factored bending moments in the beams are obtained in accordance with the provisions of ACI 13.6.5. For relatively stiff beams ($\alpha_{f1}\ell_2/\ell_1 \geq 1.0$), it is assumed that the beams attract a significant portion of the bending moments in the column strip. Thus, the beams must be designed to carry 85% of the factored column strip bending moments.

Where $\alpha_{f1}\ell_2/\ell_1 < 1.0$, ACI 13.6.5.2 permits the portion of the column strip bending moments resisted by the beams to be determined by linear interpolation between 85% and 0%, which corresponds to $\alpha_{f1}\ell_2/\ell_1 = 1.0$ and $\alpha_{f1}\ell_2/\ell_1 = 0$, respectively.

In addition to the factored bending moments from the column strip, column-line beams must be designed for the factored bending moments from loads applied directly to the beam, that is, loads located within the width of the beam web. Any load located on the slab outside of the width of the beam web needs to be distributed accordingly between the slab and beam.

	End Span			Interior Span	
	Exterior Negative	**Positive**	**Interior Negative**	**Positive**	**Interior Negative**
Flat Plate or Flat Slab					
Total moment	$0.26M_o$	$0.52M_o$	$0.70M_o$	$0.35M_o$	$0.65M_o$
Column strip	$0.26M_o$	$0.31M_o$	$0.53M_o$	$0.21M_o$	$0.49M_o$
Middle strip	0	$0.21M_o$	$0.17M_o$	$0.14M_o$	$0.16M_o$
Flat Plate or Flat Slab with Spandrel Beams ($\beta_t \geq 2.5$)*					
Total moment	$0.30M_o$	$0.50M_o$	$0.70M_o$	$0.35M_o$	$0.65M_o$
Column strip	$0.23M_o$	$0.30M_o$	$0.53M_o$	$0.21M_o$	$0.49M_o$
Middle strip	$0.07M_o$	$0.20M_o$	$0.17M_o$	$0.14M_o$	$0.16M_o$

* For $\beta_t < 2.5$, the exterior negative column strip bending moment is equal to $(0.30 - 0.03\beta_t)M_o$.

TABLE 7.3   Design Bending Moments for Flat Plates and Flat Slabs Using the Direct Design Method

Factored bending moments in middle strips   The portion of the negative and positive factored moments in the design strip that are not resisted by the column strip must be resisted by the half middle strips in that design strip (ACI 13.6.6).

Any middle strip that is adjacent and parallel to a panel edge that is supported by a wall must be designed to resist two times the factored bending moment assigned to the half middle strip corresponding to the first row of interior supports.

Table 7.3 contains bending moments for flat plates and flat slabs with and without spandrel beams based on the Direct Design Method.

Modification of Factored Moments   ACI 13.6.7 permits a reduction of 10% in the negative or positive factored bending moments calculated in accordance with the Direct Design Method, provided the total static moment in the panel in the direction of analysis is not less than M_o determined by Eq. (7.8). This provision permits a modest redistribution of bending moments in slabs designed by this method.

Factored Shear in Slab Systems with Beams   Requirements for the shear design of column-line beams in beam-supported slabs are given in ACI 13.6.8. ACI Figure R13.6.8 shows the tributary area that is to be used when determining the shear forces on an interior beam in cases where $\alpha_{f1}\ell_2/\ell_1 \geq 1.0$. Beams must be designed to resist the total factored shear forces caused by the factored loads on the tributary area shown in the figure where $\alpha_{f1}\ell_2/\ell_1 \geq 1.0$; shear forces in the slab around the column are essentially zero.

In cases where $\alpha_{f1}\ell_2/\ell_1 < 1.0$, the shear forces resisted by the beam may be obtained by linear interpolation, assuming that the beam will carry no load at $\alpha_{f1} = 0$. When the beams carry less than the total load, the portion of the shear forces that are not carried by the beams must be carried by the slab around the column. These shear forces will cause shear stresses in the slab that must be checked in the same manner as for slabs without beams (see Section 7.6).

Column-line beams must also be designed for the shear forces due to factored loads applied directly to the beam, that is, loads located within the width of the beam web.

Factored Moments in Columns and Walls   Columns and walls that support a two-way slab system must be designed to resist the appropriate negative factored bending moments transferred from the slab. In lieu of a more exact analysis, ACI Eq. (13-7) can be used to determine unbalanced moment transfer at interior supports due to gravity loads:

$$M_u = 0.07[(q_{Du} + 0.5q_{Lu})\ell_2\ell_n^2 - q'_{Du}\ell'_2(\ell'_n)^2] \qquad (7.14)$$

This equation is applicable to two adjoining spans with one span longer than the other (all of the span lengths without a "prime" are longer than those with a "prime") and with full dead load plus one-half live load on the longer span and full dead load on the shorter span. Where the longitudinal and transverse spans are equal, Eq. (7.14) reduces to the following:

$$M_u = 0.035q_{Lu}\ell_2\ell_n^2 \qquad (7.15)$$

The moment M_u obtained by Eq. (7.14) or (7.15) is distributed to the interior supporting elements above and below the slab in direct proportion to their stiffness. If the cross-sectional dimensions above and below the slab are the same, the moment is transferred on the basis of the lengths of the elements (the longer elements will resist a lesser amount of the moment than the shorter elements).

At an exterior support, the total exterior negative bending moment from the slab is transferred directly to the support. This moment is transferred to the supporting elements in proportion to their stiffness.

Example 7.4   Determine the factored bending moments at the critical sections for an interior design strip in the north-south direction, using the Direct Design Method for the flat-plate system depicted in Fig. 7.12. Assume a 9-in-thick slab, normal-weight concrete, a superimposed dead load of 20 psf, and a live load of 50 psf.

Solution   Prior to determining the factored bending moments at the critical sections, check if the Direct Design Method can be used to analyze this two-way system:

1. Three continuous spans must be present in each direction.

   There are five spans in each direction.

2. Slab panels must be rectangular with a ratio of longer to shorter span, centerline-to-centerline of supports, equal to or less than 2.

   $$\text{Longer span/shorter span} = 23.5/21.67 = 1.1 < 2.0$$

3. Successive span lengths, centerline-to-centerline of supports, in each direction must not differ by more than one-third of the longer span.

   In each direction, the span lengths are equal.

4. Columns must not be offset more than 10% of the span in the direction of offset from either axis between the centerlines of successive columns.

   No column offsets are present.

5. Loads applied to the slab must be uniformly distributed gravity loads where the ratio of the unfactored live load to the unfactored dead load is equal to or less than 2.

$$\text{Live load} = 50 \, \text{psf}$$
$$\text{Dead load of slab} = \frac{9}{12} \times 150 = 112.5 \, \text{psf}$$
$$\text{Superimposed dead load} = 20 \, \text{psf}$$
$$\text{Uniform live to dead load ratio} = 50/(112.5 + 20) = 0.4 < 2$$

6. For panels with column-line beams on all sides, ACI Eq. (13-2) must be satisfied.

   No column-line beams are present.

7. Redistribution of bending moments in accordance with ACI 8.4 is not permitted.

   Bending moments will not be redistributed in accordance with ACI 8.4.

   Therefore, the Direct Design Method can be used for gravity load analysis.
   The factored bending moments at the critical sections are determined using the steps presented earlier.

   Step 1: Determine the total factored static moment M_o in each span. The total factored static moment M_o is determined by Eq. (7.8):

$$M_o = \frac{q_u \ell_2 \ell_n^2}{8}$$

   The total factored gravity loads acting on the slab q_u are determined using the load combination of ACI Eq. (9-2) because this combination yields the maximum effects for dead and live loads (see Table 4.1 in Section 4.2):

$$q_u = 1.2 q_D + 1.6 q_L = (1.2 \times 132.5) + (1.6 \times 50) = 239 \, \text{psf}$$

   The longest clear span ℓ_n for the design strip in the direction of analysis occurs in an end span. Because the difference in the clear span lengths between the end and interior spans is relatively small, conservatively use the longest clear span to calculate M_o. This moment is used for all spans:

$$\ell_n = 21.67 - \frac{20}{2 \times 12} - \frac{24}{2 \times 12} = 19.83 \, \text{ft}$$
$$\ell_2 = 23.5 \, \text{ft}$$
$$M_o = \frac{q_u \ell_2 \ell_n^2}{8} = \frac{0.239 \times 23.5 \times 19.83^2}{8} = 276.2 \, \text{ft kips}$$

   Step 2: Distribute M_o into negative and positive bending moments in each span. The moment M_o is divided into negative and positive moments in accordance with the distribution factors given in ACI 13.6.3 (see Tables 7.2 and 7.3 of this book). A summary of the total design strip moments at the critical sections of this flat plate is given in Table 7.4.

   Step 3: Distribute the total negative and positive bending moments in the design strip to the columns strip and middle strip. The percentages of the negative and positive bending moments

End Span Moments (ft kips)			Interior Span Moments (ft kips)	
Exterior Negative	Positive	Interior Negative	Positive	Interior Negative
$0.26 M_o = -71.8$	$0.52 M_o = 143.6$	$0.70 M_o = -193.3$	$0.35 M_o = 96.7$	$0.65 M_o = -179.5$

TABLE 7.4   Summary of Total Design Strip Moments for the Flat Plate Given in Example 7.4

	End Span Moments (ft kips)			Interior Span Moments (ft kips)		
	Exterior Negative	Positive	Interior Negative	Positive	Interior Negative
Column strip	$0.26M_o = -71.8$	$0.31M_o = 85.6$	$0.53M_o = -146.4$	$0.21M_o = 58.0$	$0.49M_o = -135.3$
Middle strip	0	$0.21M_o = 58.0$	$0.17M_o = -47.0$	$0.14M_o = 38.7$	$0.16M_o = -44.2$

TABLE 7.5 Summary of Factored Bending Moments at the Critical Sections for the Flat Plate Given in Example 7.4

at the critical sections that are to be assigned to the column strips and middle strips are given in ACI 13.6.4 to 13.6.6 (see Table 7.3 of this book). A summary of the factored bending moments at the critical sections is given for this flat plate in Table 7.5.

Comments
The negative factored bending moments occur at the face of the column supports. Note that 100% of the negative bending moment at the exterior columns in the end spans must be resisted by the column strip.

Example 7.5   Determine the factored bending moments at the critical sections for an interior design strip in the north-south direction, using the Direct Design Method for the flat-slab system depicted in Fig. 7.13. Assume a 7.5-in-thick slab, an overall drop panel thickness of 9.75 in, normal-weight concrete, a superimposed dead load of 20 psf, and a live load of 80 psf.

Solution   Prior to determining the factored bending moments at the critical sections, check if the Direct Design Method can be used to analyze this two-way system:

1. Three continuous spans must be present in each direction.

   There are five spans in each direction.

2. Slab panels must be rectangular with a ratio of longer to shorter span, centerline-to-centerline of supports, equal to or less than 2.

$$\text{Longer span/shorter span} = 23.5/21.67 = 1.1 < 2.0$$

3. Successive span lengths, centerline-to-centerline of supports, in each direction must not differ by more than one-third of the longer span.

   In each direction, the span lengths are equal.

4. Columns must not be offset more than 10% of the span in the direction of offset from either axis between the centerlines of successive columns.

   No column offsets are present.

5. Loads applied to the slab must be uniformly distributed gravity loads where the ratio of the unfactored live load to the unfactored dead load is equal to or less than 2.

$$\text{Live load} = 80\,\text{psf}$$

$$\text{Dead load of slab} = \frac{7.5}{12} \times 150 = 93.8\,\text{psf}$$

$$\text{Dead load of drop panel} = 4 \times \frac{2.25}{12} \times 150 \times 8 \times 8/(23.5 \times 108.33) = 2.8\,\text{psf (weight}$$
of four drop panel projections averaged over area of design strip)

$$\text{Superimposed dead load} = 20\,\text{psf}$$

$$\text{Uniform live to dead load ratio} = 80/(93.8 + 2.8 + 20) = 0.7 < 2$$

6. For panels with column-line beams on all sides, ACI Eq. (13-2) must be satisfied.

   No column-line beams are present.

7. Redistribution of bending moments in accordance with ACI 8.4 is not permitted.

   Bending moments will not be redistributed in accordance with ACI 8.4.

Therefore, the Direct Design Method can be used for gravity load analysis.

The factored bending moments at the critical sections are determined using the steps presented earlier.

Step 1: Determine the total factored static moment M_o in each span. The total factored static moment M_o is determined by Eq. (7.8):

$$M_o = \frac{q_u \ell_2 \ell_n^2}{8}$$

The total factored gravity loads acting on the slab q_u are determined using the load combination of ACI Eq. (9-2) because this combination yields the maximum effects for dead and live loads (see Table 4.1 in Section 4.2):

$$q_u = 1.2q_D + 1.6q_L = (1.2 \times 116.6) + (1.6 \times 80) = 268\,\text{psf}$$

The longest clear span ℓ_n for the design strip in the direction of analysis occurs in an end span. Because the difference in the clear span lengths between the end and interior spans is relatively small, conservatively use the longest clear span to calculate M_o. This moment is used for all spans:

$$\ell_n = 21.67 - \frac{20}{2 \times 12} - \frac{24}{2 \times 12} = 19.83\,\text{ft}$$

$$\ell_2 = 23.5\,\text{ft}$$

$$M_o = \frac{q_u \ell_2 \ell_n^2}{8} = \frac{0.268 \times 23.5 \times 19.83^2}{8} = 309.6\,\text{ft kips}$$

Step 2: Distribute M_o into negative and positive bending moments in each span. The moment M_o is divided into negative and positive moments in accordance with the distribution factors given in ACI 13.6.3 (see Tables 7.2 and 7.3 of this book). A summary of the total design strip moments at the critical sections of this flat slab is given in Table 7.6.

Step 3: Distribute the total negative and positive bending moments in the design strip to the columns strips and middle strips. The percentages of the negative and positive bending moments at the critical sections that are to be assigned to the column strips and middle strips are given in ACI 13.6.4 to 13.6.6 (see Table 7.3 of this book).

The percentage of the total bending moment in the end spans at the face of the exterior column that is assigned to the column strip depends on the torsional stiffness β_t of the edge beam, which is determined by Eq. (7.10):

$$\beta_t = \frac{E_{cb}C}{2E_{cs}I_s}$$

The cross-sectional constant C is determined by dividing the section into separate rectangular parts and summing the values of C for each part [see Eq. (7.11) and Fig. 7.21]. Using the effective

End Span Moments (ft kips)			Interior Span Moments (ft kips)	
Exterior Negative	**Positive**	**Interior Negative**	**Positive**	**Interior Negative**
$0.30M_o = -92.9$	$0.50M_o = 154.8$	$0.70M_o = -216.7$	$0.35M_o = 108.4$	$0.65M_o = -201.2$

TABLE 7.6 Summary of Total Design Strip Moments for the Flat Slab Given in Example 7.5

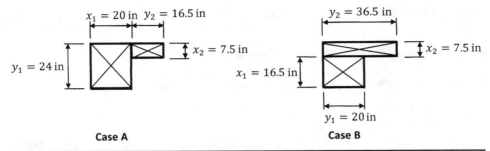

FIGURE 7.22 Calculation of C for the edge beam given in Example 7.5.

flange width $b_e = 36.5$ in, which was determined in Example 7.2, C is the larger of the following (see Fig. 7.22):

$$C_A = \left(1 - 0.63\frac{20}{24}\right)\left(\frac{20^3 \times 24}{3}\right) + \left(1 - 0.63\frac{7.5}{16.5}\right)\left(\frac{7.5^3 \times 16.5}{3}\right) = 32{,}056 \text{ in}^4 \text{ (governs)}$$

$$C_B = \left(1 - 0.63\frac{16.5}{20}\right)\left(\frac{16.5^3 \times 20}{3}\right) + \left(1 - 0.63\frac{7.5}{36.5}\right)\left(\frac{7.5^3 \times 36.5}{3}\right) = 18{,}851 \text{ in}^4$$

The moment of inertia of the slab I_s is determined by Eq. (7.2):

$$I_s = \frac{1}{12}\ell_2 h_f^3 = \frac{1}{12} \times \left(\frac{21.67 \times 12}{2} + \frac{20}{2}\right) \times 7.5^3 = 4{,}923 \text{ in}^4$$

Therefore,

$$\beta_t = \frac{E_{cb}C}{2E_{cs}I_s} = \frac{32{,}056}{2 \times 4{,}923} = 3.3$$

Because $\beta_t > 2.5$, 75% of the total bending moment is assigned to the column strip (see ACI 13.6.4.2). Thus, Table 7.3 can be used to determine the factored bending moments at the critical locations. A summary of the factored bending moments at the critical sections is provided for this flat slab in Table 7.7.

Comments
Unlike the case of the flat plate given in Example 7.4, not all of the total bending moment at the exterior column is assigned to the column strip in this flat slab; this has to do with the presence of the edge beam.

The distribution of the total moments to the column strip and middle strip at the first interior columns of the end spans and in the interior spans are the same as that for the flat plate of Example 7.4.

	End Span Moments (ft kips)			Interior Span Moments (ft kips)	
	Exterior Negative	Positive	Interior Negative	Positive	Interior Negative
Column strip	$0.23M_o = -71.2$	$0.30M_o = 92.9$	$0.53M_o = -164.1$	$0.21M_o = 65.0$	$0.49M_o = -151.7$
Middle strip	$0.07M_o = -21.7$	$0.20M_o = 61.9$	$0.17M_o = -52.6$	$0.14M_o = 43.3$	$0.16M_o = -49.5$

TABLE 7.7 Summary of Factored Bending Moments at the Critical Sections for the Flat Slab Given in Example 7.5

Example 7.6   Determine the factored bending moments at the critical sections for an interior design strip in the north-south direction, using the Direct Design Method for the beam-supported slab system depicted in Fig. 7.14. Assume a 6.5-in-thick slab, normal-weight concrete, a superimposed dead load of 20 psf, and a live load of 100 psf.

Solution   Prior to determining the factored bending moments at the critical sections, check if the Direct Design Method can be used to analyze this two-way system:

1. Three continuous spans must be present in each direction.

   There are five spans in each direction.

2. Slab panels must be rectangular with a ratio of longer to shorter span, centerline-to-centerline of supports, equal to or less than 2.

$$\text{Longer span/shorter span} = 23.5/21.67 = 1.1 < 2.0$$

3. Successive span lengths, centerline-to-centerline of supports, in each direction must not differ by more than one-third of the longer span.

   In each direction, the span lengths are equal.

4. Columns must not be offset more than 10% of the span in the direction of offset from either axis between the centerlines of successive columns.

   No column offsets are present.

5. Loads applied to the slab must be uniformly distributed gravity loads where the ratio of the unfactored live load to the unfactored dead load is equal to or less than 2.

$$\text{Live load} = 100\,\text{psf}$$
$$\text{Dead load of slab} = \frac{6.5}{12} \times 150 = 81.3\,\text{psf}$$
$$\text{Average weight of beam stem} = \frac{24 \times (24 - 6.5)}{144} \times \frac{150}{23.5} = 18.6\,\text{psf}$$
$$\text{Superimposed dead load} = 20\,\text{psf}$$
$$\text{Uniform live to dead load ratio} = 100/(81.3 + 18.6 + 20) = 0.8 < 2$$

6. For panels with column-line beams on all sides, ACI Eq. (13-2) must be satisfied.

   This requirement must be checked for both an interior and an exterior panel. The stiffness ratios α_f are determined in Example 7.3.

   Interior panel:

$$\text{North-south interior beam:} \alpha_{f1} = 6.3$$
$$\text{East-west interior beam:} \alpha_{f2} = 6.9$$
$$\frac{\alpha_{f1}\ell_2^2}{\alpha_{f2}\ell_1^2} = \frac{6.3 \times 23.5^2}{6.9 \times 21.67^2} = 1.1$$
$$0.2 < 1.1 < 5.0$$

   Exterior panel:

$$\text{North-south interior beam:} \alpha_{f1} = 6.3$$
$$\text{East-west edge beam:} \alpha_{f2} = 9.5$$
$$\frac{\alpha_{f1}\ell_2^2}{\alpha_{f2}\ell_1^2} = \frac{6.3 \times 23.5^2}{9.5 \times 21.67^2} = 0.8$$
$$0.2 < 0.8 < 5.0$$

7. Redistribution of bending moments in accordance with ACI 8.4 is not permitted.

Bending moments will not be redistributed in accordance with ACI 8.4.

Therefore, the Direct Design Method can be used for gravity load analysis.
The factored bending moments at the critical sections are determined using the steps presented earlier.

Step 1: Determine the total factored static moment M_o in each span. The total factored static moment M_o is determined by Eq. (7.8):

$$M_o = \frac{q_u \ell_2 \ell_n^2}{8}$$

The total factored gravity loads acting on the slab q_u are determined using the load combination of ACI Eq. (9-2) because this combination yields the maximum effects for dead and live loads (see Table 4.1 in Section 4.2):

$$q_u = 1.2q_D + 1.6q_L = (1.2 \times 119.9) + (1.6 \times 100) = 304 \, \text{psf}$$

The longest clear span ℓ_n for the design strip in the direction of analysis occurs in an end span. Because the difference in the clear span lengths between the end and interior spans is relatively small, conservatively use the longest clear to calculate M_o. This moment is used for all spans:

$$\ell_n = 21.67 - \frac{20}{2 \times 12} - \frac{24}{2 \times 12} = 19.83 \, \text{ft}$$

$$\ell_2 = 23.5 \, \text{ft}$$

$$M_o = \frac{q_u \ell_2 \ell_n^2}{8} = \frac{0.304 \times 23.5 \times 19.83^2}{8} = 351.2 \, \text{ft kips}$$

Step 2: Distribute M_o into negative and positive bending moments in each span. The moment M_o is divided into negative and positive moments in accordance with the distribution factors given in ACI 13.6.3 (see Table 7.2 of this book). A summary of the total design strip moments at the critical sections of this beam-supported slab is given in Table 7.8.

Step 3: Distribute the total negative and positive bending moments in the design strip to the columns strips and middle strips. The percentages of the total negative and positive bending moments at the critical sections that are to be assigned to the column strips and middle strips are given in ACI 13.6.4 to 13.6.6.

Column Strip

Equations (7.9), (7.12), and (7.13) are used to determine the percentages of negative and positive factored bending moments in the column strip.

Exterior negative moment [Eq. (7.12)]: $100 - 10\beta_t + 12\beta_t \left(\frac{\alpha_{f1}\ell_2}{\ell_1}\right)\left(1 - \frac{\ell_2}{\ell_1}\right)$

$$\frac{\alpha_{f1}\ell_2}{\ell_1} = \frac{6.3 \times 23.5}{21.67} = 6.8 > 1; \text{Therefore, use } \frac{\alpha_{f1}\ell_2}{\ell_1} = 1 \text{ in Eq. (7.12).}$$

Because the edge beam and span dimensions in this example are the same as those in Example 7.5, $\beta_t = 3.3$ (see Example 7.5 for details). However, because $\beta_t > 2.5$, use $\beta_t = 2.5$ in Eq. (7.12).

End Span Moments (ft kips)			Interior Span Moments (ft kips)	
Exterior Negative	**Positive**	**Interior Negative**	**Positive**	**Interior Negative**
$0.16M_o = -56.2$	$0.57M_o = 200.2$	$0.70M_o = -245.8$	$0.35M_o = 122.9$	$0.65M_o = -228.3$

TABLE 7.8 Summary of Total Design Strip Moments for the Beam-supported Slab Given in Example 7.6

Thus,

$$100 - 10\beta_t + 12\beta_t \left(\frac{\alpha_{f1}\ell_2}{\ell_1}\right)\left(1 - \frac{\ell_2}{\ell_1}\right) = 100 - (10 \times 2.5) + (12 \times 2.5)(1.0)\left(1 - \frac{23.5}{21.67}\right) = 72.5\%$$

Interior negative moment [Eq. (7.9)]:
$$75 + 30\left(\frac{\alpha_{f1}\ell_2}{\ell_1}\right)\left(1 - \frac{\ell_2}{\ell_1}\right) = 75 + 30(1)\left(1 - \frac{23.5}{21.67}\right) = 72.5\%$$

Positive moment [Eq. (7.13)]:
$$60 + 30\left(\frac{\alpha_{f1}\ell_2}{\ell_1}\right)\left(1.5 - \frac{\ell_2}{\ell_1}\right) = 60 + 30(1)\left(1.5 - \frac{23.5}{21.67}\right) = 72.5\%$$

Note than these percentages can also be obtained by linear interpolation of the values given in ACI 13.6.4.

Middle Strip

The half middle strips are proportioned for the percentages of the total bending moments not resisted by the column strip. Therefore, at all critical sections, the half middle strips must resist $100 - 72.5 = 27.5\%$ of the total moments.

Beams

In columns strips where $\alpha_{f1}\ell_2/\ell_1 > 1.0$, the beams are to be proportioned for 85% of the columns strip bending moments. Therefore, at all critical sections, the beams must resist $0.85 \times 72.5 = 61.6\%$ of the total moments.

A summary of the design bending moments at the critical sections of this beam-supported slab is given in Table 7.9.

The following illustrates the calculation of the interior negative bending moment in the end span.

The total moment in the column strip is equal to 72.5% of the total moment: $0.725 \times 0.70M_o = 0.51M_o = 0.51 \times 351.2 = -179.1$ ft kips.

The beam resists 85% of the column strip moment: $0.85 \times 0.51M_o = 0.43M_o = -151.0$ ft kips.

The slab resists the portion of the bending moment not resisted by the beam: $0.15 \times 0.51M_o = 0.08M_o = -28.1$ ft kips.

The middle strip resists the portion of the total bending moment not resisted by the column strip: $0.70M_o - 0.51M_o = 0.19M_o = -66.7$ ft kips.

The bending moments at the other critical sections can be determined in a similar fashion.

		End Span Moments (ft kips)			Interior Span Moments (ft kips)	
		Exterior Negative	Positive	Interior Negative	Positive	Interior Negative
Total		$0.16M_o$ $= -56.2$	$0.57M_o$ $= 200.2$	$0.70M_o$ $= -245.8$	$0.35M_o$ $= 122.9$	$0.65M_o$ $= -228.3$
Column strip	Total	$0.12M_o$ $= -42.1$	$0.41M_o$ $= 144.0$	$0.51M_o$ $= -179.1$	$0.25M_o$ $= 87.8$	$0.47M_o$ $= -165.1$
	Beam	$0.10M_o$ $= -35.1$	$0.35M_o$ $= 122.9$	$0.43M_o$ $= -151.0$	$0.21M_o$ $= 73.8$	$0.40M_o$ $= -140.5$
	Slab	$0.02M_o$ $= -7.0$	$0.06M_o$ $= 21.1$	$0.08M_o$ $= -28.1$	$0.04M_o$ $= 14.0$	$0.07M_o$ $= -24.6$
Middle strip		$0.04M_o$ $= -14.0$	$0.16M_o$ $= 56.2$	$0.19M_o$ $= -66.7$	$0.10M_o$ $= 35.1$	$0.18M_o$ $= -63.2$

TABLE 7.9 Summary of Factored Bending Moments at the Critical Sections for the Beam-supported Slab Given in Example 7.6

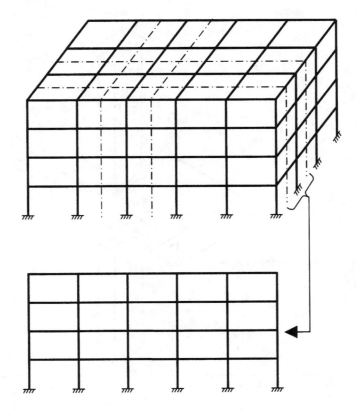

Figure 7.23
Equivalent frames
to be used in
the Equivalent
Frame Method.

Equivalent Frame Method

Introduction   Provisions for the Equivalent Frame Method, which are based on the studies reported in Refs. 6 through 8, are given in ACI 13.7. In this method, the structure is considered to be made up of equivalent frames on column lines in both the longitudinal and transverse directions. The three-dimensional building is divided into a series of two-dimensional equivalent frames in both directions, as shown in Fig. 7.23.

The two-dimensional frames, which are centered on the support lines, extend the full height of the building and consist of the columns and the portion of the slab bounded by the panel centerlines on each side of the columns. Although analysis of each equivalent frame in its entirety is permitted, a separate analysis of each floor or roof is also permitted for gravity loads (ACI 13.7.2.5). In such cases, the far ends of the columns are considered to be fixed (see Fig. 7.24).

Members of the equivalent frame are slab-beams, columns, and torsional members (see Fig. 7.25 and ACI Fig. R13.7.4). The initial step in the frame analysis is to determine

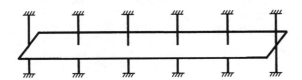

Figure 7.24
Equivalent frame
permitted in
analysis.

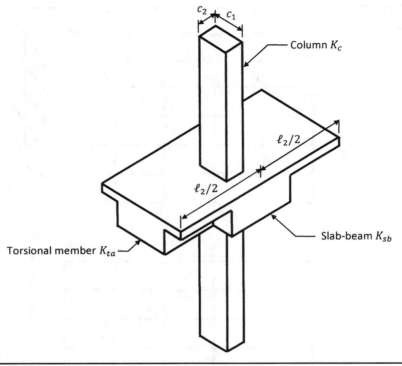

Figure 7.25   Equivalent column.

the flexural stiffness of the equivalent frame members. Once the stiffnesses and other important quantities have been determined, the moment distribution method is used to determine the factored bending moments at the critical sections.

Slab-Beams   When determining the stiffness of slab-beam members, it is permitted to determine the moment of inertia at any cross-section outside of the joints or column capitals, using the gross area of the concrete (ACI 13.7.3.1). Also, any variation in the moment of inertia along the axis of the slab-beams must be taken into account; this variation would occur, for example, where drop panels are present.

Because the analysis is based on a frame where the span lengths are measured from the centerlines of the supports, ACI 13.7.3.3 requires that the moment of inertia of the slab-beams from the center of the column to the face of the column, bracket, or capital be determined by dividing the moment of inertia of the slab-beam at the face of the support by $(1 - c_2/\ell_2)^2$, where c_2 and ℓ_2 are the widths of the column and the column-beam in the direction perpendicular to the direction of analysis, respectively.

Figure 7.26 depicts the cross-sections that are to be used when calculating the stiffness of a slab-beam in a flat plate. The stiffness of the slab-beam K_{sb} can be determined from the following equation:

$$K_{sb} = \frac{k_{AB} E_{cs} I_{sb}}{\ell_1} \qquad (7.16)$$

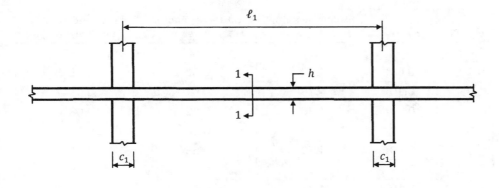

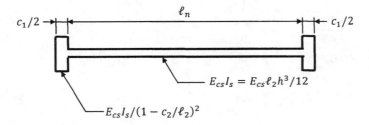

Equivalent slab-beam stiffness

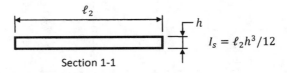

Section 1-1

FIGURE 7.26   Slab-beam stiffness in a flat plate.

In this equation, k_{AB} is the stiffness factor that takes into account the stepwise variation in the moment of inertia over the span and I_{sb} is the moment of inertia of the slab-beam away from the supports, which in this case, is the moment of inertia of the slab: $I_{sb} = I_s = \ell_2 h^3/12$. It is evident that the stiffness factor cannot be based on the assumption of uniform prismatic members.

Methods on how to determine stiffness factors and carryover factors C_{AB} can be found in numerous analysis references, including Ref. 10. Table 7.10 provides these factors for a flat plate as a function of column and span lengths, assuming $c_{A1} = c_{B1}$ and $c_{A2} = c_{B2}$. Also included in the table are fixed-end moment coefficients m_{AB}; these factors can be used to determine the fixed-end moments M_{FE} in a flat-plate system subjected to a factored uniformly distributed load q_u acting over the entire span length:

$$M_{FE} = m_{AB}q_u\ell_2\ell_1^2 \tag{7.17}$$

c_{A1}/ℓ_1	c_{A2}/ℓ_2	Stiffness Factor k_{AB}	Carryover Factor C_{AB}	Fixed-end Moment Coefficient m_{AB}
0.00	—	4.00	0.50	0.0833
0.10	0.00	4.00	0.50	0.0833
	0.10	4.18	0.51	0.0847
	0.20	4.36	0.52	0.0860
	0.30	4.53	0.54	0.0872
	0.40	4.70	0.55	0.0882
0.20	0.00	4.00	0.50	0.0833
	0.10	4.35	0.52	0.0857
	0.20	4.72	0.54	0.0880
	0.30	5.11	0.56	0.0901
	0.40	5.51	0.58	0.0921
0.30	0.00	4.00	0.50	0.0833
	0.10	4.49	0.53	0.0863
	0.20	5.05	0.56	0.0893
	0.30	5.69	0.59	0.0923
	0.40	6.41	0.61	0.0951
0.40	0.00	4.00	0.50	0.0833
	0.10	4.61	0.53	0.0866
	0.20	5.35	0.56	0.0901
	0.30	6.25	0.60	0.0936
	0.40	7.37	0.64	0.0971

TABLE 7.10 Moment Distribution Constants for Slab-beams in Flat Plates

The cross-sections that are to be used when calculating the stiffness of a slab-beam in a flat slab are shown in Fig. 7.27. Table 7.11 contains moment distribution constants for a flat slab where the following conditions are applicable: (1) $c_{A1} = c_{B1}$ and $c_{A2} = c_{B2}$; (2) the lengths of the drop panels on each side of the columns are equal to one-sixth the span length in the direction of analysis; (3) the overall thickness of the drop panel is equal 125% of the slab thickness; and (4) a factored uniformly distributed load q_u acts over the entire span length.

Reference 2 contains moment distribution factors for a wide range of geometric and loading conditions for flat-plate and flat-slab systems. Moment distribution factors for other systems can be derived using the methods given in Ref. 10.

Columns Similar to slab-beams, the stiffness of a column is based on the moment of inertia at any cross-section outside of the joints or column capitals, using the gross area of the concrete (ACI 13.7.4.1). The length of the column from the mid-depth of the slab above to that of the slab below is to be used in the calculation of the stiffness. Variation in

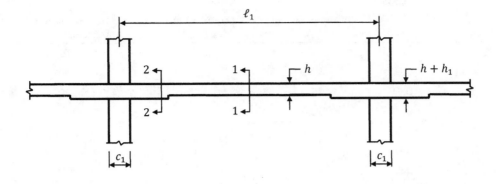

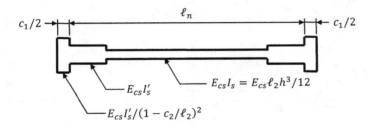

Equivalent slab-beam stiffness

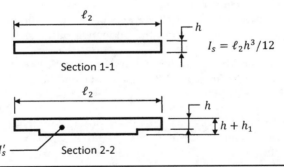

FIGURE 7.27 Slab-beam stiffness in a flat slab.

the moment of inertia along the axis of the slab-beams must also be taken into account; this variation would usually occur where column capitals are present.

The moment of inertia of the column from the center to the top or bottom face of the slab at a joint is assumed to be infinite (ACI 13.7.4.3). Thus, as with slab-beams, the stiffness factor cannot be based on the assumption of uniform prismatic members.

Figure 7.28 depicts the sections that are to be used when calculating the stiffness of a column in a flat plate or a flat slab. The stiffness of a column is

$$K_c = \frac{k_{AB} E_{cc} I_c}{L_c} \tag{7.18}$$

c_{A1}/ℓ_1	c_{A2}/ℓ_2	Stiffness Factor k_{AB}	Carryover Factor C_{AB}	Fixed-end Moment Coefficient m_{AB}
0.00	—	4.79	0.54	0.0879
0.10	0.00	4.79	0.54	0.0879
	0.10	4.99	0.55	0.0890
	0.20	5.18	0.56	0.0901
	0.30	5.37	0.57	0.0911
0.20	0.00	4.79	0.54	0.0879
	0.10	5.17	0.56	0.0900
	0.20	5.56	0.58	0.0918
	0.30	5.96	0.60	0.0936
0.30	0.00	4.79	0.54	0.0879
	0.10	5.32	0.57	0.0905
	0.20	5.90	0.59	0.0930
	0.30	6.55	0.62	0.0955

TABLE 7.11   Moment Distribution Constants for Slab-beams in Flat Slabs

In this equation, L_c is the length of the column from the mid-depth of the slab above to that of the slab below.

Table 7.12 contains moment distribution constants for columns in a flat plate where the thicknesses of the slab above and below the column are equal. Note that ℓ_c is the length of the column from the underside of the slab above to the top of the slab below.

The moment distribution constants in Table 7.13 can be used for a column in a flat slab where the thicknesses of the slab above and below the column are equal and the overall thickness of the drop panel is 125% of the slab thickness.

Torsional Members   Torsional members provide moment transfer between the slab-beams and the columns and are assumed to have a constant cross-section throughout their length (ACI 13.7.5.1). Two conditions are specified, depending on the framing members:

1. No transverse beams frame into the columns.

    In such cases, the transverse member consists of a portion of the slab having a width equal to that of the column, bracket, or capital in the direction of analysis, as illustrated in Fig. 7.29.

2. Transverse beams frame into the columns.

    For monolithic construction, T- or L-beam action is assumed, with the flanges extending on each side of the beam a distance equal to the projection of the beam above or below the slab but not greater than four times the thickness of the slab (see Fig. 7.30).

L_c/ℓ_c	Stiffness Factor k_{AB}	Carryover Factor c_{AB}
1.05	4.52	0.54
1.10	5.09	0.57
1.15	5.71	0.60
1.20	6.38	0.62
1.25	7.11	0.65
1.30	7.89	0.67
1.35	8.73	0.69
1.40	9.63	0.71
1.45	10.60	0.73
1.50	11.62	0.74

TABLE 7.12 Moment Distribution Constants for Columns in Flat Plates

FIGURE 7.28 Column stiffness for a flat plate or flat slab.

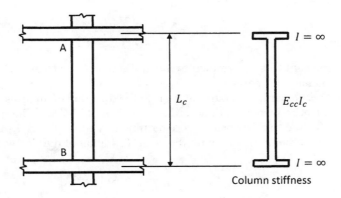

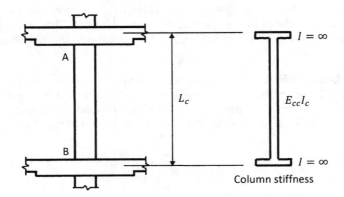

L_c/ℓ_c	Stiffness Factor k_{AB}	Carryover Factor C_{AB}
1.05	4.59	0.53
1.10	5.24	0.55
1.15	5.95	0.58
1.20	6.74	0.60
1.25	7.59	0.61
1.30	8.51	0.62
1.35	9.51	0.63
1.40	10.59	0.64
1.45	11.76	0.65
1.50	13.01	0.66

TABLE 7.13 Moment Distribution Constants for Columns in Flat Slabs

The stiffness of the torsional member K_t is calculated by the following approximate expression, which is given in ACI R13.7.5:

$$K_t = \sum \frac{9E_{cs}C}{\ell_2\,[1-(c_2/\ell_2)]^3} \qquad (7.19)$$

The constant C is determined by Eq. (7.11) where the cross-section is divided into separate rectangular parts.

Where beams frame into the support in the direction of analysis, the torsional stiffness K_t must be increased in accordance with ACI 13.7.5.2. The increased torsional stiffness K_{ta} due to these beams is determined by multiplying K_t by the ratio of the moment of inertia of the slab with the beam I_{sb} to that of the slab without the beam I_s:

$$K_{ta} = K_t\left(\frac{I_{sb}}{I_s}\right) = K_t\left(\frac{12I_{sb}}{\ell_2 h^3}\right) \qquad (7.20)$$

Analysis Procedure Once the stiffnesses of the members have been obtained in the equivalent frame, the factored moments at the critical locations can be determined using the moment distribution method. This method is convenient for analyzing partial frames such as those in the Equivalent Frame Method. The concept of an equivalent column is discussed next, and it is used to determine the distribution factors at the joints.

Equivalent column An equivalent column is assumed to consist of the actual columns above and below the slab-beam plus the attached torsional members on each side of the columns extending to the centerline of the adjacent panels (see Fig. 7.25). The stiffness of the equivalent column K_{ec} is a combination of the stiffnesses of the columns and torsional members:

$$\frac{1}{K_{ec}} = \frac{1}{\sum K_c} + \frac{1}{\sum K_{ta}} \qquad (7.21)$$

FIGURE 7.29
Torsional member
where no transverse
beams frame
into the
columns.

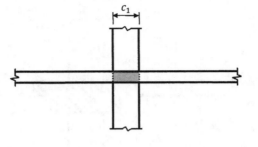

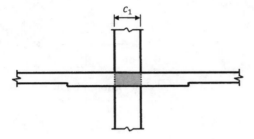

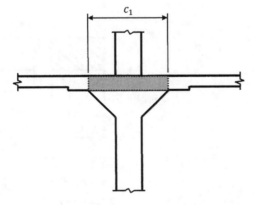

Distribution factors The generic frame shown in Fig. 7.31 is used to illustrate how distribution factors DF are determined for each member framing into a joint. Shown are the stiffnesses of the slab-beams, columns, and torsional members that are determined using the methods described previously.

The slab-beam distribution factors for both spans are given as follows:

$$DF\,(\text{span B-A}) = \frac{K_{sb1}}{K_{ec} + K_{sb1} + K_{sb2}} \tag{7.22a}$$

$$DF\,(\text{span B-C}) = \frac{K_{sb2}}{K_{ec} + K_{sb1} + K_{sb2}} \tag{7.22b}$$

These distribution factors are used in the moment distribution procedure to obtain bending moments in the slab-beams.

Figure 7.30
Torsional member
where transverse
beams frame
into the
columns.

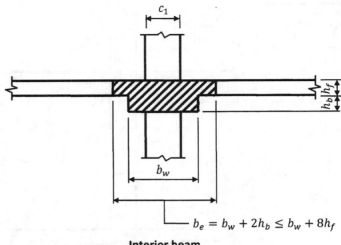

Interior beam

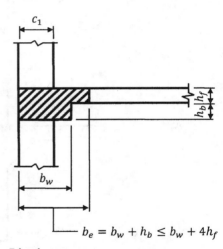

Edge beam

The equation for the distribution factor for the equivalent column is

$$DF\text{(equivalent column)} = \frac{K_{ec}}{K_{ec} + K_{sb1} + K_{sb2}} \tag{7.23}$$

This represents the unbalanced moment that is transferred to the columns from the slab-beams.

The unbalanced moment determined by Eq. (7.23) is distributed to the columns above and below the slab-beam in proportion to the actual column stiffnesses at the

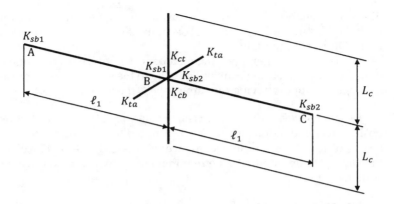

FIGURE **7.31**
Moment
distribution
factors *DF*.

joint. Thus, the portions of the unbalanced moment that are transferred to the upper and lower columns at a joint can be determined from the following equations:

$$\text{Portion of unbalanced moment transferred to upper column} = \frac{K_{ct}}{K_{cb} + K_{ct}} \quad (7.24a)$$

$$\text{Portion of unbalanced moment transferred to lower column} = \frac{K_{cb}}{K_{cb} + K_{ct}} \quad (7.24b)$$

The actual columns are designed for these bending moments.

Arrangement of live load   The arrangement of live load that will cause critical reactions is not always readily apparent. The most demanding sets of design forces must be established by investigating the effects of live load placed in various critical patterns.

ACI 13.7.6 permits the arrangement of the live load to be limited to the following conditions:

1. In cases where the unfactored live load is equal to or less than three-quarters the unfactored dead load, it is permitted to assume that maximum factored moments occur at all sections when the full factored live load acts on all of the spans at the same time.

2. In all other cases, pattern live loading must be used to obtain the maximum moments. It is permitted by the Code to use only three-quarters of the live load in such cases. The use of less than the full factored live load is based on the fact that maximum negative and positive live load moments cannot occur simultaneously and that redistribution of maximum moments is possible before failure occurs.

Critical section for factored moments   As was the case in the Direct Design Method, the critical section for negative factored moments at interior supports is at the face of rectilinear supports but not farther away than $0.175\ell_1$ from the center of the support (ACI 13.7.7.1). This limit applies in cases where there is a long, narrow support in the direction of analysis, and it helps to ensure that there is not an unwarranted reduction in the design moment. The critical sections at exterior supports are the same as those in the Direct Design Method.

In cases where the supporting member does not have a rectangular cross-section or if the sides of the rectangle are not parallel to the spans, such members are to be treated as a square support having the same area as the actual support.

Factored bending moments in column strips, middle strips, and beams   Negative and positive factored moments may be distributed to the column strips, middle strips, and beams of the slab-beam in accordance with ACI 13.6.4 to 13.6.6, which are given in the Direct Design Method. The requirement of ACI 13.6.1.6 must be satisfied for slab systems with beams.

Moment redistribution   In cases where the Equivalent Frame Method is used to analyze a slab system that meets the limitations of the Direct Design Method, the factored bending moments may be reduced so that the total static factored moment, which is equal to the sum of the average positive and negative bending moments, does not exceed M_o [see Eq. (7.8)]. This permitted reduction in design moments essentially means that the Code is not requiring the design to be based on the greater bending moments obtained from the two acceptable methods of analysis.

The flowchart shown in Fig. 7.32 can be used to determine the design bending moments, using the Equivalent Frame Method.

Example 7.7   Determine the factored bending moments at the critical sections for an interior design strip in the north-south direction, using the Equivalent Frame Method for the flat-plate system depicted in Fig. 7.33. Assume a 9-in-thick slab, normal-weight concrete, $f_c' = 4,000$ psi, a superimposed dead load of 20 psf, and a live load of 50 psf. The length of the columns from the mid-depth of the slab above to the that of the slab below is equal to 12 ft, and the same slab thickness is used on all floors.

Solution   The flowchart shown in Fig. 7.32 is used to determine the factored bending moments in this example.

Step 1: Determine the preliminary slab thickness h, using Section 7.3. A preliminary slab thickness of 9 in was given in the problem statement.

Step 2: Determine the column strips and middle strips, using Fig. 7.18. For an interior design strip, $\ell_1 = 19.5$ ft and $\ell_2 = 21.167$ ft. Because $\ell_1 < \ell_2$, the width of the column strip is equal to $\ell_1/2 = 9.75$ ft. The width of each half middle strip is equal to $(21.167 - 9.75)/2 = 5.71$ ft.

Step 3: Determine l_2 for analysis. Because this is an interior design strip that is not adjacent and parallel to an edge beam, ℓ_2 is equal to the center-to-center span length that is perpendicular to ℓ_1, which is 21.167 ft.

Step 4: Determine the modulus of elasticity of the slab-beam. For normal-weight, 4,000 psi concrete,

$$E_{cs} = w_c^{1.5} 33\sqrt{f_c'} = (150)^{1.5} \times 33\sqrt{4,000} = 3.83 \times 10^6 \text{ psi}$$

Step 5: Determine the moment of inertia of the slab-beam. Because there are no column-line beams, the moment of the inertia of the slab-beam is equal to that of the slab (see Fig. 7.26):

$$I_{sb} = \frac{\ell_2 h^3}{12} = \frac{(21.167 \times 12) \times 9^3}{12} = 15,431 \text{ in}^4$$

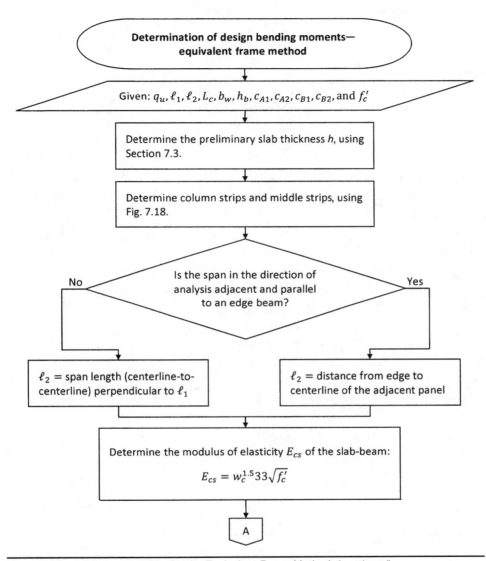

FIGURE 7.32 Analysis procedure for the Equivalent Frame Method. (*continued*)

Step 6: Determine stiffness factors k_{AB} and k_{BA} at the near and far ends of the slab-beam.

$$\frac{c_{A1}}{\ell_1} = \frac{24}{19.5 \times 12} = 0.10 \quad \text{and} \quad \frac{c_{A2}}{\ell_2} = \frac{24}{21.167 \times 12} = 0.10$$

Use Table 7.10 to obtain the following stiffness factors:

$$k_{AB} = k_{BA} = 4.18$$

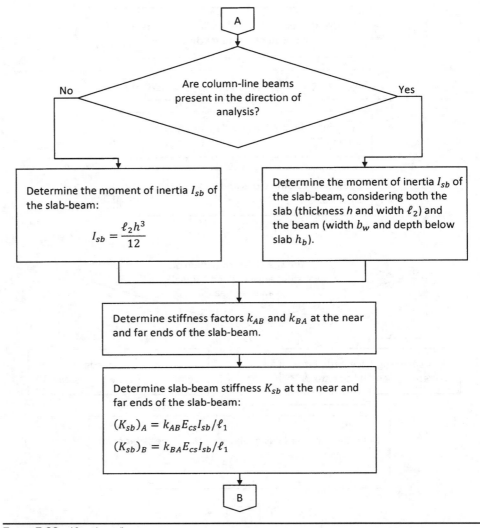

Figure 7.32   (Continued)

Step 7: Determine stiffness K_{sb} at the near and far ends of the slab-beam.

$$(K_{sb})_A = (K_{sb})_B = \frac{k_{AB}E_{cs}I_{sb}}{\ell_1} = \frac{4.18 \times 3.83 \times 10^6 \times 15{,}431}{19.5 \times 12} = 1{,}056 \times 10^6 \text{ in lb}$$

Step 8: Determine the carryover factors and fixed-end moment coefficients at the near and far ends of the slab-beam.

From Table 7.10 with $c_{A1}/\ell_1 = 0.10$ and $c_{A2}/\ell_2 = 0.10$,

$$C_{AB} = C_{BA} = 0.51$$
$$m_{AB} = m_{BA} = 0.0847$$

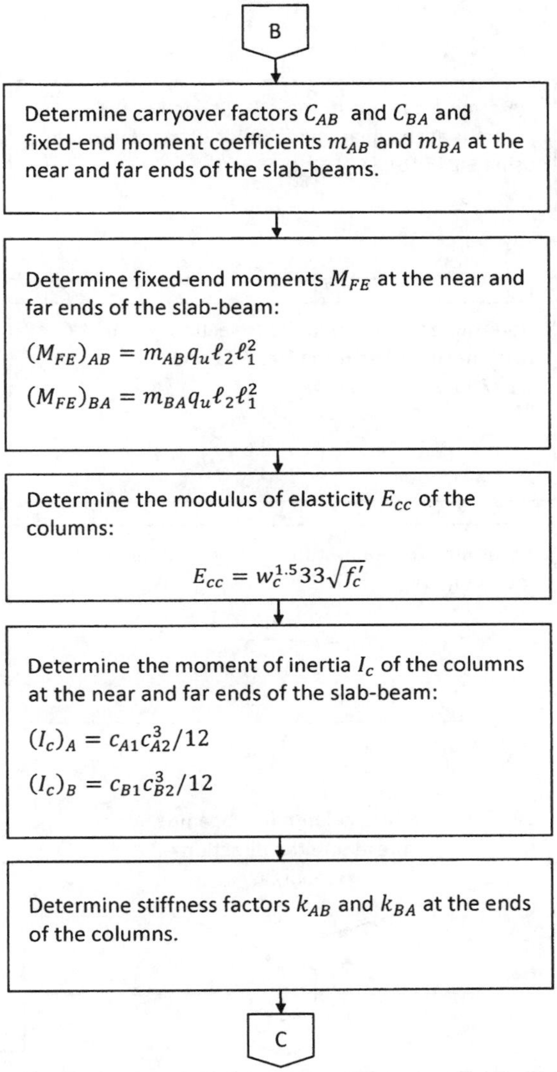

Determine carryover factors C_{AB} and C_{BA} and fixed-end moment coefficients m_{AB} and m_{BA} at the near and far ends of the slab-beams.

Determine fixed-end moments M_{FE} at the near and far ends of the slab-beam:

$$(M_{FE})_{AB} = m_{AB}q_u\ell_2\ell_1^2$$

$$(M_{FE})_{BA} = m_{BA}q_u\ell_2\ell_1^2$$

Determine the modulus of elasticity E_{cc} of the columns:

$$E_{cc} = w_c^{1.5}33\sqrt{f_c'}$$

Determine the moment of inertia I_c of the columns at the near and far ends of the slab-beam:

$$(I_c)_A = c_{A1}c_{A2}^3/12$$

$$(I_c)_B = c_{B1}c_{B2}^3/12$$

Determine stiffness factors k_{AB} and k_{BA} at the ends of the columns.

FIGURE 7.32   (*Continued*)

Step 9: Determine fixed-end moments at the near and far ends of the slab-beam.

$$\text{Live load} = 50\,\text{psf}$$
$$\text{Dead load of slab} = \frac{9}{12} \times 150 = 112.5\,\text{psf}$$
$$\text{Superimposed dead load} = 20\,\text{psf}$$

The total factored gravity loads acting on the slab q_u is determined using the load combination given in ACI Eq. (9-2), because this yields the maximum effects for dead and live loads (see

Determine column stiffness K_c at each of the columns at the near and far ends of the slab-beam, using Eq. (7.18):

$$K_c = \frac{k_{AB}E_{cc}I_c}{L_c}$$

Determine the torsional cross-section constant C of torsional member, using Fig. 7.21 and Eq. (7.11):

$$C = \sum \left(1 - 0.63\frac{x}{y}\right)\frac{x^3 y}{3}$$

Determine torsional stiffness K_t of torsional member, using Eq. (7.19):

$$K_t = \frac{9E_{cs}C}{\ell_2 \left(1 - \frac{c_2}{\ell_2}\right)^3}$$

Are column-line beams present in the direction of analysis?

No — $K_{ta} = K_t$

Yes — $K_{ta} = \dfrac{12K_t I_{sb}}{\ell_2 h^3}$

D

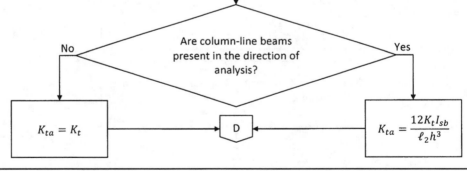

Figure 7.32   *(Continued)*

Table 4.1 in Section 4.2):

$$q_u = 1.2q_D + 1.6q_L = (1.2 \times 132.5) + (1.6 \times 50) = 239\,\text{psf}$$
$$(M_{FE})_{AB} = (M_{FE})_{BA} = m_{AB}q_u\ell_2\ell_1^2 = 0.0847 \times 0.239 \times 21.167 \times 19.5^2 = 162.9\,\text{ft kips}$$

Step 10: Determine the modulus of elasticity of the columns. For normal-weight, 4,000 psi concrete,

$$E_{cc} = w_c^{1.5}33\sqrt{f_c'} = (150)^{1.5} \times 33\sqrt{4,000} = 3.83 \times 10^6\,\text{psi}$$

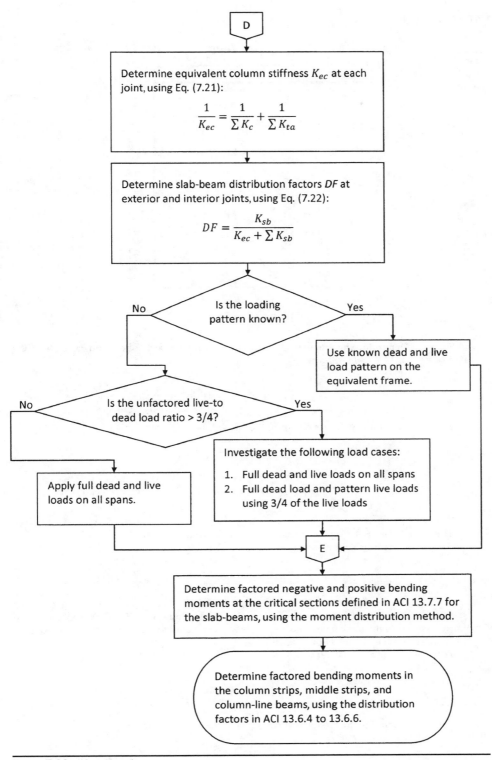

The flowchart contents:

D

Determine equivalent column stiffness K_{ec} at each joint, using Eq. (7.21):

$$\frac{1}{K_{ec}} = \frac{1}{\sum K_c} + \frac{1}{\sum K_{ta}}$$

Determine slab-beam distribution factors DF at exterior and interior joints, using Eq. (7.22):

$$DF = \frac{K_{sb}}{K_{ec} + \sum K_{sb}}$$

Is the loading pattern known?

No →

Yes → Use known dead and live load pattern on the equivalent frame.

Is the unfactored live-to dead load ratio > 3/4?

No → Apply full dead and live loads on all spans.

Yes → Investigate the following load cases:

1. Full dead and live loads on all spans
2. Full dead load and pattern live loads using 3/4 of the live loads

E

Determine factored negative and positive bending moments at the critical sections defined in ACI 13.7.7 for the slab-beams, using the moment distribution method.

Determine factored bending moments in the column strips, middle strips, and column-line beams, using the distribution factors in ACI 13.6.4 to 13.6.6.

FIGURE 7.32 (Continued)

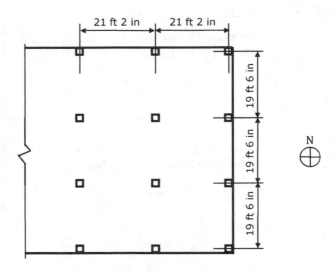

FIGURE **7.33** The
flat-plate system
given in
Example 7.7.

Columns: 24×24 in

Step 11: Determine the moment of inertia of the columns at the near and far ends of the slab-beam. For 24 × 24 in columns,

$$I_c = \frac{c_{A1}c_{A2}^3}{12} = \frac{24^4}{12} = 27{,}648 \text{ in}^4$$

Step 12: Determine stiffness factors at the ends of the columns.

$$L_c = 12 \text{ ft}$$

$$\ell_c = 12 - \frac{9}{12} = 11.25 \text{ ft}$$

$$\frac{L_c}{\ell_c} = 1.07$$

Thus, $k_{AB} = k_{BA} = 4.75$ by interpolation from Table 7.12.

Step 13: Determine column stiffness.

$$K_c = \frac{k_{AB}E_{cc}I_c}{L_c} = \frac{4.75 \times 3.83 \times 10^6 \times 27{,}648}{12 \times 12} = 3{,}493 \times 10^6 \text{ in lb}$$

Step 14: Determine the torsional constant C of the torsional member. Because there are no transverse beams, the torsional member consists of a portion of the slab having a width equal to that of the column in the direction of analysis, as illustrated in Fig. 7.29.

Therefore, using $x = 9$ in and $y = 24$ in,

$$C = \left(1 - 0.63\frac{x}{y}\right)\frac{x^3 y}{3} = \left[1 - \left(0.63 \times \frac{9}{24}\right)\right]\frac{9^3 \times 24}{3} = 4{,}454 \text{ in}^4$$

Step 15: Determine the torsional stiffness K_t of the torsional member.

$$K_t = \frac{9E_{cs}C}{\ell_2\,[1 - (c_2/\ell_2)]^3} = \frac{9 \times 3.83 \times 10^6 \times 4{,}454}{(21.167 \times 12)\{1 - [24/(21.167 \times 12)]\}^3} = 814 \times 10^6 \text{ in lb}$$

Step 16: Determine the torsional stiffness K_{ta} of the torsional member. Because there are no column-line beams, $K_{ta} = K_t = 814 \times 10^6$ in lb

Step 17: Determine the equivalent column stiffness K_{ec} at each joint. At each joint, there are two columns (one above and one below) and two torsional members (one on each side of each column).

$$\frac{1}{K_{ec}} = \frac{1}{\sum K_c} + \frac{1}{\sum K_{ta}} = \frac{1}{2 \times 3{,}493 \times 10^6} + \frac{1}{2 \times 814 \times 10^6}$$

$$K_{ec} = 1{,}320 \times 10^6 \text{ in lb}$$

Step 18: Determine slab distribution factors DF at exterior and interior joints. In general,

$$DF = \frac{K_{sb}}{K_{ec} + \sum K_{sb}}$$

At an exterior joint, there is only one slab-beam framing into the joint:

$$DF = \frac{1{,}056}{1{,}320 + 1{,}056} = 0.44$$

At an interior joint, there are two slab-beams framing into the joint:

$$DF = \frac{1{,}056}{1{,}320 + (2 \times 1{,}056)} = 0.31$$

Step 19: Determine the loading pattern.

$$\text{Uniform live to dead load ratio} = 50/(112.5 + 20) = 0.4 < 0.75$$

Because this ratio is less than 0.75, apply full dead and live loads on all spans.

Step 20: Determine the factored bending moments at the critical sections, using the moment distribution method. A summary of the computations in the moment distribution is given in Table 7.14 for the negative bending moments. It is assumed that counterclockwise bending moments are positive.

The positive factored bending moments in the end and interior spans can be determined by subtracting the average of the negative factored bending moments in the span from the total factored bending moment at the midspan for a simply supported beam:

$$M_u^+ = \frac{q_u \ell_2 \ell_1^2}{8} - \frac{M_{uL}^- + M_{uR}^-}{2}$$

For span A-B or span C-D,

$$(M_u^+)_{AB} = (M_u^+)_{CD} = \frac{0.239 \times 21.167 \times 19.5^2}{8} - \frac{95.1 + 187.4}{2} = 99.2 \text{ ft kips}$$

For span B-C,

$$(M_u^+)_{BC} = \frac{0.239 \times 21.167 \times 19.5^2}{8} - \frac{169.7 + 169.7}{2} = 70.8 \text{ ft kips}$$

The negative factored bending moments to be used in design can be taken at the faces of the supports (i.e., $24/2 = 12$ in from the centers of supports) but not at distances greater than

Joint	A	B		C		D
Member	A-B	B-A	B-C	C-B	C-D	D-C
Distribution factor DF	0.44	0.31	0.31	0.31	0.31	0.44
Carryover factor	0.51	0.51	0.51	0.51	0.51	0.51
Fixed-end moment M_{FE}	162.9	−162.9	162.9	−162.9	162.9	−162.9
Distribution	−71.7	0.0	0.0	0.0	0.0	71.7
Carryover	0.0	−36.6	0.0	0.0	36.6	0.0
Distribution	0.0	11.4	11.4	−11.4	−11.4	0.0
Carryover	5.7	0.0	−5.7	5.7	0.0	−5.7
Distribution	−2.5	1.8	1.8	−1.8	−1.8	2.5
Carryover	0.9	−1.3	−0.9	0.9	1.3	−0.9
Distribution	−0.4	0.4	0.4	−0.4	−0.4	0.4
Carryover	0.2	−0.2	−0.2	0.2	0.2	−0.2
Distribution	−0.1	0.1	0.1	−0.1	−0.1	0.1
Carryover	0.1	−0.1	−0.1	0.1	0.1	−0.1
Negative moment	95.1	−187.4	169.7	−169.7	187.4	−95.1

TABLE 7.14 Moment Distribution Computations for Negative Factored Bending Moments Given in Example 7.7

$0.175\ell_1 = 41$ in from the centers of supports. Therefore, because 12 in is less than 41 in, use the faces of supports as the critical sections for design.

The factored negative bending moments at the critical sections can be obtained from statics. For example, in the end spans, the shear force at the centerline of the exterior supports is equal to

$$(V_u)_A = (V_u)_D = \frac{[(0.239 \times 21.167 \times 19.5^2)/2] - 187.4 + 95.1}{19.5} = 44.6 \text{ kips}$$

The factored bending moment at the face of the support at A or D is then

$$(M_u)_A = (M_u)_D = \frac{0.239 \times 21.167 \times 1^2}{2} - (44.6 \times 1) + 95.1 = 53.0 \text{ ft kips}$$

The factored bending moments at the faces of the other supports can be obtained in a similar fashion. A summary of these bending moments is given in the next step.

Step 21: Determine the factored bending moments at the critical sections in the column strip and middle strip. The distribution of the design factored bending moments at the critical sections in the column strip and middle strip is performed using ACI 13.6.4 and 13.6.6 for this flat plate. The following percentages are applicable to the column strip:

- Exterior negative: 100%
- Positive: 60%
- Interior negative: 75%

The bending moments not resisted by the column strip are resisted by the middle strip. A summary of the design factored bending moments is given in Table 7.15.

	End Span Moments (ft kips)			Interior Span Moments (ft kips)	
	Exterior Negative	Positive	Interior Negative	Positive	Interior Negative
Total	53.0	99.2	135.8	70.8	122.9
Column strip	53.0	59.5	101.9	42.5	92.2
Middle strip	0	39.7	33.9	28.3	30.7

TABLE 7.15 Summary of Factored Bending Moments at the Critical Sections for the Flat Plate Given in Example 7.7

Comments

The unbalanced bending moments that occur at the joints from the slab-beams are distributed to the columns above and below the slab-beam in proportion to their stiffnesses. Because the columns have the same cross-section and length above and below the slab-beam, the stiffnesses are the same, and one-half of the unbalanced bending moment is distributed to the column above and below.

At an exterior column, the unbalanced bending moment is equal to 95.1 ft kips (see Table 7.14). The bottom of the column above the slab-beam and the top of the column below the slab-beam are subjected to a bending moment equal to $95.1/2 = 47.6$ ft kips. From Table 7.13, the carryover factor is equal to 0.54 for $L_c/\ell_c = 1.07$. Therefore, the bending moment at the top of the column above the slab-beam and at the bottom of the column below the slab-beam is equal to $0.54 \times 47.6 = 25.7$ ft kips.

At the interior columns, the unbalanced bending moment is equal to $187.4 - 169.7 = 17.7$ ft kips. Distribution of this moment to the columns above and below the slab-beam is similar to that of the exterior column.

Note that all of these bending moments are at the centerline of the slab-beam. For design purposes, the bending moments can be determined at the top and bottom of the slab-beam; however, the difference between these moments is small, and it is conservative to use the bending moment at the slab-beam centerline in the design of the column.

7.4.3 Lateral Loads

A structure utilizing two-way concrete floor and roof systems can be modeled for lateral loads, using any method that satisfies both equilibrium and compatibility. Additionally, the results obtained from such an analysis should be in reasonable agreement with test data.

Numerous analytical procedures exist for modeling frames subjected to lateral loads. The following methods have been shown to produce acceptable results[11]:

1. Finite element models
2. Effective beam width model
3. Equivalent frame model

Regardless of the analysis method that is used, stiffness of slabs, beams, columns, walls, and any other elements that are part of the lateral force–resisting system must take into account the effects of cracking so that drift caused by lateral loads is not underestimated.

Similar to beam-supported slabs, flat-plate structures behave like rigid frames when subjected to lateral loads. It has been demonstrated that only a portion of the slab is effective across its full width in resisting the effects from lateral loads. The effective beam width model will give reasonably accurate results in routine situations. In this method, the actual slab is replaced by a flexural element that has the same thickness as the slab and an effective beam width that is a fraction of the transverse width of the slab. The following equation can be used to determine the effective slab width for an interior slab–column frame[12]:

$$\text{Effective slab width} = 2c_1 + \frac{\ell_1}{3} \tag{7.25}$$

In this equation, c_1 and ℓ_1 are the column dimension and the span length in the direction of analysis, respectively. For an exterior frame, the effective slab width is equal to one-half the value determined by Eq. (7.25). Reference 12 demonstrates that this method produces an accurate estimate of elastic stiffness for regular frames.

To account for cracking, bending stiffness is typically reduced between one-half and one-quarter of the uncracked stiffness, which is a function of the slab thickness and effective slab width. When determining drifts or secondary effects in columns (see Chap. 8), lower-bound slab stiffness should be assumed in the analysis. In structures where slab–column frames interact with structural walls, a range of slab stiffnesses should be investigated in order to assess the importance of interaction.

ACI 13.5.1.3 permits combining the results of the gravity load analysis with the results from the lateral load analysis. For example, the slab can be analyzed using the Direct Design Method, and the results from that analysis can be combined with the results from the effective beam width model, using the load combinations given in ACI 9.2. The slab and other elements are subsequently designed for the effects from the critical load combinations.

7.5 Design for Flexure

7.5.1 Overview

Two-way slab systems are designed for flexure, using the basic principles of the strength design method, which are given in Chap. 5. In general, the flexural design strength ϕM_n must be equal to or greater than the required strength M_u where the factored bending moments are determined at the critical sections in the column strips, middle strips, and column-line beams (if any), using the Direct Design Method, the Equivalent Frame Method, or any other rational method of analysis.

Requirements for transfer of moment between a slab and a column must also be satisfied. In general, a fraction of the unbalanced moment at such joints must be resisted by reinforcement in a specified portion of the slab centered on the column.

7.5.2 Determining Required Reinforcement

Once the thickness of the slab has been determined on the basis of deflection and two-way shear criteria, the required amount of flexural reinforcement can be obtained in the column strips, middle strips, and any column-line beams, using $\phi M_n = M_u$.

Like in the case of beams, members in two-way slab systems should be designed as tension-controlled sections whenever possible for overall efficiency. The methods given in Section 6.2 for rectangular sections with a single layer of reinforcement can be used to determine the required reinforcement for two-way systems.

An average effective d is typically used when calculating the required flexural reinforcement. Two layers of perpendicular reinforcement are needed at both the top and the bottom of the slab. Assuming that the concrete is not exposed to weather and that stirrups will not be used for shear, the minimum clear cover to the reinforcement is 0.75 in in accordance with ACI 7.7.1(c) (see the discussion under detailing requirements for concrete protection for reinforcement and Section 7.6 for shear design). The average d for both negative and positive reinforcement in both orthogonal directions can be taken as the thickness of the slab minus 0.75 in minus one bar diameter. Thus, assuming No. 5 bars, $d = h - (0.75 + 0.5) = h - 1.25$ in. Using an average d is accurate enough in flexural calculations.

Minimum reinforcement requirements are given in ACI 13.3.1. In particular, the area of flexural reinforcement in a two-way slab system must not be less than that required in ACI 7.12.2.1 for shrinkage and temperature:

1. Slabs with Grade 40 or 50 deformed bars: $0.0020bh$

2. Slabs with Grade 60 deformed bars or welded wire reinforcement: $0.0018bh$

3. Slabs with reinforcement exceeding 60,000 psi measured at a yield strain of 0.35%: $(0.0018 \times 60{,}000)bh/f_y$

It is evident that the minimum reinforcement requirements are based on the gross area of the concrete section bh. Thus, for a column strip and middle strip, the minimum reinforcement is determined using the thickness of the slab and the width of the column strip and middle strip, respectively.

7.5.3 Transfer of Moment at Slab–Column Connections

Studies of moment transfer between slabs and columns have shown that the unbalanced moment at slab–column joints due to gravity and/or lateral load effects is transferred by a combination of flexure and eccentricity of shear (ACI 13.5.3). The requirements pertaining to the portion of the unbalanced moment transferred by flexure are discussed here; the discussion on transfer due to eccentricity of shear is given in Section 7.6.

The portion of the unbalanced moment M_u that is transferred by flexure is equal to $\gamma_f M_u$, where the fraction γ_f is determined by ACI Eq. (13-1):

$$\gamma_f = \frac{1}{1 + (2/3)\sqrt{b_1/b_2}} \tag{7.26}$$

In this equation, b_1 and b_2 are the dimensions of the critical section for two-way shear that are measured parallel and perpendicular to the direction of analysis, respectively. According to ACI 11.11.1.2, the critical section for two-way shear is located so that its perimeter b_o is a minimum. However, the perimeter need not approach closer than $d/2$ to the following:

- Edges or corners of columns, concentrated loads, or reaction areas
- Changes in slab thickness such as edges of column capitals, drop panels, or shear caps

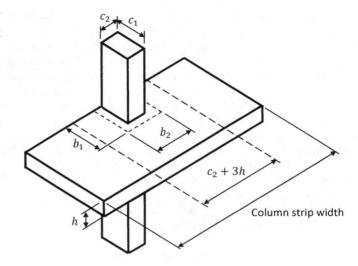

FIGURE 7.34
Transfer of moment $\gamma_f M_u$ at an edge column in a flat plate.

ACI 11.11.1.3 permits the use of a critical section with four straight edges for slabs supported by square or rectangular columns. Methods to determine b_1 and b_2 are given in Section 7.6.

It is evident from Eq. (7.26) that for a rectangular column, the portion of unbalanced moment transferred by flexure increases as the dimension of the column that is parallel to the applied moment increases.

The moment $\gamma_f M_u$ is assumed to be transferred to an effective slab width that is equal to three times the thickness of the slab or the drop panel plus the width of the column or column capital. Figure 7.34 illustrates the transfer of moment at an edge column in a flat plate. The amount of flexural reinforcement that is required within the effective slab width is calculated using the moment $\gamma_f M_u$.

This requirement is applicable primarily to two-way slab systems without beams where only reinforcement is available to resist the effects caused by the unbalanced moment. If edge beams are present, they must be designed to resist the torsional and shear stresses due to the unbalanced moment at the exterior joint.

The provisions given in ACI 13.5.3.3, which are based primarily on test results, permit an adjustment in the amount of unbalanced moment that is transferred by flexure at edge and interior slab–column connections, provided certain limitations are met.

At exterior supports where the unbalanced moment causes bending perpendicular to the edge (i.e., unbalanced moments about an axis parallel to the edge), it is permitted to take γ_f equal to 1.0, provided the factored shear force at the section V_u (excluding the shear caused by moment transfer) is equal to or less than 75% of the design shear strength ϕV_c at edge supports or 50% of ϕV_c at corner supports. The two-way shear strength provided by the concrete V_c is calculated in accordance with ACI 11.11.2.1 (see Section 7.6 of this book). Tests have shown that all of the unbalanced moment at an edge support can be transferred by flexure where the factored shear force does not exceed the fractions of the design shear strength indicated.[13,14]

At interior supports and edge columns where the unbalanced moment causes bending parallel to an edge, γ_f is permitted to be increased by up to 25% (but not greater than 1.0) provided that V_u is equal to or less than 40% of the design shear strength ϕV_c. An additional requirement that must be satisfied is that the net tensile strain ε_t in the reinforcement concentrated in the effective width defined in ACI 13.5.3.2 be equal to or greater than 0.010, based on the recommendations in Ref. 14. This requirement is not applicable to edge and corner columns.

When a frame is subjected to wind or earthquake loads, a reversal of moment can occur at the joints. In such cases, both top and bottom flexural reinforcement must be concentrated in the effective width. ACI R13.5.3.3 recommends a ratio of top to bottom reinforcement of approximately 2.

7.5.4 Detailing the Reinforcement

Concrete Protection for Reinforcement

Reinforcing bars must be placed with sufficient concrete cover to protect them from weather, fire, and other effects. ACI 7.7.1 contains the minimum requirements for non-prestressed, cast-in-place construction.

Similar to one-way slabs, concrete cover in two-way slabs without stirrups is measured from the surface of the concrete to the outermost layer of the flexural bars (see Fig. 6.7b).

For two-way slabs located in building structures that are not exposed to environmental effects, the clear cover is usually taken as three-quarters of an inch.

General Requirements

Summary of Requirements   ACI 13.3 contains detailing requirements that must be satisfied for slab reinforcement in two-way systems. The purpose of these requirements is to ensure the proper performance of the system. A summary of these provisions is given in Table 7.16.

Distribution of Flexural Reinforcement for Crack Control   In all two-way slab systems except waffle slabs, the maximum center-to-center spacing of the reinforcement at the critical sections must be equal to or less than $2h$. In addition to crack control, this limitation takes into consideration the effects that could be caused by loads concentrated on small areas of the slab. For the portions of slab over cellular spaces, such as in waffle slabs, the minimum reinforcement requirements are given in ACI 7.12.

Corner Reinforcement   The provisions of ACI 13.3.6 address exterior corners of slabs that are supported by stiff elements such as walls and edge beams. If stiff elements were not present at the exterior edges of a slab, the slab would lift when loaded. The presence of stiff elements restrains the lifting and causes additional bending moments at the exterior corners.

Corner reinforcement must be provided at both the top and the bottom of the slab, and the reinforcement in each layer in each direction must be designed for a bending moment equal to the largest positive bending moment per unit width in the slab panel. The top and bottom reinforcement must be placed parallel and perpendicular to the diagonal, respectively, as shown in Fig. 7.35 for a distance of at least one-fifth of the longer of the two span lengths in the corner panel.

Reinforcement parallel to the edges is permitted to be used instead of the diagonal bars (see Fig. 7.36).

Requirement	ACI Section Number(s)
Minimum flexural reinforcement shall not be less than that required by ACI 7.12.2.1.	13.3.1
Maximum spacing of flexural reinforcement is equal to $2h$.	13.3.2
Positive moment reinforcement perpendicular to a discontinuous edge shall extend to the edge of the slab and shall have straight or hooked embedment of at least 6 in into spandrel beams, columns, or walls.	13.3.3
Negative moment reinforcement perpendicular to a discontinuous edge shall be bent, hooked, or otherwise anchored into spandrel beams, columns, or walls. The reinforcement shall be developed at the face of the support in accordance with the provisions of ACI Chapter 12.	13.3.4
Anchorage of reinforcement shall be permitted within a slab where a slab is not supported by a spandrel beam or a wall at a discontinuous edge or where a slab cantilevers beyond a support.	13.3.5
Top and bottom reinforcement shall be provided at exterior corners in accordance with ACI 13.3.6.1 through 13.3.6.4 at exterior corners of slabs supported by edge walls or where one or more edge beams have a value of $\alpha_f > 1.0$.	13.3.6
The amount of negative reinforcement over the column of a flat slab may be reduced provided the dimensions of the drop panel conform to ACI 13.2.5. For purposes of computing the required negative reinforcement, the thickness of the drop panel below the slab shall not be assumed to be greater than one-quarter the distance from the edge of the drop panel to the face of the column or the column capital.	13.3.7
For two-way slabs without beams, reinforcement shall have minimum extensions prescribed in ACI Fig. 13.3.8. Extension of negative reinforcement is based on the longer of adjacent spans.	13.3.8.1, 13.3.8.2
Lengths of reinforcement shall be based on analysis where two-way slabs are part of the lateral force–resisting system, but shall not be less than those given in ACI Fig. 13.3.8.	13.3.8.4
All bottom bars within the column strip must be continuous or spliced with Class B tension splices or with mechanical or welded splices satisfying ACI 12.14.3. At least two bars must pass within the region bounded by the longitudinal reinforcement of the column and shall be anchored at exterior supports.	13.3.8.5

TABLE 7.16   Summary of Detailing Requirements for Two-way Slabs

Slabs without Beams   Minimum bar extensions for two-way slabs without beams are given in ACI 13.3.8 and are summarized in Fig. 7.37. It is important to note that these minimum bar lengths are based on systems subjected to uniformly distributed gravity loads only.

FIGURE 7.35
Reinforcement required at corners of slabs supported by stiff edge members.

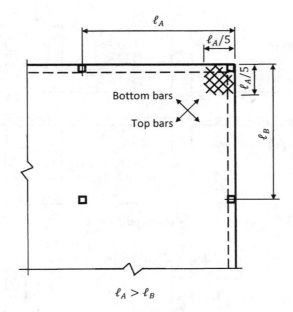

Bottom bars

Top bars

$\ell_A > \ell_B$

Applies where one or both edge beams have $\alpha_f > 1.0$

FIGURE 7.36
Alternative reinforcement layout at corners of slabs supported by stiff edge members.

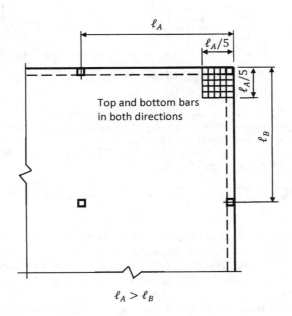

Top and bottom bars in both directions

$\ell_A > \ell_B$

Applies where one or both edge beams have $\alpha_f > 1.0$

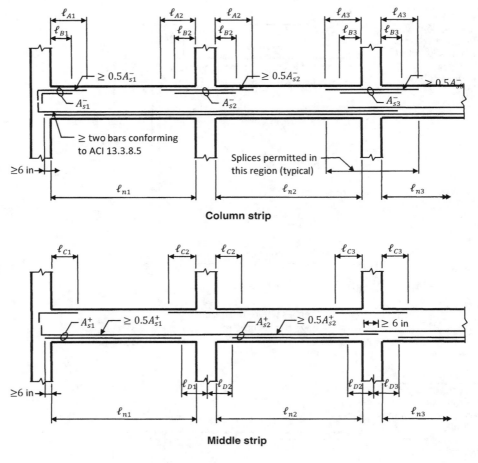

Notes:

1. $\ell_{A1} \geq 0.30\ell_{n1}$ for flat plates; $\ell_{A1} \geq 0.33\ell_{n1}$ for flat slabs
2. $\ell_{A2} \geq$ the larger of $0.30\ell_{n1}$ or $0.30\ell_{n2}$; $\ell_{A3} \geq$ the larger of $0.30\ell_{n2}$ or $0.30\ell_{n3}$
3. $\ell_{B1} \geq 0.20\ell_{n1}$; $\ell_{B2} \geq$ the larger of $0.20\ell_{n1}$ or $0.20\ell_{n2}$; $\ell_{B3} \geq$ the larger of $0.20\ell_{n2}$ or $0.20\ell_{n3}$
4. $\ell_{C1} \geq 0.22\ell_{n1}$; $\ell_{C2} \geq$ the larger of $0.22\ell_{n1}$ or $0.22\ell_{n2}$; $\ell_{C3} \geq$ the larger of $0.22\ell_{n2}$ or $0.22\ell_{n3}$
5. $\ell_{D1} \leq 0.15\ell_{n1}$; $\ell_{D2} \leq 0.15\ell_{n2}$; $\ell_{D3} \leq 0.15\ell_{n3}$

Figure 7.37   Minimum bar lengths for two-way slabs without beams.

The intent of the structural integrity requirements of ACI 13.3.8.5 is to enable two-way slab systems to span to adjacent supports should a single intermediate support be damaged or destroyed. The main purpose of the two continuous column strip bars through a support is to give the slab some residual strength after two-way shear failure at a single support.[15]

In frames that utilize the two-way slabs as part of the main lateral force–resisting system, ACI 13.3.8.4 requires that the lengths of the bars be determined by "analysis." The precise locations of inflection points cannot explicitly be found using approximate methods of analysis, because they depend on the ratio of the panel dimensions, the ratio of live to dead load, and the continuity conditions at the edges of a panel. Furthermore, there is no explicit way of determining the distribution of applied load to column strips and middle strips. A conservative approach to circumvent this problem is to make a portion of the negative reinforcing bars in the column strip continuous or to splice them using Class B splices. It is recommended that 25% of the top reinforcement in the column strip be continuous throughout the span; this is consistent with the requirement of ACI 21.3.6.4 that pertains to two-way slabs in intermediate moment frames, which are required in areas of moderate seismic risk.

7.5.5 Openings in Slab Systems

ACI 13.4.1 permits openings of any size in two-way slab systems provided that an analysis of the system with the openings that shows that all applicable strength and serviceability requirements of the Code are satisfied is performed.

For slabs without beams, such an analysis is waived when the provisions of ACI 13.4.2.1 through 13.4.2.4 are met:

1. In the area common to intersecting middle strips, openings of any size are permitted provided the total amount of reinforcement that is required for that panel without openings is maintained.

2. In the area common to intersecting column strips, the maximum permitted opening size is one-eighth the width of the column strip in either span. Also, an amount of reinforcement equivalent to that interrupted by an opening must be added on the sides of the opening.

3. In the area common to one column strip and one middle strip, the maximum permitted opening size is limited such that not more than one-quarter of the reinforcement in either strip is interrupted by openings. Also, as in the second case, an amount of reinforcement equivalent to that interrupted by an opening must be added on the sides of the opening.

The total area of reinforcement in a panel without an opening must be preserved in both directions of a panel with an opening. In other words, any reinforcement that is interrupted by an opening must be replaced on each side of the opening. Figure 7.38 illustrates these three cases.

Example 7.8 Determine the required flexural reinforcement at the critical sections for an interior design strip in the north-south direction for the flat-plate system depicted in Fig. 7.12. Assume a 9-in-thick slab, normal-weight concrete with $f_c' = 4,000$ psi, Grade 60 reinforcement, a superimposed dead load of 20 psf, and a live load of 50 psf.

Solution The required flexural reinforcement at the critical sections can be obtained using the strength design methods presented in Chap. 6 for tension-controlled, rectangular sections with a single layer of reinforcement.

Factored bending moments at the critical sections are given in Table 7.5 of Example 7.4 and were obtained using the Direct Design Method. Reinforcement is determined in the column strip and middle strip through the appropriate bending moments given in Table 7.5.

Because $\ell_1 = 21.67$ ft is less than $\ell_2 = 23.5$ ft, the width of the column strip in this interior design strip is equal to $\ell_1/2 = 10.83$ ft (see Fig. 7.18). Therefore, the width of the middle strip is equal to $23.5 - 10.83 = 12.67$ ft. These widths are used in the calculation of the flexural reinforcement.

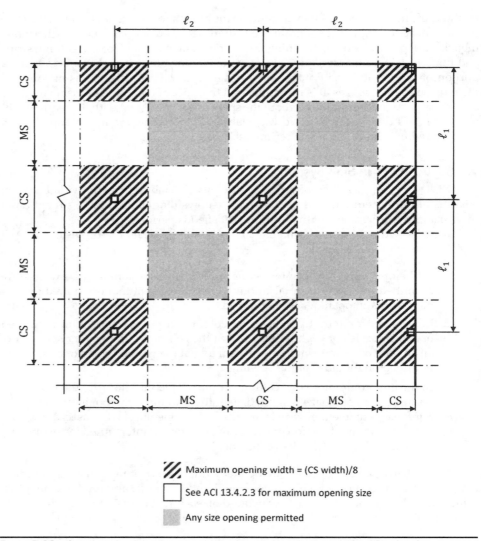

FIGURE **7.38** Openings in slab systems without beams in accordance with ACI 13.4.2.

The required flexural reinforcement at the critical sections is summarized in Table 7.17. An average $d = 9 - 1.25 = 7.75$ in was used in the calculations.

Calculations are provided for the required area of steel in the column strip at the first interior support in an end span.

The flowchart shown in Fig. 6.4, which is applicable to rectangular sections with a single layer of reinforcement, is utilized to determine A_s. It is modified as needed to satisfy the requirements for this flat plate.

Step 1: Assume tension-controlled section. Sections of flexural members, including two-way slabs, should be tension-controlled whenever possible. Thus, assume that the strength reduction factor $\phi = 0.9$.

Location			M_u (ft kips)	b (in)	A_s (in²)	Reinforcement
End span	Column strip	Exterior negative	−71.8	130	2.11*	8 No. 5†
		Positive	85.6	130	2.51	9 No. 5
		Interior negative	−146.4	130	4.33	14 No. 5
	Middle strip	Exterior negative	0.0	152	2.46*	9 No. 5†
		Positive	58.0	152	2.46*	9 No. 5†
		Interior negative	−47.0	152	2.46*	9 No. 5†
Interior span	Column strip	Positive	58.0	130	2.11*	8 No. 5†
		Negative	−135.3	130	4.02	13 No. 5
	Middle strip	Positive	38.7	152	2.46*	9 No. 5†
		Negative	−44.2	152	2.46*	9 No. 5†

* Based on minimum reinforcement requirements.
† Based on maximum spacing requirements.

TABLE 7.17 Required Slab Reinforcement for an Interior Design Strip Given in Example 7.8

Step 2: Determine the nominal strength coefficient of resistance R_n. For a rectangular section, R_n is determined by Eq. (6.5), which is a function of the factored bending moment M_u. The negative factored bending moment is given in Example 7.4 as 146.4 ft kips.

Assuming that the concrete is not exposed to weather and that stirrups will not be used as shear reinforcement, the average $d = 9 - 1.25 = 7.75$ in. Thus,

$$R_n = \frac{M_u}{\phi b d^2} = \frac{146.4 \times 12,000}{0.9 \times 130 \times 7.75^2} = 250.0 \, \text{psi}$$

Step 3: Determine the required reinforcement ratio ρ. The reinforcement ratio ρ is determined by Eq. (6.7):

$$\rho = \frac{0.85 f_c'}{f_y}\left[1 - \sqrt{1 - \frac{2R_n}{0.85 f_c'}}\right] = \frac{0.85 \times 4}{60}\left[1 - \sqrt{1 - \frac{2 \times 250.0}{0.85 \times 4,000}}\right] = 0.0043$$

Step 4: Determine the required area of tension reinforcement A_s. For a 130-in-wide column strip, the required area of negative reinforcement is

$$A_s = \rho b d = 0.0043 \times 130 \times 7.75 = 4.33 \, \text{in}^2$$

Step 5: Determine the minimum required area of reinforcement $A_{s,min}$. For two-way slabs with Grade 60 reinforcement, $A_{s,min}$ is determined in accordance with ACI 13.3.1:

$$A_{s,min} = 0.0018 bh = 0.0018 \times 130 \times 9.0 = 2.11 \, \text{in}^2 < A_s$$

Use $A_s = 4.33 \, \text{in}^2$.

Note that for the middle strip,

$$A_{s,min} = 0.0018bh = 0.0018 \times 152 \times 9 = 2.46 \, \text{in}^2$$

Step 6: Determine the depth of the equivalent rectangular stress block a.

$$a = \frac{A_s f_y}{0.85 f_c' b} = \frac{4.33 \times 60,000}{0.85 \times 4,000 \times 130} = 0.59 \, \text{in}$$

Step 7: Determine β_1. According to ACI 10.2.73, $\beta_1 = 0.85$ for $f_c' = 4,000$ psi (see Section 5.2).

Step 8: Determine the neutral axis depth c.

$$c = \frac{a}{\beta_1} = \frac{0.59}{0.85} = 0.69 \, \text{in}$$

Step 9: Determine ε_t.

$$\varepsilon_t = 0.0030 \left(\frac{d_t}{c} - 1 \right) = 0.0030 \left(\frac{7.75}{0.69} - 1 \right) = 0.0307 > 0.0040$$

Because $\varepsilon_t > 0.0040$, the maximum reinforcement requirement of ACI 10.3.5 is satisfied. Also, the section is tension-controlled because $\varepsilon_t > 0.0050$ (ACI 10.3.4), and the initial assumption that the section is tension-controlled is correct.

Step 10: Choose the size and spacing of the reinforcing bars. The required area of reinforcement is 4.33 in².

$$\text{Maximum bar spacing} = 2h = 18 \, \text{in}$$

For $b = 130$ in, $130/18 = 7.2$, say, eight, bars are needed to satisfy the maximum spacing requirements in the column strip.

Use 14 No. 5 bars ($A_s = 4.34 \, \text{in}^2$).

For $b = 152$ in, $152/18 = 8.4$, say nine, bars are needed to satisfy the maximum spacing requirements in the middle strip.

Similar calculations can be performed for the other critical sections in the column strip and middle strip.

It is evident from Table 7.17 that reinforcement based on maximum bar spacing is required in the middle strip in both the end and interior spans. This is common in flat-plate structures.

Check that the flexural reinforcement at the end support is adequate to satisfy the moment transfer requirements of ACI 13.5.3.

The total unbalanced moment at this slab–column connection is equal to 71.8 ft kips, which is the total moment in the column strip (see Table 7.17).

A fraction of this moment $\gamma_f M_u$ must be transferred over an effective width equal to $c_2 + 3h = 20 + (3 \times 9) = 47$ in.

The fraction of unbalanced moment transferred by flexure is calculated in accordance with Eq. (7.26):

$$\gamma_f = \frac{1}{1 + (2/3)\sqrt{b_1/b_2}} = \frac{1}{1 + (2/3)\sqrt{23.9/27.8}} = 0.62$$

where
$$b_1 = c_1 + \frac{d}{2} = 20 + \frac{7.75}{2} = 23.9 \, \text{in}$$
$$b_2 = c_2 + d = 20 + 7.75 = 27.8 \, \text{in}$$

FIGURE **7.39**
Reinforcement
detail at the
exterior column
given in
Example 7.8.

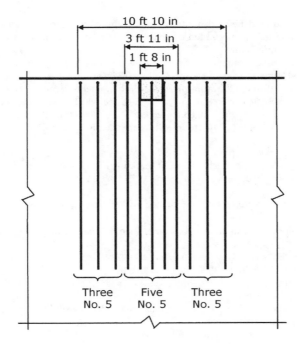

10 ft 10 in
3 ft 11 in
1 ft 8 in

Three
No. 5

Five
No. 5

Three
No. 5

For edge columns bending perpendicular to the edge, the value of γ_f computed by Eq. (7.26) may be increased to 1.0 provided that $V_u \le 0.75\phi V_c$ [ACI 13.5.3.3(a)]. No adjustment to γ_f is made in this example.

Unbalanced moment transferred by flexure $= \gamma_f M_u = 0.62 \times 71.8 = 44.5$ ft kips. The required area of steel to resist this moment in the 47-in-wide strip is $A_s = 1.32$ in^2, which is equivalent to five No. 5 bars.

$$A_{s,min} = 0.0018bh = 0.0018 \times 47 \times 9.0 = 0.76 \, in^2 < A_s$$

Provide the five No. 5 bars by concentrating five of the eight column strip bars within the 47-in width over the column (see Table 7.17). For symmetry, add another bar in the column strip and check bar spacing:

For five No. 5 bars within the 47-in width, $47/5 = 9.4$ in < 18 in.

For four No. 5 bars within the $130 - 47 = 83$-in width, $83/4 = 20.8$ in > 18 in.

Therefore, add two more No. 5 bars in the 83-in width; bar spacing $83/6 = 13.8$ in.

A total of 11 No. 5 bars are required at the end supports within the column strip, with 5 of the 11 bars concentrated within a width of 47 in centered on the column.

Reinforcement details for the top bars at the exterior column are shown in Fig. 7.39.

Similar analyses can be performed at interior columns.

Because the slab is subjected to gravity loads only, the lengths of the reinforcing bars shown in Fig. 7.37 for slabs without drop panels can be used for this flat plate.

Example 7.9   Determine the required flexural reinforcement at the critical sections for an interior design strip in the north-south direction for the flat-slab system depicted in Fig. 7.13. Assume a 7.5-in-thick slab, an overall drop panel thickness of 9.75 in, normal-weight concrete with $f'_c = 4,000$ psi, Grade 60 reinforcement, a superimposed dead load of 20 psf, and a live load of 80 psf.

Solution The required flexural reinforcement at the critical sections can be obtained using the strength design methods presented in Chap. 6 for tension-controlled, rectangular sections with a single layer of reinforcement.

Factored bending moments at the critical sections are given in Table 7.7 of Example 7.5, and were obtained using the Direct Design Method. Reinforcement is determined in the column strip and middle strip through the appropriate bending moments given in Table 7.7.

Because $\ell_1 = 21.67$ ft is less than $\ell_2 = 23.5$ ft, the width of the columns strip in this interior design strip is equal to $\ell_1/2 = 10.83$ ft (see Fig. 7.18). Therefore, the width of the middle strip is equal to $23.5 - 10.83 = 12.67$ ft. These widths are used in the calculation of the flexural reinforcement.

It was shown in Example 7.2 that the dimensions of the drop panel satisfy the requirements of ACI 13.2.5. Therefore, the effective depth d that can be used at the interior negative critical sections in the column strip can be taken as $9.75 - 1.25 = 8.5$ in. At all other locations, $d = 7.5 - 1.25 = 6.25$ in.

For the column strip at the interior negative critical sections, $A_{s,min} = 0.0018bh = 0.0018 \times 130 \times 9.75 = 2.28$ in^2.

For the column strip at all locations other than the interior negative critical sections, $A_{s,min} = 0.0018bh = 0.0018 \times 130 \times 7.5 = 1.76$ in^2.

For the middle strip, $A_{s,min} = 0.0018bh = 0.0018 \times 152 \times 7.5 = 2.05$ in^2.

Maximum bar spacing $= 2h = 2 \times 7.5 = 15.0$ in.

For $b = 130$ in, $130/15 = 8.7$, say nine, bars are needed to satisfy the maximum spacing requirements in the column strip.

For $b = 152$ in, $152/15 = 10.1$, say, 11, bars are needed to satisfy the maximum spacing requirements in the middle strip.

The required flexural reinforcement is given in Table 7.18.

Location			M_u (ft kips)	b (in)	A_s (in^2)	Reinforcement
End span	Column strip	Exterior negative	−71.2	130	2.61	9 No. 5
		Positive	92.9	130	3.43	12 No. 5
		Interior negative	−164.1	130	4.45	15 No. 5
	Middle strip	Exterior negative	−21.7	152	2.05*	11 No. 5[†]
		Positive	61.9	152	2.25	11 No. 5[†]
		Interior negative	−52.6	152	2.05*	11 No. 5[†]
Interior span	Column strip	Positive	65.0	130	2.37	9 No. 5[†]
		Negative	−151.7	130	4.10	14 No. 5
	Middle strip	Positive	43.3	152	2.05*	11 No. 5[†]
		Negative	−49.5	152	2.05*	11 No. 5[†]

* Based on minimum reinforcement requirements.
[†] Based on maximum spacing requirements.

Table 7.18 Required Slab Reinforcement for an Interior Design Strip Given in Example 7.9

The transfer of unbalanced moment at the edge columns need not be checked in this example because of the beams at the perimeter of the slab. These spandrel beams must be designed for the shear forces and torsional moments transferred from the slab.

Because the slab is subjected to gravity loads only, the lengths of the reinforcing bars shown in Fig. 7.37 for slabs with drop panels can be used for this flat slab.

Example 7.10   Determine the required flexural reinforcement at the critical sections in the slab and in the beams for an interior design strip in the north-south direction for the beam-supported slab system depicted in Fig. 7.14. Assume a 6.5-in-thick slab, normal-weight concrete with $f'_c = 4,000$ psi, Grade 60 reinforcement, a superimposed dead load of 20 psf, and a live load of 100 psf.

Solution   The required flexural reinforcement at the critical sections in the slab and beams can be obtained using the strength design methods presented in Chap. 6 for tension-controlled, rectangular sections with a single layer of reinforcement.

Factored bending moments at the critical sections are given in Table 7.9 of Example 7.6 and were obtained using the Direct Design Method. Reinforcement is determined in the column strip (slab and beams) and middle strip through the appropriate bending moments given in Table 7.9.

Because $\ell_1 = 21.67$ ft is less than $\ell_2 = 23.5$ ft, the width of the column strip in this interior design strip is equal to $\ell_1/2 = 10.83$ ft (see Fig. 7.18). Therefore, the width of the middle strip is equal to $23.5 - 10.83 = 12.67$ ft. These widths are used in the calculation of the flexural reinforcement for the slab.

For the slab, assume $d = 6.5 - 1.25 = 5.25$ in.

For the beams, assume $d = 24 - 2.5 = 21.5$ in.

For the slab in the column strip, $A_{s,min} = 0.0018bh = 0.0018 \times 130 \times 6.5 = 1.52$ in^2.

For the middle strip, $A_{s,min} = 0.0018bh = 0.0018 \times 152 \times 6.5 = 1.78$ in^2.

$$\text{Minimum reinforcement for the beams } A_{s,min} = \frac{200 b_w d}{f_y} = \frac{200 \times 24 \times 21.5}{60,000} = 1.72 \text{ in}^2$$

$$\text{Maximum bar spacing in slab} = 2h = 2 \times 6.5 = 13.0 \text{ in}$$

For $b = 130$ in, $130/13 = 10$; 11 bars are needed to satisfy the maximum spacing requirements in the column strip.

For $b = 152$ in, $152/13 = 11.7$, say, 12, bars are needed to satisfy the maximum spacing requirements in the middle strip.

The required flexural reinforcement is given in Table 7.19.

It can be shown that the beams and slabs are tension-controlled sections with the provided reinforcement and that the maximum reinforcement requirement of ACI 10.3.5 is satisfied.

The reinforcement in the beams satisfies the crack control requirements of ACI 10.6.4 and the spacing limits of ACI 7.6 and 3.3.2 (for a 24-in-wide beam, the minimum number of No. 5 bars is 3 from Table 6.2 and the maximum number is 12 from Table 6.3).

As expected, the reinforcement in the column strip slab is governed by minimum reinforcement and bar spacing requirements because stiff column-line beams are present, which carry most of the factored bending moments.

The transfer of unbalanced moment at the edge columns need not be checked in this example because of the beams at the perimeter of the slab. These spandrel beams must be designed for the shear forces and torsional moments transferred from the slab.

Because this beam-supported slab is subjected to gravity loads only, the lengths of the reinforcing bars shown in Fig. 7.37 for slabs without drop panels can be used for the slab, and the bar lengths shown in Fig. 6.25 for beams other than perimeter beams can be used for the column-line beams.

When detailing the flexural reinforcement in the edge beams, the additional structural integrity reinforcement requirements illustrated in Fig. 6.25 for perimeter beams must also be satisfied.

Location				M_u (ft kips)	b (in)	A_s (in²)	Reinforcement
End span	Column strip	Beam	Exterior negative	−35.1	24	1.72*	6 No. 5
			Positive	122.9	24	1.72*	6 No. 5
			Interior negative	−151.0	24	1.72*	6 No. 5
		Slab	Exterior negative	−7.0	130	1.52*	11 No. 4[†]
			Positive	21.1	130	1.52*	11 No. 4[†]
			Interior negative	−28.1	130	1.52*	11 No. 4[†]
	Middle strip		Exterior negative	−14.0	152	1.52*	11 No. 4[†]
			Positive	56.2	152	2.66	14 No. 4
			Interior negative	−66.7	152	2.92	15 No. 4
Interior span	Column strip	Beam	Positive	73.8	24	1.72*	6 No. 5
			Negative	−140.5	24	1.72*	6 No. 5
		Slab	Positive	14.0	130	1.52*	11 No. 4[†]
			Negative	−24.6	130	1.52*	11 No. 4[†]
	Middle strip		Positive	35.1	152	1.78*	12 No. 4[†]
			Negative	−63.2	152	2.76	14 No. 4

* Based on minimum reinforcement requirements.
† Based on maximum spacing requirements.

TABLE 7.19   Required Beam and Slab Reinforcement for an Interior Design Strip Given in Example 7.10

7.6 Design for Shear

7.6.1 Overview

The requirements of ACI 11.12 must be satisfied for shear design in two-way slabs and footings. Included are requirements for critical shear sections, nominal shear strength of concrete, and nominal shear strength of shear reinforcement.

It was shown in Section 7.5 that the shear force that needs to be resisted by the slab in two-way systems with column-line beams or walls is relatively small. In systems with stiff beams, all of the shear force is carried by the beams.

The design for shear is critical in slab systems without beams (flat plates and flat slabs), especially at exterior slab–column connections where the total exterior bending moment is transferred directly to the critical section of the exterior column. Edge beams in such cases would essentially solve the shear problems at edge and corner columns, but they add cost (additional material and time for formwork) and may not be architecturally acceptable.

Both one- and two-way shear must be considered in two-way systems supported directly on columns. One-way shear rarely governs; in most cases, two-way shear is more critical, and it has a direct impact on the required thickness of the slab. Regardless of the type of shear, the following strength equation must be satisfied:

$$V_u \leq \phi V_n = \phi(V_c + V_s) \tag{7.27}$$

In this equation, V_u is the factored shear force at the critical section for shear and V_c and V_s are the nominal shear strengths provided by the concrete and the shear reinforcement, respectively. The strength reduction factor ϕ for shear is equal to 0.75 in accordance with ACI 9.3.2.3.

Shear reinforcement is rarely used to enhance one-way shear nominal strength and is used to increase two-way shear nominal strength in situations where other measures—such as increasing the slab thickness, increasing the column size, using drop panels, or using column-line beams—are not feasible or are uneconomical.

7.6.2 One-Way Shear

Analysis for one-way shear, which is referred to as "beam action shear" in the Code, considers the slab to act as a beam spanning between columns. The critical section extends in a plane across the entire width of the slab at a distance d from the face of the support as depicted in Fig. 7.40. Except for long, narrow slabs, this type of shear is seldom critical. However, it must be checked to ensure that shear strength is not exceeded.

FIGURE 7.40 Critical section for one-way shear.

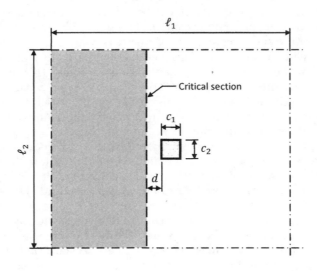

Tributary area

FIGURE 7.41 Failure surface for punching shear.

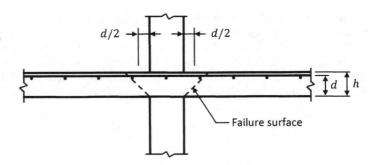

The following equation must be satisfied at the critical sections, which are located a distance d from the face of the support:

$$V_u \leq \phi V_c = \phi 2 \sqrt{f_c'} \ell d \qquad (7.28)$$

In this equation, the factored shear force at the critical section V_u is equal to the factored load on the slab q_u times the tributary area indicated in Fig. 7.40. The length ℓ is equal to the width of slab that resists the shear force. For the slab depicted in Fig. 7.40, $\ell = \ell_2$.

A 1-ft-wide section can be used to check the one-way shear strength in lieu of using the entire panel width. In such cases, the longer of the distances from the critical section to the panel edge should be used because this results in maximum V_u.

7.6.3 Two-Way Shear

Critical Shear Section

General Requirements Two-way or punching shear is generally the critical of the two types of shear in flat plates and flat slabs. At failure, a truncated cone or pyramid-shaped surface forms around the column at failure, as illustrated in Fig. 7.41. This type of failure is brittle and can occur with virtually no warning.

In slabs without shear reinforcement, the forces due to direct shear and bending are resisted by the concrete slab section around a column. This section consists of the effective slab thickness (and the overall drop panel or shear cap thickness where applicable) and the critical perimeter b_o. According to ACI 11.11.1.2, the critical section for two-way shear is located so that its perimeter b_o is a minimum but never less than $d/2$ from edges or corners of columns, concentrated loads, reaction areas, and changes in slab thickness, such as edges of column capitals, drop panels, or shear caps.

A rigorous interpretation of this definition would result in the corners of the critical section being rounded. However, the actual intent of the Code is set forth in ACI 11.11.1.3: A critical section with four straight edges is permitted for slabs supported by square or rectangular columns.

The critical section for two-way shear in a flat plate is illustrated in Fig. 7.42.

The critical sections for flat slabs are illustrated in Fig. 7.43. Shear requirements need to be checked at the critical sections located a distance $d_2/2$ from the face of the column and $d_1/2$ from the face of the drop panel, because shear failure can occur at either location. The critical sections for a system utilizing shear caps are the same as those in a system utilizing drop panels.

FIGURE 7.42 Critical
section for
two-way shear
in a flat plate.

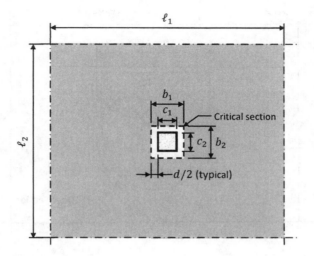

FIGURE 7.42 Critical
section for
two-way shear
in a flat plate.

Tributary area

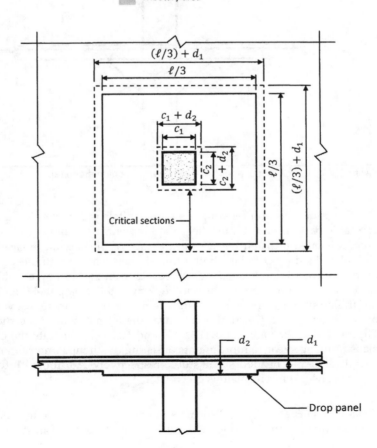

FIGURE 7.43 Critical
sections for two-way
shear in a flat slab.

Drop panel

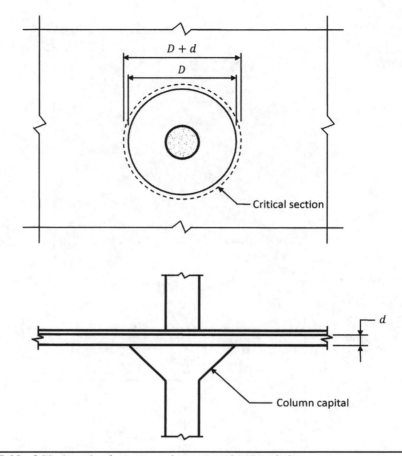

$D + d$

D

Critical section

d

Column capital

FIGURE 7.44   Critical section for two-way shear at a column capital.

The critical section for columns with capitals is depicted in Fig. 7.44. Because the capital is part of the column, shear requirements need only be checked at the critical section located a distance $d/2$ from the face of the capital. In cases where a column capital and a drop panel are both utilized, the shear strength needs to be investigated at the critical section located a distance $d/2$ from the column capital (Fig. 7.44) and at the critical section located a distance $d_1/2$ from the drop panel (Fig. 7.43).

In slab systems where shear reinforcement is used to increase overall shear capacity, an additional critical section occurs near the termination of the shear reinforcement. Figures R11.11.3(d) and (e), R11.11.4.7, and R11.11.5 illustrate the critical sections for slabs utilizing shear reinforcement consisting of stirrups (interior and edge columns), shearheads, and headed shear studs, respectively. Additional information on these types of shear reinforcement is given later.

Effect of Slab Edges   Critical shear perimeters for exterior (edge and corner) columns that are flush with the edge of a slab are clearly defined in the Code. Without any

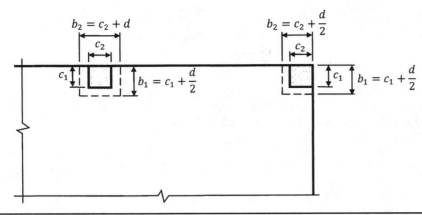

FIGURE 7.45 Critical shear perimeters for exterior columns where the columns are flush with the slab edges.

nearby openings in the slab, edge and corner columns would generally have three- and two-sided critical shear perimeters, respectively, as illustrated in Fig. 7.45.

Critical shear perimeters are not as clearly defined in the Code in cases where the slab edges cantilever beyond the face(s) of an exterior column. However, some general guidelines can be established on the basis of the provision of ACI 11.11.1.2, which requires that the perimeter of the critical section b_o be a minimum.

Consider the edge column depicted in Fig. 7.46. Depending on the length of the cantilever, the critical shear perimeter will be either three-sided or four-sided.

In the four-sided case, $b_o = 2(c_1 + c_2 + 2d)$, whereas in the general three-sided case, $b_o = 2(x_1 + c_1) + c_2 + d$. The cantilever length that results in equal perimeters is $x_1 = (c_2/2) + d$. Therefore, the following perimeters of the critical section can be used to obtain minimum b_o[14]:

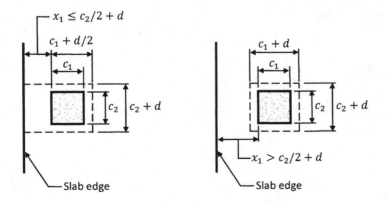

Three-sided critical perimeter   **Four-sided critical perimeter**

FIGURE 7.46 Critical section perimeters for edge columns with slab cantilevers.

- If $x_1 \leq (c_2/2) + d$, use a three-sided critical perimeter.
- If $x_1 > (c_2/2) + d$, use a four-sided critical perimeter.

Similar derivations can be performed for corner columns.

Effect of Openings   Provisions for the effect of openings in slabs on the critical shear perimeter are given in ACI 11.11.6. These provisions were originally presented in Ref. 16, and additional research has confirmed that these provisions are conservative.[17]

In general, the closer a slab opening is to a column, the greater the effect it has on the critical shear perimeter that is available to resist the shear forces. The effects of openings can be neglected where the opening is at the following locations:

1. A distance equal to or greater than 10 times the slab thickness from a concentrated load or reaction area
2. Outside of a column strip

For openings that do not satisfy these criteria and are located in slabs without shearhead reinforcement, the perimeter of the critical section is reduced by a length equal to the projection of the opening that is formed by two lines that extend from the centroid of the column, concentrated load, or reaction area and are tangent to the boundaries of the opening. Figure 7.47 illustrates the ineffective portions of b_o for both flat-plate and flat-slab systems (also see ACI Fig. 11.11.6). A reduction in b_o results in a reduction in shear strength; the relationship between b_o and the nominal shear strength provided by the concrete V_c is given later.

For slabs with shearhead reinforcement, the ineffective portion of b_o is to be taken as one-half of that defined earlier for slabs without shearhead reinforcement (ACI 11.11.6.2).

Shear Strength Provided by Concrete

The nominal shear strength provided the concrete V_c for slabs without shear reinforcement is given in ACI 11.11.2.1. In general, the shear stress provided by concrete v_c is a function of the compressive strength of the concrete and is limited to $4\lambda\sqrt{f_c'}$ for

FIGURE 7.47 Effect of openings on critical shear perimeter.

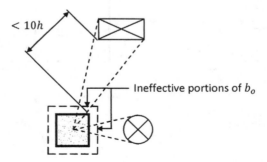

$< 10h$

Ineffective portions of b_o

Note that circular opening is located in column strip and rectangular opening is not.

square columns, where λ is the modification factor that reflects the reduced mechanical properties of lightweight concrete (see ACI 8.6.1 and Chap. 2 of this book).

The nominal shear strength V_c is obtained by multiplying the allowable stress v_c by the area of the critical section, which is equal to the perimeter of the critical section b_o times the effective depth of the slab d (see ACI Eq. 11-33):

$$V_c = 4\lambda\sqrt{f_c'}b_o d \tag{7.29}$$

Tests have indicated that the value of V_c is not conservative when the ratio β of the lengths of the long and short sides of a rectangular column or loaded area is larger than 2.0.[18] In such cases, the applied shear stress on the critical section varies from a maximum of approximately $4\lambda\sqrt{f_c'}$ around the corners of a column (or loaded area) to approximately $2\lambda\sqrt{f_c'}$ or less along the long sides of the perimeter between the two end sections. The nominal shear strength V_c is given by ACI Eq. (11-31):

$$V_c = \left(2 + \frac{4}{\beta}\right)\lambda\sqrt{f_c'}b_o d \tag{7.30}$$

It is evident from this equation that as the ratio β increases, the stress decreases linearly to a minimum of $2\lambda\sqrt{f_c'}$, which is equivalent to shear stress for one-way shear. ACI Fig. R11.11.2 illustrates the determination of β for an L-shaped column or reaction area.

Other tests have indicated that the allowable shear stress v_c decreases as the ratio b_o/d increases.[18] This is accounted for in ACI Eq. (11-32):

$$V_c = \left(\frac{\alpha_s d}{b_o} + 2\right)\lambda\sqrt{f_c'}b_o d \tag{7.31}$$

In this equation, α_s is equal to 40, 30, and 20 for interior, edge, and corner columns, respectively. Reference to these three types of columns does not suggest the actual location of a column in a building; instead, they refer to the number of sides of the critical section that are available to resist shear stress. For example, α_s is equal to 30 for the interior column depicted in ACI Fig. R11.11.6(c) with an opening on one side of the critical section, because only three sides of the critical section are available to resist shear stress caused by the external loads.

The nominal shear strength V_c for two-way shear action of slabs without shear reinforcement is the least of the values obtained by Eqs. (7.29) to (7.31).

Where shear reinforcement consisting of bars or wires and single- and multiple-leg stirrups is utilized in a slab, V_c is limited to $2\lambda\sqrt{f_c'}b_o d$ (ACI 11.11.3.1).

Shear Strength Provided by Shear Reinforcement

Shear Strength Provided by Bars, Wires, and Stirrups   Research has shown that the two-way shear strength of slabs can be increased by shear reinforcement consisting of properly anchored bars or wires and single- or multiple-leg stirrups or closed stirrups.[19-23] The use of such reinforcement is permitted provided that the effective depth of the slab is greater than 6 in but not less than 16 times the bar diameter of the shear reinforcement. ACI Fig. R11.11.3(a)–(c) illustrates three different types of this shear reinforcement.

The nominal shear strength provided by the shear reinforcement V_s is determined by the requirements of ACI 11.4. Thus, the two-way nominal shear strength V_n, which consists of the nominal shear strength provided by the concrete V_c and the nominal shear strength provided by the shear reinforcement V_s, can be determined from the following equation:

$$V_n = V_c + V_s = 2\lambda\sqrt{f_c'}b_o d + \frac{A_v f_{yt} d}{s} \leq 6\sqrt{f_c'}b_o d \qquad (7.32)$$

In this equation, A_v is the cross-sectional area of all legs of reinforcement on one peripheral line that is geometrically similar to the perimeter of the column section. For an interior column, there are four sides that contain shear reinforcement. On each side, there are two legs of reinforcement on the peripheral line, so A_v is equal to eight times the area of one stirrup leg in this case. The total shear reinforcement is determined in a similar fashion for edge and corner columns.

Shear requirements must be checked at two critical sections: The first critical section is located a distance $d/2$ from the column face, and the second is located a distance $d/2$ from the outermost line of stirrups (see Fig. 7.48). The following strength equations must be satisfied at the critical sections:

- The critical section located $d/2$ from column face:

$$V_u \leq \phi V_n = \phi(V_c + V_s) = \phi 2\lambda\sqrt{f_c'}b_o d + \frac{\phi A_v f_{yt} d}{s} \leq \phi 6\sqrt{f_c'}b_o d \qquad (7.33)$$

- The critical section located $d/2$ from the outmost peripheral line of stirrups:

$$V_u \leq \phi V_n = \phi V_c = \phi 2\lambda\sqrt{f_c'}b_o d \qquad (7.34)$$

At the critical section located $d/2$ from the face of the column, both the concrete and shear reinforcement contribute to the overall design shear strength, whereas at the critical section located $d/2$ from the last line of stirrups, only the design shear strength of the concrete is available to resist the factored shear force. The distance from the column face to the last line of stirrups (i.e., the location where the stirrups can be cut off) can be conservatively determined by setting $V_u = \phi V_c$ in Eq. (7.34).

In general, the factored shear force V_u is calculated by multiplying the total factored gravity load q_u by the net tributary area of the column, which is equal to the tributary area minus the area enclosed by the critical shear perimeter.

The spacing requirements of ACI 11.12.3.3 are summarized in Fig. 7.48 for interior, edge, and corner columns with closed stirrups. The first line of stirrups is located at a distance equal to or less than $d/2$ from the column face. The intent of this is to eliminate the possibility of shear failure between the face of the column and the innermost peripheral line of shear reinforcement. Successive lines of stirrups must be spaced perpendicular to a column face at a distance that is not to exceed $d/2$; once again, the intent is to avoid failure between consecutive peripheral lines of stirrups. The spacing between adjacent stirrup legs in the first line of stirrups in the direction parallel to a column face must be equal to or less than $2d$. These spacing limits correspond to slab shear reinforcement details that have been shown to be effective.

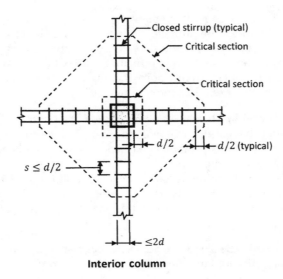

Interior column

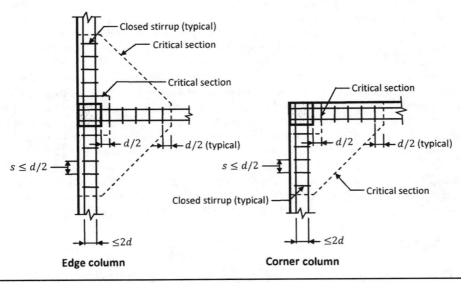

Edge column   Corner column

FIGURE 7.48 Details for closed stirrup shear reinforcement.

A symmetric distribution of the shear reinforcement should be provided around the critical section for interior columns where unbalanced moment is negligible. Even though the shear stresses are greater on the interior face of the critical section of an edge or corner column because of significant unbalanced moment (see the next section), a symmetrical distribution of the shear reinforcement is also recommended because the stirrups placed parallel to the slab edge provide some torsional strength along that edge.

It is essential that shear reinforcement (1) satisfy the requirements of ACI 12.13 for the development of web reinforcement and (2) engage longitudinal reinforcement at both the top and the bottom of the slab. When specifying shear reinforcement, it is important to keep in mind that it is very difficult to satisfy the anchorage requirements of ACI 12.13 for slabs that are less than approximately 10 in thick; it is virtually impossible to properly develop shear reinforcement in thin slabs.

Shear Strength Provided by Shearheads   Shearheads are structural steel shapes (commonly I- or channel-shaped sections) that are encased in the concrete slab immediately above the column. Like stirrups, the main purpose of the shearheads is to increase the two-way shear capacity of the slab.

Provisions for shearhead reinforcement can be found in ACI 11.11.4. Because this type of shear reinforcement is rarely used anymore (primarily because of material and labor costs), it is not covered in this book. Additional design information and an example can be found in Ref. 2.

Shear Strength Provided by Headed Shear Stud Reinforcement   Tests have shown that shear reinforcement consisting of large headed studs welded to flat steel rails are effective in resisting two-way shear in slabs.[23] Headed shear stud reinforcement can take the place of or can be used in conjunction with stirrups, drop panels, or column capitals to increase design shear strength.

A typical headed shear stud arrangement is shown in Fig. 7.49. The base rail, which is set on chairs, is nailed to the formwork around the column. The size and spacing of the studs and the length of the base rail depends on the shear requirements.

Like reinforcing bars, sufficient concrete cover must be provided to protect the base rail and head from weather, fire, and other effects. ACI 7.7.5 contains the minimum cover requirements for headed shear stud reinforcement. In particular, the concrete cover for the base rail and heads must not be less than that required for the reinforcement in the slab. ACI Fig. R7.7.5(a) illustrates these concrete cover requirements for headed shear stud reinforcement in slabs with both top and bottom bars (also see ACI 11.11.5).

Similar to stirrups, the nominal shear strength provided by the headed shear stud reinforcement V_s is determined by the requirements of ACI 11.4. The two-way nominal shear strength V_n, which consists of the nominal shear strength provided by the concrete V_c and the nominal shear strength provided by the shear reinforcement V_s, can be

Figure 7.49 Headed shear stud reinforcement.

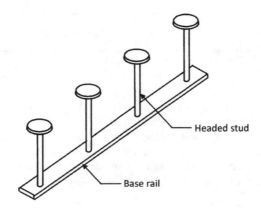

Headed stud

Base rail

determined from the following equation (see ACI 11.11.5.1):

$$V_n = V_c + V_s = 3\lambda\sqrt{f_c'}b_o d + \frac{A_v f_{yt} d}{s} \leq 8\sqrt{f_c'}b_o d \tag{7.35}$$

In this case, A_v is the cross-sectional area of all the headed studs on one peripheral line that is approximately parallel to the perimeter of the column section. This is analogous to the number of stirrup legs in slabs with closed stirrups.

ACI 11.11.5.1 also requires that $A_v f_{yt}/(b_o s)$, which is the nominal shear stress provided by the shear reinforcement, be equal to or greater than $2\sqrt{f_c'}$.

Shear requirements must be checked at two critical sections: The first critical section is located a distance $d/2$ from the column face, and the second is located a distance $d/2$ from the outermost peripheral line of headed studs (see Fig. 7.50). The following strength equations must be satisfied at the critical sections:

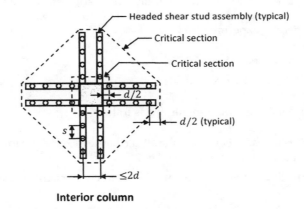

Interior column

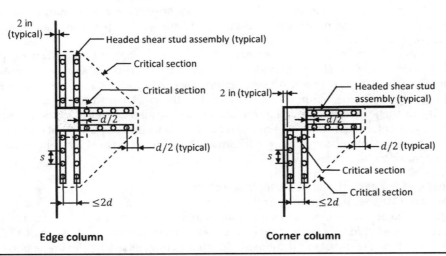

Edge column   **Corner column**

FIGURE 7.50 Details for headed shear stud reinforcement.

- The critical section located $d/2$ from column face (ACI 11.11.5.1):

$$V_u \leq \phi V_n = \phi(V_c + V_s) = \phi 3\lambda\sqrt{f_c'}b_o d + \frac{\phi A_v f_{yt} d}{s} \leq \phi 8\sqrt{f_c'}b_o d \qquad (7.36)$$

- The critical section located $d/2$ from the outmost peripheral line of headed studs (ACI 11.11.5.4):

$$V_u \leq \phi V_n = \phi V_c = \phi 2\lambda\sqrt{f_c'}b_o d \qquad (7.37)$$

The factored shear force V_u is calculated by multiplying the total factored gravity load q_u by the net tributary area of the column, which is equal to the tributary area minus the area enclosed by the critical shear perimeter. Where there is appreciable unbalanced moment, shear stresses must be computed (as described later).

The spacing s of the peripheral lines of headed studs, that is, the center-to-center spacing of the headed shear studs, must satisfy the following requirements, which have been verified by experiments[23]:

- For $v_u \leq 6\phi\sqrt{f_c'}, s \leq 0.75d$
- For $v_u > 6\phi\sqrt{f_c'}, s \leq 0.50d$

Other detailing requirements are as follows:

1. Spacing between the column face and the first peripheral line of shear reinforcement must not exceed $d/2$.
2. Spacing between adjacent reinforcement elements must not exceed $2d$.

Spacing requirements for interior, edge, and corner columns are shown in Fig. 7.50. A symmetric distribution of shear reinforcement should be provided around any column type. Overall economy can be achieved by specifying a minimum number of stud diameters and using the same spacing and base rail lengths at as many column locations as possible.

Compared with stirrups, it is evident that headed shear stud reinforcement provides larger limits for shear strength and spacing between peripheral lines of shear reinforcement. This is primarily due to performance: A stud head exhibits smaller slip than the leg of closed stirrup, which results in smaller crack widths.

The use of headed shear stud reinforcement can also alleviate reinforcement congestion at slab–column joints, as shown in Fig. 7.51. This facilitates placement of concrete in these important areas. Additional information on this type of shear reinforcement can be found in Ref. 24.

Transfer of Moment at Slab–Column Connections
Unbalanced Moment Transferred by Eccentricity of Shear   It was discussed in Section 7.5 that an unbalanced moment at a slab–column joint due to gravity and/or lateral load effects is transferred by a combination of flexure and eccentricity of shear (ACI 13.5.3). The portion of the unbalanced moment M_u that is transferred by flexure is equal to $\gamma_f M_u$, where the fraction γ_f is determined by ACI Eq. (13-1) [see Eq. (7.26)]. Design and detailing requirements pertaining to the moment $\gamma_f M_u$ are given in Section 7.5.

FIGURE 7.51   Headed shear reinforcement at a slab–column joint (*Photo courtesy of Decon USA, Inc.*).

The portion of the unbalanced moment transferred by eccentricity of shear is equal to $\gamma_v M_u$. The factored moment M_u is defined in the Code as the unbalanced moment that occurs at the centroid of the critical section. It was discussed in Section 7.4 that moments are computed at the face of supports in the Direct Design Method and at the centerline of the supports in the Equivalent Frame Method. Strictly speaking, the moments from these analysis methods would have to be calculated at the centroid of the critical section to obtain the unbalanced moment M_u. Because of the approximations that are used in the procedure to determine shear stresses, using the moments obtained directly from analysis to determine shear stresses is accurate enough; in other words, determining the moments at the centroid of the critical section is unnecessary.

The fraction γ_v is determined by ACI Eq. (13-37):

$$\gamma_v = 1 - \gamma_f = 1 - \frac{1}{1 + (2/3)\sqrt{b_1/b_2}} \tag{7.38}$$

Factored Shear Stress   The factored shear stresses v_u on the near and far faces of the critical section that is located a distance $d/2$ from the face of a column in the direction of analysis consists of two parts:

1. The shear stress caused by the shear force V_u, which acts over the area of the critical section. As noted previously, V_u is determined by multiplying the total factored gravity load q_u by the net tributary area of the column, which is equal to the tributary area minus the area enclosed by the critical shear perimeter.

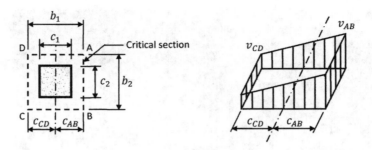

Interior column

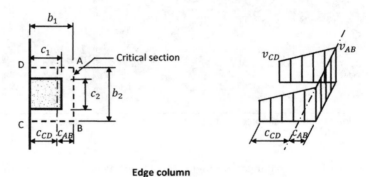

Edge column

Figure 7.52 Assumed distribution of shear stress due to direct shear and the portion of the unbalanced moment transferred by eccentricity of shear.

2. The shear stress caused by the portion of the unbalanced moment due to the eccentricity of shear $\gamma_v M_u$, which acts over the property of the critical section that is analogous to the polar moment of inertia. It is assumed that the shear stress resulting from the moment $\gamma_v M_u$ varies linearly about the critical section for two-way shear (ACI 11.11.7.2).

The shear stress distributions for an interior and edge column subjected to V_u and $\gamma_v M_u$ are illustrated in Fig. 7.52 for the critical sections located at distance $d/2$ from the face of the columns.

The total factored shear stresses on faces AB and CD of the critical section can be determined from the following equations:

$$v_{u(AB)} = \frac{V_u}{A_c} + \frac{\gamma_v M_u c_{AB}}{J_c} \tag{7.39a}$$

$$v_{u(CD)} = \frac{V_u}{A_c} - \frac{\gamma_v M_u c_{CD}}{J_c} \tag{7.39b}$$

In these equations, c_{AB} and c_{CD} are the distances from the centroid of the critical section to faces AB and CD of the critical section, respectively. The area of the critical section A_c is determined by multiplying the effective depth of the slab d by the perimeter of the critical section. The quantity J_c is the property of the critical section that is analogous to the polar moment of inertia.

Critical section properties for interior rectangular columns can be found in Fig. 7.53. Figures. 7.54 to 7.57 contain critical section properties for edge rectangular columns bending parallel to the edge, edge rectangular columns bending perpendicular to the edge, corner rectangular columns, and circular interior columns, respectively. The properties given in these figures are applicable to the critical sections located a distance $d/2$ from the faces of a column.

Appendix B contains derivations of the critical section properties for a variety of support conditions. It also contains tabulated values of the constants f_1, f_2, and f_3 that can be used to facilitate the calculation of these properties.

Appendix B of Ref. 25 contains methods to determine the properties of the nonrectangular critical section that is located $d/2$ from the outermost peripheral line of shear reinforcement. The methods presented in that reference are applicable to the critical sections of any shape regardless of the type of shear reinforcement that is used.

Requirements for Strength Design Once the maximum factored shear stress v_u has been determined, the following strength design equations must be satisfied at the applicable critical sections:

FIGURE 7.53
Properties of critical section—interior rectangular column

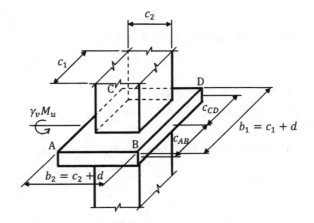

Interior rectangular column

$$c_{AB} = c_{CD} = b_1/2$$

$$A_c = f_1 d^2$$

$$J_c/c_{AB} = J_c/c_{CD} = 2f_2 d^3$$

$$f_1 = 2\left[\left(1 + \frac{c_2}{c_1}\right)\left(\frac{c_1}{d}\right) + 2\right]$$

$$f_2 = \frac{1}{6}\left[\left(1 + \frac{3c_2}{c_1}\right)\left(\frac{c_1}{d}\right)^2 + \left(5 + \frac{3c_2}{c_1}\right)\left(\frac{c_1}{d}\right) + 5\right]$$

FIGURE 7.54
Properties of
critical section—
edge rectangular
column bending
parallel to
the edge.

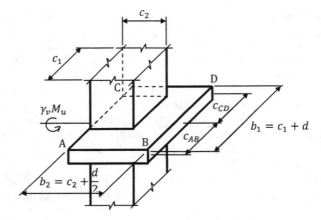

Edge rectangular column bending parallel to the edge

$$c_{AB} = c_{CD} = b_1/2$$

$$A_c = f_1 d^2$$

$$J_c/c_{AB} = J_c/c_{CD} = 2f_2 d^3$$

$$f_1 = \left(1 + \frac{2c_2}{c_1}\right)\left(\frac{c_1}{d}\right) + 2$$

$$f_2 = \frac{1}{12}\left[\left(1 + \frac{6c_2}{c_1}\right)\left(\frac{c_1}{d}\right)^2 + \left(5 + \frac{6c_2}{c_1}\right)\left(\frac{c_1}{d}\right) + 5\right]$$

- For members without shear reinforcement,

$$v_u \leq \phi v_n = \frac{\phi V_c}{b_o d} \tag{7.40}$$

The minimum value of V_c determined by Eqs. (7.29) to (7.31) is used in Eq. (7.40).

- For members with shear reinforcement (other than shearheads),

$$v_u \leq \phi v_n = \frac{\phi(V_c + V_s)}{b_o d} \tag{7.41}$$

The nominal strengths V_c and V_s that are to be used in Eq. (7.41) depend on the type of shear reinforcement that is specified and on the location of the critical section.

Requirements pertaining to two-way shear strength design are applicable primarily to slab systems without column-line beams. The stiffness of normally proportioned beams is generally large enough so that either most or all of the required shear force is resisted by the beams (see Section 7.4).

Example 7.11 Check shear strength requirements at the first interior column and an exterior column in an interior design strip for the flat-plate system depicted in Fig. 7.12. Assume a 9-in-thick slab, normal-weight concrete with $f_c' = 4{,}000$ psi, Grade 60 reinforcement, a superimposed dead load of

FIGURE 7.55
Properties of
critical section—
edge rectangular
column bending
perpendicular to
the edge.

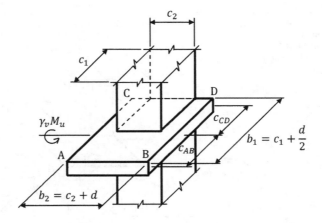

Edge rectangular column bending perpendicular to the edge

$$c_{AB} = [f_3/(f_2 + f_3)][c_1 + (d/2)]$$

$$c_{CD} = [f_2/(f_2 + f_3)][c_1 + (d/2)]$$

$$A_c = f_1 d^2$$

$$J_c/c_{AB} = 2f_2 d^3$$

$$J_c/c_{CD} = 2f_3 d^3$$

$$f_1 = 2 + \frac{c_1}{d}\left(2 + \frac{c_2}{c_1}\right)$$

$$f_2 = \frac{\left(\frac{c_1}{d} + \frac{1}{2}\right)^2 \left[\frac{c_1}{d}\left(1 + \frac{2c_2}{c_1}\right) + \frac{5}{2}\right] + \left[\frac{c_1}{d}\left(1 + \frac{c_2}{2c_1}\right) + 1\right]}{6\left(\frac{c_1}{d} + \frac{1}{2}\right)}$$

$$f_3 = \frac{\left(\frac{c_1}{d} + \frac{1}{2}\right)^2 \left[\frac{c_1}{d}\left(1 + \frac{2c_2}{c_1}\right) + \frac{5}{2}\right] + \left[\frac{c_1}{d}\left(1 + \frac{c_2}{2c_1}\right) + 1\right]}{6\left[\frac{c_1}{d}\left(1 + \frac{c_2}{c_1}\right) + \frac{3}{2}\right]}$$

20 psf, and a live load of 50 psf. Also assume that the Direct Design Method can be used to compute bending moments in the slab.

Solution Both one- and two-way shear requirements must be checked at both the first interior column and an edge column in an interior design strip.

First Interior Column

One-way shear The critical section for one-way shear is located a distance $d = 9 - 1.25 = 7.75$ in from the face of the column (see Fig. 7.40).

The total factored gravity loads acting on the slab q_u is determined using the load combination given in ACI Eq. (9-2), because this yields the maximum effects for dead and live loads (see Table 4.1 in Section 4.2 and Example 7.4):

$$q_u = 1.2q_D + 1.6q_L = (1.2 \times 132.5) + (1.6 \times 50) = 239 \, \text{psf}$$

FIGURE **7.56**
Properties of
critical section—
corner rectangular
column.

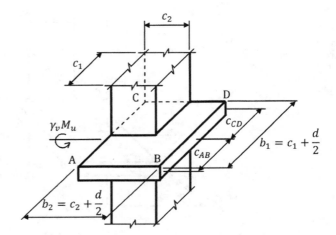

Corner rectangular column bending perpendicular to the edge

$$c_{AB} = [f_3/(f_2 + f_3)][c_1 + (d/2)]$$

$$c_{CD} = [f_2/(f_2 + f_3)][c_1 + (d/2)]$$

$$A_c = f_1 d^2$$

$$J_c/c_{AB} = 2f_2 d^3$$

$$J_c/c_{CD} = 2f_3 d^3$$

$$f_1 = 1 + \frac{c_1}{d}\left(1 + \frac{c_2}{c_1}\right)$$

$$f_2 = \frac{\left(\frac{c_1}{d} + \frac{1}{2}\right)^2 \left[\frac{c_1}{d}\left(1 + \frac{4c_2}{c_1}\right) + \frac{5}{2}\right] + \left[\frac{c_1}{d}\left(1 + \frac{c_2}{c_1}\right) + 1\right]}{12\left(\frac{c_1}{d} + \frac{1}{2}\right)}$$

$$f_3 = \frac{\left(\frac{c_1}{d} + \frac{1}{2}\right)^2 \left[\frac{c_1}{d}\left(1 + \frac{4c_2}{c_1}\right) + \frac{5}{2}\right] + \left[\frac{c_1}{d}\left(1 + \frac{c_2}{c_1}\right) + 1\right]}{12\left[\frac{c_1}{d}\left(1 + \frac{2c_2}{c_1}\right) + \frac{3}{2}\right]}$$

The maximum factored shear force at the critical section is

$$V_u = 0.239 \times \left(\frac{23.5}{2} - \frac{24}{2 \times 12} - \frac{7.75}{12}\right) \times 21.67 = 52.3\,\text{kips}$$

Design shear strength is computed by Eq. (7.28):

$$\phi V_c = \phi 2\lambda\sqrt{f_c'}\,\ell d = 0.75 \times 2 \times 1.0\sqrt{4,000} \times (21.67 \times 12) \times 7.75/1,000 = 191.2\,\text{kips}$$

Because $V_u < \phi V_c$, one-way shear strength requirements are satisfied.

Two-way shear The total factored shear stress is the sum of the direct shear stress plus the shear stress due to the fraction of the unbalanced moment transferred by eccentricity of shear.

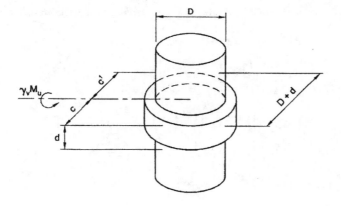

Circular interior column

$$c = c' = (D + d)/2$$

$$A_c = f_1 d^2$$

$$J_c/c = J_c/c' = 2f_2 d^3$$

$$f_1 = \pi \left(\frac{D}{d} + 1 \right)$$

$$f_2 = \frac{\pi}{8} \left(\frac{D}{d} + 1 \right)^2 + \frac{1}{6}$$

FIGURE 7.57 Properties of critical section—circular interior column.

The critical section for two-way shear is located a distance $d/2 = 3.9$ in from the face of the column (see Fig. 7.42).

At the first interior column, the factored shear force due to gravity loads is

$$V_u = q_u(A_t - b_1 b_2)$$

$$= 0.239 \left[(23.5 \times 21.67) - \left(\frac{31.75}{12} \right) \left(\frac{31.75}{12} \right) \right] = 120.0 \text{ kips}$$

where $b_1 = b_2 = 24 + 7.75 = 31.75$ in

The total unbalanced moment is equal to the difference between the total interior negative moments on both sides of the column (see Table 7.4):

$$M_u = 193.3 - 179.5 = 13.8 \text{ ft kips}$$

Determine γ_v by Eq. (7.38):

$$\gamma_v = 1 - \gamma_f = 1 - \frac{1}{1 + (2/3)\sqrt{b_1/b_2}} = 1 - \frac{1}{1 + (2/3)\sqrt{31.75/31.75}} = 0.40$$

The section properties of the critical section are determined using Fig. 7.53 for an interior column:

$$c_{AB} = b_1/2 = 31.75/2 = 15.9 \text{ in}$$

$$f_1 = 2\left[\left(1 + \frac{c_2}{c_1}\right)\left(\frac{c_1}{d}\right) + 2\right] = 2\left[(1+1)\left(\frac{24}{7.75}\right) + 2\right] = 16.39$$

$$A_c = f_1 d^2 = 16.39 \times 7.75^2 = 984.4 \text{ in}^2$$

$$f_2 = \frac{1}{6}\left[\left(1 + \frac{3c_2}{c_1}\right)\left(\frac{c_1}{d}\right)^2 + \left(5 + \frac{3c_2}{c_1}\right)\left(\frac{c_1}{d}\right) + 5\right]$$

$$= \frac{1}{6}\left[(1+3)\left(\frac{24}{7.75}\right)^2 + (5+3)\left(\frac{24}{7.75}\right) + 5\right] = 11.36$$

$$J_c/c_{AB} = 2f_2 d^3 = 2 \times 11.36 \times 7.75^3 = 10,576 \text{ in}^3$$

These section properties can also be obtained using the tables given in Appendix B. The total factored shear stress is determined by Eq. (7.39):

$$v_{u(AB)} = \frac{V_u}{A_c} + \frac{\gamma_v M_u c_{AB}}{J_c}$$

$$= \frac{120,000}{984.4} + \frac{0.4 \times 13.8 \times 12,000}{10,576} = 121.9 + 6.3 = 128.2 \text{ psi}$$

The allowable stress for a square column is obtained by Eq. (7.29):

$$\phi v_c = \frac{\phi V_c}{b_o d} = \phi 4\lambda\sqrt{f_c'} = 0.75 \times 4 \times 1.0\sqrt{4,000} = 189.7 \text{ psi} > 128.6 \text{ psi}$$

Edge Column

One-way shear The critical section for one-way shear is located a distance $d = 9 - 1.25 = 7.75$ in from the face the column (see Fig. 7.40).

The maximum factored shear force at the critical section is

$$V_u = 0.239 \times \left(\frac{21.67}{2} - \frac{20}{2 \times 12} - \frac{7.75}{12}\right) \times 23.5 = 52.6 \text{ kips}$$

Design shear strength is computed by Eq. (7.28):

$$\phi V_c = \phi 2\lambda\sqrt{f_c'}\ell d = 0.75 \times 2 \times 1.0\sqrt{4,000} \times (23.5 \times 12) \times 7.75/1,000 = 207.3 \text{ kips}$$

Because $V_u < \phi V_c$, one-way shear strength requirements are satisfied.

Two-way shear The total factored shear stress is the sum of the direct shear stress plus the shear stress due to the fraction of the unbalanced moment transferred by eccentricity of shear.

The critical section for two-way shear is located a distance $d/2 = 3.9$ in from the face of the column (see Fig. 7.42).

At the edge column, the factored shear force due to gravity loads is

$$V_u = q_u(A_t - b_1 b_2)$$

$$= 0.239\left[\left(\frac{21.67}{2} + \frac{20}{2 \times 12}\right)(23.5) - \left(\frac{23.88}{12}\right)\left(\frac{27.75}{12}\right)\right] = 64.4 \text{ kips}$$

where $b_1 = 20 + \dfrac{7.75}{2} = 23.88$ in

$b_2 = 20 + 7.75 = 27.75$ in

Because the Direct Design Method was used to compute the moments, ACI 13.6.3.6 requires that the unbalanced moment at the edge column that is transferred by eccentricity of shear be $0.3M_o = 0.3 \times 276.2 = 82.9$ ft kips (see Example 7.4).

Determine γ_v by Eq. (7.38):

$$\gamma_v = 1 - \gamma_f = 1 - \frac{1}{1 + (2/3)\sqrt{b_1/b_2}} = 1 - \frac{1}{1 + (2/3)\sqrt{23.88/27.75}} = 0.38$$

The section properties of the critical section are determined using Fig. 7.55 for an edge column bending perpendicular to the edge:

$$f_1 = 2 + \frac{c_1}{d}\left(2 + \frac{c_2}{c_1}\right) = 2 + \frac{20}{7.75}\left(2 + \frac{20}{20}\right) = 9.74$$

$$f_2 = \frac{[(c_1/d) + (1/2)]^2\,\{(c_1/d)\,[1 + (2c_2/c_1)] + (5/2)\} + \{(c_1/d)\,[1 + (c_2/2c_1)] + 1)\}\left[\frac{c_1}{d}\left(1 + \frac{c_2}{2c_1}\right) + 1\right]}{6[(c_1/d) + (1/2)]}$$

$$= \frac{[(20/7.75) + (1/2)]^2\{(20/7.75)[1 + (2 \times 20)/20] + (5/2)\} + \{(20/7.75)[1 + 20/(2 \times 20)] + 1\}}{6[(20/7.75) + (1/2)]} = 5.52$$

$$f_3 = \frac{[(c_1/d) + (1/2)]^2\,\{(c_1/d)\,[1 + (2c_2/c_1)] + (5/2)\} + \{(c_1/d)\,[1 + (c_2/2c_1)] + 1)\}\left[\frac{c_1}{d}\left(1 + \frac{c_2}{2c_1}\right) + 1\right]}{6\{(c_1/d)[1 + (c_2/c_1)] + (3/2)\}}$$

$$= \frac{[(20/7.75) + (1/2)]^2\{(20/7.75)[1 + (2 \times 20)/20] + (5/2)\} + \{(20/7.75)[1 + 20/(2 \times 20)] + 1\}}{6\{(20/7.75)[1 + (20/20)] + (3/2)\}} = 2.55$$

$$c_{AB} = \frac{f_3}{f_2 + f_3}\left(c_1 + \frac{d}{2}\right) = \frac{2.55}{5.52 + 2.55}\left(20 + \frac{7.75}{2}\right) = 7.5 \text{ in}$$

$$A_c = f_1 d^2 = 9.74 \times 7.75^2 = 585.0 \text{ in}^2$$

$$J_c/c_{AB} = 2f_2 d^3 = 2 \times 5.52 \times 7.75^3 = 5139 \text{ in}^3$$

These section properties can also be obtained using the tables given in Appendix B. The total factored shear stress is determined by Eq. (7.39):

$$v_{u(AB)} = \frac{V_u}{A_c} + \frac{\gamma_v M_u c_{AB}}{J_c}$$

$$= \frac{64,400}{585.0} + \frac{0.38 \times 82.9 \times 12,000}{5,139} = 110.1 + 73.6 = 183.7 \text{ psi}$$

The allowable stress for a square column is obtained from Eq. (7.29):

$$\phi v_c = \frac{\phi V_c}{b_o d} = \phi 4\lambda\sqrt{f_c'} = 0.75 \times 4 \times 1.0\sqrt{4,000} = 189.7 \text{ psi} > 183.7 \text{ psi}$$

Shear strength requirements are satisfied at both the first interior column and the edge column.

Example 7.12 Check shear strength requirements at the first interior column and an exterior column in an interior design strip for the flat-slab system depicted in Fig. 7.13. Assume a 7.5-in-thick slab, an overall drop panel thickness of 9.75 in, normal-weight concrete with $f'_c = 4,000$ psi, Grade 60 reinforcement, a superimposed dead load of 20 psf, and a live load of 80 psf.

Solution Both one- and two-way shear requirements must be checked at both the first interior column and an edge column in an interior design strip.

First Interior Column

One-way shear The critical section for one-way shear is located a distance d from the face of the column and traverses overall slab thicknesses of 9.75 in within the drop panel region and 7.5 in outside of the drop panel region. In the calculations for one-way shear, conservatively use $d = 7.5 - 1.25 = 6.25$ in. The critical section located a distance d from the face of the drop panel is not critical and need not be checked for one-way shear.

The total factored gravity loads acting on the slab q_u is determined using the load combination given in ACI Eq. (9-2), because this yields the maximum effects for dead and live loads (see Table 4.1 in Section 4.2 and Example 7.5):

$$q_u = 1.2q_D + 1.6q_L = (1.2 \times 116.6) + (1.6 \times 80) = 268\,\text{psf}$$

The maximum factored shear force at the critical section is

$$V_u = 0.268 \times \left(\frac{23.5}{2} - \frac{24}{2 \times 12} - \frac{6.25}{12}\right) \times 21.67 = 59.4\,\text{kips}$$

Design shear strength is computed by Eq. (7.28):

$$\phi V_c = \phi 2\lambda\sqrt{f'_c}\,\ell d = 0.75 \times 2 \times 1.0\sqrt{4,000} \times (21.67 \times 12) \times 6.25/1,000 = 154.2\,\text{kips}$$

Because $V_u < \phi V_c$, one-way shear strength requirements are satisfied.

Two-way shear The total factored shear stress is the sum of the direct shear stress plus the shear stress due to the fraction of the unbalanced moment transferred by eccentricity of shear.

Two critical sections must be checked when drop panels are utilized: the critical section located a distance $d/2 = (9.75 - 1.25)/2 = 4.25$ in from the face of the column and the critical section located a distance $d/2 = (7.5 - 1.25)/2 = 3.13$ in from the face of the drop panel (see Fig. 7.43).

Critical section located at d/2 from face of column At the first interior column, the factored shear force due to gravity loads is

$$V_u = q_u(A_t - b_1 b_2)$$

$$= 0.268\left[(23.5 \times 21.67) - \left(\frac{32.5}{12}\right)\left(\frac{32.5}{12}\right)\right] = 134.5\,\text{kips}$$

where $b_1 = b_2 = 24 + 8.5 = 32.5$ in

The total unbalanced moment is equal to the difference between the total interior negative moments on both sides of the column (see Table 7.6):

$$M_u = 216.7 - 201.2 = 15.5 \, \text{ft kips}$$

Determine γ_v by Eq. (7.38):

$$\gamma_v = 1 - \gamma_f = 1 - \frac{1}{1 + (2/3)\sqrt{b_1/b_2}} = 1 - \frac{1}{1 + (2/3)\sqrt{32.5/32.5}} = 0.40$$

The section properties of the critical section are determined using Fig. 7.53 for an interior column:

$$c_{AB} = b_1/2 = 32.5/2 = 16.3 \, \text{in}$$

$$f_1 = 2\left[\left(1 + \frac{c_2}{c_1}\right)\left(\frac{c_1}{d}\right) + 2\right] = 2\left[(1 + 1)\left(\frac{24}{8.5}\right) + 2\right] = 15.29$$

$$A_c = f_1 d^2 = 15.29 \times 8.5^2 = 1,104.7 \, \text{in}^2$$

$$f_2 = \frac{1}{6}\left[\left(1 + \frac{3c_2}{c_1}\right)\left(\frac{c_1}{d}\right)^2 + \left(5 + \frac{3c_2}{c_1}\right)\left(\frac{c_1}{d}\right) + 5\right]$$

$$= \frac{1}{6}\left[(1 + 3)\left(\frac{24}{8.5}\right)^2 + (5 + 3)\left(\frac{24}{8.5}\right) + 5\right] = 9.91$$

$$J_c/c_{AB} = 2 f_2 d^3 = 2 \times 9.91 \times 8.5^3 = 12,172 \, \text{in}^3$$

These section properties can also be obtained using the tables given in Appendix B. The total factored shear stress is determined by Eq. (7.39):

$$v_{u(AB)} = \frac{V_u}{A_c} + \frac{\gamma_v M_u c_{AB}}{J_c}$$

$$= \frac{134,500}{1,104.7} + \frac{0.4 \times 15.5 \times 12,000}{12,172} = 121.8 + 6.1 = 127.9 \, \text{psi}$$

The allowable stress for a square column is obtained from Eq. (7.29):

$$\phi v_c = \frac{\phi V_c}{b_o d} = \phi 4\lambda\sqrt{f_c'} = 0.75 \times 4 \times 1.0\sqrt{4,000} = 189.7 \, \text{psi} > 127.9 \, \text{psi}$$

Critical section located at d/2 from face of drop panel The factored shear force due to gravity loads at this location is

$$V_u = q_u(A_t - b_1 b_2)$$

$$= 0.268\,[(23.5 \times 21.67) - (8.5)(8.5)] = 117.0 \, \text{kips}$$

where $b_1 = b_2 = 8 + \dfrac{6.25}{12} = 8.5 \, \text{ft}$

The section properties of the critical section are determined using Fig. 7.53 for an interior column:

$$c_{AB} = b_1/2 = 8.5/2 = 4.3 \, \text{ft}$$

In the equations for f_1 and f_2, use $c_1 = c_2 = 8 \times 12 = 96$ in.

$$f_1 = 2\left[\left(1 + \frac{c_2}{c_1}\right)\left(\frac{c_1}{d}\right) + 2\right] = 2\left[(1 + 1)\left(\frac{96}{6.25}\right) + 2\right] = 65.44$$

$$A_c = f_1 d^2 = 65.44 \times 6.25^2 = 2{,}556.3 \text{ in}^2$$

$$f_2 = \frac{1}{6}\left[\left(1 + \frac{3c_2}{c_1}\right)\left(\frac{c_1}{d}\right)^2 + \left(5 + \frac{3c_2}{c_1}\right)\left(\frac{c_1}{d}\right) + 5\right]$$

$$= \frac{1}{6}\left[(1 + 3)\left(\frac{96}{6.25}\right)^2 + (5 + 3)\left(\frac{96}{6.25}\right) + 5\right] = 178.6$$

$$J_c/c_{AB} = 2f_2 d^3 = 2 \times 178.6 \times 6.25^3 = 87{,}207 \text{ in.}^3$$

The total factored shear stress is determined by Eq. (7.39):

$$v_{u(AB)} = \frac{V_u}{A_c} + \frac{\gamma_v M_u c_{AB}}{J_c}$$

$$= \frac{117{,}000}{2{,}556.3} + \frac{0.4 \times 15.5 \times 12{,}000}{87{,}207} = 45.8 + 1.0 = 46.8 \text{ psi}$$

It is evident that the factored shear stress due to moment transfer by eccentricity of shear is small compared with that from direct shear at the outermost critical section.

For large rectangular critical sections that typically occur around drop panels, the shear strength given in Eq. (7.31) will usually govern:

$$\phi v_c = \frac{\phi V_c}{b_o d} = \phi\left(\frac{\alpha_s d}{b_o} + 2\right)\lambda\sqrt{f_c'}$$

Because the critical section is four-sided, $\alpha_s = 40$.
Also, $b_o = 2(b_1 + b_2) = 2(8.5 + 8.5) \times 12 = 408$ in.
Thus,

$$\phi v_c = \phi\left(\frac{40 \times 6.25}{408} + 2\right)\lambda\sqrt{f_c'} = \phi 2.6\lambda\sqrt{f_c'} \text{ psi}$$

As expected, this design shear strength is less than that from Eq. (7.29), which is equal to $\phi 4\lambda\sqrt{f_c'}$, and that from Eq. (7.30), which is equal to $\phi[2 + (4/4)]\lambda\sqrt{f_c'} = \phi 3\lambda\sqrt{f_c'}$.

Therefore,

$$\phi v_c = 0.75 \times 2.6 \times 1.0\sqrt{4{,}000} = 123.3 \text{ psi} > 46.8 \text{ psi}$$

These calculations clearly show the significant effect that relatively large values of b_o/d can have on the design shear strength.

Edge Column

One- and two-way shear requirements need not be checked in the slab at the edge column because of the spandrel beams. These beams must be designed for the bending moments, shear forces, and torsional moments that are transferred from the slab.

Shear strength requirements are satisfied at both the first interior column and the edge column.

Example 7.13 Check shear strength requirements in an interior design strip for the beam-supported slab system depicted in Fig. 7.14. Assume a 6.5-in-thick slab, normal-weight concrete with $f_c' = 4{,}000$ psi, Grade 60 reinforcement, a superimposed dead load of 20 psf, and a live load of 100 psf.

Solution   According to ACI 13.6.8.1, beams with $\alpha_{f1}\ell_2/\ell_1 \geq 1.0$ are to be proportioned for the entire shear force caused by the factored loads on the tributary area defined in ACI Fig. R13.6.8.

Determine $\alpha_{f1}\ell_2/\ell_1$ for the beams in the interior design strip.

From Example 7.3, it was determined that $\alpha_f = 6.3$ for the north-south interior beams. Thus,

$$\alpha_{f1}\ell_2/\ell_1 = 6.3 \times 23.5/21.67 = 6.8 > 1.0$$

Therefore, the column-line beams in the interior design strip must be designed to resist the entire shear force.

The total factored gravity loads acting on the slab q_u are determined using the load combination given in ACI Eq. (9-2), because this yields the maximum effects for dead and live loads (see Table 4.1 in Section 4.2 and Example 7.6):

$$q_u = 1.2q_D + 1.6q_L = (1.2 \times 119.9) + (1.6 \times 100) = 304 \, \text{psf}$$

The factored load per unit length at the center of a beam is equal to (see ACI Fig. R13.6.8)

$$w_u = 0.304 \times 21.67 = 6.6 \, \text{kips/ft}$$

The load decreases linearly to zero at the centerlines of the columns.

The factored shear force V_u in the beam at the critical section located a distance d from face of support can be determined from statics, and the required shear reinforcement can be obtained using the methods given in Chap. 6.

Example 7.14   Check two-way shear strength requirements at the edge column in an interior design strip for the flat-plate system depicted in Fig. 7.58. Assume a 10-in-thick slab, normal-weight concrete with $f_c' = 4,000$ psi, Grade 60 reinforcement, a superimposed dead load of 20 psf, and a live load of 80 psf. Also assume that the Direct Design Method can be used to determine moments in the slab. Provide shear reinforcement if needed.

FIGURE 7.58 The flat-plate system given in Example 7.14.

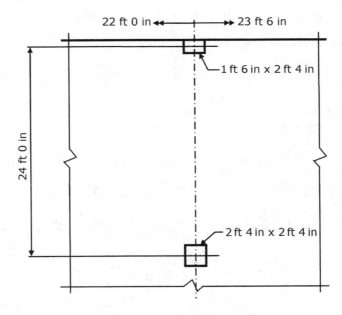

Solution   The total factored shear stress is the sum of the direct shear stress plus the shear stress due to the fraction of the unbalanced moment transferred by eccentricity of shear.

The total factored gravity loads acting on the slab q_u is determined using the load combination given in ACI Eq. (9-2), because this yields the maximum effects for dead and live loads (see Table 4.1 in Section 4.2).

$$q_D = \left(\frac{10}{12} \times 150\right) + 20 = 145\,\text{psf}$$

$$q_u = 1.2q_D + 1.6q_L = (1.2 \times 145) + (1.6 \times 80) = 302\,\text{psf}$$

The critical section for two-way shear is located a distance $d/2 = (10 - 1.25)/2 = 4.4$ in from the face of the column (see Fig. 7.42).

At the edge column, the factored shear force due to gravity loads is

$$V_u = q_u(A_t - b_1 b_2)$$

$$= 0.302\left[\left(\frac{24}{2} + \frac{18}{2 \times 12}\right)\left(\frac{22.0 + 23.5}{2}\right) - \left(\frac{22.38}{12}\right)\left(\frac{36.75}{12}\right)\right] = 85.9\,\text{kips}$$

where $b_1 = 18 + \dfrac{8.75}{2} = 22.38$ in

$b_2 = 28 + 8.75 = 36.75$ in

Because it is assumed that the Direct Design Method can be used to compute the moments, ACI 13.6.3.6 requires that the unbalanced moment at the edge column that is transferred by eccentricity of shear is equal to $0.3M_o$, where M_o is

$$M_o = \frac{q_u \ell_2 \ell_n^2}{8} = \frac{0.302 \times [(22.0 + 23.5)/2] \times \{24 - [18/(2 \times 12)] - [28/(2 \times 12)]\}^2}{8} = 418.8\,\text{ft kips}$$

Thus,

$$M_u = 0.3 \times 418.8 = 125.6\,\text{ft kips}$$

Determine γ_v by Eq. (7.38):

$$\gamma_v = 1 - \gamma_f = 1 - \frac{1}{1 + (2/3)\sqrt{b_1/b_2}} = 1 - \frac{1}{1 + (2/3)\sqrt{22.38/36.75}} = 0.34$$

The section properties of the critical section are determined using Fig. 7.55 for an edge column bending perpendicular to the edge:

$$f_1 = 2 + \frac{c_1}{d}\left(2 + \frac{c_2}{c_1}\right) = 2 + \frac{18}{8.75}\left(2 + \frac{28}{18}\right) = 9.31$$

$$f_2 = \frac{[(c_1/d) + (1/2)]^2\{(c_1/d)[1 + (2c_2/c_1)] + (5/2)\} + \{(c_1/d)[1 + (2c_2/c_1)] + 1\}}{6[(c_1/d) + (1/2)]}$$

$$= \frac{[(18/8.75) + (1/2)]^2\{(18.8.75)[1 + (2 \times 28)/18] + (5/2)\} + \{(18/18.75)[1 + 28/(2 \times 18)] + 1\}}{6[(18/8.75) + (1/2)]} = 4.97$$

$$f_3 = \frac{[(c_1/d) + (1/2)]^2\{(c_1/d)[1 + (2c_2/c_1)] + (5/2)\} + \{(c_1/d)[1 + (c_2/2c_1)] + 1\}}{6\{(c_1/d)[1 + (c_2/c_1)] + (3/2)\}}$$

$$= \frac{[(18/8.75) + (1/2)]^2\{(18.8.75)[1 + (2 \times 28)/18] + (5/2)\} + \{(18/18.75)[1 + 28/(2 \times 18)] + 1\}}{6\{(18/8.75)[1 + (28/18)] + (1/2)\}} = 1.88$$

$$c_{AB} = \frac{f_3}{f_2 + f_3}\left(c_1 + \frac{d}{2}\right) = \frac{1.88}{4.97 + 1.88}\left(18 + \frac{8.75}{2}\right) = 6.1 \text{ in}$$

$$A_c = f_1 d^2 = 9.31 \times 8.75^2 = 712.8 \text{ in}^2$$

$$J_c/c_{AB} = 2f_2 d^3 = 2 \times 4.97 \times 8.75^3 = 6,659 \text{ in}^3$$

These section properties can also be obtained using the tables given in Appendix B. The total factored shear stress is determined by Eq. (7.39):

$$v_{u(AB)} = \frac{V_u}{A_c} + \frac{\gamma_v M_u c_{AB}}{J_c}$$

$$= \frac{85,900}{712.8} + \frac{0.34 \times 125.6 \times 12,000}{6,659} = 120.5 + 77.0 = 197.5 \text{ psi}$$

The allowable stress is the smallest of the values obtained from Eqs. (7.29) to (7.31):

$$\phi v_c = \frac{\phi V_c}{b_o d} = \phi 4\lambda\sqrt{f_c'} = 0.75 \times 4 \times 1.0\sqrt{4,000} = 189.7 \text{ psi (governs)}$$

$$\phi v_c = \frac{\phi V_c}{b_o d} = \phi\left(2 + \frac{4}{\beta}\right)\lambda\sqrt{f_c'} = 0.75 \times \left(2 + \frac{4}{1.56}\right) \times 1.0\sqrt{4,000} = 216.5 \text{ psi}$$

$$\phi v_c = \frac{\phi V_c}{b_o d} = \phi\left(\frac{\alpha_s d}{b_o} + 2\right)\lambda\sqrt{f_c'} = 0.75 \times \left[\frac{30 \times 8.75}{(2 \times 22.38) + 36.75} + 2\right] \times 1.0\sqrt{4,000} = 330.2 \text{ psi}$$

Because $v_u > \phi v_c$, shear strength requirements are not satisfied at the edge column. Increase shear strength by providing (1) closed stirrups and (2) headed shear studs.

Increase Shear Strength by Closed Stirrups

Step 1: Determine if stirrups can be utilized in this slab in accordance with ACI 11.11.3. Assuming No. 4 stirrups ($d_b = 0.50$ in),

$$\text{Average } d = 8.75 \text{ in} > \begin{cases} 6.0 \text{ in} \\ 16d_b = 16 \times 0.50 = 8.0 \text{ in} \end{cases}$$

Step 2: Check the maximum shear strength permitted with stirrups in accordance with ACI 11.11.3.2.

$$v_{u(AB)} = 197.5 \text{ psi} < \phi 6\sqrt{f_c'} = 0.75 \times 6\sqrt{4,000} = 284.6 \text{ psi}$$

Step 3: Determine the design shear strength of the concrete with stirrups in accordance with ACI 11.11.3.1.

$$\phi v_c = \phi 2\lambda\sqrt{f_c'} = 0.75 \times 2 \times 1.0\sqrt{4,000} = 94.9 \text{ psi}$$

Step 4: Determine the required area of stirrups. The required area of stirrups A_v at the critical section located a distance $d/2$ from the face of the column can be determined from Eq. (7.33). Assuming a stirrup spacing s equal to 4 in, which is less than the maximum spacing of $d/2 = 4.4$ in (ACI 11.11.3.3), the total area of stirrups A_v required on the three sides of the column is

$$A_v = \frac{(v_{u(AB)} - \phi v_c)b_o s}{\phi f_y}$$

$$= \frac{(197.5 - 94.9)[(2 \times 22.38) + 36.75] \times 4}{0.75 \times 60,000} = 0.74 \text{ in}^2$$

The required area of stirrups per side is

$$A_v \text{ (per side)} = \frac{0.74}{3} = 0.25 \text{ in}^2$$

Use No. 4 stirrups spaced at 4.0 in on center [A_v (per side) $= 0.40$ in^2].

Step 5: Determine the distance from faces of column where stirrups can be terminated. Stirrups can be terminated where the design strength of the concrete can resist the factored shear stress without stirrups.

At the critical section located a distance $d/2$ from the outermost peripheral line of stirrups, the design shear strength of the concrete is determined by Eq. (7.34):

$$\phi V_n = \phi V_c = \phi 2\lambda \sqrt{f_c'} b_o d$$

Assume that the stirrups are terminated 24 in from the faces of the column. Therefore, the outermost critical section is located $24 + 4.4 = 28.4$ in from the faces of the column.

The perimeter of the critical section is (see Fig. 7.48)

$$b_o = (2 \times 18) + 28 + (2\sqrt{2} \times 28.4) = 144.3 \text{ in}$$

Thus,

$$\phi V_c = 0.75 \times 2 \times 1.0\sqrt{4,000} \times 144.3 \times 8.75/1,000 = 119.8 \text{ kips}$$

The factored shear force at the face of the critical section located $d/2$ from the face of the column was determined earlier as 85.9 kips. At the critical section located $d/2$ from the outermost stirrups, V_u is significantly less than 85.9 kips, and the shear stress due to the portion of the unbalanced moment transferred by eccentricity of shear is negligible. Therefore, because ϕV_c is greater than V_u at the critical section located $d/2$ from the outermost stirrups, shear strength requirements are adequate at that section.

Use six No. 4 closed stirrups spaced at 4.0 in along the three sides of the column.

The first peripheral line of stirrups must be located not farther than $d/2$ from the face of the column. Thus, locate the first line of stirrups 4.0 in from the faces of the column. See Fig. 7.48 for other detailing requirements.

Increase Shear Strength by Headed Shear Stud Reinforcement

Step 1: Check the maximum shear strength permitted with headed shear studs in accordance with ACI 11.11.5.1.

$$v_{u(AB)} = 197.5 \text{ psi} < \phi 8\sqrt{f_c'} = 0.75 \times 8\sqrt{4,000} = 379.5 \text{ psi}$$

Step 2: Determine the design shear strength of the concrete with headed shear studs in accordance with ACI 11.11.5.1.

$$\phi v_c = \phi 3\lambda\sqrt{f_c'} = 0.75 \times 3 \times 1.0\sqrt{4,000} = 142.3 \text{ psi}$$

Step 3: Determine the required spacing of headed shear studs. The required spacing of headed shear studs s at the critical section located a distance $d/2$ from the face of the column can be determined from Eq. (7.35).

In order to satisfy the requirements of ACI 11.11.5.3 related to spacing between adjacent shear reinforcement elements (i.e., spacing between elements must not exceed $2d = 17.5$ in), three lines of headed studs are provided on each column face.

Assuming $1/2$-in diameter studs ($A_{stud} = 0.196$ in.^{2}), the required spacing is

$$s = \frac{A_v f_{yt}}{(v_{u(AB)} - \phi v_c)b_o}$$

$$= \frac{(9 \times 0.196) \times 51,000}{(197.5 - 142.3)[(2 \times 22.38) + 36.75]} = 20.0 \text{ in}$$

In this equation, A_v is the cross-sectional area of all the headed shear studs on one peripheral line that is parallel to the perimeter of the column section, and the minimum specified yield strength of the headed studs is 51,000 psi (see ACI R3.5.5).

Because $v_{u(AB)} < \phi 6\sqrt{f_c'}$, maximum stud spacing $= 0.75d = 6.6$ in.

Assuming a 6-in spacing, check the requirement of ACI 11.11.5.1:

$$\frac{A_v f_{yt}}{b_o s} = \frac{(9 \times 0.196) \times 51,000}{[(2 \times 22.38) + 36.75] \times 6} = 184.0 \text{ psi} > 2\sqrt{f_c'} = 126.5 \text{ psi}$$

Step 4: Determine distance from faces of column where headed shear studs can be terminated. Headed shear studs can be terminated where the design strength of the concrete can resist the factored shear stress without the headed shear studs.

At the critical section located a distance $d/2$ from the outermost peripheral line of headed studs, the design shear strength of the concrete is determined by Eq. (7.37):

$$\phi V_n = \phi V_c = \phi 2\lambda \sqrt{f_c'} b_o d$$

Assume that the headed studs are terminated 22 in from the faces of the column. Therefore, the outermost critical section is located $22 + 4.4 = 26.4$ in from the faces of the column.

The perimeter of the critical section is (see Fig. 7.50)

$$b_o = (2 \times 18) + 28 + (2\sqrt{2} \times 26.4) = 138.7 \text{ in}$$

Thus,

$$\phi V_c = 0.75 \times 2 \times 1.0\sqrt{4,000} \times 138.7 \times 8.75/1,000 = 115.1 \text{ kips}$$

The factored shear force at the face of the critical section located $d/2$ from the face of the column was determined earlier as 85.9 kips. At the critical section located $d/2$ from the outermost line of headed shear studs, V_u is significantly less than 85.9 kips, and the shear stress due to the portion of the unbalanced moment transferred by eccentricity of shear is negligible. Therefore, because ϕV_c is greater than V_u at the critical section located $d/2$ from the outermost line of studs, shear strength requirements are adequate at that section.

Use $1/2$-in-diameter headed studs spaced at 6.0 in on center along the three sides of the column.

The first peripheral line of headed shear studs must be located not farther than $d/2$ from the face of the column. Thus, locate the first line of stirrups 4.0 in from the faces of the column. See Fig. 7.50 for other detailing requirements.

7.7 Design Procedure

The following design procedure can be used in the design of two-way slab systems. Included is the information presented in the previous sections on how to analyze, design, and detail a two-way system for flexure and shear.

Step 1: Determine the preliminary slab thickness (Section 7.3). The first step in the design procedure is to determine a preliminary slab thickness. A minimum slab thickness must be provided to satisfy serviceability requirements. For slab systems without beams, it is also advisable at this stage to run a preliminary investigation of the shear strength of the slab in the vicinity of the columns or other support locations.

For overall economy, the slab thickness for the entire floor plate should be based on the minimum thickness for the panel that requires the largest thickness. Varying the amount of flexural reinforcement, and not the thickness of the slab, produces the most economical solution.

Step 2: Determine bending moments at the critical sections (Section 7.4). Once a preliminary slab thickness has been obtained, the next step is to determine the factored bending moments in the column strips and middle strips at the critical locations along the span of a design strip. The Direct Design Method can be used to determine bending moments due to gravity loads, provided that the limitations of ACI 13.6.1 are satisfied. The Equivalent Frame Method can be used in cases where the Direct Design Method cannot be used.

The Code also permits other analysis methods to determine bending moments, as long as equilibrium and geometric compatibility are satisfied.

Step 3: Determine the required flexural reinforcement (Section 7.5). The required flexural reinforcement is determined at the critical locations in both the column strips and the middle strips. It is advantageous to design the slab as a tension-controlled section with the strength reduction factor equal to 0.9. The methods presented in Chap. 6 for rectangular sections with a single layer of tension reinforcement can be used to determine the required area of flexural reinforcement in slabs.

Minimum reinforcement requirements are given in ACI 13.3.1. The area of reinforcement at any critical section must not be less than that required for temperature and shrinkage in accordance with ACI 7.12.2.2. For Grade 60 reinforcement, the minimum reinforcement ratio is 0.0018, which is based on the gross concrete area. According to ACI 13.3.2, spacing of the flexural reinforcement is limited to two times the thickness of the slab.

Flexural strength requirements with respect to unbalanced moments at slab–column joints must also be satisfied. The portion of the unbalanced moment transferred by flexure is resisted by the reinforcement in the effective slab width defined in ACI 13.5.3.2.

Step 4: Check shear strength requirements at columns (Section 7.6). Both one- and two-way shear must be checked at the critical sections around columns. The total factored two-way shear stress at a critical section around a support consists of the shear stress due to direct shear force and the shear stress due to the portion of the unbalanced moment that is transferred by eccentricity of shear.

The total factored shear stress must be equal to or less than the design shear strength. In cases where the design shear strength of the concrete is not sufficient, shear reinforcement can be utilized to increase the total design shear strength.

Typically, proportioned column-line beams usually resist most, if not all, of the shear force in a column strip.

Step 5: Detail the reinforcement (Sections 7.5 and 7.6). Detailing requirements for slab reinforcement in two-way systems are given in ACI 13.3. ACI Fig. 13.3.8 provides minimum bar extensions for reinforcement in slabs without beams on the basis of uniformly distributed gravity loads.

For bending moments resulting from combined gravity and lateral loads, the minimum lengths given in this figure may not be sufficient. Bar lengths must be determined by the development requirements of ACI 12.10 to 12.12 for general load cases. The provided bar lengths must not be taken less than those prescribed in ACI Fig. 13.3.8.

According to ACI 13.3.6, special top and bottom reinforcement must be provided at the exterior corners of a slab with spandrel beams that have a stiffness α_f greater than 1.0. The reinforcement must be designed for a moment equal to the largest positive moment per unit width in the panel and must be placed in a band parallel to the diagonal at the top of the slab and a band perpendicular to the diagonal at the bottom of the slab. Additionally, the reinforcement must extend at least one-fifth of the longer span in each direction from the corner. Alternatively, the reinforcement may be placed in two layers parallel to the edges of the slab at both the top and the bottom of the slab.

For slabs with shear reinforcement, the detailing requirements of ACI 11.11 must be satisfied for the particular type of reinforcement that is used in the slab.

Example 7.15 Design the interior strip for the flat-plate system depicted in Fig. 7.59.

The flat plate is part of the lateral force–resisting system, and the interior design strip is subjected to the following bending moments:

End span: ±27 ft kips at exterior column, ±25 ft kips at first interior column

Interior span: ±20 ft kips at interior column

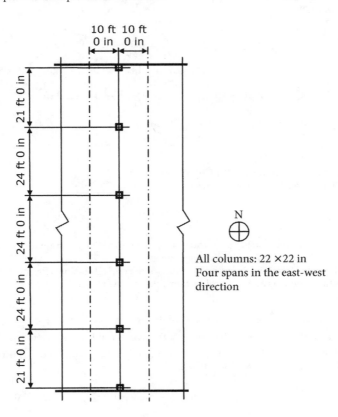

Figure 7.59 The flat-plate system given in Example 7.15.

10 ft 10 ft
0 in   0 in

21 ft 0 in

24 ft 0 in

24 ft 0 in

24 ft 0 in

21 ft 0 in

N

All columns: 22 ×22 in
Four spans in the east-west direction

Assume normal-weight concrete with $f_c' = 4{,}000$ psi, Grade 60 reinforcement, a superimposed dead load of 15 psf, and a live load of 80 psf.

Solution   The design procedure outlined earlier will be used to design this flat plate.

Step 1: Determine the preliminary slab thickness. For slabs without interior beams spanning between the supports, the provisions of ACI 9.5.3.2 must be used to determine the minimum slab thickness based on serviceability requirements.

The minimum slab thickness is determined from ACI Table 9.5(c) for slab systems without drop panels and edge beams and with Grade 60 reinforcement:

$$\text{Exterior panel: longest } \ell_n = (21 \times 12) - 22 = 230 \text{ in}$$

$$h_{min} = \frac{\ell_n}{30} = \frac{230}{30} = 7.7 \text{ in}$$

$$\text{Interior panel: longest } \ell_n = (24 \times 12) - 22 = 266 \text{ in}$$

$$h_{min} = \frac{\ell_n}{33} = \frac{266}{33} = 8.1 \text{ in}$$

Thus, the minimum slab thickness is governed by an interior panel.

Use Fig. 7.11 to estimate minimum slab thickness based on two-way shear strength at an edge column, assuming a 9-in-thick slab.

The total factored gravity loads acting on the slab q_u is determined using the load combination of ACI Eq. (9-2) because this yields the maximum effects for dead and live loads (see Table 4.1 in Section 4.2):

$$q_D = \left(\frac{9}{12} \times 150 \right) + 15 = 127.5 \text{ psf}$$

$$q_u = 1.2 q_D + 1.6 q_L = (1.2 \times 127.5) + (1.6 \times 80) = 281 \text{ psf}$$

$$\text{Tributary area } A = \frac{1}{2} \left(21 + \frac{22}{12} \right) \times 20 = 228.33 \text{ ft}^2$$

$$\text{Area of edge column } c_1^2 = 22^2/144 = 3.36 \text{ ft}^2$$

$$A/c_1^2 = 228.33/3.36 = 68.0$$

From Fig. 7.11, obtain $d/c_1 \cong 0.25$ for $q_u = 281$ psf. Therefore, $d = 0.25 \times 22 = 5.5$ in.

$$h = 5.5 + 1.25 = 6.75 \text{ in.} < 8.1 \text{ in}$$

Try a 9-in-thick slab. This thickness is greater than the minimum thickness of 5 in prescribed in ACI 9.5.3.2(a) for slabs without drop panels.

Step 2: Determine bending moments at the critical sections. Prior to determining the factored bending moments at the critical sections, check if the Direct Design Method can be used to analyze this two-way system:

1. Three continuous spans must be present in each direction.

   There are five spans in the north-south direction and four spans in the east-west direction.

2. Slab panels must be rectangular with a ratio of the longer to shorter span, centerline-to-centerline of supports, equal to or less than 2.

$$\text{Longer span/shorter span} = 24/20 = 1.2 < 2.0$$

3. Successive span lengths, centerline-to-centerline of supports, in each direction must not differ by more than one-third of the longer span.

   In the north-south direction, $21/24 = 0.875 > 0.67$.

4. Columns must not be offset more than 10% of the span in the direction of offset from either axis between the centerlines of successive columns.

   No column offsets are present.

5. Loads applied to the slab must be uniformly distributed gravity loads where the ratio of the unfactored live load to the unfactored dead load is equal to or less than.

$$\text{Live load} = 80\,\text{psf}$$
$$\text{Dead load} = 127.5\,\text{psf}$$
$$\text{Uniform live to dead load ratio} = 80/127.5 = 0.63 < 2$$

6. For panels with column-line beams on all sides, ACI Eq. (13-2) must be satisfied.

   No column-line beams are present.

7. Redistribution of bending moments in accordance with ACI 8.4 is not permitted.

   Bending moments will not be redistributed in accordance with ACI 8.4.

   Therefore, the Direct Design Method can be used for gravity load analysis.

   The steps presented in Section 7.4 will be used to determine the bending moments at the critical sections. Instead of using factored gravity loads, service dead and live load moments will be computed because these moments must be combined with those due to the effects from wind, using the load combinations of ACI 9.2.

   Step 2A: Determine the static moment M_o in each span. The static moment M_o is determined by Eq. (7.8):

$$M_o = \frac{q\ell_2\ell_n^2}{8}$$

In this equation, q is the uniformly distributed service dead or live load.

Calculate M_o for the end and interior spans.

End span:

$$\ell_n = 21 - \frac{22}{12} = 19.17\,\text{ft}$$
$$\ell_2 = 20.0\,\text{ft}$$
$$(M_o)_D = \frac{q_D\ell_2\ell_n^2}{8} = \frac{0.128 \times 20.0 \times 19.17^2}{8} = 117.6\,\text{ft kips}$$
$$(M_o)_L = \frac{q_L\ell_2\ell_n^2}{8} = \frac{0.080 \times 20.0 \times 19.17^2}{8} = 73.5\,\text{ft kips}$$

Interior span:

$$\ell_n = 24 - \frac{22}{12} = 22.17\,\text{ft}$$
$$\ell_2 = 20.0\,\text{ft}$$
$$(M_o)_D = \frac{q_D\ell_2\ell_n^2}{8} = \frac{0.128 \times 20.0 \times 22.17^2}{8} = 157.3\,\text{ft kips}$$
$$(M_o)_L = \frac{q_L\ell_2\ell_n^2}{8} = \frac{0.080 \times 20.0 \times 22.17^2}{8} = 98.3\,\text{ft kips}$$

Step 2B: Distribute M_o into negative and positive bending moments in each span. The moment M_o is divided into negative and positive moments in accordance with distribution factors

End Span Moments (ft kips)			Interior Span Moments (ft kips)	
Exterior Negative	**Positive**	**Interior Negative**	**Positive**	**Interior Negative**
$0.26(M_o)_D =$ -30.6	$0.52(M_o)_D =$ 61.2	$0.70(M_o)_D =$ -82.3	$0.35(M_o)_D =$ 55.1	$0.65(M_o)_D =$ -102.3
$0.26(M_o)_L =$ -19.1	$0.52(M_o)_L =$ 38.2	$0.70(M_o)_L =$ -51.5	$0.35(M_o)_L =$ 34.4	$0.65(M_o)_L =$ -63.9

TABLE 7.20   Summary of Dead and Live Load Moments for the Flat Plate Given in Example 7.15

given in ACI 13.6.3 (see Tables 7.2 and 7.3 of this book). A summary of the dead and live load moments at the critical sections is given in Table 7.20.

Step 2C: Distribute the dead and live load moments to the column strip and middle strip. The percentages of the negative and positive bending moments at the critical sections that are assigned to the column strips and middle strips are given in ACI 13.6.4 to 13.6.6 (see Table 7.3 of this book). A summary of the dead and live load moments at the critical section is given for this flat plate in Table 7.21.

According to ACI 13.5.1.3, the results of the gravity load analysis are combined with the moments due to wind, using the applicable load combinations given in ACI 9.2.

Equation (7.25) is used to determine the effective width of the slab to resist the effects from the wind loads.

For the end span,

$$\text{Effective slab width} = 2c_1 + \frac{\ell_1}{3} = (2 \times 22) + \frac{21 \times 12}{3} = 128 \text{ in}$$

For the interior span,

$$\text{Effective slab width} = 2c_1 + \frac{\ell_1}{3} = (2 \times 22) + \frac{24 \times 12}{3} = 140 \text{ in}$$

The width of the column strip for the end and interior spans is equal to $\ell_2/2 = (20 \times 12)/2 = 120$ in.

Therefore, because the effective slab width in the end span and the column strip width are almost equal, conservatively assume that the column strip resists all of the effects from the wind loads.

A summary of the factored load combinations is given in Table 7.22 for an end span and a typical interior span. Two values of bending moments are given at the supports in the column strip: The

	End Span Moments (ft kips)						Interior Span Moments (ft kips)			
	Exterior Negative		**Positive**		**Interior Negative**		**Positive**		**Interior Negative**	
Column strip	$0.26M_o$		$0.31M_o$		$0.53M_o$		$0.21M_o$		$0.49M_o$	
	Dead	-30.6	Dead	36.5	Dead	-62.3	Dead	33.0	Dead	-77.1
	Live	-19.1	Live	22.8	Live	-39.0	Live	20.6	Live	-48.2
Middle strip	0		$0.21M_o$		$0.17M_o$		$0.14M_o$		$0.16M_o$	
			Dead	24.7	Dead	-20.0	Dead	22.0	Dead	-25.2
			Live	15.4	Live	-12.5	Live	13.8	Live	-15.7

TABLE 7.21   Summary of Dead and Live Load Moments at the Critical Sections for the Flat Plate Given in Example 7.15

Table 7.22 Summary of Design Bending Moments (ft kips) for the Interior Design Strip Given in Example 7.15

Load Combination	End Span						Interior Span			
	Column Strip			Middle Strip			Column Strip		Middle Strip	
	Exterior Negative	Positive	Interior Negative	Exterior Negative	Positive	Interior Negative	Positive	Negative	Positive	Negative
1.4D	−42.8	51.1	−87.2	0.0	34.6	−28.0	46.2	−107.9	30.8	−35.3
1.2D + 1.6L	−67.3	80.3	−137.2	0.0	54.3	−44.0	72.6	−169.6	48.5	−55.4
1.2D + 1.6L + 0.8W	−45.7	80.3	−157.2	0.0	54.3	−44.0	72.6	−153.6	48.5	−55.4
	−88.9		−117.2					−185.6		
1.2D + 0.5L + 1.6W	−3.1	55.2	−134.3	0.0	37.3	30.3	49.9	−84.6	33.3	−38.1
	−89.5		−54.3					−148.6		
0.9D + 1.6W	15.7	32.9	−96.1	0.0	22.2	−18.0	29.7	−37.4	19.8	−22.7
	−70.7		−16.1					−101.4		

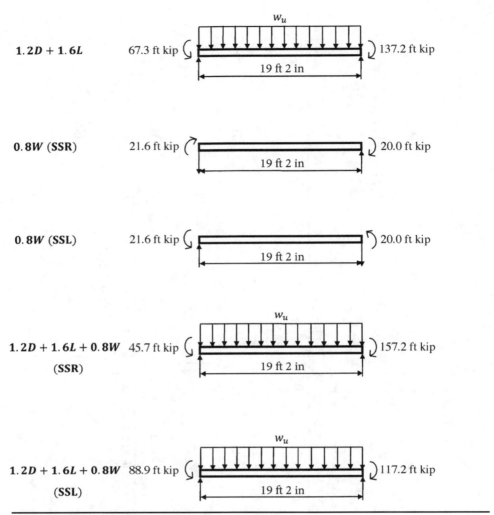

Figure 7.60   Combination of gravity and wind load effects.

top value corresponds to wind blowing from left to right (sidesway right), and the bottom value corresponds to wind blowing from right to left (sidesway left). Figure 7.60 illustrates the proper combination of the effects due to gravity and wind loads in the end span for the load combination $1.2D + 1.6L + 0.8W$ for sidesway to the right (SSR) and sidesway to the left (SSL).

Note that at the exterior column in the end span, a positive moment occurs at the face of the column due to reversal of moments caused by the wind load effects.

Step 3: Determine the required flexural reinforcement. The required flexural reinforcement at the critical sections can be obtained using the strength design methods presented in Chap. 6 for tension-controlled, rectangular sections with a single layer of reinforcement.

It was determined in step 2 that the width of the column strip is equal to 120 in. Therefore, the width of the middle strip is equal to $240 - 120 = 120$ in. These widths are used in the calculation of the flexural reinforcement.

A summary of the required flexural reinforcement at the critical sections in the column strip and middle strip is given in Table 7.23.

Location			M_u (ft kips)	b (in)	A_s (in^2)	Reinforcement
End span	Column strip	Exterior negative	−89.5	120	2.63	9 No. 5
		Positive	80.3	120	2.35	8 No. 5
		Interior negative	−157.2	120	4.72	16 No. 5
	Middle strip	Exterior negative	0.0	120	1.94*	7 No. 5
		Positive	54.3	120	1.94*	7 No. 5
		Interior negative	−44.0	120	1.94*	7 No. 5
Interior span	Column strip	Positive	72.6	120	2.12	7 No. 5
		Negative	−185.6	120	5.62	19 No. 5
	Middle strip	Positive	48.5	120	1.94*	7 No. 5
		Negative	−55.4	120	1.94*	7 No. 5

*Based on minimum reinforcement requirements: $A_{s,min} = 0.0018bh = 1.94$ in^2.

TABLE 7.23   Required Slab Reinforcement for the Flat Plate Given in Example 7.15

Maximum spacing requirements:

Maximum bar spacing $= 2h = 18$ in.

For $b = 120$ in, $120/18 = 6.7$, say, seven bars are needed to satisfy maximum spacing requirements.

Note that the eight No. 5 bars that are provided at the bottom of the slab in the column strip in the end span are adequate to resist the 15.7 ft kip positive bending moment at the face of the exterior column due to the reversal of wind load effects. Additional positive reinforcement would have been required at that location if that moment were greater than the positive moment near the midspan.

Check that the flexural reinforcement at the end support is adequate to satisfy the moment transfer requirements of ACI 13.5.3.

The total unbalanced moment at this slab–column connection is equal to 89.5 ft kips, which is the total moment in the column strip (see Table 7.23).

A fraction of this moment $\gamma_f M_u$ must be transferred over an effective width equal to $c_2 + 3h = 22 + (3 \times 9) = 49$ in.

The fraction of unbalanced moment transferred by flexure is calculated in accordance with Eq. (7.26):

$$\gamma_f = \frac{1}{1 + (2/3)\sqrt{b_1/b_2}} = \frac{1}{1 + (2/3)\sqrt{25.88/29.75}} = 0.62$$

where $b_1 = c_1 + \dfrac{d}{2} = 22 + \dfrac{7.75}{2} = 25.88$ in

$b_2 = c_2 + d = 22 + 7.75 = 29.75$ in

For edge columns bending perpendicular to the edge, the value of γ_f computed by Eq. (7.26) may be increased to 1.0 provided that $V_u \le 0.75\phi V_c$ [ACI 13.5.3.3(a)]. No adjustment to γ_f is made in this example.

Unbalanced moment transferred by flexure $= \gamma_f M_u = 0.62 \times 89.5 = 55.5$ ft kips. The required area of steel to resist this moment in the 49-in-wide strip is $A_s = 1.66$ in^2, which is equivalent to six No. 5 bars.

$$A_{s,min} = 0.0018bh = 0.0018 \times 49 \times 9.0 = 0.79 \, \text{in}^2 < A_s$$

Provide the six No. 5 bars by concentrating six of the nine column strip bars (see Table 7.23) within the 49-in width over the column. For symmetry, add another bar in the column strip and check bar spacing:

For six No. 5 bars within the 49-in width, $49/6 = 8.2$ in < 18 in.

For four No. 5 bars within the $120 - 49 = 71$-in width, $71/4 = 17.8$ in < 18 in.

A total of 10 No. 5 bars are required at the end supports within the column strip, with 6 of the 10 bars concentrated within a width of 49 in centered on the column.

Similar calculations can be performed for the first interior column and an interior column.

Step 4: Check the shear strength requirements at columns (Section 7.6). Both one- and two-way shear requirements must be checked at the first interior column and at an edge column in an interior design strip. Similar calculations can be performed for an interior column.

First Interior Column

One-way shear   The critical section for one-way shear is located a distance $d = 9 - 1.25 = 7.75$ in from the face of the column (see Fig. 7.40).

The total factored gravity load acting on the slab q_u is determined using the load combination of ACI Eq. (9-2) because this yields the maximum effects for dead and live loads (see Table 4.1 in Section 4.2):

$$q_u = 1.2q_D + 1.6q_L = (1.2 \times 127.5) + (1.6 \times 80) = 281 \, \text{psf}$$

The maximum factored shear force at the critical section is

$$V_u = 0.281 \times \left(\frac{24}{2} - \frac{22}{2 \times 12} - \frac{7.75}{12} \right) \times 20 = 58.7 \, \text{kips}$$

Design shear strength is computed by Eq. (7.28):

$$\phi V_c = \phi 2\lambda \sqrt{f_c'} \ell d = 0.75 \times 2 \times 1.0 \sqrt{4,000} \times (20 \times 12) \times 7.75/1,000 = 176.5 \, \text{kips}$$

Because $V_u < \phi V_c$, one-way shear strength requirements are satisfied.

Two-way shear   The total factored shear stress is the sum of the direct shear stress plus the shear stress due to the fraction of the unbalanced moment transferred by eccentricity of shear.

The critical section for two-way shear is located a distance $d/2 = 3.9$ in from the face of the column (see Fig. 7.42).

Factored shear stresses must be checked for gravity load combinations and gravity plus wind load combinations. It is not readily apparent which of these combinations produces the greatest combined shear stress.

Gravity loads: $1.2D + 1.6L$ At the first interior column, the factored shear force due to gravity loads is

$$V_u = q_u(A_t - b_1 b_2)$$
$$= 0.281 \left[(22.5 \times 20) - \left(\frac{29.75}{12} \right) \left(\frac{29.75}{12} \right) \right] = 124.7 \, \text{kips}$$

where $b_1 = b_2 = 22 + 7.75 = 29.75$ in

The total unbalanced moment is equal to the difference between the total interior negative moments on both sides of the column (see Table 7.20):

$$M_u = [(1.2 \times 102.3) + (1.6 \times 63.9)] - [(1.2 \times 82.3) + (1.6 \times 51.5)] = 43.8 \text{ ft kips}$$

Determine γ_v by Eq. (7.38):

$$\gamma_f = \frac{1}{1 + (2/3)\sqrt{b_1/b_2}} = \frac{1}{1 + (2/3)\sqrt{29.75/29.75}} = 0.6$$

$$\gamma_v = 1 - \gamma_f = 1 - 0.6 = 0.4$$

The section properties of the critical section are determined using Fig. 7.53 for an interior column:

$$c_{AB} = b_1/2 = 29.75/2 = 14.9 \text{ in}$$

$$f_1 = 2\left[\left(1 + \frac{c_2}{c_1}\right)\left(\frac{c_1}{d}\right) + 2\right] = 2\left[(1 + 1)\left(\frac{22}{7.75}\right) + 2\right] = 15.36$$

$$A_c = f_1 d^2 = 15.36 \times 7.75^2 = 922.6 \text{ in}^2$$

$$f_2 = \frac{1}{6}\left[\left(1 + \frac{3c_2}{c_1}\right)\left(\frac{c_1}{d}\right)^2 + \left(5 + \frac{3c_2}{c_1}\right)\left(\frac{c_1}{d}\right) + 5\right]$$

$$= \frac{1}{6}\left[(1 + 3)\left(\frac{22}{7.75}\right)^2 + (5 + 3)\left(\frac{22}{7.75}\right) + 5\right] = 9.99$$

$$J_c/c_{AB} = 2f_2 d^3 = 2 \times 9.99 \times 7.75^3 = 9{,}300 \text{ in}^3$$

These section properties can also be obtained using the tables given in Appendix B. The total factored shear stress is determined by Eq. (7.39):

$$v_{u(AB)} = \frac{V_u}{A_c} + \frac{\gamma_v M_u c_{AB}}{J_c}$$

$$= \frac{124{,}700}{922.6} + \frac{0.4 \times 43.8 \times 12{,}000}{9{,}300} = 135.2 + 22.6 = 157.8 \text{ psi}$$

The allowable stress for a square column is obtained from Eq. (7.29):

$$\phi v_c = \frac{\phi V_c}{b_o d} = \phi 4\lambda\sqrt{f_c'} = 0.75 \times 4 \times 1.0\sqrt{4{,}000} = 189.7 \text{ psi} > 157.8 \text{ psi}$$

Gravity plus wind loads
1. $1.2D + 1.6L + 0.8W$

From previous calculations, $q_u = 281$ psf and $V_u = 124.7$ kips.

The maximum shear force due to wind loads is determined from statics by dividing the sum of the wind moments in the span by the span length: $(27 + 25)/21 = 2.5$ kips.

Thus, the total factored shear force at the first interior column is

$$V_u = 124.7 + (0.8 \times 2.5) = 126.7 \text{ kips}$$

The total unbalanced moment is equal to the sum of the unbalanced moments due to gravity and wind loads:

$$M_u = 43.8 + [0.8 \times (25 + 20)] = 79.8 \text{ ft kips}$$

Note that the unbalanced moment due to the wind loads is the sum of the moments acting on opposite sides of the column.

Therefore, the total factored shear stress is

$$v_{u(AB)} = \frac{126{,}700}{922.6} + \frac{0.4 \times 79.8 \times 12{,}000}{9{,}300} = 137.3 + 41.2 = 178.5 \, \text{psi} < 189.7 \, \text{psi}$$

2. $1.2D + 0.5L + 1.6W$

$$q_u = 1.2q_D + 0.5q_L = (1.2 \times 127.5) + (0.5 \times 80) = 193 \, \text{psf}$$

$$V_u = 0.193 \left[(22.5 \times 20) - \left(\frac{29.75}{12}\right)\left(\frac{29.75}{12}\right) \right] = 85.7 \, \text{kips}$$

The total factored shear force due to gravity plus wind loads is

$$V_u = 85.7 + (1.6 \times 2.5) = 89.7 \, \text{kips}$$

The total unbalanced moment is equal to the sum of the unbalanced moments due to gravity and wind loads:

$$M_u = [(1.2 \times 102.3) + (0.5 \times 63.9)] - [(1.2 \times 82.3) + (0.5 \times 51.5)] + [1.6 \times (25 + 20)]$$
$$= 30.2 + 72.0 = 102.2 \, \text{ft kips}$$

Therefore, the total factored shear stress is

$$v_{u(AB)} = \frac{89{,}700}{922.6} + \frac{0.4 \times 102.2 \times 12{,}000}{9{,}300} = 97.2 + 52.8 = 150.0 \, \text{psi} < 189.7 \, \text{psi}$$

Therefore, shear strength requirements are satisfied at the first interior column.

Edge Column

One-way shear   The critical section for one-way shear is located a distance $d = 9 - 1.25 = 7.75$ in from the face of the column (see Fig. 7.40).

The maximum factored shear force at the critical section is

$$V_u = 0.281 \times \left(\frac{21}{2} - \frac{22}{2 \times 12} - \frac{7.75}{12} \right) \times 20 = 50.2 \, \text{kips}$$

Design shear strength is computed by Eq. (7.28):

$$\phi V_c = \phi 2\lambda \sqrt{f_c'}\, \ell d = 0.75 \times 2 \times 1.0\sqrt{4{,}000} \times (20 \times 12) \times 7.75/1{,}000 = 176.5 \, \text{kips}$$

Because $V_u < \phi V_c$, one-way shear strength requirements are satisfied.

Two-way shear   Like at the interior column, factored shear stresses must be checked for gravity load combinations and gravity plus wind load combinations.

Gravity loads: 1.2D + 1.6L   At the edge column, the factored shear force due to gravity loads is

$$V_u = q_u(A_t - b_1 b_2)$$
$$= 0.281 \left[\left(\frac{21}{2} + \frac{22}{2 \times 12} \right) (20) - \left(\frac{25.88}{12}\right)\left(\frac{29.75}{12}\right) \right] = 62.7 \, \text{kips}$$

where $b_1 = 22 + \dfrac{7.75}{2} = 25.88$ in

$b_2 = 22 + 7.75 = 29.75$ in

Because the Direct Design Method was used to compute the moments, ACI 13.6.3.6 requires that the unbalanced moment at the edge column that is transferred by eccentricity of shear be (see step 2)

$$M_u = 0.3M_o = 0.3 \times [(1.2 \times 117.6) + (1.6 \times 73.5)] = 77.6 \text{ ft kips}$$

Determine γ_v by Eq. (7.38):

$$\gamma_v = 1 - \gamma_f = 1 - \frac{1}{1 + (2/3)\sqrt{b_1/b_2}} = 1 - \frac{1}{1 + (2/3)\sqrt{25.88/29.75}} = 0.38$$

The properties of the critical section are determined using Fig. 7.55 for an edge column bending perpendicular to the edge:

$$f_1 = 2 + \frac{c_1}{d}\left(2 + \frac{c_2}{c_1}\right) = 2 + \frac{22}{7.75}(2 + 1) = 10.52$$

$$f_2 = \frac{[(c_1/d) + (1/2)]^2\{(c_1/d)[1 + (2c_2/c_1)] + (5/2)\} + \{(c_1/d)[1 + (c_2/2c_1)] + 1\}}{6[(c_1/d) + (1/2)]}$$

$$= \frac{[(22/7.75) + (1/2)]^2 [(22/7.75)(1 + 2) + (5/2)] + \{(22/7.75)[1 + (1/2)] + 1\}}{6[(22/7.75) + (1/2)]} = 6.39$$

$$f_3 = \frac{[(c_1/d) + (1/2)]^2\{(c_1/d)[1 + (2c_2/c_1)] + (5/2)\} + \{(c_1/d)[1 + (c_2/2c_1)] + 1\}}{6\{(c_1/d)[1 + (c_2/c_1)] + (1/2)\}}$$

$$= \frac{[(22/7.75) + (1/2)]^2 [(22/7.75)(1 + 2) + (5/2)] + \{(22/7.75)[1 + (1/2)] + 1\}}{6[(22/7.75)(1 + 1) + (3/2)]} = 2.97$$

$$c_{AB} = \frac{f_3}{f_2 + f_3}\left(c_1 + \frac{d}{2}\right) = \frac{2.97}{6.39 + 2.97}\left(22 + \frac{7.75}{2}\right) = 8.2 \text{ in}$$

$$A_c = f_1 d^2 = 10.52 \times 7.75^2 = 631.9 \text{ in}^2$$

$$J_c/c_{AB} = 2f_2 d^3 = 2 \times 6.39 \times 7.75^3 = 5,949 \text{ in}^3$$

These section properties can also be obtained using the tables given in Appendix B. The total factored shear stress is determined by Eq. (7.39):

$$v_{u(AB)} = \frac{V_u}{A_c} + \frac{\gamma_v M_u c_{AB}}{J_c}$$

$$= \frac{62,700}{631.9} + \frac{0.38 \times 77.6 \times 12,000}{5,949} = 99.2 + 59.5 = 158.7 \text{ psi} < 189.7 \text{ psi}$$

Gravity plus wind loads
1. $1.2D + 1.6L + 0.8W$

From previous calculations, $q_u = 281$ psf and $V_u = 62.7$ kips.

The maximum shear force due to wind loads is determined from statics by dividing the sum of the wind moments in the span by the span length: $(27 + 25)/21 = 2.5$ kips.

Thus, the total factored shear force at the edge column is

$$V_u = 62.7 + (0.8 \times 2.5) = 64.7 \, \text{kips}$$

When lateral loads are considered, shear stress calculations can be based on the actual unbalanced moment rather than that based on the provision of ACI 13.6.3.6, which, as shown earlier, requires the unbalanced moment to be $0.3M_o$. Thus, using Table 7.22, the total unbalanced moment is equal to the sum of the unbalanced moments due to gravity and wind loads:

$$M_u = 67.3 + (0.8 \times 27) = 88.9 \, \text{ft kips}$$

Therefore, the total factored shear stress is

$$v_{u(AB)} = \frac{64,700}{631.9} + \frac{0.38 \times 88.9 \times 12,000}{5,949} = 102.4 + 68.1 = 170.5 \, \text{psi} < 189.7 \, \text{psi}$$

2. $1.2D + 0.5L + 1.6W$

$$q_u = 1.2q_D + 0.5q_L = (1.2 \times 127.5) + (0.5 \times 80) = 193 \, \text{psf}$$

$$V_u = 0.193 \left[\left(\frac{21}{2} + \frac{22}{2 \times 12} \right)(20) - \left(\frac{25.88}{12} \right)\left(\frac{29.75}{12} \right) \right] = 43.0 \, \text{kips}$$

The total factored shear force is

$$V_u = 43.0 + (1.6 \times 2.5) = 47.0 \, \text{kips}$$

From Table 7.22, the unbalanced moment at the exterior column for this load combination is $M_u = 89.5$ ft kips.

Therefore, the total factored shear stress is

$$v_{u(AB)} = \frac{47,000}{631.9} + \frac{0.38 \times 89.5 \times 12,000}{5,949} = 74.4 + 68.6 = 143.0 \, \text{psi} < 189.7 \, \text{psi}$$

Therefore, shear strength requirements are satisfied at the edge column.

Step 5: Detail the reinforcement. The flexural reinforcement must be developed at the critical sections in the column strip and middle strip. Because the slab is subjected to the effects from wind, the minimum bar extensions given in Fig. 7.37 may not be adequate.

Conservatively provide 25% of the top bars in the column strip continuous over the span. This eliminates the need to locate inflection points for the various load combinations. The remaining bars in the column strip and the bars in the middle can be cut off at the locations indicated in Fig. 7.37.

References

1. American Concrete Institute (ACI), Committee 435. 1995. *Control of Deflection in Concrete Structures*, ACI 435. ACI, Farmington Hills, MI.
2. Portland Cement Association (PCA). 2008. *Notes on ACI 318-08 Building Code Requirements for Structural Concrete*. PCA, Skokie, IL.
3. American Concrete Institute (ACI), Committee 209. 1992 (Reapproved 1997). *Prediction of Creep, Shrinkage, and Temperature Effects in Concrete Structures*, ACI 209. ACI, Farmington Hills, MI.
4. International Code Council (ICC). 2009. *International Building Code*. ICC, Washington, DC.

5. Jirsa, J. O., Sozen, M. A., and Siess, C. P. 1969. Pattern loadings on reinforced concrete floor slabs. *Proceedings, ASCE* 95(ST6):1117-1137.
6. Corley, W. G., Sozen, M. A., and Siess, C. P. 1961. *Equivalent-frame Analysis for Reinforced Concrete Slabs*. Structural Research Series No. 218. Civil Engineering Studies, University of Illinois, Urbana, IL, 166 pp.
7. Jirsa, J. O., Sozen, M. A., and Siess, C. P. 1963. *Effects of Pattern Loadings on Reinforced Concrete Floor Slabs*. Structural Research Series No. 269. Civil Engineering Studies, University of Illinois, Urbana, IL.
8. Corley, W. G., and Jirsa, J. O. 1970. Equivalent frame analysis for slab design. *ACI Journal* 67(11):875-884.
9. Gamble, W. L. 1972. Moments in beam supported slabs. *ACI Journal* 69(3):149-157.
10. Portland Cement Association (PCA). 1958. *Handbook of Frame Constants*, EB034. PCA, Skokie, IL, 34 pp.
11. Vanderbilt, M. D., and Corley, W. G. 1983. Frame analysis of concrete buildings. *Concrete International: Design and Construction* 5(12):33-43.
12. Hwang, S. -J., and Moehle, J. P. 2000. Models for laterally loaded slab–column frames. *ACI Structural Journal* 97(2):345-352.
13. Moehle, J. P. 1988. Strength of slab–column edge connections. *ACI Structural Journal* 85(1):89-98.
14. Joint ACI-ASCE Committee 352. 1989 (Reapproved 2004). *Recommendations for Design of Slab–column Connections in Monolithic Reinforced Concrete Structures*, ACI 352.1. ACI, Farmington Hills, MI.
15. Mitchell, D., and Cook, W. D. 1984 Preventing progressive collapse of slab structures. *Journal of Structural Engineering* 110(7):1513-1532.
16. Joint ACI-ASCE Committee 326 (now 426). 1962 Shear and diagonal tension. *ACI Journal* 59(1):1-30; (2):277-334; (3):352-396.
17. Joint ACI-ASCE Committee 426. 1974 The shear strength of reinforced concrete members-slabs. *Proceedings*, ASCE 100(ST8):1543-1591.
18. Vanderbilt M. D. 1972 Shear strength of continuous plates. *Journal of the Structural Division, ASCE* 98(ST5):961-973.
19. Hawkins, N. M. 1974 *Shear Strength of Slabs with Shear Reinforcement, Shear in Reinforced Concrete*, Sp-42, Vol. 2. American Concrete Institute, Farmington Hills, MI.
20. Broms, C. E. 1990 Shear reinforcement for deflection ductility of flat plates. *ACI Structural Journal* 87(6):696-705.
21. Yamada, T., Nanni, A., and Endo, K. 1991. Punching shear resistance of flat slabs: influence of reinforcement type and ratio. *ACI Structural Journal* 88(4):555-563.
22. Hawkins, N. M., Mitchell, D., and Hannah, S. N. 1975 The effects of shear reinforcement on reversed cyclic loading behavior of flat plate structures. *Canadian Journal of Civil Engineering* 2:572-582.
23. Joint ACI-ASCE Committee 421. 1999 (Reapproved 2006). *Shear Reinforcement for Slabs*. American Concrete Institute, Farmington Hills, MI.
24. Joint ACI-ASCE Committee 421. 2008 *Guide to Shear Reinforcement for Slabs*, ACI 421.1. American Concrete Institute, Farmington Hills, MI.

Problems

7.1. Determine the slab thickness to satisfy the deflection criteria of ACI 9.5.3 for a flat-plate floor system with six bays in the north-south direction and five in the east-west direction. The center-to-center span lengths are 21 ft 9 in in the north-south direction and 18 ft 2 in in the east-west direction. All interior columns are 26 × 26 in, and all perimeter columns are 24 × 18 in where the 24-in dimension is parallel to the east-west direction. Assume Grade 60 reinforcement.

7.2.   Determine the slab thickness to satisfy the deflection criteria of ACI 9.5.3 for a two-way concrete floor system with eight bays in the north-south direction and six in the east-west direction. The center-to-center span lengths are 22 ft 2 in in the north-south direction and 19 ft 8 in in the east-west direction. All interior columns are 22 × 22 in, and all perimeter columns are 18 × 18 in. Perimeter beams are 18 in wide and 24 in deep, and interior beams are 22 in wide and 24 in deep. Beams are centered on all of the column lines. Preliminary calculations indicate that a 6-in-thick slab can be utilized. Assume Grade 60 reinforcement.

7.3.   Given the two-way slab system in Problem 7.2, determine bending moments at the critical sections in an interior design strip in the north-south direction. Assume a 6-in-thick slab, a superimposed service dead load of 10 psf, and a service live load of 80 psf. Also assume that the Direct Design Method can be used to determine bending moments.

7.4.   Given the two-way slab system in Problem 7.3, determine bending moments at the critical sections in an edge design strip in the east-west direction.

7.5.   Determine the maximum factored one-way shear force at the critical section of an interior column for the interior panel of a flat-plate floor system that has center-to-center span lengths of 21 and 19 ft. Columns are 22 × 18 in with the 22-in side parallel to the 21-ft span. Assume a slab thickness of 8 in, a superimposed service dead load of 15 psf, and a live load of 100 psf. Also assume normal-weight concrete with a compressive strength of 4,000 psi and Grade 60 reinforcement.

7.6.   Given the two-way system in Problem 7.5, determine the factored two-way shear force at the critical section of an interior column.

7.7.   Given the two-way slab system in Problem 7.5, determine the allowable one- and two-way shear strength at an interior column.

7.8.   Determine the required flexural reinforcement at the critical sections in the end bay of a flat-plate floor system. The end bay is square with center-to-center span lengths of 22 ft. The edge column is 16 × 16 in, and the interior column is 20 × 20 in. Assume an 8-in-thick slab (lightweight with a compressive strength of 4,000 psi) and Grade 60 reinforcement. Also assume a superimposed service dead load of 20 psf and a service live load of 50 psf. The Direct Design method can be used to determine bending moments.

7.9.   Given the flat-plate system in Example 7.8, design and detail an edge design strip in the east-west direction.

7.10.   Given the flat-slab system in Example 7.9, design and detail an edge design strip in the east-west direction.

7.11.   Given the flat-plate system in Example 7.15, determine the following at the edge column: (a) drop panel dimensions, (b) shear cap dimensions, (c) size and spacing of closed stirrups, and (d) size and spacing of headed stud shear reinforcement. Assume a service live load of 125 psf.

CHAPTER 8

Columns

8.1 Introduction

Columns are structural elements that support axial loads from the roof and floors. They are usually oriented vertically in a building, and a typical cross-section is rectangular or circular. However, virtually any orientation and shape can be provided as needed.

ACI 2.2 provides the following definition of a column:

> Member with a ratio of height-to-least lateral dimension exceeding 3 used primarily to support axial compressive load. For a tapered member, the least lateral dimension is the average of the top and bottom dimensions of the smaller side.

In a typical column stack, the loads are collected at each floor level and are transmitted to the column below; the columns in the lowest level of the building transfer the loads from above to the foundations. Columns may also be supported by beams or walls at any level above ground. Transfer beams are members that support one or more columns from above, and are generally used at locations where open, column-free space (like a lobby) is needed below. These beams are supported by one or more columns or walls.

In addition to axial loads, columns may be subjected to bending moments. Gravity loads can cause unbalanced moments at column–beam or column–slab joints, especially at the perimeter of a structure. These unbalanced moments are transmitted from the roof or floor system to the columns. Columns that are in a frame that is part of the lateral force–resisting system of a structure must be designed to resist axial loads, bending moments, and shear forces due to the combined effects of gravity and lateral forces caused by wind or earthquakes.

The axial loads on a column are usually compressive. As such, columns are often referred to as *compression members*. However, if the effects from lateral loads are large enough, it is possible for a column to be subjected to a net tensile axial load under one or more load combinations. Columns are also described as *members subjected to combined axial load and bending*, as are similar members such as walls.

Pedestals are also members that support primarily axial compressive loads. A pedestal is defined as a member having a ratio of height to least horizontal lateral dimension equal to or less than 3.

In addition to sectional and material properties, the design strength of a column depends on the type of lateral reinforcement in the member. The primary types are ties and spirals. Tied columns have rectangular or circular ties that enclose the longitudinal reinforcement in the column. Spiral reinforcement consists of an essentially continuously wound reinforcing bar that is in the form of a cylindrical helix that encloses the longitudinal reinforcement at a specified pitch. It is usually used in circular columns and provides a higher degree of lateral confinement than that provided by ties.

Design strength also depends on the slenderness of a column. Use of high-strength concrete has resulted in smaller column cross-sectional dimensions, which increases the likelihood of slenderness (secondary) effects being included in the design of the column.

Design requirements are presented in this chapter for columns subjected to axial loads, combined flexure and axial loads, and shear. Methods are presented on the following:

1. Sizing the cross-section

2. Determining the required amount of reinforcement

3. Detailing the reinforcement

The effects of slenderness, which depend on the geometric properties (cross-sectional dimensions and length) and end-support conditions, are also covered.

8.2 Preliminary Column Sizing

Preliminary sizes of typical columns in a structure are needed for a variety of reasons, including frame analysis and initial cost estimation.

In the early stages of design, it is common practice to obtain preliminary column sizes, utilizing axial gravity loads only. The axial load in a column at a particular floor level is obtained by multiplying the dead and live loads at that level by the area that is tributary to the column. The tributary area depends on the sizes of the bay, which are frequently dictated by architectural and functional requirements. The type of floor system also influences column spacing. For example, a flat-plate floor system usually requires columns to be spaced closer than those supporting a beam-supported floor system (see Section 7.2). Live loads at the roof and at the floor levels can be reduced in accordance with the applicable provisions of the governing building code (see Section 3.2). In general, larger tributary areas translate into greater permitted reductions in live load.

The total axial gravity loads in the first-story columns are calculated by summing the loads at each floor level over the height of the building.

Factored axial loads P_u are determined by the load combinations given in ACI 9.2. The load factors can be applied at each floor level, and the summation can be performed using factored loads. Equivalently, the service loads can be summed over the column stack, and the load factors can be applied later.

A preliminary column size is obtained by setting the total factored axial load P_u equal to the design axial load strength $\phi P_{n,max}$ given by ACI Eq. (10-1) or (10-2). These equations are applicable to members with spiral reinforcement and tie reinforcement, respectively (see Section 5.5). The appropriate equation is subsequently solved for the gross area of the column A_g, assuming practical values for the total area of longitudinal reinforcement A_{st}, the compressive strength of the concrete f_c', and the yield strength of the reinforcement f_y.

It is evident from the preceding discussion that a preliminary column size is often obtained by assuming that the effects from bending moments are relatively small and that the column section is nonslender; that is, secondary effects are negligible. The first of these assumptions is usually valid for columns that are not part of the lateral force–resisting system. However, unbalanced gravity load moments due to unequal spans are just one example where gravity loads cause bending moments on a column. Methods to determine when slenderness effects need to be considered are given in Section 8.5. For many columns, slenderness effects are not an issue.

A preliminary column size should be determined using a low percentage of longitudinal reinforcement. This allows reinforcement to be added in the final design stage without having to change the column dimensions. Additional longitudinal reinforcement may also be required to account for the effects of axial load in combination with bending moments and/or slenderness effects. Columns that have longitudinal reinforcement ratios A_{st}/A_g in the range of 1% to 2% are usually the most economical because concrete carries axial compressive loads more cost-effectively than reinforcing steel. Generally, it is usually more economical to use larger column sizes with less longitudinal reinforcement.

The design chart shown in Fig. 8.1 can be used to obtain a preliminary size of a nonslender, tied column with Grade 60 longitudinal reinforcement that is loaded at an

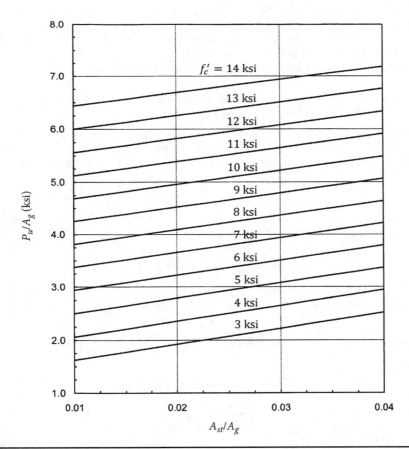

FIGURE 8.1 Design chart for nonslender column with tie reinforcement.

eccentricity of no more than 10% of the column dimension (i.e., the bending moments are zero or relatively small). The information provided in the chart was derived for various concrete compressive strengths f'_c and longitudinal reinforcement ratios A_{st}/A_g by setting the factored axial load P_u equal to the design axial load strength $\phi P_{n,max}$ given in ACI Eq. (10-2). Similar design charts can be generated for other column sizes and shapes and other material strengths.

The dimensions of a column can be influenced by architectural and functional requirements. One or both dimensions of a rectangular column may be limited, which could result in a column that is slender.

Columns must be sized not only for strength but also for constructability. To ensure proper concrete placement and consolidation, column dimensions and bar sizes must be selected to minimize reinforcement congestion, especially at beam–column or slab–column joints. A smaller number of larger bars usually improve constructability. Section 8.7 contains information and design aids that facilitate the selection of longitudinal bars that adequately fit within a column section.

Significant cost savings are often realized where column forms can be reused from story to story. In low-rise buildings, it is generally more economical to use the same column size over the full height of the building and to vary the amount of longitudinal reinforcement as required. In taller buildings, the size of the column should change over the height, but the number of changes should be kept to a minimum. The same column size can be used over a number of stories by judiciously varying the amount of longitudinal reinforcement and the strength of the concrete. In any building, it is economically unsound to vary column size to suit the load at each story level.

Example 8.1 Determine a preliminary column size for a tied reinforced concrete column that is subjected to a factored axial load of 1,200 kips. Assume $f'_c = 7,000$ psi and Grade 60 reinforcement.

Solution Because no additional information is provided, initially assume that bending moments and second-order effects are negligible. As such, ACI Eq. (10-2) can be used to determine a preliminary column size by setting the factored axial load P_u equal to the design axial load strength $\phi P_{n,max}$:

$$P_u = \phi P_{n,max} = \phi 0.80[0.85 f'_c(A_g - A_{st}) + f_y A_{st}]$$

This equation can be rewritten in the following form:

$$\frac{P_u}{A_g} = \phi 0.80 \left[0.85 f'_c \left(1 - \frac{A_{st}}{A_g} \right) + f_y \left(\frac{A_{st}}{A_g} \right) \right]$$

Substituting the known quantities into this equation and rearranging terms results in

$$\frac{1,200}{A_g} = (0.65 \times 0.80) \left[(0.85 \times 7) \left(1 - \frac{A_{st}}{A_g} \right) + 60 \left(\frac{A_{st}}{A_g} \right) \right]$$

$$A_g = \frac{1,200}{3.09[1 - (A_{st}/A_g)] + 31.2(A_{st}/A_g)}$$

Columns with a longitudinal reinforcement ratio A_{st}/A_g between 1% and 2% are usually the most economical. Table 8.1 provides a summary of required column areas A_g for 1%, 1.5%, and 2% reinforcement ratios.

A_{st}/A_g	A_g (in.^{2})
0.010	356.0
0.015	341.7
0.020	328.6

TABLE 8.1 Preliminary Size of the Column Given in Example 8.1

The required column area can also be determined from Fig. 8.1. For a longitudinal reinforcement ratio of 1% and $f_c' = 7$ ksi, obtain from the figure a value of P_u/A_g equal to approximately 3.375. Thus, $A_g = 1{,}200/3.375 = 355.6$ in^2, which essentially matches the value given in Table 8.1.

A 20 × 20 in cross-section with a gross area of 400 in^2 is satisfactory for all three reinforcement ratios. Column dimensions may be dictated by architectural or other requirements.

Once a preliminary column section has been established, it can be used in a structural model, and the analysis will yield refined values of axial loads, bending moments, and shear forces on the member. Prior to final design, it must be determined if the effects of slenderness need to be considered in the design of the column.

8.3 Analysis and Design Methods

8.3.1 Analysis Methods

Methods of analysis for reinforced concrete structures are presented in ACI 8.3 and are summarized in Section 3.3 of this book. A discussion of the provisions relevant to columns follows.

Frames are permitted to be analyzed by a number of different methods. A first-order frame analysis is an elastic analysis that does not include the internal force effects resulting from the overall lateral deflection of the frame (i.e., it is assumed that secondary effects are negligible). In such cases, it is permitted to fix the far ends of the column when computing gravity load moments. Bending moments at a beam–column or slab–column joint are distributed to the columns above and below the joint in accordance with the relative column stiffnesses and the restraint conditions at the ends of the column. The stiffness of a column is proportional to the modulus of elasticity of the concrete and the moment of inertia of the cross-section and is inversely proportional to the length of the column.

A second-order analysis considers the effects of deflections on geometry and axial flexibility. Second-order effects need to be considered in the design of certain columns in order to obtain the correct amplified moments for design. Column slenderness and its effects on design moments are covered in Section 8.5.

8.3.2 Design Methods

Regardless of the method of analysis, columns must be designed for the most critical combinations of factored axial loads and bending moments due to the applied loads. The effects of unbalanced floor or roof loads as well as any eccentric loading must be considered in the design of all columns.

Derivation of the nominal axial strength and the flexural strength of compression members is given in Section 5.7 and is based on the general principles of the strength design method. Nominal strengths are determined by a strain compatibility analysis for given strain distribution. Design strengths are obtained by multiplying the nominal strengths by the strength reduction factor defined in ACI 9.3, which depends on the magnitude of the strain in the reinforcing bars that are farthest from the extreme compression face of the section.

The following equations must be satisfied in the design of any column:

$$\phi P_n \geq P_u \tag{8.1}$$

$$\phi M_n \geq M_u \tag{8.2}$$

The factored axial load and bending moment acting on a reinforced concrete section must be equal to or less than the corresponding design value in order for the section to satisfy strength requirements.

Interaction diagrams are usually used to determine the adequacy of a reinforced concrete column subjected to axial loads and bending moments. As noted in Section 5.7, these diagrams are a collection of design strength values that are determined using strain compatibility analyses. The cross-sectional dimensions of the column, the amount and distribution of longitudinal reinforcement in the section, the compression strength of the concrete, and the yield strength of the longitudinal reinforcement are all used in the construction of interaction diagrams.

Section 8.4 contains methods to construct interaction diagrams for rectangular and circular reinforced concrete sections. Slenderness effects and their impact on the design of columns are covered in Section 8.5.

8.4 Interaction Diagrams

8.4.1 Overview

For concrete members subjected to combined flexure and axial load, it is convenient to construct interaction diagrams. In general terms, an interaction diagram shows the relationship between axial load and bending moment at failure. The results from strain compatibility analyses for a number of strain distributions are summarized in an interaction diagram.

An interaction diagram is commonly used in establishing the adequacy of a section that is subjected to a combination of factored axial loads and bending moments, which are determined by the load combinations in ACI 9.2. Methods to construct interactions diagrams for rectangular and circular sections are presented next.

8.4.2 Rectangular Sections

Nominal Strength

The general principles and assumptions of the strength design method can be applied to reinforced concrete sections subjected to axial compressive load and bending (see Section 5.7). Figure 5.39 contains a step-by-step procedure that can be used to determine the nominal axial strength P_n and the nominal flexural strength M_n for a particular strain distribution in a rectangular section. Failure of the section is assumed to occur when the applied axial load and bending moment attain these values.

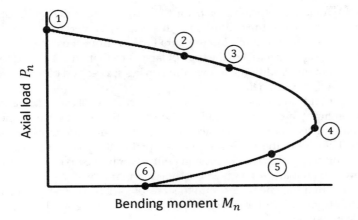

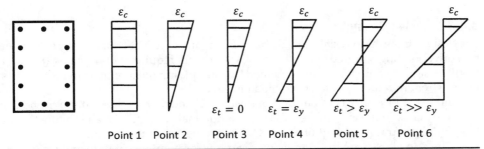

Point 1   Point 2   Point 3   Point 4   Point 5   Point 6

FIGURE 8.2   Strain distributions related to key points on an interaction diagram.

A number of different strain distributions that correspond to key points on an interaction diagram are illustrated in Fig. 8.2.

Point 1 corresponds to the case of pure compression. The strain over the entire depth of the section in this case is equal to the ultimate strain in the concrete ($\varepsilon_c = 0.0030$). The strength of the column under pure axial load is equal to P_o [see Eq. (5.35) in Section 5.5]:

$$P_o = 0.85 f_c'(A_g - A_{st}) + f_y A_{st} \tag{8.3}$$

Recall that the Code reduces this axial load to account for minimum eccentricities, based on the type of lateral reinforcement in the section [see Eqs. (5.36) and (5.37) in Section 5.5 for members with spiral reinforcement and tie reinforcement, respectively].

Point 2 corresponds to crushing of the concrete at the compression face of the section and zero stress at the other face. Because the tensile strength of the concrete is taken as zero, which is the fourth design assumption of the strength design method (see Section 5.2), this point represents the onset of cracking at the face of the section farthest from the compression face. All points on the interaction diagram that fall below this point represent cases in which the section is partially cracked. At this point, and at other points similar to this one, the column fails as soon as the maximum compressive strain reaches

0.0030. There are no large deformations prior to failure because the reinforcement has not yielded in tension. Thus, the column fails in a brittle manner.

At Point 3, the strain in the reinforcing bars farthest from the compression face is equal to zero. This point is related to the type of lap splice that must be used for the longitudinal reinforcement. Additional information on lap splices and other reinforcement details are provided in Section 8.7.

Point 4 corresponds to balanced failure in which crushing of the concrete and yielding of the reinforcing steel occur simultaneously: The compressive strain in the concrete is equal to 0.0030 at the compression face of the section, and the tensile strain in the reinforcing steel closest to the tension face ε_t is equal to the yield strain ε_y. This point also represents the change from compression failures for higher axial loads and tension failures for lower axial loads for a given bending moment.

At Point 5, the reinforcement closest to the tension face has been strained to several times the yield strain before the concrete reaches its crushing strain of 0.0030. This implies ductile behavior.

Point 6 represents the case of pure bending. Under pure bending, the strain ε_t in the reinforcing bars closest to the tension face is many times greater than ε_y.

8.4.3 Design Strength

The design strengths for axial load and bending moment must be equal to or greater than the corresponding required strengths to satisfy strength requirements [see Eqs. (8.1) and (8.2)]. Design strength values are obtained by multiplying the nominal strength values by the strength reduction factor ϕ defined in ACI 9.3.

The value of ϕ that is to be used in design depends on the magnitude of the net tensile strain ε_t in the extreme tension steel at nominal strength, which is a function of the particular strain distribution (see ACI 9.3 and Fig. 4.2 of this book).

Shown in Fig. 8.3 are the interaction diagrams for nominal strength and for design strength for a typical tied, rectangular reinforced concrete column with a symmetrical distribution of longitudinal reinforcement.

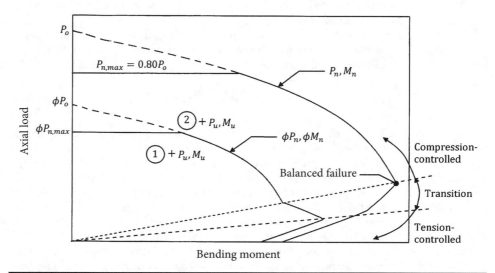

FIGURE 8.3 Nominal and design strength interaction diagrams.

The maximum allowable axial load $P_{n,max}$ for a tied column is shown in the figure and is equal to 80% of the strength of the column under pure axial load P_o where P_o is determined by Eq. (8.3).

Also identified in Fig. 8.3 are the portions of the diagrams that correspond to compression-controlled sections, tension-controlled sections, and sections in the transition region.

Compression-controlled sections are those with axial load–bending moment combinations above and to the left of the balanced failure point. The strength reduction factor ϕ for compression-controlled sections with other than spiral reinforcement is 0.65 (ACI 9.3.2.2). Thus, the design strength interaction diagram in this region has values that are 65% of the values on the nominal diagram. For axial loads at or near the nominal axial compressive strength, it is possible for the depth to the neutral axis and the depth of the stress block to be greater than the depth of the section. In such cases, the depth of the stress block a should be taken as the depth of the section, and the strain compatibility analysis should be performed using that value.

Tensioned-controlled sections are those with axial load–bending moment combinations in the lower portion of the interaction diagram indicated in Fig. 8.3. In accordance with ACI 9.3.2.1, the strength reduction factor for these sections is equal to 0.9.

Between the compression- and tension-controlled sections are sections in the transition region. As indicated previously, the ϕ-factor varies linearly in this region (see ACI 9.3.2.2 and Fig. 4.2 of this book).

For purposes of design, factored axial load–bending moment combinations that fall on or within the boundaries of the design strength interaction diagram can be safely carried by the column section For example, the column section is adequate for the factored axial load–bending moment combination denoted by Point 1 in Fig. 8.3. No modifications need to be made to the concrete column in such cases.

In contrast, the column dimensions, amount of longitudinal steel, and/or compressive strength of the concrete must be increased when factored axial load–bending moment combinations fall outside the boundaries of the design strength interaction diagram (see Point 2 in Fig. 8.3).

Example 5.10 illustrates the determination of the nominal strengths P_n and M_n corresponding to balanced failure for a rectangular reinforced concrete column. Other key points on the interaction diagram for that column are determined in the following example.

Example 8.2   Determine the following points on the design strength interaction diagram for the rectangular column shown in Fig. 8.4:

1. Maximum axial load capacity

2. Crushing of the concrete at the compression face and zero stress at the other face

3. Strain in the reinforcing bars farthest from the compression face ε_t equal to zero

4. Balanced failure

5. Strain in the reinforcing bars farthest from the compression face ε_t equal to $2\varepsilon_y$

6. Pure bending

Assume that the extreme compression fiber occurs at the top of the section and that ties are utilized as the lateral reinforcement. Also assume normal-weight concrete with $f_c' = 7,000$ psi and Grade 60 reinforcement ($f_y = 60,000$ psi).

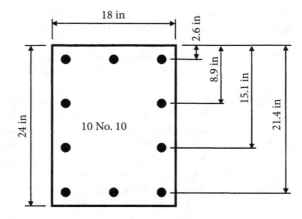

FIGURE 8.4 The rectangular column given in Example 8.2.

Solution

Point 1—Maximum Axial Load Capacity

The design maximum axial load capacity $\phi P_{n,max}$ is determined by ACI Eq. (10-2) for a tied column [see Eq. (5.39)]:

$$\phi P_{n,max} = 0.80\phi[0.85f_c'(A_g - A_{st}) + f_y A_{st}]$$
$$= 0.80 \times 0.65[0.85 \times 7 \times (432 - 12.7) + (60 \times 12.7)] = 1{,}694 \text{ kips}$$

Point 2—Crushing of the Concrete at the Compression Face and Zero Stress at the Other Face

The strain in the reinforcing steel farthest from the compression face can be determined from similar triangles (see Figs. 8.2 and 8.4):

$$\frac{0.003}{24} = \frac{\varepsilon_t}{24 - 21.4}$$

$$\varepsilon_t = \frac{0.0030 \times 2.6}{24} = 0.00033$$

The flowchart shown in Fig. 5.39 is utilized to determine P_n and M_n for this strain distribution.

Step 1: Check the minimum and maximum longitudinal reinforcement limits. The minimum and maximum amounts of longitudinal reinforcement permitted in a compression member are specified in ACI 10.9.1:

$$\text{Minimum } A_{st} = 0.01A_g = 0.01 \times 18 \times 24 = 4.32 \text{ in}^2$$
$$\text{Maximum } A_{st} = 0.08A_g = 0.08 \times 18 \times 24 = 34.6 \text{ in}^2$$

The provided area of longitudinal reinforcement $A_{st} = 10 \times 1.27 = 12.7 \text{ in}^2$ falls between the minimum and maximum limits.

Step 2: Determine the neutral axis depth c. In this case, the neutral axis depth is equal to the depth of the section, which is 24 in.

Step 3: Determine β_1.

$$\beta_1 = 1.05 - 0.00005f_c' = 1.05 - (0.00005 \times 7{,}000) = 0.70 \quad \text{for} \quad f_c' = 7{,}000 \text{ psi}$$

(see ACI 10.2.7.3 and Section 5.2 of this book)

Step 4: Determine the depth of the equivalent stress block *a*.

$$a = \beta_1 c = 0.70 \times 24.0 = 16.8 \text{ in}$$

Step 5: Determine *C*. The concrete compression resultant force *C* is determined by Eq. (5.42):

$$C = 0.85 f'_c a b = 0.85 \times 7 \times 16.8 \times 18 = 1{,}799.3 \text{ kips}$$

Step 6: Determine ε_{si}. The strain in the reinforcement ε_{si} at the various layers is determined by similar triangles where compression strains are positive (see Fig. 8.2):

- Layer 1 ($d_1 = 2.6$ in):

$$\varepsilon_{s1} = \frac{0.0030(24 - 2.6)}{24} = 0.0027$$

- Layer 2 ($d_2 = 8.9$ in):

$$\varepsilon_{s2} = \frac{0.0030(24 - 8.9)}{24} = 0.0019$$

- Layer 3 ($d_3 = 15.1$ in):

$$\varepsilon_{s3} = \frac{0.0030(24 - 15.1)}{24} = 0.0011$$

- Layer 4 ($d_4 = 21.4$ in):

$$\varepsilon_{s4} = \frac{0.0030(24 - 21.4)}{24} = 0.00033 \text{ (checks)}$$

It is evident that all of the layers of reinforcement are in compression. Also, the layer of reinforcement closest to the extreme compression fiber yields (i.e., $\varepsilon_{s1} > \varepsilon_y = 0.0020$).

Step 7: Determine f_{si}. The stress in the reinforcement f_{si} at the various layers is determined by multiplying ε_{si} by the modulus of elasticity of the steel E_s:

- Layer 1: $f_{s1} = 0.0027 \times 29{,}000 = 78.3$ ksi > 60 ksi; use $f_{s1} = 60$ ksi
- Layer 2: $f_{s2} = 0.0019 \times 29{,}000 = 55.1$ ksi
- Layer 3: $f_{s3} = 0.0011 \times 29{,}000 = 31.9$ ksi
- Layer 4: $f_{s4} = 0.00033 \times 29{,}000 = 9.6$ ksi

Step 8: Determine F_{si}. The force in the reinforcement F_{si} at the various layers is determined by Eq. (5.44) or (5.45), which depends on the location of the steel layer:

- Layer 1 ($d_1 = 2.6$ in $< a = 16.8$ in): $F_{s1} = [60 - (0.85 \times 7)] \times 3 \times 1.27 = 205.9$ kips
- Layer 2 ($d_2 = 8.9$ in $< a = 16.8$ in): $F_{s2} = [55.1 - (0.85 \times 7)] \times 2 \times 1.27 = 124.8$ kips
- Layer 3 ($d_3 = 15.1$ in $< a = 16.8$ in): $F_{s3} = [31.9 - (0.85 \times 7)] \times 2 \times 1.27 = 65.9$ kips
- Layer 4: $F_{s4} = 9.6 \times 3 \times 1.27 = 36.6$ kips

Note that the compression steel in the top three layers fall within the depth of the equivalent stress block; thus, Eq. (5.45) is used to determine the forces in the reinforcement in those layers.

Step 9: Determine P_n and M_n. The nominal axial strength P_n and nominal flexural strength M_n of the section are determined by Eqs. (5.46) and (5.47), respectively:

$$P_n = C + \sum F_{si} = 1{,}799.3 + (205.9 + 124.8 + 65.9 + 36.6) = 2{,}232.5 \text{ kips}$$

$$\begin{aligned} M_n &= 0.5C(h-a) + \sum F_{si}(0.5h - d_i) \\ &= [0.5 \times 1{,}799.3 \times (24 - 16.8)] + [205.9(12 - 2.6) + 124.8(12 - 8.9) \\ &\quad + 65.9(12 - 15.1) + 36.6(12 - 21.4)] \\ &= 6{,}477.5 + 1{,}774.0 = 8{,}251.5 \text{ in kips} = 687.6 \text{ ft kips} \end{aligned}$$

The design axial load and bending moment are obtained by multiplying P_n and M_n by the strength reduction factor ϕ. Because $\varepsilon_t = \varepsilon_{s4} = 0.00033$, which is less than the compression-controlled strain limit of 0.0020 for sections with Grade 60 reinforcement (ACI 10.3.3), the section is compression-controlled and $\phi = 0.65$ (see ACI 9.3.2.2 and Fig. 4.2 of this book).

Therefore,

$$\phi P_n = 0.65 \times 2{,}232.5 = 1{,}451.1 \text{ kips}$$

$$\phi M_n = 0.65 \times 687.6 = 446.9 \text{ ft kips}$$

Point 3—Strain in the Reinforcing Bars Farthest from the Compression Face Is Equal to Zero

Step 1: Check the minimum and maximum longitudinal reinforcement limits. The provided area of longitudinal reinforcement falls between the minimum and maximum limits (see the calculations under Point 2).

Step 2: Determine the neutral axis depth c. In this case, the neutral axis depth is equal to the depth from the compression face to the reinforcing bars farthest from the compression face, which is 21.4 in.

Step 3: Determine β_1.

$$\beta_1 = 1.05 - 0.00005 f_c' = 1.05 - (0.00005 \times 7{,}000) = 0.70 \quad \text{for} \quad f_c' = 7{,}000 \text{ psi}$$

(see ACI 10.2.7.3 and Section 5.2 of this book)

Step 4: Determine the depth of the equivalent stress block a.

$$a = \beta_1 c = 0.70 \times 21.4 = 15.0 \text{ in}$$

Step 5: Determine C. The concrete compression resultant force C is determined by Eq. (5.42):

$$C = 0.85 f_c' a b = 0.85 \times 7 \times 15.0 \times 18 = 1{,}606.5 \text{ kips}$$

Step 6: Determine ε_{si}. The strain in the reinforcement ε_{si} at the various layers is determined by similar triangles where compression strains are positive (see Fig. 8.2):

- Layer 1 ($d_1 = 2.6$ in):

$$\varepsilon_{s1} = \frac{0.0030(21.4 - 2.6)}{21.4} = 0.0026$$

- Layer 2 ($d_2 = 8.9$ in):

$$\varepsilon_{s2} = \frac{0.0030(21.4 - 8.9)}{21.4} = 0.0018$$

- Layer 3 ($d_3 = 15.1$ in):

$$\varepsilon_{s3} = \frac{0.0030(21.4 - 15.1)}{21.4} = 0.0009$$

- Layer 4 ($d_4 = 21.4$ in):

$$\varepsilon_{s4} = \frac{0.0030(21.4 - 21.4)}{21.4} = 0 \text{ (checks)}$$

It is evident that all of the layers of reinforcement are in compression, except for layer 4 where it was given that the strain is zero. Also, the layer of reinforcement closest to the extreme compression fiber yields (i.e., $\varepsilon_{s1} > \varepsilon_y = 0.0020$).

Step 7: Determine f_{si}. The stress in the reinforcement f_{si} at the various layers is determined by multiplying ε_{si} by the modulus of elasticity of the steel E_s:

- Layer 1: $f_{s1} = 0.0026 \times 29,000 = 75.4$ ksi > 60 ksi; use $f_{s1} = 60$ ksi
- Layer 2: $f_{s2} = 0.0018 \times 29,000 = 52.2$ ksi
- Layer 3: $f_{s3} = 0.0009 \times 29,000 = 26.1$ ksi
- Layer 4: $f_{s4} = 0$ ksi

Step 8: Determine F_{si}. The force in the reinforcement F_{si} at the various layers is determined by Eq. (5.44) or (5.45), which depends on the location of the steel layer:

- Layer 1 ($d_1 = 2.6$ in $< a = 15.0$ in): $F_{s1} = [60 - (0.85 \times 7)] \times 3 \times 1.27 = 205.9$ kips
- Layer 2 ($d_2 = 8.9$ in $< a = 15.0$ in): $F_{s2} = [52.2 - (0.85 \times 7)] \times 2 \times 1.27 = 117.5$ kips
- Layer 3: $F_{s3} = 26.1 \times 2 \times 1.27 = 66.3$ kips
- Layer 4: $F_{s4} = 0$ kips

Note that the compression steel in the top two layers fall within the depth of the equivalent stress block; thus, Eq. (5.45) is used to determine the forces in the reinforcement in those layers.

Step 9: Determine P_n and M_n. The nominal axial strength P_n and nominal flexural strength M_n of the section are determined by Eqs. (5.46) and (5.47), respectively:

$$P_n = C + \sum F_{si} = 1,606.5 + (205.9 + 1,17.5 + 66.3 + 0) = 1,996.2 \text{ kips}$$

$$M_n = 0.5C(h - a) + \sum F_{si}(0.5h - d_i)$$
$$= [0.5 \times 1,606.5 \times (24 - 15.0)] + [205.9(12 - 2.6) + 117.5(12 - 8.9)$$
$$+ 66.3(12 - 15.1) + 0]$$
$$= 7,229.3 + 2,094.2 = 9,323.5 \text{ in kips} = 777.0 \text{ ft kips}$$

The design axial load and bending moment are obtained by multiplying P_n and M_n by the strength reduction factor ϕ. Because $\varepsilon_t = \varepsilon_{s4} = 0$, which is less than the compression-controlled strain limit of 0.0020 for sections with Grade 60 reinforcement (ACI 10.3.3), the section is compression-controlled and $\phi = 0.65$ (see ACI 9.3.2.2 and Fig. 4.2 of this book).

Therefore,

$$\phi P_n = 0.65 \times 1,996.2 = 1,297.5 \text{ kips}$$
$$\phi M_n = 0.65 \times 777.0 = 505.1 \text{ ft kips}$$

Point 4—Balanced Failure

The design strength values for this strain distribution are determined in Example 5.10:

$$\phi P_n = 0.65 \times 955.6 = 621.1 \text{ kips}$$

$$\phi M_n = 0.65 \times 965.5 = 627.6 \text{ ft kips}$$

At balanced failure, $\varepsilon_t = \varepsilon_{s4} = 0.0020$, which is the limit for compression-controlled sections.

Point 5—Strain in the Reinforcing Bars Farthest from the Compression Face Is Equal to Two Times the Yield Strength

Step 1: Check the minimum and maximum longitudinal reinforcement limits. The provided area of longitudinal reinforcement falls between the minimum and maximum limits (see the calculations under Point 2).

Step 2: Determine the neutral axis depth c. The strain in reinforcing bars farthest from the compression face is given as two times the yield strain: $\varepsilon_{s4} = 2\varepsilon_y = 2 \times 0.0020 = 0.0040$.

The neutral axis depth is determined by Eq. (5.41):

$$c = \frac{0.0030 d_t}{\varepsilon_t + 0.0030} = \frac{0.0030 \times 21.4}{0.0040 + 0.0030} = 9.2 \text{ in}$$

Step 3: Determine β_1.

$$\beta_1 = 1.05 - 0.00005 f_c' = 1.05 - (0.00005 \times 7{,}000) = 0.70 \quad \text{for} \quad f_c' = 7{,}000 \text{ psi}$$

(see ACI 10.2.7.3 and Section 5.2 of this book)

Step 4: Determine the depth of the equivalent stress block a.

$$a = \beta_1 c = 0.70 \times 9.2 = 6.4 \text{ in}$$

Step 5: Determine C. The concrete compression resultant force C is determined by Eq. (5.42):

$$C = 0.85 f_c' a b = 0.85 \times 7 \times 6.4 \times 18 = 685.4 \text{ kips}$$

Step 6: Determine ε_{si}. The strain in the reinforcement ε_{si} at the various layers is determined by Eq. (5.43) where compression strains are positive:

• Layer 1 ($d_1 = 2.6$ in):

$$\varepsilon_{s1} = \frac{0.0030(9.2 - 2.6)}{9.2} = 0.0022$$

• Layer 2 ($d_2 = 8.9$ in):

$$\varepsilon_{s2} = \frac{0.0030(9.2 - 8.9)}{9.2} = 0.0001$$

• Layer 3 ($d_3 = 15.1$ in):

$$\varepsilon_{s3} = \frac{0.0030(9.2 - 15.1)}{9.2} = -0.0019$$

• Layer 4 ($d_4 = 21.4$ in):

$$\varepsilon_{s4} = \frac{0.0030(9.2 - 21.4)}{9.2} = -0.0040 \text{ (checks)}$$

It is evident that the top two layers of reinforcement are in compression and that the bottom two layers are in tension. Also, the layer of reinforcement closest to the extreme compression fiber and the layer farthest from the extreme compression fiber yield (i.e., ε_{s1} and $\varepsilon_{s4} > \varepsilon_y = 0.0020$).

Step 7: Determine f_{si}. The stress in the reinforcement f_{si} at the various layers is determined by multiplying ε_{si} by the modulus of elasticity of the steel E_s:

- Layer 1: $f_{s1} = 0.0022 \times 29{,}000 = 63.8$ ksi > 60 ksi; use $f_{s1} = 60$ ksi
- Layer 2: $f_{s2} = 0.0001 \times 29{,}000 = 2.9$ ksi
- Layer 3: $f_{s3} = -0.0019 \times 29{,}000 = -55.1$ ksi
- Layer 4: $f_{s4} = -0.0040 \times 29{,}000 = -116.0$ ksi > -60 ksi; use $f_{s4} = -60$ ksi

Step 8: Determine F_{si}. The force in the reinforcement F_{si} at the various layers is determined by Eq. (5.44) or (5.45), which depends on the location of the steel layer:

- Layer 1 ($d_1 = 2.6$ in $< a = 6.4$ in): $F_{s1} = [60 - (0.85 \times 7)] \times 3 \times 1.27 = 205.9$ kips
- Layer 2: $F_{s2} = 2.9 \times 2 \times 1.27 = 7.4$ kips
- Layer 3: $F_{s3} = -55.1 \times 2 \times 1.27 = -140.0$ kips
- Layer 4: $F_{s4} = -60 \times 3 \times 1.27 = -228.6$ kips

Note that the compression steel in the top layer falls within the depth of the equivalent stress block; thus, Eq. (5.45) is used to determine the forces in the reinforcement in that layer.

Step 9: Determine P_n and M_n. The nominal axial strength P_n and nominal flexural strength M_n of the section are determined by Eqs. (5.46) and (5.47), respectively:

$$P_n = C + \sum F_{si} = 685.4 + (205.9 + 7.4 - 140.0 - 228.6) = 530.1 \text{ kips}$$

$$M_n = 0.5C(h - a) + \sum F_{si}(0.5h - d_i)$$

$$= [0.5 \times 685.4 \times (24 - 6.4)] + [205.9(12 - 2.6) + 7.4(12 - 8.9)$$

$$+ (-140.0)(12 - 15.1) + (-228.6)(12 - 21.4)]$$

$$= 6{,}031.5 + 4{,}541.2 = 10{,}572.7 \text{ in kips} = 881.1 \text{ ft kips}$$

The design axial load and bending moment are obtained by multiplying P_n and M_n by the strength reduction factor ϕ. Because $\varepsilon_t = \varepsilon_{s4} = 0.0040$ falls between the limits of compression- and tension-controlled sections (0.0020 and 0.0050, respectively), the section is in the transition region, and ϕ can be calculated by the following equation (see ACI 9.3.2.2 and Fig. 4.2 of this book):

$$\phi = 0.65 + (\varepsilon_t - 0.0020)\left(\frac{250}{3}\right) = 0.65 + (0.0040 - 0.0020)\left(\frac{250}{3}\right) = 0.82$$

Therefore,

$$\phi P_n = 0.82 \times 530.1 = 434.7 \text{ kips}$$

$$\phi M_n = 0.82 \times 881.1 = 722.5 \text{ ft kips}$$

Point 6—Pure Bending

For sections with multiple layers of reinforcement, there is no easy way of determining M_n for the case of pure bending. A trial-and-error procedure is usually utilized using various neutral axis depths c. A strain compatibility analysis is performed using the assumed value of c, and the nominal axial strength P_n is calculated using Eq. (5.46). The iterations can end after a value of zero (or a value close to zero) is found for P_n.

For the case of pure bending, the neutral axis should be located within the upper half of the section. After several iterations, determine P_n, assuming $c = 5.225$ in. Also, $a = \beta_1 c = 3.658$ in.

The concrete compression resultant force C is determined by Eq. (5.42):

$$C = 0.85 f'_c ab = 0.85 \times 7 \times 3.658 \times 18 = 391.8 \text{ kips}$$

The strain in the reinforcement ε_{si} at the various layers is determined by Eq. (5.43) where compression strains are positive:

- Layer 1 ($d_1 = 2.6$ in):

$$\varepsilon_{s1} = \frac{0.0030(5.225 - 2.6)}{5.225} = 0.0015$$

- Layer 2 ($d_2 = 8.9$ in):

$$\varepsilon_{s2} = \frac{0.0030(5.225 - 8.9)}{5.225} = -0.0021$$

- Layer 3 ($d_3 = 15.1$ in):

$$\varepsilon_{s3} = \frac{0.0030(5.225 - 15.1)}{5.225} = -0.0057$$

- Layer 4 ($d_4 = 21.4$ in):

$$\varepsilon_{s4} = \frac{0.0030(5.225 - 21.4)}{5.225} = -0.009$$

It is evident that the top layer of reinforcement is in compression and that the bottom three layers are in tension. Also, the reinforcement in all three of the layers that are in tension yield (i.e., ε_{s2}, ε_{s3}, and $\varepsilon_{s4} > \varepsilon_y = 0.0020$).

The stress in the reinforcement f_{si} at the various layers is determined by multiplying ε_{si} by the modulus of elasticity of the steel E_s:

- Layer 1: $f_{s1} = 0.0015 \times 29,000 = 43.5$ ksi
- Layer 2: $f_{s2} = -60$ ksi
- Layer 3: $f_{s3} = -60$ ksi
- Layer 4: $f_{s4} = -60$ ksi

The force in the reinforcement F_{si} at the various layers is determined by Eq. (5.44) or (5.45), which depends on the location of the steel layer:

- Layer 1 ($d_1 = 2.6$ in $< a = 3.658$ in): $F_{s1} = [43.5 - (0.85 \times 7)] \times 3 \times 1.27 = 143.1$ kips
- Layer 2: $F_{s2} = -60 \times 2 \times 1.27 = -152.4$ kips
- Layer 3: $F_{s3} = -60 \times 2 \times 1.27 = -152.4$ kips
- Layer 4: $F_{s4} = -60 \times 3 \times 1.27 = -228.6$ kips

Note that the compression steel in the top layer falls within the depth of the equivalent stress block; thus, Eq. (5.45) is used to determine the forces in the reinforcement in that layer.

The nominal axial strength P_n is determined by Eq. (5.46):

$$P_n = C + \sum F_{si} = 391.8 + (143.1 - 152.4 - 152.4 - 228.6) = 1.5 \text{ kips} \cong 0$$

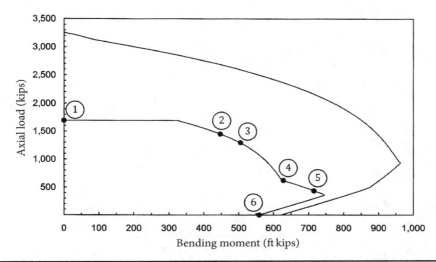

FIGURE 8.5 Nominal and design strength interaction diagrams about the major axis of the column given in Example 8.2.

The nominal flexural strength M_n of the section is determined by Eq. (5.47):

$$M_n = 0.5C(h-a) + \sum F_{si}(0.5h - d_i)$$
$$= [0.5 \times 391.8 \times (24 - 3.658)] + [143.1(12 - 2.6) + (-152.4)(12 - 8.9)$$
$$+ (-152.4)(12 - 15.1) + (-228.6)(12 - 21.4)]$$
$$= 3{,}985.0 + 3{,}494.0 = 7{,}479.0 \text{ in kips} = 623.3 \text{ ft kips}$$

Because the section is tension-controlled ($\varepsilon_t = \varepsilon_{s4} > 0.0050$), $\phi = 0.9$ (see ACI 9.3.2.2 and Fig. 4.2 of this book).
Therefore,

$$\phi M_n = 0.9 \times 623.3 = 561.0 \text{ ft kips}$$

Both the nominal and design strength interaction diagrams for this column are shown in Fig. 8.5. Points 1 through 6 are indicated on the figure as well.

Comments
The interaction diagram has been constructed for bending about the major axis of the reinforced concrete section. The procedure outlined earlier can be used to construct the interaction diagram of the column about the minor axis. The nominal and design strength interaction diagrams about the minor axis of the column are shown in Fig. 8.6. As expected, the design uniaxial compressive strength is the same, and the design flexural strength is less.

Example 8.3 Check the adequacy of the column in Example 8.2 for the factored load combinations given in Table 8.2, assuming that the column resists the bending moments about its major axis.

Solution A summary of the service axial loads and bending moments is given in Table 8.2. Also given in the table are the factored load combinations in accordance with ACI 9.2 (see Table 4.1 of this book).

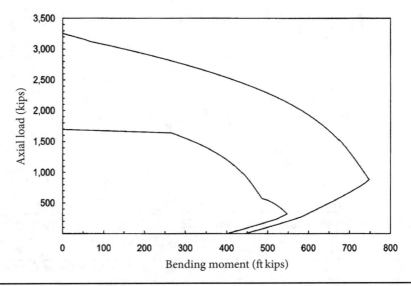

FIGURE 8.6 Nominal and design strength interaction diagrams about the minor axis of the column given in Example 8.2.

The factored load effects are added together in all of the load combinations except in load combination 6, which corresponds to ACI Eq. (9-6). The factored wind load effects are subtracted from the factored dead load effects in load combination 6 because the dead load counteracts the effects from wind; this produces a more critical effect on the column (see the discussion in Section 4.2).

The six load combinations in Table 8.2 are plotted in Fig. 8.7, which contains the design strength interaction diagram that was constructed for this column in Example 8.2.

Load Case	Axial Loads (kips)	Bending Moment (ft kips)
Dead (D)	675	15
Roof live (L_r)	18	—
Live (L)	270	10
Wind (W)	±40	±150
Load Combination		
1 $1.4D$	945	21
2 $1.2D + 1.6L + 0.5L_r$	1,251	34
3 $1.2D + 1.6L_r + 0.5L$	974	23
4 $1.2D + 1.6L_r + 0.8W$	871	138
5 $1.2D + 1.6W + 0.5L + 0.5L_r$	1,018	263
6 $0.9D - 1.6W$	544	227

TABLE 8.2 Summary of Axial Loads and Bending Moments for the Column Given in Example 8.3

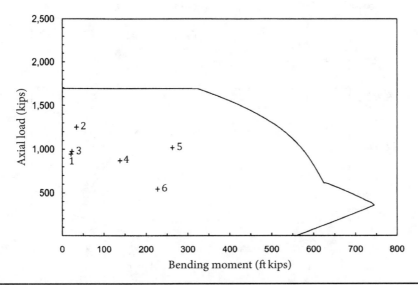

FIGURE 8.7 Design strength interaction diagram about the major axis of the column given in Example 8.3.

It is clear from Fig. 8.7 that all of the load combination points fall within the interaction diagram; therefore, the 18 × 24 in column reinforced with 10 No. 10 bars is adequate.

Example 8.4 Check the adequacy of the column given in Example 8.2, using the axial loads and bending moments in Example 8.3, assuming that the column resists the bending moments about its minor axis.

Solution The six load combinations given in Table 8.2 are plotted in Fig. 8.8, which contains the design strength interaction diagram that was constructed for this column in Example 8.2.
It is clear from Fig. 8.8 that all of the load combination points fall within the interaction diagram; therefore, the 18 × 24 in column reinforced with 10 No. 10 bars is adequate for bending about the minor axis as well.

8.4.4 Circular Sections

The procedure outlined earlier for rectangular sections, which is based on strain compatibility analyses, can also be used to construct interaction diagrams for circular sections.
The main difference between the analyses of a rectangular and a circular section pertains to the shape of the compression zone: For a rectangular section, the shape of the compression zone is rectangular, whereas for a circular section, the shape is related to a segment of a circle.
Two cases are possible for circular sections, which are based on the depth of the compression area a. Nominal strength equations for axial load and bending moment are derived for both cases.

- Case 1: $a \leq h/2$

    In the first case, the depth of the compression zone is equal to or less than the radius of the circular section $h/2$ (see Fig. 8.9).

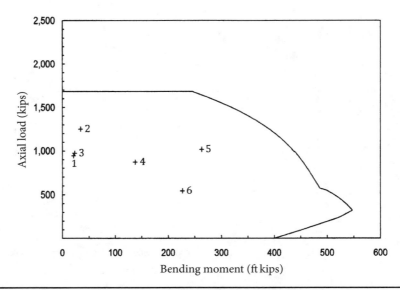

FIGURE 8.8 Design strength interaction diagram about the minor axis of the column given in Example 8.4.

The compression zone in this case is a segment of a circle of depth a. Because the compressive force and its moment about the centroid of the section are needed to calculate the nominal axial load and bending moment, the area of the segment and the location of its centroid must be determined. The area A of the circular segment can be calculated by the following equation:

$$A = \frac{h^2}{8}(\theta - \sin\theta) \tag{8.4}$$

FIGURE 8.9 Circular column where the depth of the compression zone a is equal to or less than $h/2$.

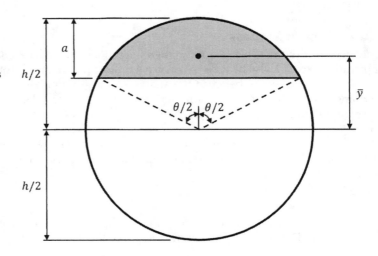

The angle θ, which is expressed in radians, can be determined from trigonometry (see Fig. 8.9):

$$\theta = 2 \cos^{-1}\left(1 - \frac{2a}{h}\right) \tag{8.5}$$

The centroid $\bar{y}$ of the circular segment is located at the following distance from the center of the circle:

$$\bar{y} = \frac{2h \sin^3(\theta/2)}{3(\theta - \sin\theta)} \tag{8.6}$$

The resultant compressive force C in the concrete is obtained by multiplying the stress $0.85 f_c'$ by the area A:

$$C = \frac{0.85 f_c' h^2(\theta - \sin\theta)}{8} \tag{8.7}$$

This compressive force is added to the forces in the reinforcing bars to obtain the nominal axial strength P_n:

$$P_n = C + \sum F_{si} \tag{8.8}$$

Recall that the magnitude of the force F_{si} depends on whether the steel is located in the compression zone or not (see the discussion in Section 5.7).

The nominal flexural strength M_n is determined by summing the moments about the centroid of the column:

$$M_n = C\bar{y} + \sum F_{si}(0.5h - d_i) \tag{8.9}$$

In this equation, $\bar{y}$ is determined by Eq. (8.6) and d_i is the distance from the compression face of the section to the centroid of the reinforcing bar(s) at level i.

- Case 2: $a > h/2$

  Illustrated in Fig. 8.10 is the case where the depth of the compression zone a is greater than the radius $h/2$. Similar to case 1, both the area and the centroid of the shaded area are needed for nominal strength calculations.

  The area of the compression zone A can be determined by subtracting the area of the segment below a from the area of the circle:

$$A = \pi \left(\frac{h}{2}\right)^2 - \frac{h^2}{8}(\bar{\theta} - \sin\bar{\theta}) \tag{8.10}$$

The angle $\bar{\theta}$ is determined from trigonometry (see Fig. 8.10):

$$\bar{\theta} = 2 \cos^{-1}\left(\frac{2a}{h} - 1\right) \tag{8.11}$$

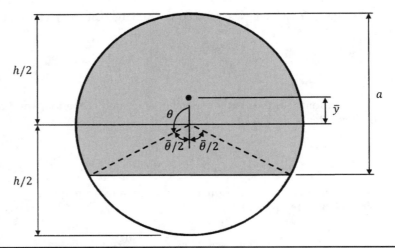

Figure 8.10   Circular column where the depth of the compression zone a is greater than $h/2$.

The following equation for A is obtained by first substituting $\bar{\theta} = 2(\pi - \theta)$ into Eq. (8.11) and solving for θ and then substituting θ into Eq. (8.10):

$$A = \frac{h^2}{4}(\theta - \sin\theta\cos\theta) \qquad (8.12)$$

The distance $\bar{y}$ from the center of the section to the centroid of the shaded area is

$$\bar{y} = \frac{h\sin^3\theta}{3(\theta - \sin\theta\cos\theta)} \qquad (8.13)$$

Equations (8.8) and (8.9) can be utilized to determine the nominal axial load and bending moment, using the appropriate quantities derived earlier.

Example 8.5   Determine the design strengths ϕP_n and ϕM_n corresponding to balanced failure for the tied circular column shown in Fig. 8.11. Assume normal-weight concrete with $f_c' = 5{,}000$ psi and Grade 60 reinforcement.

Solution   The flowchart shown in Fig. 5.39 is utilized to determine ϕP_n and ϕM_n for this compression member; some of the steps are modified to account for the differences pertaining to the circular section.

It is important to note in this example that the design axial load strength and bending moment strength for bending about the x-axis are different from the values for bending about the y-axis because of the different bar orientations relative to the direction of the neutral axis. Design strengths are determined in both cases.

Case 1—Bending About the x-axis

Step 1: Check the minimum and maximum longitudinal reinforcement limits. The minimum and maximum amounts of longitudinal reinforcement permitted in a compression member are

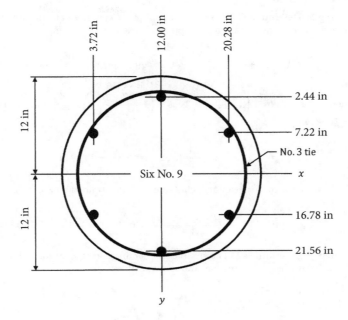

Figure 8.11 The circular column given in Example 8.5.

specified in ACI 10.9.1:

$$\text{Minimum } A_{st} = 0.01A_g = 0.01 \times \pi \times (24/2)^2 = 4.52 \text{ in}^2$$

$$\text{Maximum } A_{st} = 0.08A_g = 0.08 \times \pi \times (24/2)^2 = 36.2 \text{ in}^2$$

The provided area of longitudinal reinforcement $A_{st} = 6 \times 1.0 = 6.00$ in² falls between the minimum and maximum limits.

Step 2: Determine the neutral axis depth c. Balanced failure occurs when crushing of the concrete and yielding of the reinforcing steel occur simultaneously (see Section 5.3). The balanced failure point also represents the change from compression failures for higher axial loads and tension failures for lower axial loads for a given bending moment. ACI 10.3.3 permits the yield strain of the reinforcement to be taken as 0.0020 for Grade 60 reinforcement; thus, $\varepsilon_{s4} = \varepsilon_t = 0.0020$.

The neutral axis depth is determined by Eq. (5.41):

$$c = \frac{0.0030d_t}{\varepsilon_t + 0.0030} = \frac{0.0030 \times 21.56}{0.0020 + 0.0030} = 12.9 \text{ in}$$

Step 3: Determine β_1.

$$\beta_1 = 1.05 - 0.00005f_c' = 1.05 - (0.00005 \times 5,000) = 0.80 \quad \text{for} \quad f_c' = 5,000 \text{ psi}$$
$$\text{(see ACI 10.2.7.3 and Section 5.2 of this book)}$$

Step 4: Determine the depth of the compression zone a.

$$a = \beta_1 c = 0.80 \times 12.9 = 10.3 \text{ in}$$

Step 5: Determine C. Because the depth of the compression zone a is less than the radius of the section, the area of the compression zone is determined by Eq. (8.4):

$$A = \frac{h^2}{8}(\theta - \sin\theta)$$

where $\theta = 2\cos^{-1}\left(1 - \frac{2a}{h}\right) = 2\cos^{-1}\left(1 - \frac{2 \times 10.3}{24}\right) = 2.86$ rad [see Eq. (8.5)]

Thus,

$$A = \frac{24^2}{8}(2.86 - \sin 2.86) = 185.9 \text{ in}^2$$

The concrete compression resultant force C is determined by Eq. (8.7):

$$C = \frac{0.85 f'_c h^2(\theta - \sin\theta)}{8} = 0.85 \times 5 \times 185.9 = 790.1 \text{ kips}$$

Step 6: Determine ε_{si}. The strain in the reinforcement ε_{si} at the various layers is determined by Eq. (5.43) where compression strains are positive:

• Layer 1 ($d_1 = 2.44$ in):

$$\varepsilon_{s1} = \frac{0.0030(12.9 - 2.44)}{12.9} = 0.0024$$

• Layer 2 ($d_2 = 7.22$ in):

$$\varepsilon_{s2} = \frac{0.0030(12.9 - 7.22)}{12.9} = 0.0013$$

• Layer 3 ($d_3 = 16.78$ in):

$$\varepsilon_{s3} = \frac{0.0030(12.9 - 16.78)}{12.9} = -0.0009$$

• Layer 4 ($d_4 = 21.56$ in):

$$\varepsilon_{s4} = \frac{0.0030(12.9 - 21.56)}{12.9} = -0.0020 \text{ (checks)}$$

It is evident that the top two layers of reinforcement are in compression and that the bottom two layers are in tension. Also, the layers of reinforcement closest to and farthest from the extreme compression fiber yield.

Step 7: Determine f_{si}. The stress in the reinforcement f_{si} at the various layers is determined by multiplying ε_{si} by the modulus of elasticity of the steel E_s:

• Layer 1: $f_{s1} = 0.0024 \times 29,000 = 69.6$ ksi > 60 ksi; use $f_{s1} = 60$ ksi
• Layer 2: $f_{s2} = 0.0013 \times 29,000 = 37.7$ ksi
• Layer 3: $f_{s3} = -0.0009 \times 29,000 = -26.1$ ksi
• Layer 4: $f_{s4} = -60$ ksi

Step 8: Determine F_{si}. The force in the reinforcement F_{si} at the various layers is determined by Eq. (5.44) or (5.45), which depends on the location of the steel layer:

- Layer 1 ($d_1 = 2.44$ in $< a = 10.3$ in): $F_{s1} = [60 - (0.85 \times 5)] \times 1 \times 1.00 = 55.8$ kips
- Layer 2 ($d_2 = 7.22$ in $< a = 10.3$ in): $F_{s2} = [37.7 - (0.85 \times 5)] \times 2 \times 1.00 = 66.9$ kips
- Layer 3: $F_{s3} = -26.1 \times 2 \times 1.00 = -52.2$ kips
- Layer 4: $F_{s4} = -60 \times 1 \times 1.00 = -60.0$ kips

Note that the compression steel in the top two layers fall within the depth of the compression zone; thus, Eq. (5.45) is used to determine the forces in the reinforcement in those layers.

Step 9: Determine P_n and M_n. The nominal axial strength P_n is determined by Eq. (8.8):

$$P_n = C + \sum F_{si} = 790.1 + (55.8 + 66.9 - 52.2 - 60.0) = 800.6 \text{ kips}$$

The nominal flexural strength M_n is determined by Eq. (8.9) where the centroid $\bar{y}$ of the circular segment is determined by Eq. (8.6):

$$\bar{y} = \frac{2h \sin^3 (\theta/2)}{3(\theta - \sin\theta)} = \frac{2 \times 24 \times \sin^3 (2.86/2)}{3(2.86 - \sin 2.86)} = 6.0 \text{ in}$$

$$\begin{aligned} M_n &= C\bar{y} + \sum F_{si}(0.5h - d_i) \\ &= (790.1 \times 6.0) + [55.8(12 - 2.44) + 66.9(12 - 7.22) \\ &\quad + (-52.2)(12 - 16.78) + (-60.0)(12 - 21.56)] \\ &= 4,740.6 + 1,676.4 = 6,417.0 \text{ in kips} = 534.8 \text{ ft kips} \end{aligned}$$

This section is compression-controlled because ε_t is equal to the compression-controlled strain limit of 0.0020 (see ACI 10.3.3). Thus, in accordance with ACI 9.3.2.2, the strength reduction factor ϕ is equal to 0.65 for a compression-controlled section with lateral reinforcement consisting of ties (or, equivalently, a compression-controlled section without spiral reinforcement, conforming to ACI 10.9.3). Therefore, the design axial strength ϕP_n and design flexural strength ϕM_n are

$$\phi P_n = 0.65 \times 800.6 = 520.4 \text{ kips}$$
$$\phi M_n = 0.65 \times 534.8 = 347.6 \text{ ft kips}$$

Case 2—Bending About the y-axis

Step 1: Check the minimum and maximum longitudinal reinforcement limits. It was shown in Case 1 that the provided area of longitudinal reinforcement falls between the minimum and maximum limits.

Step 2: Determine the neutral axis depth c. The neutral axis depth is determined by Eq. (5.41):

$$c = \frac{0.0030 d_t}{\varepsilon_t + 0.0030} = \frac{0.0030 \times 20.28}{0.0020 + 0.0030} = 12.2 \text{ in}$$

Step 3: Determine β_1.

$$\beta_1 = 1.05 - 0.00005 f_c' = 1.05 - (0.00005 \times 5,000) = 0.80 \quad \text{for} \quad f_c' = 5,000 \text{ psi}$$
(see ACI 10.2.7.3 and Section 5.2 of this book)

Step 4: Determine the depth of the compression zone a.

$$a = \beta_1 c = 0.80 \times 12.2 = 9.8 \text{ in}$$

Step 5: Determine C. Because the depth of the compression zone a is less than the radius of the section, the area of the compression zone is determined by Eq. (8.4):

$$A = \frac{h^2}{8}(\theta - \sin\theta)$$

where $\theta = 2\cos^{-1}\left(1 - \frac{2a}{h}\right) = 2\cos^{-1}\left(1 - \frac{2 \times 9.8}{24}\right) = 2.77 \text{ rad}$ [see Eq. (8.5)]

Thus,

$$A = \frac{24^2}{8}(2.77 - \sin 2.77) = 173.3 \text{ in}^2$$

The concrete compression resultant force C is determined by Eq. (8.7):

$$C = \frac{0.85 f_c' h^2 (\theta - \sin\theta)}{8} = 0.85 \times 5 \times 173.3 = 736.5 \text{ kips}$$

Step 6: Determine ε_{si}. The strain in the reinforcement ε_{si} at the various layers is determined by Eq. (5.43) where compression strains are positive:

- Layer 1 ($d_1 = 3.72$ in):

$$\varepsilon_{s1} = \frac{0.0030(12.2 - 3.72)}{12.2} = 0.0021$$

- Layer 2 ($d_2 = 12.0$ in):

$$\varepsilon_{s2} = \frac{0.0030(12.2 - 12.0)}{12.2} = 0.00005$$

- Layer 3 ($d_3 = 20.28$ in):

$$\varepsilon_{s3} = \frac{0.0030(12.2 - 20.28)}{12.2} = -0.0020 \text{ (checks)}$$

It is evident that the top two layers of reinforcement are in compression and that the bottom layer is in tension. Also, the layers of reinforcement closest to and farthest from the extreme compression fiber yield.

Step 7: Determine f_{si}. The stress in the reinforcement f_{si} at the various layers is determined by multiplying ε_{si} by the modulus of elasticity of the steel E_s:

- Layer 1: $f_{s1} = 0.0021 \times 29,000 = 60.9 \text{ ksi} > 60 \text{ ksi}$; use $f_{s1} = 60 \text{ ksi}$
- Layer 2: $f_{s2} = 0.00005 \times 29,000 = 1.5 \text{ ksi}$
- Layer 3: $f_{s3} = -60 \text{ ksi}$

Step 8: Determine F_{si}. The force in the reinforcement F_{si} at the various layers is determined by Eq. (5.44) or (5.45), which depends on the location of the steel layer:

- Layer 1 ($d_1 = 3.72$ in $< a = 9.8$ in): $F_{s1} = [60 - (0.85 \times 5)] \times 2 \times 1.00 = 111.5 \text{ kips}$
- Layer 2: $F_{s2} = 1.5 \times 2 \times 1.00 = 3.0 \text{ kips}$
- Layer 3: $F_{s3} = -60.0 \times 2 \times 1.00 = -120.0 \text{ kips}$

Note that the compression steel in the top layer falls within the depth of the compression zone; thus, Eq. (5.45) is used to determine the forces in the reinforcement in that layer.

Step 9: Determine P_n and M_n. The nominal axial strength P_n is determined by Eq. (8.8):

$$P_n = C + \sum F_{si} = 736.5 + (111.5 + 3.0 - 120.0) = 731.0 \text{ kips}$$

The nominal flexural strength M_n is determined by Eq. (8.9) where the centroid $\bar{y}$ of the circular segment is determined by Eq. (8.6):

$$\bar{y} = \frac{2h \sin^3 (\theta/2)}{3(\theta - \sin\theta)} = \frac{2 \times 24 \times \sin^3 (2.77/2)}{3(2.77 - \sin 2.77)} = 6.3 \text{ in}$$

$$M_n = C\bar{y} + \sum F_{si}(0.5h - d_i)$$
$$= (736.5 \times 6.3) + [111.5(12 - 3.72) + 3.0(12 - 12) + (-120.0)(12 - 20.28)]$$
$$= 4,640.0 + 1,916.8 = 6,556.8 \text{ in kips} = 546.4 \text{ ft kips}$$

Because the section is compression-controlled, the strength reduction factor ϕ is equal to 0.65, and the design axial strength ϕP_n and design flexural strength ϕM_n are

$$\phi P_n = 0.65 \times 731.0 = 475.2 \text{ kips}$$

$$\phi M_n = 0.65 \times 546.4 = 355.2 \text{ ft kips}$$

Comments
Other points on the interaction diagram can be determined using the method outlined earlier. For strain distributions that are primarily compressive, a large portion of the section is under compression, and Eqs. (8.12) and (8.13) would be utilized in the determination of the design strengths.

As illustrated in this example, the design axial load and moment strengths depend on the bar arrangement for reinforced concrete columns with six longitudinal bars. Because the designer has no control over the arrangement of the bars when they are placed in the formwork, the column must be checked for the most critical cases. Utilizing eight or more longitudinal bars in a section essentially eliminates the need to determine different design strengths based on bar arrangement.

8.4.5   Design Aids

A number of design aids that assist in the design of columns subjected to axial load and bending moment are available. The intent of the design aids is to eliminate the routine and repetitive calculations that are required in the construction of an interaction diagram.

Reference 1 contains nondimensionalized nominal strength interaction diagrams for rectangular and circular sections with a variety of bar arrangements. Numerous tables for rectangular and circular columns are contained in Ref. 2, which cover a wide range of cross-sectional dimensions, concrete compressive strength, longitudinal reinforcement ratios, and bar arrangements. These tables contain values corresponding to key points on the interaction diagram.

8.5 Slenderness Effects

8.5.1 Overview

It is important to determine whether the effects of slenderness need to be considered or not early in the design of any column, because second-order effects can have a significant influence on design strength. In very simple terms, a column is slender if its applicable cross-sectional dimension is small in comparison with its length.

The term "short column" is often used to indicate a column that has a strength equal to that computed for its cross-section. In such cases, strength can be represented by an interaction diagram, which is constructed on the basis of the geometric and material properties of the section (see Section 8.4). If a column does not deflect laterally, actual failure is theoretically represented by any point along the nominal strength interaction curve. In other words, any combination of axial load and bending moment that falls outside of the interaction curve is assumed to cause failure. This is commonly referred to as a *material failure*. For purposes of design, a column that is not slender has adequate strength when all of the factored combinations of axial load and bending moment that are obtained from an elastic first-order analysis of the frame fall within or on the design strength interaction curve.

A *slender column* is defined as a column whose strength is reduced by second-order deformations due to horizontal displacements. Consider the column shown in Fig. 8.12 that is subjected to an axial load P and a bending moment M. The moment M can be expressed as the axial load P times an eccentricity e.

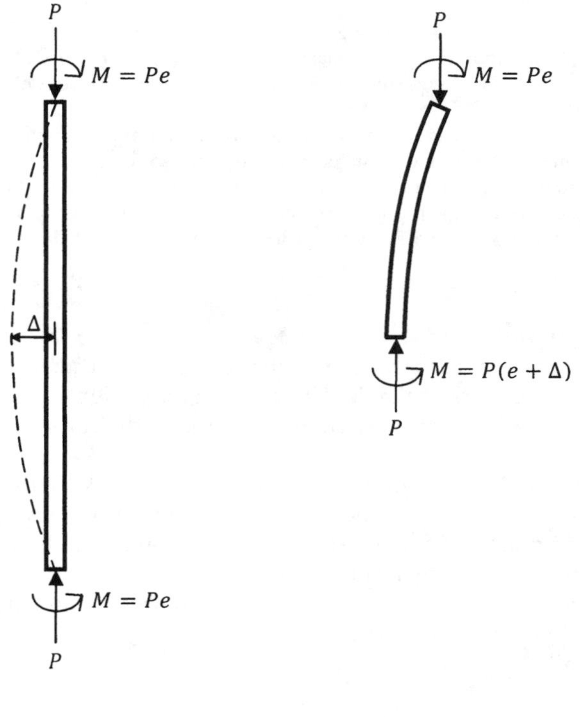

FIGURE 8.12 P-delta effects in a column.

The column has a horizontal deflection Δ along the span because of the applied loading. This deflection, in turn, causes an additional (or secondary) moment, as shown in the free-body diagram of the deflected shape of the column. Thus, the total moment in the column is equal to the applied moment Pe due to the external loading plus the moment $P\Delta$ due to the horizontal deflection of the member. Secondary effects caused by horizontal deflection are commonly referred to as *P-delta effects*.

Secondary effects have a relatively small impact on the design of columns that are not slender. In contrast, the deflection Δ due to the applied loading can increase and become unstable with an increase in P for columns that are slender. When this occurs, the column buckles under the effects of the applied loads. This type of failure is known as a *stability failure* or *elastic instability*, and it generally occurs at a load that is less than that corresponding to material failure of the section.

It is important to keep in mind that a column with a given slenderness ratio may be considered a short column for design under one set of constraints and a slender column under another set. This is discussed in more detail later.

The effects of slenderness can be neglected in compression members with slenderness ratios that are equal to or less than the limiting values given in ACI 10.10.1. These limiting values are given for members not braced against sidesway and those braced against sidesway. The Code requirements on what constitutes a braced member are discussed next.

In the Code, slenderness ratio is defined as the effective length factor k times the unsupported length of the column ℓ_u divided by the radius of gyration of the column cross-section r, which is equal to $\sqrt{I/A_g}$, where I and A_g are the moment of inertia in the direction of analysis and the gross area of the cross-section, respectively. More information on these quantities is given after the discussion on braced and unbraced frames.

8.5.2   Compression Members Braced and Unbraced Against Sidesway

Sway in buildings due to lateral or other loads can have a dramatic influence on second-order effects in columns. It is common for secondary effects to increase with increasing sway.

Distinguishing between members that are braced against sidesway and those that are not can usually be done by comparing the total lateral stiffness of the columns in a story with that of the bracing elements. Moment frame buildings are typically laterally flexible, and the columns are more susceptible to secondary effects than those in a frame with more rigid bracing elements, such as walls.

The Code provides information that can be used where it is not readily apparent whether a frame is braced against sidesway or not. ACI 10.10.1 permits compression members to be considered braced against sidesway when the bracing elements in the structure have a total lateral stiffness in the direction of analysis of at least 12 times the gross stiffness of all the columns within a given story. For moment frames, the bracing elements are the beams and columns that are part of the lateral force–resisting system. In buildings with walls, the walls alone or in combination with specifically identified frames are the bracing elements.

It is evident from the preceding discussion that entire stories in a building are either braced or not braced against sidesway. Depending on a number of factors, it is possible for a column in one story to be braced and for the same column immediately above or below that story to be unbraced.

ACI 10.10.5.1 and 10.10.5.2 give two additional ways of determining whether a story or member is braced or not (the terms that are used in these Code sections that refer to frames or members that are braced and not braced are *nonsway frames* and *sway frames*, respectively).

According to ACI 10.10.5.1, a column or story can be considered nonsway when the end moments due to second-order effects are equal to or less than 1.05 times the end moments determined by a first-order analysis. It has been demonstrated that this 5% limitation provides a reasonable limit on nonsway conditions.[3]

An alternate method is given in ACI 10.10.5.2, which utilizes the stability index Q. A story in a structure is assumed to be nonsway when Q is equal to or less than 0.05 [see ACI Eq. (10-10)]:

$$Q = \frac{\sum P_u \Delta_o}{V_{us} \ell_c} \leq 0.05 \tag{8.14}$$

In this equation $\sum P_u$ is the total factored axial load in the story of the building that is being evaluated (i.e., the sum of all the factored loads on the columns and walls in that story); Δ_o is the first-order relative lateral deflection between the top and the bottom of the story; V_{us} is the factored horizontal shear in the story; and ℓ_c is the length of the columns in the story measured center-to-center of the joints in the frame. This method should not be used in cases where the factored horizontal shear force V_{us} is equal to zero. Also, the sum of the factored axial loads $\sum P_u$ must correspond to the lateral loading case for which this sum is the greatest.

Equation (8.14) is commonly used in a P-delta analysis. When the externally applied horizontal story shear V_{us} is applied to a story in a frame, the story deflects an amount equal to Δ_o (see Fig. 8.13). The sum of the factored axial loads in the story $\sum P_u$ is displaced horizontally by this amount.

The numerator of Eq. (8.14) is the moment in the story due to the axial loads being displaced by the amount Δ_o, and the denominator is the overall moment in the story due to V_{us}. This equation essentially means that second-order (or P-delta) effects need not be considered when the overall story moment due to the axial loads is equal to or less than 5% of the overall moment due to the horizontal shear.

FIGURE 8.13
Definition of
stability index Q.

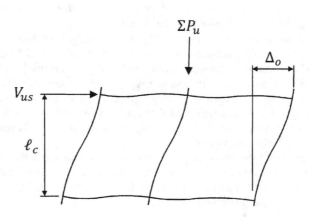

In cases where service loads are used to calculate the lateral deflection of a frame, the following equation can be used to determine Q (see the discussion in ACI R10.10.5):

$$Q = \frac{1.2 \sum (P_D + P_L)(1.43\Delta_S)}{V_s \ell_c} \leq 0.05 \tag{8.15}$$

In this equation, $\Sigma(P_D + P_L)$ is the sum of the total service dead and live axial loads in a story; Δ_S is the first-order, service-level interstory deflection; and V_s is the service-level story shear.

Stories in a building that do not meet the criteria of ACI 10.10.5.1 or 10.10.5.2 are considered sway stories. As noted previously, a frame may contain both nonsway and sway stories.

Example 8.6   The following quantities have been determined for a story in a multistory building by a first-order elastic analysis: $\Sigma P_u = 20{,}000$ kips, $\Delta_o = 0.40$ in, and $V_{us} = 350$ kips. The story height $\ell_c = 12$ ft. Determine if the story is nonsway or sway.

Solution   Because factored loads are given, Eq. (8.14) is used to determine if this story is nonsway or sway:

$$Q = \frac{\sum P_u \Delta_o}{V_{us} \ell_c} = \frac{20{,}000 \times 0.40}{350 \times (12 \times 12)} = 0.16 > 0.05$$

Because the stability index for the story is greater than 0.05, the story is sway.

8.5.3   Consideration of Slenderness Effects

Introduction

Consideration of column slenderness for nonsway and sway frames depends on the slenderness ratio of the column $k\ell_u/r$. Different slenderness limits are prescribed in the Code for nonsway and sway frames.

Prior to discussing these limits, information is presented on the quantities that make up the slenderness ratio.

Unsupported Length ℓ_u

According to ACI 10.10.1, the unsupported length of a compression member ℓ_u shall be taken as the clear distance between floor slabs, beams, or other members capable of providing lateral support in the direction of analysis. The beam shown in Fig. 8.14 provides lateral support to the column in the direction parallel to the x-axis. That is why the unsupported length $(\ell_u)_1$ is equal to the distance from the bottom of the beam to the top of the slab in that direction. The beam does not provide lateral support to the column parallel to the y-direction, and the unsupported length $(\ell_u)_2$ is the distance from the bottom of the slab at the top of the column to the top of the slab at the bottom of the column.

Where column capitals or haunches are present, ℓ_u is measured to the lower extremity of the capital or haunch in the plane considered, as illustrated in Fig. 8.15 for a column capital.

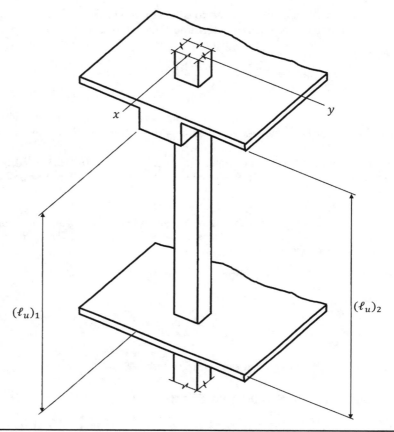

Figure 8.14 Unsupported column length with beams and slab.

Radius of Gyration *r*

The radius of gyration r is defined as $\sqrt{I/A_g}$, where I is the moment of inertia of the section in the direction of analysis and A_g is the gross cross-sectional area of the section. For a rectangular section where the dimension of the column parallel to the direction of analysis is h_1 and the dimension perpendicular to the direction of analysis is h_2, the radius of gyration is

$$r = \sqrt{\frac{h_2 h_1^3/12}{h_1 h_2}} = 0.29 h_1$$

Note that ACI 10.10.1.2 permits r to be taken as $0.3h_1$ for rectangular sections.

Similarly, r can be taken as $0.25h$ for a circular member where h is the diameter of the section.

Effective Length Factor *k*

The effective length factor k for a compression member depends on rotational and lateral restraints at the ends of a column and on the type of frame (nonsway or sway).

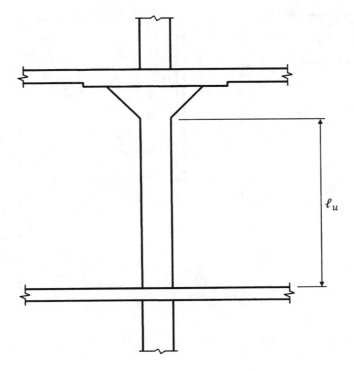

ℓ_u

In general, the effective column length is equal to $k\ell_u$; this corresponds to the length of the column between inflection points on the deflected shape of the column at the onset of buckling.

For the column on the left in Fig. 8.16, which is pinned at both ends and laterally braced against sidesway, the shape of the deflected column at the onset of buckling is a half-sine wave. In this case, the inflection points occur at the ends of the member, which means that the effective length of the column is equal to the unbraced column length ℓ_u. Thus, k is equal to 1.0 in this case.

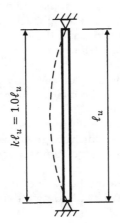

$k\ell_u = 1.0\ell_u$

ℓ_u

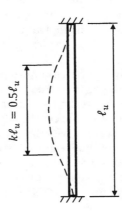

$k\ell_u = 0.5\ell_u$

ℓ_u

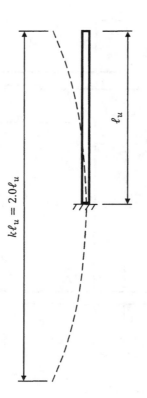

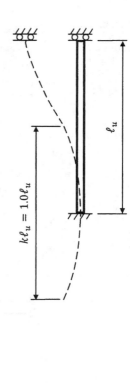

FIGURE 8.17
Effective length
factors *k* for
columns in sway
frames.

Similarly, for the column on the right in Fig. 8.16, which is fixed at both ends in a nonsway frame, the effective length of the column is one-half of ℓ_u (i.e., the length between the inflection points is equal to one-half of the unbraced length). Therefore, *k* is equal to 0.5 for these end conditions.

Figure 8.17 illustrates the idealized effective length factors for columns in a sway frame. The column on the left represents the case of a cantilevered column that is fully fixed against rotation and translation at the base and is free to rotate and deflect at the top. In this situation, the distance between the points of inflection is equal to twice the unbraced length of the column ℓ_u; thus, the effective length factor *k* = 2.0. The column on the right is fixed against rotation at both ends, and the top end is free to deflect horizontally. The distance between the points of inflection is equal to the unbraced length ℓ_u, and *k* = 1.0.

Rarely does a column have end supports that are perfectly pinned or fixed. In general, the degree of end restraint depends on the relative stiffnesses of the flexural members and columns that frame into the joint.

In lieu of a more exact analysis, the most commonly used design aid for estimating *k* is the Jackson and Moreland Alignment Charts, which are given in ACI Fig. R10.10.1.1. These charts allow a graphical determination of *k* for columns of constant cross-section in both nonsway and sway frames.[4,5]

The effective length factor *k* is determined on the basis of the stiffness ratio Ψ at the ends of the column. This ratio is determined by dividing the sum of the stiffnesses of the columns at a joint by the sum of the stiffnesses of the flexural members framing into that joint in the direction of analysis:

$$\Psi = \frac{\sum EI/\ell_c}{\sum EI/\ell} \qquad (8.16)$$

In this equation, ℓ_c is the length of the columns measured center-to-center of the joints in the frame and ℓ is the span length of the flexural members measured center-to-center of the joints.

The charts are based on the following equations[6]:

For nonsway frames,

$$\Psi = \frac{-2k}{\pi} \tan \frac{\pi}{2k} \qquad (8.17)$$

For sway frames,

$$\Psi = \frac{6k}{\pi} \cot \frac{\pi}{2k} \qquad (8.18)$$

A value of the stiffness ratio must be determined at both the top and the bottom of the column. The charts can be used to graphically obtain k by drawing a line from the stiffness ratio Ψ_A at the top of the column to the stiffness ratio Ψ_B at the bottom of the column. The value of k is obtained from the chart at the location where the line crosses the k-axis. For example, consider a column in a nonsway frame where $\Psi_A = 3.0$ and $\Psi_B = 0.7$. From the chart shown in ACI Fig. R10.10.1.1(a) for nonsway frames, graphically obtain $k \approx 0.8$.

Nonsway Frames

Slenderness effects may be neglected in nonsway frames when ACI Eq. (10-7) is satisfied:

$$\frac{k\ell_u}{r} \leq 34 - 12 \left(\frac{M_1}{M_2} \right) \leq 40 \qquad (8.19)$$

In this equation, M_1 and M_2 are the smaller factored end moment and the larger factored end moment, respectively, that are obtained from an elastic analysis of the frame. The ratio M_1/M_2 is positive if the column is bent in single curvature and negative if the column is bent in double curvature. Because columns are more stable when bent in double curvature, the limiting slenderness ratio is larger than that for columns bent in single curvature. The negative value of M_1/M_2 allows a wider range of columns to be treated as short columns. In no case shall the slenderness limit exceed 40.

The effective length factor k can be taken as 1.0 for all nonsway frames regardless of the stiffness factors at both ends of a column; this will produce conservative results in frames where the actual stiffnesses result in k-values less than 1.0.

Sway Frames

For sway frames, slenderness effects can be neglected when $k\ell_u/r$ is less than 22 [see ACI 10.10.1(a) and ACI Eq. (10-6)]. In such cases, the required cross-sectional dimensions, longitudinal reinforcement, and material properties can be determined for the governing factored load combinations based on the results from a first-order elastic analysis.

The effective length factor k must always be greater than 1.0 for columns in a sway frame [see ACI Fig. R10.10.1.1(b)]. Depending on the relative stiffnesses of the columns and flexural members framing into the joint, values of k equal to or greater than 2.0 are possible.

8.5.4   Methods of Analysis

Overview

ACI 10.10.2 outlines three methods that are permitted to analyze compression members where slenderness effects cannot be neglected:

1. Nonlinear second-order analysis
2. Elastic second-order analysis
3. Moment magnification procedure

In general, a second-order analysis yields results that are more reasonable than those obtained from an approximate method like the moment magnification procedure. Thus, the second-order procedures will usually result in a more economical design, especially for sway frames. As expected, approximate methods tend to give more conservative results.

Regardless of the analysis method that is used, it is important to limit the overall total moment in the structural members of a frame to ensure that the structure is stable. According to ACI 10.10.2.1, the total moment, which includes the second-order (or P-delta) effects due to slenderness, shall not exceed 140% of the moments obtained from a first-order analysis. It has been demonstrated by analytical research that the probability of stability failure increases significantly when the stability index Q exceeds 0.2.[3] When Q is equal to 0.2, the ratio of the secondary moment to the primary moment is equal to $1/(1 - Q) = 1.25$. In Section 12.8.7 of ASCE/SEI 7-05, which addresses P-delta effects in structures subjected to earthquake loads, the maximum value of the stability coefficient θ is given as 0.25 (θ is similar to the Code stability factor Q). This corresponds to a ratio of the secondary moment to the primary moment equal to 1.33. The 1.4 limit in ACI 10.10.2.1 is based on both of these findings.

Nonlinear Second-order Analysis

This analysis method is the most comprehensive and usually the most accurate in predicting the behavior of slender columns. Material nonlinearity, member curvature and lateral drift, duration of loads, shrinkage and creep, and interaction with the foundation must all be included in the analysis.

The methodology for a nonlinear second-order analysis is beyond the scope of this book. More information on how to undertake such an analysis is given in ACI R10.10.3.

Elastic Second-order Analysis

As the name implies, this method of analysis is performed in the elastic range, using member stiffnesses that represent those immediately prior to failure. In addition to using a strength reduction factor ϕ that accounts for variability in cross-sectional strength, a

Member	Moment of Inertia
Compression members	
Columns	$0.70I_g$
Walls—uncracked	$0.70I_g$
Walls—cracked	$0.35I_g$
Flexural members	
Beams	$0.35I_g$
Flat plates and flat slabs	$0.25I_g$

TABLE 8.3 Moments of Inertia to Use in an Elastic Second-order Analysis

stiffness reduction factor that accounts for variability of member stiffness (i.e., cracked regions) along the length of the member is used. The influence of axial loads and the effects of load duration must also be accounted for in this type of analysis.

The section properties of the members in the frame must be determined prior to analysis. The effects of cracking and the other effects outlined earlier must be included when establishing the section properties. In lieu of a more sophisticated analysis, ACI 10.10.4.1 permits the use of reduced moments of inertia for purposes of analysis. These values are obtained by multiplying the reduced moments of inertia derived in Ref. 3 for nonprestressed members by a stiffness reduction factor of 0.875. Reduced values of the moment of inertia based on these reduction factors result in an overestimation of the second-order deflections obtained from analysis on the order of 1.20 to 1.25. A summary of the reduced moments of inertia is given in Table 8.3.

In addition to reduced moments of inertia, ACI 10.10.4.1 requires that the modulus of elasticity of the concrete determined by ACI 8.5.1 and the gross area of the section be used in the analysis.

The reduced moments of inertia in Table 8.3 are based on an analysis where strength-level (factored) loads are used. An analysis using service-level loads is needed, for example, when determining lateral deflections or when calculating the period of a building. ACI R10.10.4.1 notes that it is satisfactory to perform a service-level analysis using 1.43 times the tabulated values of the reduced moments of inertia in Table 8.3.

More refined values of I are given in ACI 10.10.4.1 for compression members and flexural members that can be used in the analysis instead of the corresponding values given in Table 8.3[7,8]:

For compression members,

$$0.35I_g \leq I = \left(0.80 + \frac{25A_{st}}{A_g}\right)\left(1 - \frac{M_u}{P_u h} - \frac{0.5P_u}{P_o}\right)I_g \leq 0.875I_g \qquad (8.20)$$

For flexural members,

$$I = (0.10 + 25\rho)\left(1.2 - \frac{0.2b_w}{d}\right)I_g \leq 0.50I_g \qquad (8.21)$$

In Eq. (8.20), P_u and M_u are the factored axial load and bending moment on the compression member for the load combination under consideration and P_o is the maximum concentric axial load that can be carried by a short column, which is determined by Eq. (5.35).

Equations (8.20) and (8.21) were derived using a stiffness reduction factor comparable to the one used in deriving the reduced moments of inertia given in Table 8.3. These equations are also applicable for all levels of loading, including ultimate and service, even though they are presented in terms of factored loads. For service-level analysis, P_u and M_u in the equations should be replaced by a corresponding service-level axial load and bending moment.

When analyzing a system with walls, it is customary to initially assume that the walls are uncracked in flexure, using a reduced moment of inertia of $0.70I_g$. If the analysis indicates that the walls will crack in flexure based on the modulus of rupture, the analysis should be repeated using a reduced moment of inertia of $0.35I_g$.

In systems with beams, the moment of inertia of the beams should be determined using the web of the beam and the effective flange width defined in ACI 8.12. It is permitted to take I as the average of the values obtained from Eq. (8.21) for the critical positive and negative moment sections. Note that I need not be taken less than 25% of the gross moment of inertia.

Sustained lateral loads are not commonly encountered in building design because wind and earthquake loads are not sustained loads. However, a sustained load can occur, for example, where unequal earth pressures act on two sides of a building. In such cases, the moment of inertia I for compression members must be divided by $(1 + \beta_{ds})$, where β_{ds} is the ratio of the maximum sustained shear within a story to the maximum factored shear in that story associated with the same load combination (ACI 10.10.4.2). The magnitude of β_{ds} must be equal to or less than 1.0.

Moment Magnification Procedure

Overview　The moment magnification procedure is an approximate method that accounts for slenderness effects by magnifying the moments obtained from a first-order analysis. In general, a slender column must be designed to resist the combined effects from factored axial compressive loads and magnified bending moments, which are obtained by multiplying the first-order bending moments by a moment magnifier that is a function of the factored axial load and the critical buckling load of the column. The column is adequate when all of the factored load combinations for combined axial load and bending fall within or on the design strength interaction diagram, which is obtained from a strain compatibility analysis of the section (see Section 8.4).

Requirements for nonsway and sway frames are given in ACI 10.10.6 and 10.10.7, respectively, and are discussed next.

Nonsway Frames　Provisions for slender columns in nonsway frames are given in ACI 10.10.6. Columns must be designed for combinations of factored axial loads P_u and factored magnified moments M_c where M_c is determined by ACI Eq. (10-11):

$$M_c = \delta_{ns} M_2 \tag{8.22}$$

In this equation, M_2 is the larger of the two factored end moments and δ_{ns} is the moment magnification factor that accounts for second-order effects [see ACI Eq. (10-12)]:

$$\delta_{ns} = \frac{C_m}{1 - (P_u / 0.75 P_c)} \geq 1.0 \qquad (8.23)$$

The factor C_m relates the actual moment diagram to an equivalent uniform moment diagram. The moment magnification procedure assumes that the maximum moment is at or near the midheight of the column. If the maximum moment occurs at one end of the column instead of at the midheight, the design must be based on an equivalent uniform moment equal to $C_m M_2$. This would lead to the same maximum moment when magnified.

Two cases related to transverse loads between the supports must be examined. If transverse loads do not act between the supports of a compression member, the value of C_m is determined by ACI Eq. (10-16):

$$C_m = 0.6 + 0.4 \frac{M_1}{M_2} \qquad (8.24)$$

In this equation, the ratio M_1 / M_2 of the smaller factored end moment to the larger factored end moment is positive if the column is bent in single curvature and negative if the column is bent in double curvature.

For compression members that are subject to transverse loads between their supports, it is possible that the maximum moment will occur at a section away from the end of the member. In such cases, the largest calculated moment occurring anywhere along the member should be used for M_2, and this moment is magnified by δ_{ns}. The factor C_m is to be taken as 1.0 in this case.

The critical buckling load P_c is determined by ACI Eq. (10-13):

$$P_c = \frac{\pi^2 EI}{(k \ell_u)^2} \qquad (8.25)$$

In this equation, which is essentially Euler's equation for column buckling, k is the effective length factor, which can be taken equal to 1.0 for nonsway frames; ℓ_u is the unsupported length of the compression member (see Figs. 8.14 and 8.15); and EI is the stiffness of the column in the direction of analysis.

The stiffness EI must account for variations due to cracking, creep, and nonlinearity of the concrete stress–strain curve (see Chap. 2). Three methods to determine EI are presented in ACI 10.10.6.1. In the first method, EI is calculated by ACI Eq. (10-14):

$$EI = \frac{0.2 E_c I_g + E_s I_{se}}{1 + \beta_{dns}} \qquad (8.26)$$

The term I_{se} is the moment of inertia of the longitudinal reinforcement about the centroidal axis of the cross-section in the direction of analysis and β_{dns} is defined in ACI 10.10.6.2 as the ratio of the maximum factored axial sustained load to the maximum factored axial load associated with the same load combination. This factor, which is typically equal to the factored dead load divided by the total factored load in a column, accounts for the increase in lateral deflections and corresponding moments due to creep. As discussed in Chap. 2, creep of concrete transfers some of the axial load from the concrete to the longitudinal reinforcing bars over time. In columns with relatively small amounts of longitudinal reinforcement, the additional stress due to creep can eventually lead to compression failure in the bars and a corresponding reduction in stiffness. In no case shall β_{dns} be taken greater than 1.0.

Equation (8.26) was originally derived for small eccentricity ratios $M_u/P_u h$ and large axial load ratios P_u/P_o and represents the lower limit of the practical range of stiffness values, especially for columns with larger longitudinal reinforcement ratios. Under these conditions, axial load effects and, thus, slenderness effects are most pronounced.

In the second method, EI is determined by ACI Eq. (10-15):

$$EI = \frac{0.4 E_c I_g}{1 + \beta_{dns}} \tag{8.27}$$

This equation is a simplified version of Eq. (8.26) and does not directly account for the amount of longitudinal reinforcement in the section. As such, this equation is not as accurate as Eq. (8.26), especially in columns with greater amounts of longitudinal reinforcement (in such cases, the effects of the reinforcement are greatly underestimated).

In the third method, EI is determined by the following equation:

$$EI = \frac{E[0.80 + (25 A_{st}/A_g)][1 - (M_u/P_u h) - (0.5 P_u/P_o)]I_g}{1 + \beta_{dns}} \tag{8.28}$$

Equation (8.28), which is obtained by dividing ACI Eq. (10-8) by $1 + \beta_{dns}$, generally yields more accurate values of EI than those determined by Eqs. (8.26) and (8.27).

The factor 0.75 in the denominator of Eq. (8.23) is a stiffness reduction factor similar to the one derived for the reduced moments of inertia, which is equal to 0.875. Additional information on the development of this stiffness factor can be found in ACI R10.10.6.

If the factored column moments from the structural analysis are very small or are equal to zero, the Code requires that the minimum moment defined by ACI Eq. (10-17) be used in the design of the column for slenderness:

$$M_{2,min} = P_u(0.6 + 0.03h) \tag{8.29}$$

In situations where $M_{2,min} > M_2$, the factored column end moments M_1 and M_2 from the structural analysis are to be used in Eq. (8.24) to determine C_m. Alternatively, C_m can be taken equal to 1.0. In either case, this eliminates what would otherwise be a discontinuity between columns with computed moments less than $M_{2,min}$ and columns with computed moments equal to or greater than $M_{2,min}$.

Example 8.7   A typical floor plan in a five-story building is given in Fig. 8.18.

Determine the required longitudinal reinforcement in the first story of column C4 for analysis in the north-south direction. Assume normal-weight concrete with $f'_c = 4,000$ psi and Grade 60 reinforcement. Also assume (1) that the walls provide all of the resistance to wind loads in the north-south direction and (2) that the first-order relative lateral deflection Δ_o between the top and the bottom of the first story due a total factored wind load V_{us} of 246 kips is equal to 0.05 in. The 16-ft story height for the first story is measured from the mid-depth of the first elevated slab to the top of the foundation.

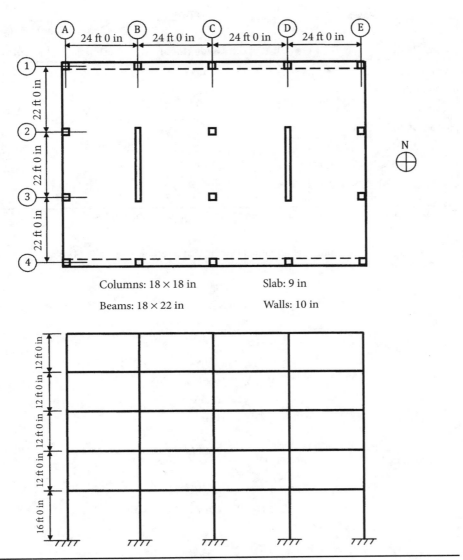

FIGURE 8.18 The typical floor plan of building given in Example 8.7.

Solution

Step 1: Determine the factored axial loads and bending moments for column C4 in the first story. A summary of the service axial loads and bending moments on the column is given in Table 8.4. These values were obtained from an elastic analysis of the frame, using the appropriate reduced moments of inertia given in ACI 10.10.4.1. Because the walls are designed to resist all of the effects from wind, column C4 is not part of the lateral force–resisting system and can be designed for the effects of gravity loads only. The applicable load combinations of ACI 9.2 are also given in Table 8.4.

Step 2: Determine if the frame in the first story is nonsway or sway. From inspection, it would appear that the frame is a nonsway frame in the north-south direction, owing to the relatively stiff

Load Case	Axial Loads (kips)	Bending Moment (ft kips)	
		Top	Bottom
Dead (D)	218.8	27.2	13.7
Snow (S)	17.2	—	—
Live (L)	109.9	17.0	8.6
Load Combination			
1 1.4D	306.3	38.1	19.2
2 1.2D + 1.6L + 0.5S	447.0	59.8	30.2
3 1.2D + 1.6S + 0.5L	345.0	41.1	20.7

TABLE 8.4   Summary of Axial Loads and Bending Moments on Column C4

walls that brace the frame over its entire height in that direction. Use Eq. (8.14) to confirm that the frame is a nonsway frame:

$$Q = \frac{\sum P_u \Delta_o}{V_{us} \ell_c} \le 0.05$$

The following total loads at the first floor level were determined from the analysis:

$$P_D = 5{,}251 \text{ kips}$$
$$P_L = 2{,}636 \text{ kips}$$
$$P_S = 412 \text{ kips}$$
$$P_W = 0 \text{ kips}$$

The total factored axial load $\sum P_u$ must correspond to the lateral loading case for which it is a maximum. In this example, ACI Eq. (9-4) produces the largest $\sum P_u$:

$$\sum P_u = 1.2 P_D + 1.6 P_W + 0.5 P_L + 0.5 P_S$$
$$= (1.2 \times 5{,}251) + (1.6 \times 0) + (0.5 \times 2{,}636) + (0.5 \times 412)$$
$$= 7{,}825 \text{ kips}$$

Therefore,

$$Q = \frac{\sum P_u \Delta_o}{V_{us} \ell_c} = \frac{7{,}825 \times 0.05}{246 \times 16 \times 12} = 0.01 < 0.05$$

Because $Q < 0.05$, the frame in the first story is a nonsway frame, as expected.

Step 3: Determine if slenderness effects need to be considered. Because the frame is a nonsway frame, slenderness effects need not be considered where Eq. (8.19) is satisfied:

$$\frac{k\ell_u}{r} \le 34 - 12\left(\frac{M_1}{M_2}\right) \le 40$$

For the 18-in square column, $r = 0.3 \times 18 = 5.4$ in.

Also, the unsupported length ℓ_u of the column in the north-south direction is the distance between the bottom of the slab and the top of the foundation:

$$\ell_u = (16 \times 12) - (9/2) = 187.5 \text{ in}$$

Using an effective length factor $k = 1.0$ for a nonsway frame results in the following slenderness ratio:

$$\frac{k\ell_u}{r} = \frac{1.0 \times 187.5}{5.4} = 34.7 < 40.0$$

Because the column is subjected to the effects of gravity loads only, it is bent in single curvature. From Table 8.4, it is evident that the ratio M_1/M_2 is approximately 0.5 for all of the load combinations. Thus,

$$\frac{k\ell_u}{r} = 34.7 > 34 - 12\left(\frac{M_1}{M_2}\right) = 34 - (12 \times 0.5) = 28$$

Therefore, slenderness effects must be considered in the design of this column.

Step 4: Determine the magnified moments in the column, using the moment magnification procedure. In lieu of an elastic or nonlinear second-order analysis, the moment magnification procedure is used to determine the magnified factored moments due to second-order effects.

Calculations are provided for load combination 2 of Table 8.4.

From Eq. (8.22):

$$M_c = \delta_{ns} M_2$$

where

$$\delta_{ns} = \frac{C_m}{1 - (P_u/0.75P_c)} \geq 1.0$$

Because there are no transverse loads between the supports of this column, use Eq. (8.24) to determine C_m:

$$C_m = 0.6 + 0.4\frac{M_1}{M_2} = 0.6 + 0.4\left(\frac{30.2}{59.8}\right) = 0.8$$

The critical buckling load P_c is determined by Eq. (8.25):

$$P_c = \frac{\pi^2 EI}{(k\ell_u)^2}$$

Compare the stiffness values determined by Eqs. (8.26) to (8.28).
The EI from Eq. (8.26) is

$$EI = \frac{0.2E_c I_g + E_s I_{se}}{1 + \beta_{dns}}$$

The modulus of elasticity of the concrete is determined by ACI 8.5.1:

$$E_c = w_c^{1.5}33\sqrt{f_c'} = (150)^{1.5} \times 33\sqrt{4,000} = 3,834,254\,\text{psi}$$

$$I_g = \frac{18^4}{12} = 8,748\,\text{in}^4$$

$$E_s = 29,000,000 \text{ psi (see ACI 8.5.2)}$$

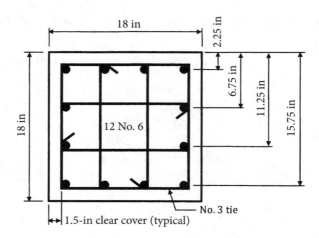

The moment of inertia I_{se} of the longitudinal reinforcement about the centroidal axis of the cross-section in the direction of analysis is determined for a specific reinforcement layout. Assume the layout depicted in Fig. 8.19. The provided area of steel is equal to $12 \times 0.44 = 5.28\ \text{in}^2$, which is greater than the lower limit of $0.01 \times 18^2 = 3.24\ \text{in}^2$ and is less than the upper limit of $0.08 \times 18^2 = 25.92\ \text{in}^2$.

Thus,

$$I_{se} = 2[(4 \times 0.44)(15.75 - 9)^2 + (2 \times 0.44)(11.25 - 9)^2] = 169.3\ \text{in}^4$$

Because the dead load is the only sustained load, β_{dns} is determined as follows:

$$\beta_{dns} = \frac{1.2D}{1.2D + 1.6L + 0.6S} = \frac{1.2 \times 218.8}{447.0} = 0.59$$

Therefore, the EI from Eq. (8.26) is

$$EI = \frac{0.2E_cI_g + E_sI_{se}}{1 + \beta_{dns}} = \frac{(0.2 \times 3{,}834 \times 8{,}748) + (29{,}000 \times 169.3)}{1 + 0.59} = 7.31 \times 10^6\ \text{kip in}^2$$

The EI from Eq. (8.27) is

$$EI = \frac{0.4E_cI_g}{1 + \beta_{dns}} = \frac{0.4 \times 3{,}834 \times 8{,}748}{1 + 0.59} = 8.44 \times 10^6\ \text{kip in}^2$$

The EI from Eq. (8.28) is

$$EI = \frac{E[0.80 + (25A_{st}/A_g)]\,[1 - (M_u/P_uh) - (0.5P_u/P_o)]\,I_g}{1 + \beta_{dns}}$$

The maximum concentric axial load P_o is determined by Eq. (5.35):

$$P_o = 0.85f_c'(A_g - A_{st}) + f_yA_{st}$$
$$= [0.85 \times 4 \times (18^2 - 5.28)] + (60 \times 5.28) = 1{,}400.5\ \text{kips}$$

Load Combination	β_{dns}	$EI \times 10^6$ (kips in^2)	P_c (kips)	δ_{ns}	$M_{2,min}$ (ft kips)	M_c (ft kips)
1	1.00	5.81	1,631	1.07	29.1	40.8
2	0.59	7.31	2,052	1.13	42.5	67.6
3	0.76	6.60	1,852	1.07	32.8	44.0

TABLE 8.5 Summary of Results from Moment Magnification Procedure for Column C4

Therefore, the EI from Eq. (8.28) is

$$EI = \frac{3,834 \times \{0.80 + [(25 \times 5.28)/18^2]\}\{1 - [(59.8 \times 12)/(447.0 \times 18)] - [(0.5 \times 447.0)/1,400.5]\} \times 8,748}{1 + 0.59}$$

$$= 19.1 \times 10^6 \text{ kip in}^2$$

The smallest value of EI will result in the smallest value of P_c, which in turn, will result in the largest moment magnification factor δ_{ns}. Therefore, use the EI determined by Eq. (8.26) to calculate P_c:

$$P_c = \frac{\pi^2 EI}{(k\ell_u)^2} = \frac{\pi^2 \times 7.31 \times 10^6}{(1.0 \times 187.5)^2} = 2,052.2 \text{ kips}$$

The moment magnification factor δ_{ns} is equal to the following:

$$\delta_{ns} = \frac{C_m}{1 - (P_u/0.75P_c)} = \frac{0.8}{1 - [447.0/(0.75 \times 2,052.2)]} = 1.13 > 1.0$$

Thus,

$$M_c = \delta_{ns} M_2 = 1.13 \times 59.8 = 67.6 \text{ ft kips}$$

Check the minimum factored bending moment requirement [see Eq. (8.29)]:

$$M_{2,min} = P_u(0.6 + 0.03h) = 447.0 \times [0.6 + (0.03 \times 18)] = 509.6 \text{ in kips} = 42.5 \text{ ft kips} < M_c$$

Therefore, the factored axial load and moment in load combination 2 become $P_u = 447.0$ kips and $M_u = M_c = 67.6$ ft kips.

Similar calculations can be performed for the other two load combinations.

A summary of the results from the moment magnification procedure for the three governing load combinations is given in Table 8.5.

Step 5: Check the adequacy of the section for combined axial load and bending. The design strength interaction diagram for column C4 reinforced with 12 No. 6 bars is given in Fig. 8.20. Also shown in the figure are the factored axial loads and magnified bending moments for the three load combinations.

Because all three load combination points fall within the design strength interaction diagram, the column is adequate for analysis in the north-south direction.

Sway Frames Provisions of the moment magnification procedure for sway frames are given in ACI 10.10.7. The design procedure essentially consists of two steps.

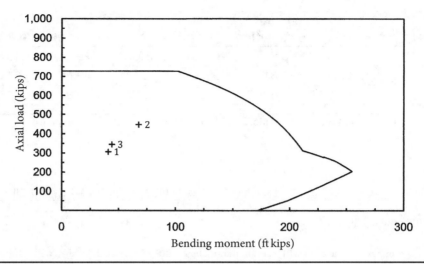

Figure 8.20 Design strength interaction diagram for column C4.

Step 1: Determine the magnified sway moments. In the first step, the magnified sway moments $\delta_s M_{1s}$ and $\delta_s M_{2s}$ at the ends of the columns are calculated. The moments M_{1s} and M_{2s} are due to loads that cause appreciable sidesway (e.g., wind and seismic loads) and are determined using a first-order elastic analysis of the frame.

The Code recognizes two methods of determining the moment magnification factor δ_s for sway frames. In the first method, δ_s is determined using the stability index Q for the story [see ACI Eq. (10-20)]:

$$\delta_s = \frac{1}{1-Q} \geq 1 \tag{8.30}$$

This equation is the solution of the infinite series that represents a general iterative P-delta analysis for second-order moments.[3] It has been demonstrated that this equation closely predicts the magnitudes of second-order moments in a sway frame for δ_s equal to or less than 1.5.[9] If it is found that δ_s exceeds 1.5, then δ_s must be calculated using a second-order elastic analysis in accordance with ACI 10.10.3 or 10.10.4 or the method given in ACI 10.10.7.4.

In the derivation of Eq. (8.30), it is assumed that the second-order moments are caused by equal and opposite forces of $P\Delta/\ell_c$, applied at the top and bottom of the story. This produces a second-order moment diagram that is a straight line. The actual moment diagram is curved, and it leads to displacements that are approximately 15% greater than those from the approximate second-order moment diagram. This effect can be included in the analysis by changing the denominator in Eq. (8.30) to $(1-1.15Q)$. However, the Code permits the simpler form of Eq. (8.30) without this modification.

The stability index Q is determined by Eq. (8.14) and is based on deflections that are calculated using an elastic first-order analysis and reduced section properties in accordance with Table 8.3.

In the second method, δ_s is permitted to be determined by ACI Eq. (10-21), which was used in the magnified moment procedure that appeared in earlier editions of the Code:

$$\delta_s = \frac{1}{1 - (\Sigma P_u / 0.75 \Sigma P_c)} \geq 1 \tag{8.31}$$

In this equation, the factored axial loads P_u are summed over the entire story, and the critical buckling loads P_c are summed over the sway-resisting columns in that story. This equation reflects the interaction of all of the columns in a story of a sway frame. In the absence of torsional displacements about a vertical axis, the lateral displacements of all the columns in a sway frame should be the same. A three-dimensional second-order analysis should be used in situations where there is a significant torsional displacement of the structure due to lateral loads; the moment magnification in the columns farthest from the center of twist can be underestimated in such cases by the approximate moment magnification procedure.

Like in the case of nonsway frames, the stiffness EI that is used in the calculation of P_c is determined by ACI 10.10.6.1. The term β_{ds}, which is defined in ACI 10.10.4.2, is used in the denominator of the equations for EI instead of β_{dns}. As noted previously, β_{ds} will normally be zero for a sway frame because wind and seismic loads are not sustained loads. In certain cases, sustained loads may act on a structure (such as lateral earth pressure) and β_{ds} will not be zero.

In the case of very slender columns, it is possible for the midheight deflection to be substantial even though the column is adequately braced by other columns in the frame. ACI R10.10.7.4 recommends that the adequacy of such columns be checked using the nonsway provisions given in ACI 10.10.6.

Step 2: Determine total moments at the ends of the column. The magnified sway moments determined in Step 1 are added to the unmagnified moments M_{ns} at each end of the column. The moments M_{ns} are due to loads that cause no appreciable sidesway (such as gravity loads) and are computed using a first-order elastic analysis.

The following equations can be used to determine the minimum and maximum total moments M_1 and M_2 at the ends of the column [see ACI Eqs. (10-18) and (10-19)]:

$$M_1 = M_{1ns} + \delta_s M_{1s} \tag{8.32}$$

$$M_2 = M_{2ns} + \delta_s M_{2s} \tag{8.33}$$

In these equations, M_{1ns} and M_{1s} are the factored ends moments due to loads that cause no appreciable sidesway and due to those that cause appreciable sidesway, respectively, at the end of the column at which M_1 acts. Similarly, M_{2ns} and M_{2s} are the factored ends moments due to loads that cause no appreciable sidesway and due to those that cause appreciable sidesway, respectively, at the end of the column at which M_2 acts. By definition, M_1 and M_2 are the smaller and larger end moments, respectively. The magnitude of M_1 is taken as positive if a column is bent in single curvature and negative if bent in double curvature, and M_2 is taken as the largest moment in the member if transverse loading occurs between the supports.

In lieu of determining the magnified sway moments, using the two methods presented earlier, $\delta_s M_{s1}$ and $\delta_s M_{2s}$ can be determined by an elastic or nonlinear second-order analysis in accordance with ACI 10.10.4 and 10.10.3, respectively.

Load Case	Axial Loads (kips)	Bending Moment (ft kips)			
		Top	Bottom	M_1	M_2
Dead (D)	218.8	0.0	0.0	0.0	0.0
Snow (S)	17.2	—	—	—	—
Live (L)	109.9	−15.5	−7.8	−7.8	−15.5
Wind (W)	0	±50.6	±421.5	50.6	421.5
Load Combination					
1 $1.4D$	306.3	0.0	0.0	0.0	0.0
2 $1.2D + 1.6L + 0.5S$	447.0	−24.8	−12.5	−12.5	−24.8
3 $1.2D + 1.6S + 0.5L$	345.0	−7.8	−3.9	−3.9	−7.8
4 $1.2D + 1.6S + 0.8W$	290.1	40.5	337.2	40.5	337.2
5 $1.2D + 1.6S − 0.8W$	290.1	−40.5	−337.2	−40.5	−337.2
6 $1.2D + 1.6W + 0.5L + 0.5S$	326.1	73.2	670.5	73.2	670.5
7 $1.2D − 1.6W + 0.5L + 0.5S$	326.1	−88.7	−678.3	−88.7	−678.3
8 $0.9D + 1.6W$	196.9	81.0	674.4	81.0	674.4
9 $0.9D − 1.6W$	196.9	−81.0	−674.4	−81.0	−674.4

TABLE 8.6   Summary of Axial Loads and Bending Moments on Column C4

ACI 10.10.7.1 requires that the magnified moments at the ends of a column must be considered in the design of the beams framing into the column. The stiffnesses of the beams in a sway frame play a key role in the stability of the columns, and this provision ensures that the beams will have adequate strength to resist the magnified column moments.

Example 8.8   Check the adequacy of column C4 of Example 8.7 for analysis in the east-west direction. Assume that the first-order relative lateral deflection Δ_o between the top and the bottom of the first story due a total factored wind load V_{us} of 169 kips is equal to 0.55 in.

Solution
   Step 1: Determine the factored axial loads and bending moments for column C4 in the first story. A summary of the service axial loads and bending moments on the column is given in Table 8.6. These values were obtained from an elastic analysis of the frame, using the appropriate reduced moments of inertia given in ACI 10.10.4.1. The moments M_1 and M_2, which correspond to the smaller and larger end moments, respectively, are also given in the table.
   Column C4 is in the moment frame along line 4 that forms part of the lateral force–resisting system in the east-west direction; therefore, it is subjected to the effects of gravity and wind loads. The live load bending moments at the top and bottom of the column are due to pattern live loading on the spans adjacent to the column.
   Also included in Table 8.6 are the applicable load combinations of ACI 9.2. The corresponding factored moments M_1 and M_2 are also given in the table.
   The bending moments must be categorized as those that do and those that do not cause appreciable sway. Because the frame is symmetrical, the gravity loads do not cause appreciable sidesway.

Load Combination	Bending Moments (ft kips)					
	M_1	M_2	M_{1ns}	M_{2ns}	M_{1s}	M_{2s}
1	0.0	0.0	0.0	0.0	—	—
2	−12.5	−24.8	−12.5	−24.8	—	—
3	−3.9	−7.8	−3.9	−7.8	—	—
4	40.5	337.2	0.0	0.0	40.5	337.2
5	−40.5	−337.2	0.0	0.0	−40.5	−337.2
6	73.2	670.5	−7.8	−3.9	81.0	674.4
7	−88.7	−678.3	−7.8	−3.9	−81.0	−674.4
8	81.0	674.4	0.0	0.0	81.0	674.4
9	−81.0	−674.4	0.0	0.0	−81.0	−674.4

TABLE 8.7   Summary of Nonsway and Sway Moments for Column C4

Therefore, the gravity load moments are designated as M_{ns}. The wind loads cause appreciable sidesway, and these moments are designated as M_s. A summary of the nonsway and sway moments for each load combination is given in Table 8.7.

It is important to note that the moments M_{1ns} and M_{1s} are the factored nonsway and sway moments, respectively, at the end of the column at which M_1 acts. Similarly, M_{2ns} and M_{2s} are the factored nonsway and sway moments, respectively, at the end of the column at which M_2 acts. For example, in load combination 6, $M_1 = 73.2$ ft kips, and it acts at the top of the column (see Table 8.6). Therefore, the moments M_{1ns} and M_{1s} are determined using the service load moments at the top of the column because they correspond to M_1:

$$M_{1ns} = 1.2M_D + 0.5M_L$$
$$= (1.2 \times 0.0) + [0.5 \times (-15.5)] = -7.8 \text{ ft kips}$$
$$M_{1s} = 1.6M_W = 1.6 \times 50.6 = 81.0 \text{ ft kips}$$

The moments M_{2ns} and M_{2s} are determined using the service loads at the bottom of the column because they correspond to M_2:

$$M_{2ns} = 1.2M_D + 0.5M_L$$
$$= (1.2 \times 0.0) + [0.5 \times (-7.8)] = -3.9 \text{ ft kips}$$
$$M_{2s} = 1.6M_W = 1.6 \times 421.5 = 674.4 \text{ ft kips}$$

The other moments given in Table 8.7 can be obtained in a similar fashion.

Step 2: Determine if the frame in the first story is nonsway or sway. From inspection, it would appear that the frame is a sway frame in the east-west direction, owing to the relatively flexible moment frames that are used to brace the frame over the height of the building in that direction. Use Eq. (8.14) to confirm that the frame is a sway frame:

$$Q = \frac{\sum P_u \Delta_o}{V_{us} \ell_c} > 0.05$$

The following total loads at the first floor level were determined from the analysis:

$$P_D = 5{,}251 \text{ kips}$$
$$P_L = 2{,}636 \text{ kips}$$
$$P_S = 412 \text{ kips}$$
$$P_W = 0 \text{ kips}$$

The total factored axial load ΣP_u must correspond to the lateral loading case for which it is a maximum. In this example, ACI Eq. (9-4) produces the largest ΣP_u:

$$\begin{aligned}
\Sigma P_u &= 1.2 P_D + 1.6 P_W + 0.5 P_L + 0.5 P_S \\
&= (1.2 \times 5{,}251) + (1.6 \times 0) + (0.5 \times 2{,}636) + (0.5 \times 412) \\
&= 7{,}825 \text{ kips}
\end{aligned}$$

The length of the column ℓ_c in this direction of analysis is equal to the distance from the midheight of the beam to the top of the foundation (recall from Example 8.7 that the 16-ft story height for the first story is measured from the mid-depth of the first elevated slab to the top of the foundation):

$$\ell_c = 16 - \frac{9}{2 \times 12} - \frac{22 - 9}{12} + \frac{22}{2 \times 12} = 15.46 \text{ ft}$$

Therefore,

$$Q = \frac{\Sigma P_u \Delta_o}{V_{us} \ell_c} = \frac{7{,}825 \times 0.55}{169 \times 15.46 \times 12} = 0.14 > 0.05$$

Because $Q > 0.05$, the frame in the first story is a sway frame, as expected.

Step 3: Determine if slenderness effects need to be considered. Because the frame is a sway frame, slenderness effects need not be considered where ACI Eq. (10-6) is satisfied:

$$\frac{k \ell_u}{r} \le 22$$

For the 18-in square column, $r = 0.3 \times 18 = 5.4$ in.

Because the column-line beam provides lateral support to the top of the column in the east-west direction, the unsupported length ℓ_u of the column in this direction is the distance between the bottom of the beam and the top of the foundation:

$$\ell_u = 16 - \frac{9}{2 \times 12} - \frac{22 - 9}{12} = 14.54 \text{ ft}$$

The effective length factor k is determined using the alignment chart shown in ACI Fig. R10.10.1.1(b) for sway frames. At the bottom of the column, which is essentially fixed against rotation, use a stiffness ratio $\Psi_B = 1.0$.

The stiffness ratio Ψ_A at the top of the column is determined using the ratio of the stiffness of the columns to the stiffness of the beams.

The moments of inertia of the columns and beams are calculated using the reduction factors given in ACI 10.10.4.1.

$$I_{col} = 0.7 \times \frac{1}{12} \times 18^4 = 6{,}124 \text{ in}^4$$
$$E_c = 33(w_c)^{1.5} \sqrt{f_c'} = 33 \times (150)^{1.5} \times \sqrt{4{,}000}/1{,}000 = 3{,}834 \text{ ksi}$$

For the column below the first elevated level,

$$\frac{E_c I_{col}}{\ell_c} = \frac{3,834 \times 6,124}{15.46 \times 12} = 127 \times 10^3 \text{ in kips}$$

For the column above the first elevated level,

$$\frac{E_c I_{col}}{\ell_c} = \frac{3,834 \times 6,124}{12 \times 12} = 163 \times 10^3 \text{ in kips}$$

$$I_{beam} = 0.35 \times \frac{1}{12} \times 18 \times 22^3 = 5,590 \text{ in}^4$$

$$\frac{E_c I_{beam}}{\ell} = \frac{3,834 \times 5,590}{24 \times 12} = 74 \times 10^3 \text{ in kips}$$

Thus, Ψ_A at the top of the column is determined by Eq. (8.16):

$$\Psi_A = \frac{\sum EI/\ell_c}{\sum EI/\ell} = \frac{127 + 163}{2 \times 74} = 2.0$$

From the alignment chart shown in ACI Fig. R10.10.1.1(b) for sway frames with $\Psi_A = 2.0$ and $\Psi_B = 1.0, k \approx 1.45$.

The slenderness ratio for column C4 is

$$\frac{k\ell_u}{r} = \frac{1.45 \times 14.54 \times 12}{5.4} = 47 > 22$$

Because, $k\ell_u/r > 22$, slenderness effects must be considered.

Step 4: Determine the total moment M_2 in the column including slenderness effects. The total moment M_2 in the column is determined by Eq. (8.33) for each of the load combinations:

$$M_2 = M_{2ns} + \delta_s M_{2s}$$

This moment is used in the design of the column because it is greater than the moment M_1.

The nonsway and sway moments M_{2ns} and M_{2s}, respectively, are given in Table 8.7 for all of the load combinations.

Two methods are given in the Code to determine the magnification factor δ_s for sway frames. Both methods are examined next.

Determination of δ_s by ACI 10.10.7.3

In ACI 10.10.7.3, δ_s is determined using the stability index Q for the story:

$$\delta_s = \frac{1}{1 - Q} \geq 1$$

where

$$Q = \frac{\sum P_u \Delta_o}{V_{us} \ell_c}$$

In the equation for Q, the only constant is ℓ_c; all of the other quantities depend on the load combination.

Table 8.8 contains a summary of the calculations for δ_s and M_2, using ACI 10.10.7.3.

Load Combination	ΣP_u (kips)	Δ_o (in.)	V_{us} (kips)	Q	δ_s	M_2 (ft kips)
1	7,351	—	—	—	—	0.0
2	10,725	—	—	—	—	−24.8
3	8,278	—	—	—	—	−7.8
4	6,960	0.27	84.5	0.12	1.14	384.4
5	6,960	0.27	84.5	0.12	1.14	−384.4
6	7,825	0.55	169.0	0.14	1.16	774.5
7	7,825	0.55	169.0	0.14	1.16	−786.2
8	4,726	0.55	169.0	0.08	1.09	735.1
9	4,726	0.55	169.0	0.08	1.09	−735.1

TABLE 8.8 Summary of Slenderness Calculations for Column C4, Using ACI 10.10.7.3

The calculations are illustrated for load combination number 7. Similar calculations can be performed for the other load combinations.

$$\Sigma P_u = 1.2 P_D - 1.6 P_W + 0.5 P_L + 0.5 P_S$$

$$= (1.2 \times 5{,}251) - (1.6 \times 0) + (0.5 \times 2{,}636) + (0.5 \times 412)$$

$$= 7{,}825 \text{ kips}$$

$$Q = \frac{\sum P_u \Delta_o}{V_{us} \ell_c} = \frac{7{,}825 \times 0.55}{169 \times 15.46} = 0.14$$

$$\delta_s = \frac{1}{1-Q} = \frac{1}{1-0.14} = 1.16$$

$$M_2 = M_{2ns} + \delta_s M_{2s}$$

$$= -3.9 + [1.16 \times (-674.4)] = -786.2 \text{ ft kips}$$

Determination of δ_s by ACI 10.10.7.4

In ACI 10.10.7.4, δ_s is determined using Eq. (8.31):

$$\delta_s = \frac{1}{1 - (\Sigma P_u / 0.75 \Sigma P_c)} \geq 1$$

In this equation, the sum of the critical buckling loads P_c for all of the sway-resisting columns in the first story is determined by Eq. (8.25):

$$P_c = \frac{\pi^2 EI}{(k \ell_u)^2}$$

The stiffness EI is calculated in accordance with ACI 10.10.6.1, and the effective length factor k is determined in accordance with ACI 10.10.7.2 for sway frames.

It is evident from the design strength interaction diagram of this column shown in Fig. 8.20 that the section reinforced with 12 No. 6 bars is not adequate for the total bending moment combined with a factored axial load of 326.1 kips. Therefore, because the reinforcement is not known at this point in the analysis, use Eq. (8.27) to determine EI:

$$EI = \frac{0.4E_cI_g}{1 + \beta_{ds}} = \frac{0.4 \times 3{,}834 \times 18^4}{12(1 + 0)} = 13.42 \times 10^6 \text{ in}^2 \text{ kips}$$

The term β_{ds} is equal to zero in this equation because the wind load is not a sustained lateral load.

This value of EI is applicable to all of the columns in the moment frames (sway-resisting columns) along lines 1 and 4 because the sizes of these columns are all the same.

The effective length factor k was determined in step 3 for column C4 and is equal to 1.45. This value of k is applicable to columns B1, C1, D1, B4, C4, and D4 because they all have two beams framing into the top of the columns. Therefore, P_c for these columns is equal to the following:

$$P_c = \frac{\pi^2 EI}{(k\ell_u)^2} = \frac{\pi^2 \times 13.42 \times 10^6}{(1.45 \times 14.54 \times 12)^2} = 2{,}069 \text{ kips}$$

Columns A1, E1, A4, and E4 have only one beam framing into the top of the columns, and the ratio Ψ_A is determined using Eq. (8.16):

$$\Psi_A = \frac{\sum EI/\ell_c}{\sum EI/\ell} = \frac{127 + 163}{74} = 3.9$$

With $\Psi_A = 3.9$ and $\Psi_B = 1.0$, the effective length factor k from ACI Fig. R10.10.1.1(b) for sway frames is approximately equal to 1.61.

Therefore, P_c for these columns is equal to the following:

$$P_c = \frac{\pi^2 EI}{(k\ell_u)^2} = \frac{\pi^2 \times 13.42 \times 10^6}{(1.61 \times 14.54 \times 12)^2} = 1{,}679 \text{ kips}$$

The summation of P_c for all of the sway-resisting columns in the first story is equal to

$$\sum P_c = (6 \times 2{,}069) + (4 \times 1{,}679) = 19{,}130 \text{ kips}$$

Table 8.9 contains a summary of the calculations for δ_s and M_2, using ACI 10.10.7.4.

Load Combination	$\sum P_u$ (kips)	V_{us} (kips)	δ_s	M_2 (ft kips)
1	7,351	—	—	0.0
2	10,725	—	—	−24.8
3	8,278	—	—	−7.8
4	6,960	84.5	1.94	654.9
5	6,960	84.5	1.94	−654.9
6	7,825	169.0	2.20	1,479.6
7	7,825	169.0	2.20	−1,487.4
8	4,726	169.0	1.49	1,005.6
9	4,726	169.0	1.49	−1,005.6

TABLE 8.9 Summary of Slenderness Calculations for Column C4, Using ACI 10.10.7.4

It is evident from Table 8.9 that the magnification factor δ_s is greater than the limiting value of 1.4 given in ACI 10.10.2.1. As such, this approximate method, which is generally conservative, should not be used. The results from Table 8.8 are used to design the column.

Step 5: Check the adequacy of the section for combined axial load and bending. The column reinforced with 12 No. 6 bars is not adequate for the critical load combinations given in Table 8.8. Initially, try increasing the amount of longitudinal reinforcement and the compressive strength of the concrete.

It can be shown that this column is not adequate even when the reinforcement ratio is increased to approximately 6% and the compressive strength is increased to 7,000 psi.

Therefore, the dimensions of the column must be increased. This entails performing another first-order elastic analysis of the structure for combined gravity and wind loads. The results of this analysis and subsequent design of the column are not shown here.

It may be possible to determine a column size so that slenderness effects can be neglected. In this example, the beam stiffness at the top of the column has a significant influence on the degree of slenderness for this sway frame. A few iterations using different combinations of column and beam sizes can result in a design where the effects from slenderness need not be considered.

8.6 Biaxial Loading

Axial load in combination with biaxial bending moments must be considered in the design of corner columns and other columns where moments occur simultaneously about both principal axes.

As was discussed in Section 8.4, an interaction diagram for a reinforced concrete section represents the strength of the section for combinations of axial load and bending moment about one of the principal axes. These interaction diagrams facilitate the design of such members.

The strength of an axially loaded column subjected to bending moments about both principal axes can be represented by a biaxial strength interaction surface, as shown in Fig. 8.21.

This surface is formed by a series of uniaxial interaction diagrams that are drawn radially from the vertical axis. Intermediate interaction diagrams between the angle θ equal to 0 degrees (uniaxial bending about the y-axis) and 90 degrees (uniaxial bending about the x-axis) are obtained by varying the angle of the neutral axis for assumed strain configurations.

In general, a biaxial strength interaction surface is obtained by performing a series of strain compatibility analyses that are involved and time-consuming. Designing a column for combined axial load and biaxial bending is a very lengthy process without the use of a computer program.

A number of simplified methods that have produced satisfactory results have been developed through the years. A summary of these methods can be found in ACI R10.3.6 and R10.3.7.

The reciprocal load method provides a simple and conservative estimate of the strength of a member under biaxial loading conditions.[10] The nominal axial load strength P_{ni} corresponding to eccentricities about both axes of a section can be obtained from the following equation:

$$\frac{1}{P_{ni}} = \frac{1}{P_{nx}} + \frac{1}{P_{ny}} - \frac{1}{P_o} \tag{8.34}$$

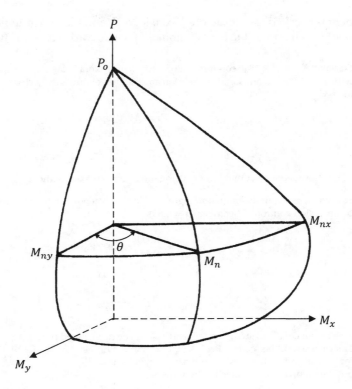

FIGURE 8.21 Biaxial strength interaction surface.

where P_{nx} = nominal axial load strength when the member is subjected to the uniaxial moment M_{nx}

P_{ny} = nominal axial load strength when the member is subjected to the uniaxial moment M_{ny}

P_o = nominal axial load strength at zero eccentricity

Determination of the quantities that make up P_{ni} is relatively straightforward. The nominal axial load strengths P_{nx} and P_{ny} can be obtained from uniaxial nominal strength strain compatibility analyses, and P_o is determined by Eq. (5.35).

The design axial load strength ϕP_{ni} is obtained by multiplying P_{ni} from Eq. (8.34) by the appropriate strength reduction factor ϕ based on the strain in the reinforcing bars farthest from the compression face of the section. This design strength must be equal to or greater than the factored axial load P_u in order to satisfy the requirements of the strength design method.

The reciprocal load method produces reasonably accurate results when flexure does not govern the design of the section, that is, when P_{nx} and P_{ny} are greater than the axial load P_b corresponding to balanced failure. In cases where flexure governs, one of the other methods referenced in ACI R10.3.6 and R10.3.7 should be utilized.

Slenderness effects must also be considered in the design of columns subjected to biaxial bending. The moment about each axis is magnified separately on the basis of the restraint conditions corresponding to that axis. The adequacy of the compression member is checked for the factored axial load P_u and the maximum magnified bending moment for all applicable load combinations in accordance with ACI 9.2.

Example 8.9   A 28 × 28 in column is reinforced with 12 No. 10 bars. Assume normal-weight concrete with $f'_c = 6,000$ psi and Grade 60 reinforcement. The column is subjected to the following factored loads:

$$P_u = 2,000 \text{ kips}$$
$$M_{ux} = 400 \text{ ft kips}$$
$$M_{uy} = 200 \text{ ft kips}$$

Check the adequacy of the column, using the reciprocal load method.

Solution   The required nominal strength axial load and bending moments are determined assuming compression-controlled behavior ($\phi = 0.65$):

$$P_n = \frac{2,000}{0.65} = 3,077 \text{ kips}$$

$$M_{nx} = \frac{400}{0.65} = 615.4 \text{ ft kips}$$

$$M_{ny} = \frac{200}{0.65} = 307.7 \text{ ft kips}$$

The uniaxial load strengths P_{nx} and P_{ny} can be determined from the nominal strength interaction diagram of the section. Because the section is symmetrical, the same diagram can be utilized for bending about the x-axis and y-axis (see Fig. 8.22).
The following uniaxial load strengths are obtained from Fig. 8.22:

- For $M_{nx} = 615.4$ ft kips, $P_{nx} = 4,195$ kips
- For $M_{ny} = 307.7$ ft kips, $P_{ny} = 4,505$ kips

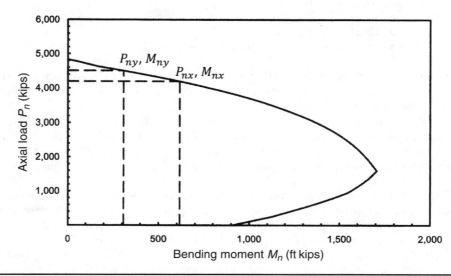

Figure 8.22   Nominal strength interaction diagram for the column given in Example 8.9.

Note that both P_{nx} and P_{ny} are greater than the axial load at balanced failure, which confirms that the section is compression-controlled. Thus, the reciprocal load method can be used. The nominal axial load strength at zero eccentricity P_o is determined by Eq. (5.35):

$$P_o = 0.85 f_c'(A_g - A_{st}) + f_y A_{st}$$
$$= 0.85 \times 6 \times [28^2 - (12 \times 1.27)] + (60 \times 12 \times 1.27) = 4{,}835 \text{ kips}$$

The nominal axial load strength P_{ni} is determined by Eq. (8.34):

$$\frac{1}{P_{ni}} = \frac{1}{P_{nx}} + \frac{1}{P_{ny}} - \frac{1}{P_o} = \frac{1}{4{,}195} + \frac{1}{4{,}505} - \frac{1}{4{,}835}$$

or $P_{ni} = 3{,}944$ kips.

Because the section is compression-controlled,

$$\phi P_{ni} = 0.65 \times 3{,}944 = 2{,}564 \text{ kips} > P_u = 2{,}000 \text{ kips}$$

Thus, the section is adequate for combined axial load and biaxial bending.

8.7 Reinforcement Details

8.7.1   Overview

Longitudinal and lateral reinforcement for compression members must satisfy the requirements given in ACI Chaps. 7 and 10. Limitations are provided on the size and spacing of both types of reinforcement.

The longitudinal and lateral bars must be spaced far enough apart so that concrete can flow easily between the bars without honeycombing. Minimum bar spacing is especially critical at splice locations. The lateral reinforcement must be spaced close enough to provide adequate lateral support to the longitudinal reinforcement and to provide sufficient shear strength where needed.

Splice requirements for longitudinal reinforcement in columns are given in ACI 12.17. The type of lap splice that must be provided is based on the stress in the reinforcing bars under factored loads.

8.7.2   Limits for Reinforcement

Longitudinal Reinforcement
Minimum and Maximum Areas of Longitudinal Reinforcement   The minimum and maximum areas of longitudinal reinforcement are prescribed in ACI 10.9.1 (see Section 5.5 of this book). The following limits are applicable regardless of the type of lateral reinforcement that is used in the member:

- Minimum $A_{st} = 0.01 A_g$
- Maximum $A_{st} = 0.08 A_g$

In these equations, A_g is the gross cross-sectional area of the member.

The 1% lower limit is meant to provide resistance to any bending moments that are not accounted for in the analysis because of, for example, construction tolerances or misalignments. This lower limit is also meant to help reduce creep and shrinkage in the concrete under sustained compressive stresses. After time, a portion of the sustained compressive stress is transferred to the longitudinal reinforcement, and having a minimum amount of such reinforcement helps in resisting these stresses.

In order for the concrete to be properly placed and consolidated, the size and number of longitudinal reinforcing bars must be chosen to minimize reinforcement congestion. The upper limit on the longitudinal reinforcement is meant to help achieve these goals. The maximum area of reinforcement must not exceed 4% of the gross column area at sections where lap splices are utilized. As noted previously, economy is achieved in column design when the ratio of the longitudinal reinforcement is between 1% and 2% of the gross area of the section.

ACI 10.8.2 through 10.8.4 permit a column to be designed of sufficient size to carry the required factored loads and to add additional concrete to the section without having to increase the minimum longitudinal reinforcement to satisfy ACI 10.9.1. The additional concrete should not be considered to carry any load; however, it should be considered in analysis because it increases the stiffness of the section.

The provision given in ACI 10.8.4 is commonly employed in the design of columns in the upper floors of a building. It is economical to use the same column size over a number of floors, so at the top of the building where the load requirements are usually less, the column cross-section is typically larger than required for loading. ACI 10.8.4 permits the minimum reinforcement to be based on a reduced effective area that is equal to or greater than 50% of the total gross area. For example, assume that a column has a required axial strength $P_u = 250$ kips. Also assume that the cross-section and area of longitudinal reinforcement have been determined on the basis of a minimum reinforcement ratio of 1% and that $\phi P_n = 450$ kips. The ratio of the required strength to the design strength is $P_u/\phi P_n = 250/450 = 0.56$. Therefore, according to ACI 10.8.4, the minimum longitudinal reinforcement ratio may be taken as $0.56 \times 1.0 = 0.56\%$ instead of 1%. If $\phi P_n = 600$ kips instead of 450 kips, $P_u/\phi P_n = 250/600 = 0.42$, which is less than 0.50. In this case, the minimum longitudinal reinforcement ratio may be taken as 0.50% instead of 1%.

The reduction of the minimum area of longitudinal reinforcement according to ACI 10.8.4 is not applicable to special moment frames or special structural walls designed in accordance with the seismic provisions given in ACI Chap. 21.

Minimum Number of Longitudinal Bars   A minimum of four longitudinal bars are required in compression members where rectangular or circular ties are used as lateral reinforcement (ACI 10.9.2). For bars enclosed by spirals conforming to ACI 10.9.3, a minimum of six longitudinal bars are required.

For members that have a circular arrangement of longitudinal bars, the orientation of the bars will affect the moment strength of the column where the number of longitudinal bars is less than eight. Calculations that illustrate this are given in Example 8.5. Because it is impossible to know the orientation of the bars placed in the field, the design of the column must be based on the most critical bar orientation, that is, the orientation that results in the lowest strength.

A minimum of three longitudinal bars are required in sections that utilize triangular ties. A tied triangular column needs three longitudinal bars located at each apex of the triangular ties.

Spacing of Longitudinal Bars   The longitudinal bars in reinforced concrete columns must be spaced at a sufficient distance so that concrete can flow easily between the bars and between the bars and the formwork. The provisions given in ACI 7.6.3 are intended to satisfy this objective.

In spiral or tie columns, the minimum clear distance that is to be provided between longitudinal bars is equal to the largest of the following (see ACI 7.6.3):

1. 1.5 times the diameter of the bar

2. 1.5 in

3. 1.33 times the maximum aggregate size

The third criterion is related to the maximum aggregate size and is provided to ensure that the concrete fully encases the reinforcing bars without honeycombing.

Minimum clear distance requirements are also applicable to the clear distance between a contact lap splice and any adjacent bars or splices.

Figure 8.23 contains the minimum face dimension of rectangular tied columns with normal lap splices based on the requirements presented earlier. The column face dimensions have been rounded to the nearest inch and have been determined using a 1.5-in clear cover to No. 4 ties.

The following equations were utilized to generate the information given in Fig. 8.23:

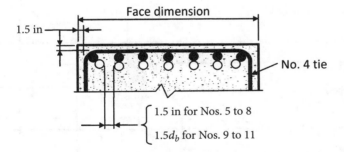

Bar	Number of bars per face												
size No.	2	3	4	5	6	7	8	9	10	11	12	13	14
5	8	10	12	14	17	19	21	23	25	27	29	31	34
6	9	11	13	15	18	20	22	24	27	29	31	33	36
7	9	11	14	16	18	21	23	26	28	30	33	35	37
8	9	12	14	17	19	22	24	27	29	32	34	37	39
9	10	13	16	18	21	24	27	30	33	35	38	41	44
10	11	14	17	20	23	27	30	33	36	39	42	46	49
11	11	15	18	22	25	29	32	36	40	43	47	50	54

FIGURE 8.23   Minimum face dimension (inches) of rectangular tied columns with normal lap splices.

- For Nos. 5 through 8 longitudinal bars,

$$\text{Minimum face dimension} = 2(\text{cover} + d_{tie}) + nd_b + 1.5(n-1)$$
$$+ [(3 + 2d_b) \times \cos\theta - 0.586d_b - 3] \quad (8.35)$$

- For Nos. 9 through 11 longitudinal bars,

$$\text{Minimum face dimension} = 2(\text{cover} + d_{tie}) + nd_b + 1.5(n-1) + 1.38d_b \quad (8.36)$$

In these equations,

$$n = \text{number of longitudinal bars per face}$$
$$d_{tie} = \text{diameter of tie bars}$$
$$d_b = \text{diameter of longitudinal bars}$$
$$\theta = \arcsin\frac{(1 - 1/\sqrt{2})d_b}{1.5 + d_b} \quad \text{(see Fig. 8.24)}$$

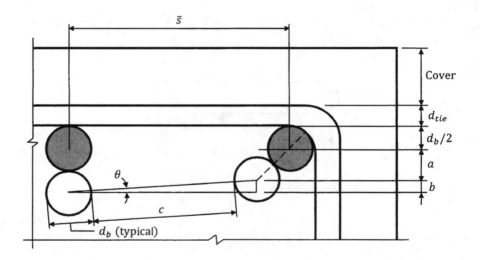

$$a = d_b/\sqrt{2}$$

$$b = (1 - 1/\sqrt{2})d_b$$

$$c = \begin{cases} 1.5 \text{ in for Nos. 3 through 8 bars} \\ 1.5d_b \text{ for Nos. 9 through 11 bars} \end{cases}$$

$$\theta = \arcsin[b/(c + d_b)]$$

FIGURE 8.24 Distance between bars at the corner of a column with normal lap splices.

These equations can also be used to determine minimum column face dimensions for other tie sizes and cover.

Example 8.10   Determine the minimum face dimension of a rectangular column to accommodate seven No. 8 bars. Assume a 1.5-in clear cover to No. 4 ties.

Solution   Equation (8.35) is used to determine the minimum face dimension because No. 8 bars are specified:

Minimum face dimension $= 2(\text{cover} + d_{tie}) + nd_b + 1.5(n-1) + [(3 + 2d_b)\cos\theta - 0.586d_b - 3]$

$$= 2(1.5 + 0.50) + (7 \times 1.0) + 1.5(7-1) + [(3 + (2 \times 1.0))\cos 11.3 - (0.586 \times 1.0) - 3] = 21.3\,\text{in}$$

where

$$\theta = \arcsin\frac{(1 - 1/\sqrt{2})d_b}{1.5 + d_b} = \arcsin\frac{(1 - 1/\sqrt{2}) \times 1.0}{1.5 + 1.0} = 11.3\,\text{degrees}$$

Round the minimum face dimension up to 22 in; this matches the minimum face dimension shown in Fig. 8.23.

Figure 8.25 contains the maximum number of bars in a circular or square column that has longitudinal bars arranged in a circle with normal lap splices. The information given in this figure satisfies the minimum clear distance requirements of ACI 7.6.3 and the reinforcement limits of ACI 10.9.1. The number of bars have been rounded to the nearest whole number and were determined using a 1.5-in clear cover to No. 4 spirals or ties.

The following equation was utilized to generate the information given in Fig. 8.25 (see Fig. 8.26):

$$n = \frac{180}{\arcsin\left[(\bar{s}/2)/a\right]} \tag{8.37}$$

In this equation,

$$n = \text{maximum number of longitudinal bars}$$
$$\bar{s} = \text{minimum clear space between longitudinal bars}$$
$$= 1.5\,\text{in} + d_b \text{ for Nos. 5 though 8 bars}$$
$$= 1.5d_b + d_b = 2.5d_b \text{ for Nos. 9 through 11 bars}$$
$$a = h/2 - [\text{cover} + (d_s\text{ or }d_{tie})] - 1.5d_b$$
$$d_s = \text{diameter of spiral bar}$$
$$d_{tie} = \text{diameter of tie bars}$$
$$d_b = \text{diameter of longitudinal bars}$$

This equation can also be used to determine the maximum number of bars for other spiral or tie sizes and cover.

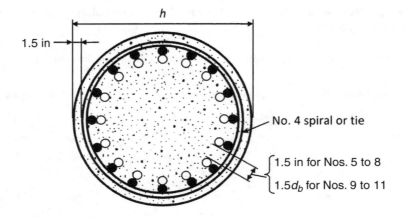

h	Bar size (No.)						
(in)	**5**	**6**	**7**	**8**	**9**	**10**	**11**
12	8	7	6	6	4*	—	—
14	11	10	9	8	7	5*	4*
16	14	13	12	11	9	7	6
18	17	16	14	13	11	9	8
20	20	19	17	16	13	11	10
22	23	21	20	18	16	13	12
24	26	24	22	21	18	15	13
26	29	27	25	23	20	17	15
28	32	30	28	26	22	19	17
30	35	33	30	28	25	21	19
32	38	35	33	31	27	23	21
34	41	38	36	33	29	25	22
36	44	41	38	36	31	27	24

*Applicable to circular tied columns only.

FIGURE 8.25   Maximum number of bars in columns having longitudinal bars arranged in a circle and normal lap splices.

Example 8.11   Determine the maximum number of bars that can be accommodated in an 18-in-diameter column with No. 7 bars. Assume a 1.5-in clear cover to No. 4 spirals.

Solution   Use Eq. (8.37) to determine the maximum number of bars n:

$$n = \frac{180}{\arcsin\left[(\bar{s}/2)/a\right]}$$

$$= \frac{180}{\arcsin\left[(2.375/2)/5.69\right]}$$

$$= 14.9$$

FIGURE 8.26
Distance between
bars in a column
with bars arranged
in a circle and
normal lap splices.

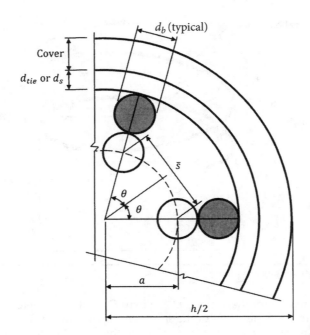

$$a = h/2 - [\text{cover} + (d_{tie} \text{ or } d_s)] - 1.5d_b$$

$$\theta = \frac{1}{2}\left(\frac{360}{n}\right) = \arcsin\left(\frac{\bar{s}/2}{a}\right)$$

where $\bar{s} = 1.5 + d_b = 1.5 + 0.875 = 2.375$ in for No. 7 bars

$\quad a = h/2 - (\text{cover} + d_s) - 1.5d_b$

$\quad\quad = (18/2) - (1.5 + 0.5) - (1.5 \times 0.875) = 5.69$ in

Round the number of bars down to 14; this matches the maximum number of bars shown in Fig. 8.25.

Minimum Spiral Reinforcement   The minimum amount of spiral reinforcement that must be provided in a spiral column is given by ACI Eq. (10-5):

$$\rho_s = 0.45\left(\frac{A_g}{A_{ch}} - 1\right)\frac{f_c'}{f_{yt}} \tag{8.38}$$

The volumetric spiral reinforcement ratio is equal to the volume of the spiral reinforcement divided by the volume of the concrete core measured to the outside edges of the spiral reinforcement.

The volume of the spiral reinforcement is determined by multiplying the area of the spiral bar A_{bs} by the length of one 360-degree loop of the spiral; this length is equal

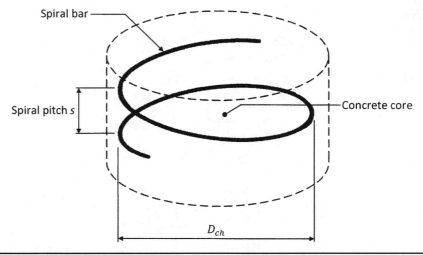

Spiral bar

Spiral pitch s

Concrete core

D_{ch}

FIGURE 8.27   Spiral reinforcement.

to the circumference of the spiral πD_{ch}, where D_{ch} is the diameter of the column core measured to the outside edges of the spiral reinforcement (see Fig. 8.27).

The volume of the concrete core is equal to the area of the core $A_{ch} = \pi D_{ch}^2/4$ times the center-to-center spacing s of the spirals (the center-to-center spacing of the spirals is commonly referred to as spiral pitch).

Thus, ρ_s can be expressed as

$$\rho_s = \frac{A_{bs}\pi D_{ch}}{(\pi D_{ch}^2/4)\, s} = \frac{4 A_{bs}}{D_{ch}s} \tag{8.39}$$

Substituting Eq. (8.39) into Eq. (8.38) and solving for s results in

$$s = \frac{8.9 A_{bs}}{D_{ch}[(A_g/A_{ch}) - 1]\left(f_c'/f_{yt}\right)} \tag{8.40}$$

Thus, for a given spiral bar size, the center-to-center spacing of the spirals s must be less than or equal to that given by Eq. (8.40) in order to satisfy the minimum requirements of ACI 10.9.3. The clear spacing between spirals shall not exceed 3 in or be less than 1 in (ACI 7.10.4.3). Additionally, the spacing must satisfy the requirements of ACI 3.3.2. These limitations must also be considered when choosing the spiral pitch.

The yield strength of the spiral reinforcement f_{yt} that can be used in determining the minimum amount of spiral reinforcement must be equal to or less than 100,000 psi. Research has confirmed that 100,000 psi reinforcement can be used for confinement.[11–13] However, lap splices determined in accordance with ACI 7.10.4.5(a) are not permitted in cases where f_{yt} exceeds 60,000 psi.

Spiral reinforcement increases the strength of the concrete that is within the core of the column after the concrete shell outside of the core spalls off caused by load and deformation. The purpose of minimum spiral reinforcement is to provide additional load-carrying strength for concentrically loaded columns that is equal to or slightly greater than the strength that is lost when the outer concrete shell spalls off. It has been shown that concrete columns that contain the minimum amount of spiral reinforcement required by the section exhibit considerable toughness and ductility.

Example 8.12   Determine the maximum allowable spiral pitch for a 24-in-diameter column, assuming $f'_c = 5,000$ psi with a 3/4-in maximum aggregate size and a 1.5-in clear cover to No. 4 spirals (Grade 60).

Solution   Equation (8.40) is used to determine the maximum spiral pitch s:

$$s = \frac{8.9\,A_{bs}}{D_{ch}[(A_g/A_{ch}) - 1]\,(f'_c/f_{yt})}$$

$$= \frac{8.9 \times 0.20}{(24 - 3)\{[(\pi \times 12^2)/(\pi \times 10.5^2)] - 1\} \times (5/60)} = 3.3 \text{ in}$$

Try a 3-in spiral pitch.
Check the spacing requirements of ACI 7.10.4.3:

Clear spacing $= 3 - 0.5 = 2.5$ in, which is less than 3 in and greater than 1 in

Minimum clear spacing in accordance with ACI 3.3.2 $= 1.33$ (maximum aggregate size) $= 1.33 \times 0.75 = 1.0$ in < 3 in

Therefore, use a maximum spiral pitch of 3 in.

Lateral Reinforcement
Spiral Reinforcement   Requirements for compression members with spiral reinforcement are given in ACI 7.10.4 and are summarized in Table 8.10.

Standard spiral sizes are Nos. 3 to 5 bars. Minimum sizes are based on practical considerations in cast-in-place construction.

The main purposes of the spiral reinforcement are as follows:

1. To hold the longitudinal reinforcement in the proper position when the concrete is placed

2. To prevent the longitudinal reinforcement from buckling outward through the relatively thin layer of concrete cover

Spirals are permitted to be terminated at the level of the lowest horizontal reinforcement of the members framing into the column. In cases where one or more sides of the column are not enclosed by beams or brackets, ties are required from the termination of the spiral to the bottom of the slab, drop panel, or shear cap. These ties must enclose the longitudinal column reinforcement and the portion of the bars from the beams that are bent into the column for anchorage.

The proper pitch and alignment of the spirals must be maintained during concrete placement. Prior to the 1989 Code, spacers were required to hold the spiral cage in place. Information on the number of recommended spacers based on spiral bar size and core diameter is given in ACI R7.10.4. Note that any method of installation is acceptable provided the spirals are held firmly in place.

Tie Reinforcement   Requirements for compression members with tie reinforcement are given in ACI 7.10.5 and are summarized in Table 8.11.

Requirement	ACI Section Number
Spirals shall consist of evenly spaced continuous bar or wire of such size and so assembled to permit handling and placing without distortion from designed dimensions.	7.10.4.1
Minimum spiral diameter = 3/8 in.	7.10.4.2
Clear spacing between spirals shall not exceed 3 in or be less than 1 in. Also see ACI 3.3.2.	7.10.4.3
Anchorage of spiral reinforcement shall be provided by 1-1/2 turns of spiral bar or wire at each end of a spiral unit.	7.10.4.4
Spiral reinforcement shall be spliced using lap splices conforming to ACI 7.10.4.5(a) or mechanical or welded splices conforming to ACI 7.10.4.5(b).	7.10.4.5
Spirals shall extend from the top of the footing or slab in any story to the level of the lowest horizontal reinforcement in the members supported above.	7.10.4.6
Ties shall extend above the termination of the spiral to the bottom of a slab, drop panel, or shear cap where beams or brackets do not frame into all sides of a column.	7.10.4.7
In columns with capitals, spirals shall extend to a level at which the diameter or width of the capital is 2 times that of the column.	7.10.4.8
Spirals shall be held firmly in place and true to line.	7.10.4.9

TABLE 8.10 Summary of Requirements for Spiral Reinforcement

Standard hook dimensions for ties, which are the same as those for stirrups, are given in ACI 7.1.3.

The main purposes of the ties are the same as those of spirals: to hold the longitudinal reinforcement in the proper position when the concrete is placed and to prevent the longitudinal bars from buckling outward through the relatively thin layer of concrete cover. The maximum spacing criteria in ACI 7.10.5.2 are intended to achieve these purposes.

ACI Fig. R7.10.5 illustrates the requirements of ACI 7.10.5.3 pertaining to tie arrangement and maximum clear spacing between laterally supported bars. The corner bars and every other longitudinal bar must have lateral support in cases where the center-to-center spacing of the longitudinal bars on a side is equal to or less than 6 in plus the diameter of the longitudinal bar. If the spacing is greater than that, lateral support must be provided for the intermediate bars as well.

Crossties are commonly used in columns to provide lateral support for intermediate bars. A crosstie is a continuous reinforcing bar that has a seismic hook on one end and a hook not less than 90 degrees with at least a six-diameter extension on the other. A seismic hook is defined as a hook with a bend that is not less than 135 degrees and an extension that is not less than 6 bar diameters or 3 in, whichever is greater.

Figure 8.28 illustrates crossties in a column where the clear spacing of the longitudinal bars on a side is equal to or less than 6 in. Note that all of the hooks engage the peripheral longitudinal bars and that the 90-degree hooks of successive crossties alternate from one side of the column to the other.

Requirement	ACI Section Number
At least No. 3 ties shall enclose No. 10 or smaller longitudinal bars, and at least No. 4 ties shall enclose Nos. 11, 14, and 18 bars and bundled longitudinal bars.	7.10.5.1
Vertical spacing of ties shall not exceed the smaller of the following: 1. 48 tie bar diameters 2. 16 longitudinal bar diameters 3. The least dimension of the compression member	7.10.5.2
Ties shall be arranged such that every corner or alternate longitudinal bar has lateral support provided by the corner of a tie with an included angle of not more than 135 degrees. No bar shall be farther than 6 in clear on each side along the tie from a laterally supported bar. A complete circular tie is also permitted.	7.10.5.3
Ties shall be located vertically not more than one-half a tie spacing above the top of a footing or slab in any story and should be spaced to not more than one-half a tie spacing below the lowest horizontal reinforcement in a slab, drop panel, or shear cap above.	7.10.5.4
Where beams or brackets frame from four directions into a column, it is permitted to terminate the ties not more than 3 in below the lowest reinforcement in the shallowest of the beams or brackets.	7.10.5.5
Where anchor bolts are placed in the top of a column or pedestal, the bolts must be enclosed by the lateral reinforcement that also surrounds at least four vertical bars of the column or pedestal. The lateral reinforcement must be distributed within 5 in of the top of the column or pedestal and must consist of at least two No. 4 or three No. 3 bars.	7.10.5.6

TABLE 8.11   Summary of Requirements for Tie Reinforcement

The tie detail requirements of ACI 7.10.5 are illustrated in Fig. 8.29. Also illustrated in the figure are the offset bends that must be provided in the longitudinal bars immediately below the splice locations. Offset longitudinal bars must conform to the requirements of ACI 7.8.1. Additional ties must be provided within 6 in of the offset to help resist the horizontal component of the force transmitted through the inclined portion of the bars (see ACI 7.8.1.3).

Lap splice locations for the longitudinal bars are also depicted in Fig. 8.29. Information on the different types of splices for columns is given next.

8.7.3   Splices

Overview

Splice requirements for column are given in ACI 12.17. Provisions are provided for lap splices, mechanical or welded splices, and end-bearing splices.

FIGURE 8.28
Crossties in
a column.

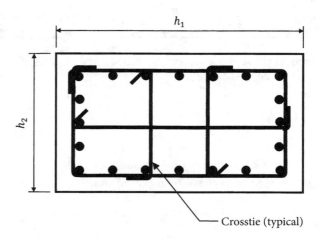

Crosstie (typical)

In general, column splices must satisfy the requirements for all load combinations. Gravity load combinations, which produce compressive stresses in all of the longitudinal reinforcing bars, may govern the design of a column (cross-sectional dimensions and longitudinal reinforcement). However, load combinations that include wind or seismic effects may produce significant tensile stresses in some of the longitudinal bars. In such cases, the column splice must be designed for tension even though the design of the column is governed by gravity loads only.

Where the longitudinal bar stress due to factored loads is compressive, all of the splice types listed earlier may be used. Lap splices and mechanical or welded splices are permitted when the longitudinal bar stress is tensile; end-bearing splices are not permitted in such cases.

Lap Splices

Lap splices are the most popular type of splices used in columns. The type of lap splice—compressive or tensile—that must be used depends on the stress in the longitudinal bars due to the factored load combinations. Figure 8.30 illustrates the lap splice requirements for columns. The three zones depicted in the figure are covered later.

Lap splices are permitted to occur immediately above the top of the slab, as shown in Fig. 8.29. This location facilitates the overall construction of the structure. The longitudinal bars from the column below extend above the slab a distance equal to or greater than the required lap splice length. The longitudinal bars for the column above are tied to these bars after the floor below has been constructed.

Zone 1 Zone 1 corresponds to the portion of the interaction diagram where all of the longitudinal bars in the column are in compression. If all of the factored load combinations fall within Zone 1, a compression lap splice conforming to ACI 12.6 may be used.

The minimum compression lap splice length is given in ACI 12.16.1:

$$0.0005 f_y d_b \geq 12 \text{ in for } f_y \leq 60,000 \text{ psi} \tag{8.41a}$$

$$(0.0009 f_y - 24) d_b \geq 12 \text{ in for } f_y > 60,000 \text{ psi} \tag{8.41b}$$

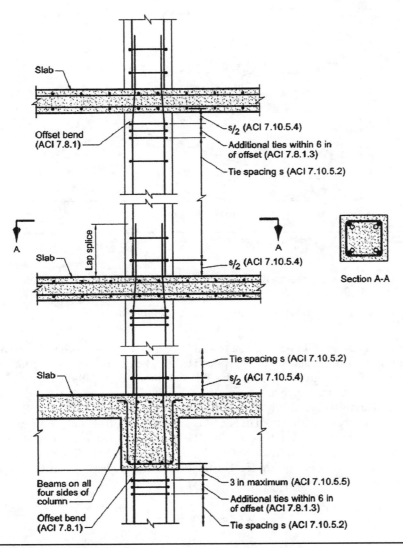

Slab

Offset bend
(ACI 7.8.1)

$s/2$ (ACI 7.10.5.4)

Additional ties within 6 in
of offset (ACI 7.8.1.3)

Tie spacing s (ACI 7.10.5.2)

A

Lap splice

Slab

A

$s/2$ (ACI 7.10.5.4)

Section A-A

Tie spacing s (ACI 7.10.5.2)

Slab

$s/2$ (ACI 7.10.5.4)

Beams on all
four sides of
column

3 in maximum (ACI 7.10.5.5)

Additional ties within 6 in
of offset (ACI 7.8.1.3)

Offset bend
(ACI 7.8.1)

Tie spacing s (ACI 7.10.5.2)

FIGURE 8.29   Tie and splice details.

In cases where the compressive strength of the concrete is less than 3,000 psi, the lap splice must be increased by one-third.

A minimum tensile strength is required for all compression splices. This accounts for misalignments and other types of situations that can introduce tensile stresses in the bars. A compression lap splice length determined in accordance with ACI 12.16 has a tensile strength of at least $0.25A_b f_y$.

Longitudinal bar sizes are typically reduced at designated locations over the height of a column for overall economy. At such locations, the splice length must be equal to or greater than the larger of the following:

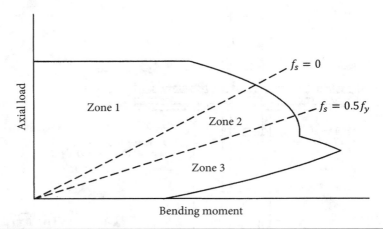

FIGURE 8.30   Lap splice requirements for columns.

- The development length in compression ℓ_{dc} of the larger bar, determined in accordance with ACI 12.3
- The compression lap splice length of the smaller bar

Lap splices are prohibited for Nos. 14 and 18 bars (see ACI 12.14.2.1). However, ACI 12.16.2 permits such bar sizes to be lap spliced to No. 11 and smaller bars in compression.

The lap splice length determined by Eq. (8.41) may be multiplied by 0.83 in tied columns where ties that have an effective area equal to or greater than $0.0015hs$ in both directions of analysis are provided over the lap splice length (ACI 12.17.2.4).

Consider the column depicted in Fig. 8.28. In the direction of analysis perpendicular to the column dimension h_1, there are four tie bars with an area per bar of A_b. In the direction of analysis perpendicular to h_2, there are three tie bars. The compression lap splice length determined by Eq. (8.41) may be multiplied by 0.83 when both of the following equations are satisfied:

$$4A_b \geq 0.0015h_1s$$

$$3A_b \geq 0.0015h_2s$$

In these equations, s is the vertical spacing of the ties along the splice length. Note that the reduced lap splice length must not be taken less than 12 in.

Because spiral reinforcement provides increased resistance to splitting, lap splice lengths in columns with spiral reinforcement may be taken as 0.75 times the length determined by Eq. (8.41). The spiral reinforcement must conform to ACI 7.10.4 and 10.9.3, and in no case shall the lap splice length be taken less than 12 in.

Zone 2   Zone 2 corresponds to the portion of the interaction diagram where the stress in the longitudinal bars on the tension face of the member is tensile and does not exceed 50% of the yield strength of the reinforcement.

A tension lap splice conforming to the requirements of ACI 12.15 must be used in columns where one or more of the factored load combinations fall within Zone 2. Details on how to determine tension lap splice lengths can be found in Section 6.2.

If half or fewer of the longitudinal bars are spliced at any section and alternate lap splices are staggered a distance not less than the tension development length ℓ_d, a Class A tension lap splice may be used. A Class B tension lap splice is required where more than one-half of the longitudinal bars are spliced at one section. In typical column construction, all of the bars are spliced above the slab, so a Class B splice is required.

Zone 3   Zone 3 corresponds to the portion of the interaction diagram where the stress in the longitudinal bars on the tension face of the member is tensile and exceeds 50% of the yield strength of the reinforcement.

A Class B tension lap splice conforming to the requirements of ACI 12.15 must be used in columns where one or more of the factored load combinations fall within Zone 3.

Example 8.13   Determine the required lap splice length for a 20-in square tied column reinforced with 12 No. 6 bars. Assume normal-weight concrete with $f_c' = 4{,}000$ psi and Grade 60 reinforcement. Also assume a 1.5-in clear cover to No. 3 ties and that all of the bars will be spliced at the same location.

The governing factored load combinations determined in accordance with ACI 9.2 are given in Table 8.12.

Solution   Figure 8.31 contains the design strength interaction diagram for this column along with the load points corresponding to the four load combinations.

It is evident from the figure that load combinations 3 and 4 produce tensile stresses in the longitudinal bars closest to the tension face of the section, which exceed 50% of the yield strength of the reinforcement. Because all bars are to be spliced at the same location, a Class B tension lap splice must be provided in accordance with ACI 12.15.1:

$$\text{Class B splice length} = 1.3\ell_d \geq 12 \text{ in}$$

The tension development length ℓ_d is determined by ACI Eq. (12-1):

$$\ell_d = \left(\frac{3}{40} \frac{f_y}{\lambda\sqrt{f_c'}} \frac{\psi_t\psi_e\psi_s}{[(c_b + K_{tr})/d_b]} \right) d_b \geq 12 \text{ in}$$

Load Combination	Axial Load (kips)	Bending Moment (ft kips)
1 1.4D	220	16
2 1.2D + 1.6L + 0.5S	242	31
3 1.2D + 0.5L + 0.5S + 1.6W	213	197
4 0.9D − 1.6W	136	167

TABLE 8.12 Summary of Factored Axial Loads and Bending Moments for the Column Given in Example 8.13

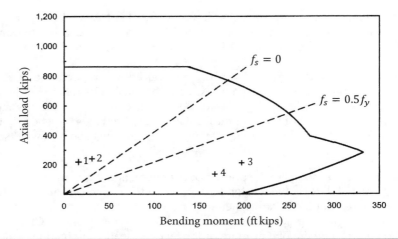

FIGURE 8.31   Design strength interaction diagram for the column given in Example 8.13.

where $\lambda = 1.0$ for normal-weight concrete
$\psi_t = 1.0$ for bars other than top bars
$\psi_e = 1.0$ for uncoated reinforcement
$\psi_s = 0.8$ for No. 6 bars

$$c_b = 1.5 + 0.375 + \frac{0.75}{2} = 2.3 \text{ in (governs)}$$

$$= \frac{20 - 2(1.5 + 0.375) - 0.75}{2 \times 3} = 2.6 \text{ in}$$

The required spacing of the No. 3 ties is the smallest of the following:

- 16 (longitudinal bar diameters) $= 16 \times 0.75 = 12.0$ in (governs)
- 48 (tie bar diameters) $= 48 \times 0.375 = 18.0$ in
- Least column dimension $= 20$ in

Use No. 3 ties spaced 12.0 in on center.

Clear space between the longitudinal bars $= \dfrac{20 - 2(1.5 + 0.375) - 0.75}{3} - 0.75 = 4.4$ in

Because the clear space between the bars is less than 6 in, one of the longitudinal bars on each face does not need lateral support. However, for symmetry, provide a No. 3 crosstie on each of the interior bars (i.e., provide two crossties in each direction).

$$K_{tr} = \frac{40 A_{tr}}{sn} = \frac{40 \times 4 \times 0.11}{20 \times 4} = 0.2$$

$$\frac{c_b + K_{tr}}{d_b} = \frac{2.3 + 0.2}{0.75} = 3.3 > 2.5; \text{ use } 2.5$$

Therefore,

$$\ell_d = \left(\frac{3}{40} \frac{60,000}{1.0\sqrt{4,000}} \frac{1.0 \times 1.0 \times 0.8}{2.5} \right) \times 0.75 = 17.1 \text{ in} = 1.4 \text{ ft}$$

Class B splice length $= 1.3 \times 1.4 = 1.8$ ft

Use a splice length of 2 ft 0 in with the splice located just above the slab.
Reinforcement details for this column are similar to those shown in Fig. 8.29.

Example 8.14 Determine the required compression lap splice length for a column where 12 No. 7 bars are spliced to 12 No. 9 bars. Assume normal-weight concrete with $f_c' = 6,000$ psi and Grade 60 reinforcement. Also assume a 1.5-in clear cover to the No. 4 ties and that all of the bars will be spliced at the same location.

Solution ACI 12.16.2 requires that the compression splice length be taken as the larger of the following:

- The development length in compression ℓ_{dc} of the larger bar, determined in accordance with ACI 12.3
- The compression lap splice length of the smaller bar

The development length ℓ_{dc} of the No. 9 bars is the larger of the following (see ACI 12.3.2):

- $$\frac{0.02 f_y d_b}{\lambda \sqrt{f_c'}} = \frac{0.02 \times 60,000 \times 1.128}{1.0 \sqrt{6,000}} = 17.5 \text{ in}$$
- $0.0003 f_y d_b = 0.0003 \times 60,000 \times 1.128 = 20.3$ in (governs)

The compression lap splice length of the No. 7 bars is determined by Eq. (8.41a) for Grade 60 reinforcement:

$$0.0005 f_y d_b = 0.0005 \times 60,000 \times 0.875 = 26.2 \text{ in}$$

Therefore, the splice length must be equal to a minimum of 26.2 in.
Use a splice length of 2 ft 4 in.

Mechanical or Welded Splices

Mechanical or welded splices are permitted in columns and must meet the requirements of ACI 12.14.3.2 and 12.14.3.4, respectively. As noted previously, these types of splices may be used in both compression and tension.

A mechanical splice is defined in Ref. 14 as a "complete assembly of a coupler, a coupling sleeve, or an end-bearing sleeve, including any additional intervening material or other components required to accomplish the splicing of reinforcing bars." More information on the different types of mechanical splices can be found in Ref. 14.

A variety of proprietary mechanical devices are available that can be used to splice the longitudinal reinforcing bars in a column. They are commonly specified at locations where long lap splices would be required or at locations where lap splices would cause congestion, such as at beam–column joints.

A full mechanical splice must be able to develop in tension or compression 125% of the yield strength of the reinforcing bar. The 25% increase above the yield strength is considered an economical way of achieving a ductile failure (rather than a brittle failure) in the mechanical splice.

A full welded splice must also be able to develop 125% of the yield strength of the reinforcing bar. These types of splices are primarily intended for No. 6 bars and larger. The requirement of $1.25 f_y$ is intended to provide sound welding that is adequate in both tension and compression.

End-Bearing Splices

End-bearing splices transmit compressive stresses from one longitudinal bar to another by bearing of square cut ends held in concentric contact by a suitable device. Tolerances for the bar ends are given in ACI 12.16.4.2.

These types of splices are almost exclusively used in columns with vertical longitudinal bars. ACI 12.16.4.3 requires that end-bearing splices be used only in members where closed ties, closed stirrups, or spiral reinforcement is provided. This requirement ensures that a minimum shear resistance is provided in members containing these types of splices.

End-bearing splices must have a minimum tensile strength of 25% of the yield strength of the longitudinal reinforcement area on each face of a column. This is typically achieved by either staggering the location of the end-bearing splices or adding additional longitudinal reinforcement through the splice locations.

8.8 Shear Requirements

Shear strength requirements for columns are essentially the same as those for flexural members. The following equation must be satisfied at all sections along the length of a column:

$$\phi V_n \geq V_u \tag{8.42}$$

The required shear strength V_u is determined by the applicable load combinations in ACI 9.2, based on an elastic analysis of the structure. The nominal shear strength V_n is the sum of the nominal shear strength provided by the concrete V_c and that provided by the lateral reinforcement V_s. The strength reduction factor ϕ is equal to 0.75 for shear (ACI 9.3.2.3).

ACI 11.2 provides two methods to determine V_c for members subjected to axial compression. In the first method, V_c is determined by ACI Eq. (11-4):

$$V_c = 2\left(1 + \frac{N_u}{2{,}000 A_g}\right)\lambda\sqrt{f_c'}b_w d \tag{8.43}$$

In this equation, N_u is the factored axial compressive load determined by the applicable load combinations given in ACI 9.2. The magnitude of N_u is taken as positive for compressive loads. The effective depth d should be determined from a strain compatibility analysis for each applicable load combination.

The second method to determine V_c is more complex than the first. In this method, ACI Eq. (11-5) is used to calculate V_c where M_m, which is determined by ACI Eq. (11-6), is substituted for M_u in that equation. Also, the ratio $V_u d / M_u$ is not limited to 1 as required by ACI 11.2.2.1. Thus, V_c can be determined by the following:

$$V_c = \left(1.9\lambda\sqrt{f_c'} + 2{,}500\rho_w \frac{V_u d}{M_m}\right)b_w d$$

$$\leq 3.5\lambda\sqrt{f_c'}b_w d\sqrt{1 + \frac{N_u}{500 A_g}} \tag{8.44}$$

In this equation, $M_m = M_u - N_u \left[(4h - d)/8\right] > 0$ and $\rho_w = A_s/b_w d$. The reinforcement ratio ρ_w is determined using the area of the tensile reinforcement A_s in the section. If $M_m \leq 0$, then V_c is determined by the upper limit given in Eq. (8.44). Derivations of these equations along with comparisons with test data are given in Ref. 15.

In the rare occasions where axial tension acts on a column, V_c can be taken equal to zero (ACI 11.2.1.3), or it can be determined by ACI Eq. (11-8):

$$V_c = 2\left(1 + \frac{N_u}{500 A_g}\right)\lambda\sqrt{f_c'}b_w d \geq 0 \tag{8.45}$$

In this equation, the factored axial tension load N_u is taken as negative, and the tensile stress N_u/A_g is expressed in pounds per square inch.

A comparison of shear strength equations for members subjected to axial load can be found in ACI Fig. R11.2.2.2.

For circular columns, ACI 11.2.3 permits the area $b_w d$ that is used to compute V_c to be taken as the product of the diameter of the section and the effective depth d. In lieu of determining d by a strain compatibility analysis, it can be taken as 80% of the diameter of the section.

The nominal shear strength of the lateral reinforcement V_s is determined by ACI Eq. (11-15):

$$V_s = \frac{A_v f_{yt} d}{s} \tag{8.46}$$

Typically, V_s is initially calculated using the lateral reinforcement required by ACI 7.10. If additional shear strength is needed, the amount and spacing of the lateral reinforcement can be adjusted appropriately or the strength of the concrete can be increased. In cases where the shear demand is very high, the size of the column may need to be increased as well.

Example 8.15 Determine the design shear strength ϕV_n for the column given in Example 8.13.

Solution Because no information is provided on the factored shear forces V_u for the four load combinations, determine the nominal shear strength of the concrete V_c, using Eq. (8.43).

It can be shown for all four load combinations that the layer of reinforcement closest to the compression face of the section is in compression and that the other three layers are in tension; that is, the neutral axis c in all four cases lies between the layer of steel that is closest to the compression face and the adjoining layer of steel. Therefore, the effective depth d is equal to the distance from the extreme compression fiber to the centroid of the three layers of longitudinal reinforcement below the neutral axis.

The distance from the tension face of the section to the centroid of these three layers of tension reinforcement can be determined as follows (see Fig. 8.32):

$$\bar{y} = \frac{[(4 \times 0.44) \times 2.25] + [(2 \times 0.44) \times 7.42] + [(2 \times 0.44) \times 12.58]}{8 \times 0.44} = 6.1 \text{ in}$$

Thus, $d = 20 - 6.1 = 13.9$ in.

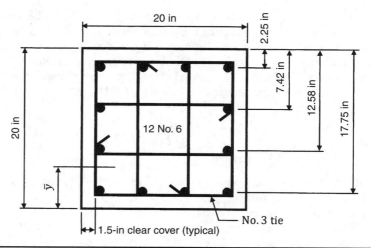

FIGURE 8.32 Location of the centroid of tensile reinforcement given in Example 8.15.

The nominal shear strength of the concrete V_c is determined using the smallest factored axial load given in Table 8.12:

$$V_c = 2\left(1 + \frac{N_u}{2,000\,A_g}\right)\lambda\sqrt{f_c'}\,b_w d$$

$$= 2\left(1 + \frac{136,000}{2,000 \times 20^2}\right) \times 1.0\sqrt{4,000} \times 20 \times 13.9/1,000 = 41.1 \text{ kips}$$

The nominal shear strength of the lateral reinforcement V_s is determined by Eq. (8.46), using No. 3 ties with four legs spaced at 12 in on center (see Example 8.13):

$$V_s = \frac{A_v f_{yt} d}{s}$$

$$= \frac{(4 \times 0.11) \times 60 \times 13.9}{12} = 30.6 \text{ kips}$$

Therefore, the design shear strength ϕV_n is

$$\phi V_n = \phi(V_c + V_s) = 0.75\,(41.1 + 30.6) = 53.8 \text{ kips}$$

Example 8.16 Check the adequacy of the column in Example 8.13, assuming that the factored shear force V_u is equal to 39 kips for load combination 3.

Solution Equation (8.44) will be used to determine V_c.

It was shown in Example 8.15 that the effective depth d of the section is equal to 13.9 in for all four load combinations.

The moment M_m is calculated from the following equation for the third load combination (see Table 8.12):

$$M_m = M_u - N_u\frac{4h - d}{8}$$

$$= (197 \times 12) - 213\left[\frac{(4 \times 20) - 13.9}{8}\right] = 604.1 \text{ in kips}$$

Also, the reinforcement ratio ρ_w is determined using the area of tensile reinforcement in the section:

$$\rho_w = \frac{A_s}{b_w d} = \frac{8 \times 0.44}{20 \times 13.9} = 0.0127$$

The nominal shear strength of the concrete V_c is

$$V_c = \left(1.9\lambda\sqrt{f_c'} + 2{,}500\rho_w \frac{V_u d}{M_m}\right) b_w d$$

$$= \left[1.9 \times 1.0\sqrt{4{,}000} + \left(2{,}500 \times 0.0127 \times \frac{39 \times 13.9}{604.1}\right)\right] \times 20 \times 13.9/1{,}000$$

$$= 41.3 \text{ kips}$$

Check the upper limit on V_c:

$$V_c = 3.5\lambda\sqrt{f_c'}b_w d\sqrt{1 + \frac{N_u}{500A_g}}$$

$$= 3.5 \times 1.0\sqrt{4{,}000} \times 20 \times 13.9\sqrt{1 + \frac{213{,}000}{500 \times 20^2}}/1{,}000$$

$$= 88.4 \text{ kips} > 41.3 \text{ kips}$$

Since $V_u = 39$ kips $> \phi V_c = 31$ kips, the maximum spacing of the ties is equal to $d/2 = 13.9/2 = 6.95$ in (see ACI 11.4.5.1). Assume the ties are spaced 6 in on center.

Therefore, the design shear strength ϕV_n is

$$\phi V_n = \phi(V_c + V_s) = 0.75\,(41.3 + 61.2) = 76.9 \text{ kips} > V_u = 39 \text{ kips}$$

Comments

The nominal shear strength of the concrete V_c determined by Eq. (8.43) is

$$V_c = 2\left(1 + \frac{N_u}{2{,}000A_g}\right)\lambda\sqrt{f_c'}b_w d$$

$$= 2\left(1 + \frac{213{,}000}{2{,}000 \times 20^2}\right) \times 1.0\sqrt{4{,}000} \times 20 \times 13.9/1{,}000 = 44.5 \text{ kips}$$

In this example, this value of V_c determined by Eq. (8.43) is essentially the same as the value determined by Eq. (8.44).

8.9 Design Procedure

The following design procedure can be used in the design of columns. Included is the information presented in the previous sections on analysis, design, and detailing.

Step 1: Determine the preliminary column size (Section 8.2). The first step in the design procedure is to select a preliminary column size and amount of longitudinal reinforcing steel in the column.

Unless column size is dictated by architectural requirements, a preliminary column size should be based on a low percentage of longitudinal reinforcement, considering only axial load and temporarily ignoring the effects from bending moments and slenderness. This allows the column capacity to be increased in the final design stages by adding reinforcement rather than by increasing the column dimensions.

Figure 8.1 can be used to quickly select a preliminary column size for nonslender, tied columns loaded at an eccentricity of no more than approximately 10% of the overall column thickness, that is, with zero or small bending moments.

When choosing the amount of longitudinal reinforcement in a compression member, it is important to keep in mind the limits that are prescribed in ACI 10.9.1. The area of longitudinal steel A_{st} must be between 1% and 8% of the gross area of the section. The reinforcement ratio should not exceed 4% if column bars are lap spliced. Columns with reinforcement ratios in the range of 1% to 2% are usually the most cost-effective.

Step 2: Choose the reinforcing bar size and layout (Section 8.7). The size of the reinforcing bars and their layout are chosen on the basis of the preliminary area of longitudinal reinforcement determined in Step 1. For proper concrete placement and consolidation, bar sizes that minimize reinforcement congestion, especially at beam–column joints, must be chosen. A smaller number of larger bars usually improve constructability.

The minimum number of longitudinal bars in compression members is four for bars within rectangular or circular ties, three for bars within triangular ties, and six for bars enclosed by spirals conforming to ACI 10.9.3.

Figure 8.23 can be used to determine the number of longitudinal bars that can be accommodated on the face of a rectangular, tied column with normal lap splices, based on the provisions of ACI 7.6.3 and assuming a 1.5-in clear cover to No. 4 ties. Similarly, Fig. 8.25 contains the maximum number of bars that can be accommodated in circular columns or square columns with longitudinal reinforcement arranged in a circle with normal lap splices, assuming a No. 4 spiral or tie.

Step 3: Acquire an interaction diagram (Section 8.4). For columns subjected to bending moments primarily about one axis, interaction diagrams (or equivalent) facilitate the determination of adequacy for columns subjected to combined axial loads and bending moments.

An interaction diagram for a specific column size and longitudinal bar arrangement can be constructed using the methods given in Section 8.4. The design aids referenced in that section can also be utilized.

For biaxial bending, the approximate method discussed in Section 8.6 can be used to determine the adequacy of a column.

Step 4: Determine whether the frame is a nonsway or sway frame (Section 8.5). The methods given in Section 8.5 can be used to determine whether a frame is a nonsway or sway frame. Consideration of slenderness effects depends on whether the frame is nonsway or sway.

Step 5: Determine whether slenderness effects need to be considered (Section 8.5). The limiting slenderness ratios for both nonsway and sway frames are presented in Section 8.5.

If slenderness effects need not be considered, all that is left to be done is as follows: detail the transverse reinforcement; check the shear strength; and provide splices for the longitudinal reinforcement. These items are discussed later.

Where slenderness effects must be considered, continue to step 6.

Step 6: Determine the magnified moments in cases where slenderness effects must be considered (Section 8.5). In cases where slenderness effects must be considered, the bending moments acquired from the first-order elastic analysis must be magnified in accordance with ACI 10.10.3, 10.10.4, or 10.10.6 for nonsway frames and ACI 10.10.3, 10.10.4, or 10.10.7 for sway frames.

For compression members subjected to biaxial bending, the moment about each axis is magnified separately on the basis of the restraint conditions corresponding to that axis. The adequacy of the compression member is checked for the factored axial load P_u and the maximum magnified bending moment for all applicable load combinations in accordance with ACI 9.2.

At this point, the size of the column, the amount of longitudinal reinforcement, and/or the compressive strength of the concrete may have to be adjusted in order to satisfy strength requirements.

Step 7: Detail the lateral reinforcement (Section 8.7). Once the combined flexural and axial load strength requirements have been satisfied for all load combinations, lateral (or transverse) reinforcement in the column must be detailed in accordance with ACI 7.10.

Spiral reinforcement must conform to ACI 10.9.3 and 7.10.4, and tie reinforcement must satisfy the requirements of ACI 7.10.5 (see Fig. 8.29).

Step 8: Check shear strength requirements (Section 8.8). The design provisions for shear that are applicable to columns are given in ACI Chap. 11. The required shear strength V_u must be equal to or less than the design shear strength ϕV_n, which consists of the design shear strength of the concrete ϕV_c and that of the lateral reinforcement ϕV_s.

ACI 11.2 gives two methods to determine the nominal concrete strength V_c for members subjected to axial compression. The nominal shear strength of the lateral reinforcement V_s is determined by ACI Eq. (11-15).

In cases where lateral reinforcement is required for shear, it is common to check shear strength requirements based on the minimum transverse reinforcement obtained from Step 7. Additional transverse reinforcement above the minimum required may be needed in certain cases.

Step 9: Provide splices for the longitudinal reinforcement (Section 8.7). ACI 12.17 contains requirements for lap splices, mechanical splices, welded splices, and end-bearing splices. End-bearing splices are permitted only where the longitudinal bars are in compression.

Lap splices are the most commonly used type of splices in columns. The type of lap splice (compression, Class A tension, or Class B tension) that is required depends on the factored load combinations and the magnitude of stress in the longitudinal bars (see Fig. 8.30). Lap splices can occur immediately above the top of the slab, as illustrated in Fig. 8.29.

References

1. American Concrete Institute (ACI). 2009. *ACI Design Handbook*, SP-17(09). ACI, Farmington Hills, MI.
2. Concrete Reinforcing Steel Institute (CRSI). 2008. *CRSI Design Handbook*, 10th ed. CRSI, Schaumburg, IL.
3. MacGregor, J. G., and Hage, S. E. 1977. Stability analysis and design concrete. *Proceedings, ASCE* 103(ST 10).

4. American Concrete Institute (ACI), Committee 340. 1997. *ACI Design Handbook*. ACI 340R-97, SP-17(97). ACI, Farmington Hills, MI.

5. Concrete Research Council. 1966. *Guide to Design Criteria for Metal Compression Members*, 2nd ed. Fritz Engineering laboratory, Lehigh University, Bethlehem, PA.

6. Kavanagh, T. C. 1962. Effective length of framed columns. *Transactions, ASCE* 127:81–101.

7. Khuntia, M., and Ghosh, S. K. 2004. Flexural stiffness of reinforced concrete columns and beams: Analytical approach. *ACI Structural Journal* 101(3):351–363.

8. Khuntia, M., and Ghosh, S. K. 2004. Flexural stiffness of reinforced concrete columns and beams: Experimental verification. *ACI Structural Journal* 101(3):364–374.

9. Lai, S. M. A., and MacGregor, J. G. 1983. Geometric nonlinearities in unbraced multistory frames. *Journal of Structural Engineering* 109(11):2528–2545.

10. Bresler, B. 1960. Design criteria for reinforced concrete columns under axial load and biaxial bending. *ACI Journal* 57(5):481–490.

11. Saatcioglu, M., and Razvi, S. R. 2002. Displacement-based design of reinforced concrete columns for confinement. *ACI Structural Journal* 99(1):3–11.

12. Pessiki, S., Graybeal, B., and Mudlock, M. 2001. Proposed design of high-strength spiral reinforcement in compression members. *ACI Structural Journal* 98(6):799–810.

13. Richart, F. E., Brandzaeg, A., and Brown, R. L. 1929. *The Failure of Plain and Spirally Reinforced Concrete in Compression*, Bulletin No. 190. University of Illinois Engineering Experiment Station, Urbana-Champaign, IL, 74 pp.

14. American Concrete Institute (ACI), Committee 439. 2007. *Types of Mechanical Splices for Reinforcing Bars*, ACI 439.3. ACI, Farmington Hills, MI.

15. MacGregor, J. G., and Hanson, J. M. 1969. Proposed changes in shear provisions for reinforced and prestressed concrete beams. *ACI Journal* 66(4):276–288.

Problems

8.1. Determine the design axial strength $\phi P_{n,max}$ of a 28-in-diameter column reinforced with nine No. 9 bars. The column has spiral reinforcement conforming to ACI 7.10.4 and 10.9.3. Assume normal-weight concrete with $f_c' = 6,000$ psi and Grade 60 reinforcement.

8.2. Determine the design axial load strength ϕP_n corresponding to a strain limit at the limit for a compression-controlled section of a 22×22 in column reinforced with eight No. 10 bars that are uniformly distributed in the cross-section. The clear cover to No. 4 ties is 1.5 in. Assume normal-weight concrete with $f_c' = 5,000$ psi and Grade 60 reinforcement.

8.3. Given the column in Problem 8.2, determine the design axial moment strength ϕM_n corresponding to a balanced strain condition.

8.4. Given the column in Problem 8.2, determine the design axial load strength ϕP_n corresponding to a strain limit at the limit for a tension-controlled section.

8.5. Determine the magnified moment M_c of a 24×24 in column reinforced with 12 No. 9 bars that are uniformly distributed in the cross-section. The clear cover to No. 3 ties is 1.5 in. Assume normal-weight concrete with $f_c' = 5,000$ psi and Grade 60 reinforcement. The column is part of a nonsway frame where $k = 1.0$ and $\beta_{dns} = 0.95$. The factored loads are $P_u = 1,000$ kips, $M_1 = 10$ ft kips, and $M_2 = 25$ ft kips.

8.6. Given the column in Problem 8.5, determine the design moment strength ϕM_n corresponding to a design axial load $\phi P_n = 650$ kips.

8.7. Determine the effective length factor k of a 28×28 in column in the first story of a sway frame. The column is essentially fixed at its base, and at the top of the column, 28- $\times$ 20-in- beams frame into the column at both sides. The lengths of the column and beams are 16 and 24 ft, respectively. Assume normal-weight concrete with $f_c' = 5,000$ psi for the column and $f_c' = 4,000$ for the beams.

8.8.   Given the column in Problem 8.7, determine the moment magnification factor δ_s. Assume $\Sigma P_u = 25{,}000$ kips, $\Delta_o = 0.35$ in, $V_{us} = 300$ kips, and $k = 1.6$.

8.9.   Determine the required vertical spacing of ties for a 24×18 in tied column reinforced with eight No. 9 bars. Assume normal-weight concrete with $f_c' = 5{,}000$ psi and Grade 60 reinforcement. Also assume that the clear cover to No. 3 ties is 1.5 in. The column is subjected to $N_u = 675$ kips and $V_u = 125$ kips acting parallel to the short side of the column.

8.10.   Determine the required Class B lap splice length of a 36×36 in column reinforced with 20 No. 10 bars that are uniformly distributed in the cross-section. The clear cover to No. 4 ties is 1.5 in. Assume normal-weight concrete with $f_c' = 9{,}000$ psi and Grade 60 reinforcement.

CHAPTER 9

Walls

9.1 Introduction

A wall is defined in ACI 2.2 as a member, usually vertical, that is used to enclose or separate spaces in a building or structure. There are many different types of walls, but they are typically categorized as non–load-bearing and load-bearing. A non–load-bearing wall supports primarily its own weight. In contrast, a load-bearing wall supports dead and live loads from the floor and roof systems in addition to its own weight. IBC 202 contains more precise definitions of these wall types.

Because of their relatively large in-plane lateral stiffness, walls can attract a significant portion of the effects due to wind or earthquakes. They are used alone or in combination with moment frames to resist these load effects. Because a wall is much stiffer in the direction parallel to the plane of the wall than perpendicular to this plane, it is commonly assumed that only the walls that are oriented parallel to the direction of the lateral loads resist the lateral load effects in that direction. Such walls must be designed for combinations of axial loads, bending moments about their strong axis, and shear forces, and they are referred to as *structural walls* in ACI 318.

Walls must also be designed for any bending moments about their minor axis, caused by lateral loads applied perpendicular to the plane of the wall (e.g., a wall that is situated at the perimeter of a building is subjected to wind loads perpendicular to its face) or by axial loads acting at an eccentricity from the centroid of the wall.

Basement walls and retaining walls are subjected to lateral earth pressure perpendicular to the plane of the wall. A cantilever retaining wall is designed for flexure in accordance with the strength design method presented in Chaps. 5 and 6.

Like columns, load-bearing walls are designed for the effects of axial loads in combination with bending moments and, thus, are referred to as *members subjected to combined axial load and bending*. They are also identified as *shear walls* because they usually resist most, if not all, of the shear forces generated by the horizontal loads in the direction parallel to the length of the wall.

Properly proportioned walls can reduce lateral displacements of a building frame, which can result in the frame being designated as nonsway. This can have a significant impact on the design of the columns in the frame (especially in regard to second-order effects) and on the overall performance of the structure. Section 8.5 contains methods on how to determine if a frame is nonsway or sway.

This chapter focuses on the analysis and design of walls that are subjected to axial loads, combined axial loads and bending, and shear. In general, provisions are presented on the following:

1. Sizing the cross-section

2. Determining the required amount of reinforcement

3. Detailing the reinforcement.

The requirements for reinforced concrete walls are given in ACI Chap. 14. These requirements are applicable to the design of cast-in-place, precast, and tilt-up wall systems.

Cast-in-place concrete walls are cast on site, utilizing formwork that is also built on site. After the forms have been erected, the required reinforcing bars are set in the forms at the proper location. The concrete is subsequently deposited into the forms.

Precast concrete walls are manufactured in a precasting plant under controlled conditions and are subsequently shipped to the site for erection. Such walls are reinforced with either nonprestressed or prestressed reinforcement.

Tilt-up concrete walls are cast in a horizontal position at the jobsite and then tilted up into their final position in the structure. According to the Code, tilt-up concrete construction is a form of precast concrete.

This chapter covers design methods that are applicable for these three types of walls after they are in their final positions within the structure. In the case of precast and tilt-up concrete walls, design methods for handling or erection are not covered, as they are beyond the scope of this book.

9.2 Design Methods for Axial Loads and Flexure

9.2.1 Overview

Chapter 14 of the Code gives three methods that can be utilized to design reinforced concrete walls: walls designed as compression members (ACI 14.4), the empirical design method (ACI 14.5), and an alternate design method for slender walls (ACI 14.8). The limitations and provisions of each method are covered later.

Regardless of the analysis method that is used in the design of a wall, the minimum reinforcement requirements of ACI 14.3 must be satisfied. A summary of these requirements is given in Table 9.1.

9.2.2 Walls Designed as Compression Members

According to ACI 14.4, walls subjected to axial loads or axial loads and bending moments are permitted to be designed in accordance with the following provisions:

- ACI 10.2 (design assumptions of the strength design method)
- ACI 10.3 (general principles and requirements of the strength design method)
- ACI 10.10 (slenderness effects in compression members)
- ACI 10.11 (axially loaded members supporting a slab system)
- ACI 10.14 (bearing strength)
- ACI 14.2 (general design requirements for walls)
- ACI 14.3 (minimum reinforcement requirements for walls)

Requirement	ACI Section Number
Minimum ratio of the vertical reinforcement area to the gross concrete area ρ_ℓ: 　For deformed bars not larger than No. 5 bars with 　　$f_y \geq 60{,}000$ psi, $\rho_\ell = 0.0012$. 　For other deformed bars, $\rho_\ell = 0.0015$. 　For welded wire reinforcement not larger than W31 or D31, 　　$\rho_\ell = 0.0012$.	14.3.2
Minimum ratio of the horizontal reinforcement area to the gross concrete area ρ_t: 　For deformed bars not larger than No. 5 bars with 　　$f_y \geq 60{,}000$ psi, $\rho_t = 0.0020$. 　For other deformed bars, $\rho_t = 0.0025$. 　For welded wire reinforcement not larger than W31 or D31, 　　$\rho_t = 0.0020$.	14.3.3
Walls that are more than 10 in thick (except basement walls) must have two layers of reinforcement placed in each direction according to the following: 　One layer must consist of not less than one-half and not more than two-thirds of the total reinforcement required for each direction. This layer must be placed not less than 2 in or more than one-third the thickness of the wall from the exterior surface. 　The other layer, which consists of the balance of the reinforcement in that direction, must be placed not less than three-quarters of an inch or more than one-third the thickness of the wall from the interior surface.	14.3.4
The maximum center-to-center spacing of the vertical reinforcement is three times the wall thickness or 18 in, whichever is smaller.	14.3.5
Vertical reinforcement need not be enclosed by lateral ties where the following conditions are met: 　The vertical reinforcement area is equal to or less than 1% of the gross concrete area. 　The vertical reinforcement is not required as compression reinforcement.	14.3.6
Additional reinforcement must be provided around window, door, and similar-sized openings in a wall.	14.3.7

TABLE 9.1　Minimum Reinforcement Requirements for Walls

Any wall may be designed using the general principles of the strength design method. However, walls that do not meet the limitations of ACI 14.5 or 14.8 must be designed as compression members by this method.

The following equations must be satisfied in the design of any wall subjected to combined axial load and bending:

$$\phi P_n \geq P_u \tag{9.1}$$

$$\phi M_n \geq M_u \tag{9.2}$$

The factored axial load P_u and bending moment M_u acting on a reinforced concrete section must be equal to or less than the corresponding design values ϕP_n and ϕM_n in order for strength requirements to be satisfied.

A design strength interaction diagram facilitates the design of a wall section. Details on how to construct such a diagram are given in Chap. 8.

Like columns, slenderness effects must be considered in the design of walls where required. In the direction parallel to the length of the wall, slenderness effects can usually be neglected because the radius of gyration of the wall in that direction is relatively large; this results in a slenderness ratio less than the limits given in ACI 10.10.1. However, in the direction parallel to the thickness of a wall, the slenderness ratio is usually greater than the prescribed slenderness limits. As such, the factored bending moments about the minor axis of the wall must be magnified to account for slenderness effects.

The three methods of analysis that are permitted for analyzing compression members where slenderness effects cannot be neglected are (1) nonlinear second-order analysis (ACI 10.10.3), (2) elastic second-order analysis (ACI 10.10.4), and (3) moment magnification procedure (ACI 10.10.5). More often than not, the frame in the direction of analysis can be considered nonsway where walls are used to resist the lateral loads.

ACI Eqs. (10-14) and (10-15) that are given in the moment magnification procedure to determine the stiffness EI were not originally derived for members, like walls, that have a single layer of reinforcement. Reference 1 contains the following equation of EI for members with a single layer of reinforcement:

$$\frac{0.1 E_c I_g}{\beta} \leq EI = \frac{E_c I_g}{\beta}\left(0.5 - \frac{e}{h}\right) \leq \frac{0.4 E_c I_g}{\beta} \tag{9.3}$$

In this equation, E_c is the modulus of elasticity of the concrete; I_g is the gross moment of inertia of the wall section in the direction of analysis; e is the eccentricity of the axial loads and lateral loads for all applicable load combinations; h is the overall thickness of the wall section; and β is a term related to the sustained axial loads and the area of vertical reinforcement:

$$\beta = 0.9 + 0.5\beta_d^2 - 12\rho \geq 1.0 \tag{9.4}$$

The term β_d is the ratio of the sustained load to the total load on the wall section and ρ is the ratio of the area of vertical reinforcement to the gross concrete area of the wall section.

Example 9.1 Check the adequacy of the wall depicted in Fig. 9.1 for the factored load combinations given in Table 9.2, which occur at the base of the wall. Assume that the wall resists the bending moments about its major axis. Also assume normal-weight concrete with $f'_c = 5,000$ psi and Grade 60 reinforcement ($f_y = 60,000$ psi) and that the frame is nonsway.

Solution

Step 1: Determine the factored load combinations. The service-level axial loads and bending moments were obtained from an elastic analysis of the frame, using the appropriate reduced moments of inertia given in ACI 10.10.4.1. The applicable load combinations are determined in accordance with ACI 9.2. A summary of the axial loads and bending moments is given in Table 9.2.

Load Case	Axial Load (kips)	Bending Moment (ft kips)
Dead (D)	500	0
Roof live (L_r)	8	0
Live (L)	400	0
Wind (W)	0	±3,004
Load Combination		
1 $1.4D$	700	0
2 $1.2D + 1.6L + 0.5L_r$	1,244	0
3 $1.2D + 1.6L_r + 0.5L$	813	0
4 $1.2D + 1.6L_r + 0.8W$	613	2,403
5 $1.2D + 1.6W + 0.5L + 0.5L_r$	804	4,806
6 $0.9D - 1.6W$	450	−4,806

TABLE 9.2   Summary of Axial Loads and Bending Moments on the Wall Given in Example 9.1

FIGURE 9.1 Wall section of Example 9.1.

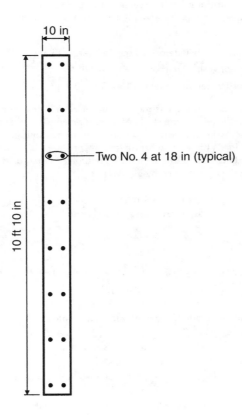

10 in

10 ft 10 in

Two No. 4 at 18 in (typical)

Step 2: Determine if slenderness effects need to be considered. Because the frame is a nonsway frame, slenderness effects need not be considered where Eq. (8-19) is satisfied:

$$\frac{k\ell_u}{r} \le 34 - 12\left(\frac{M_1}{M_2}\right) \le 40$$

For the 130-in wall in the direction of analysis, $r = 0.3 \times 130 = 39.0$ in.

The bending moment at the top of the wall at this level is equal to approximately 98% of the bending moment at the base, and the wall is bent in single curvature. Therefore,

$$\frac{k\ell_u}{r} = \frac{1.0 \times 10 \times 12}{39.0} = 3.1 \le 34 - (12 \times 0.98) = 22.2$$

Thus, slenderness effects may be neglected in the design of the wall in this direction.

Step 3: Determine the design strength interaction diagram for bending about the strong axis. The interaction diagram for this member is determined using strain compatibility analyses for bending about the strong axis (see Chap. 8).

Construction of the interaction diagram is illustrated by calculating the nominal strengths P_n and M_n corresponding to balanced failure. The flowchart shown in Fig. 5.39 is utilized to determine P_n and M_n for this compression member; it is modified to account for the requirements in wall design.

Step 3A: Check the minimum reinforcement limit. The minimum ratio ρ_ℓ of the vertical reinforcement area to the gross concrete area for a wall section is specified in ACI 14.3.2 (see Table 9.1):

Minimum $\rho_\ell = 0.0012$ for Grade 60 deformed bars not larger than No. 5 bars.

Thus, the minimum area of vertical reinforcement $= 0.0012 \times 10 \times 130 = 1.6$ in^2.

The provided area of vertical reinforcement $= 16 \times 0.2 = 3.2$ in$^2 > 1.6$ in^2.

Step 3B: Determine the neutral axis depth c. Balanced failure occurs when crushing of the concrete and yielding of the reinforcing steel occur simultaneously (see Section 5.3). The balanced failure point also represents the change from compression failures for higher axial loads and tension failures for lower axial loads for a given bending moment. ACI 10.3.3 permits the yield strain of the reinforcement to be taken as 0.0020 for Grade 60 reinforcement; thus, $\varepsilon_{s4} = \varepsilon_t = 0.0020$.

The neutral axis depth is determined by Eq. (5.41):

$$c = \frac{0.0030 d_t}{\varepsilon_t + 0.0030} = \frac{0.0030 \times 128}{0.0020 + 0.0030} = 76.8 \text{ in}$$

Step 3C: Determine β_1.

$$\beta_1 = 1.05 - 0.00005 f_c' = 1.05 - (0.00005 \times 5,000) = 0.80 \quad \text{for} \quad f_c' = 5,000 \text{ psi}$$

(see ACI 10.2.7.3 and Section 5.2 of this book)

Step 3D: Determine the depth of the equivalent stress block a.

$$a = \beta_1 c = 0.80 \times 76.8 = 61.4 \text{ in}$$

Step 3E: Determine C. The concrete compression resultant force C is determined by Eq. (5.42):

$$C = 0.85 f_c' a b = 0.85 \times 5 \times 61.4 \times 10 = 2,609.5 \text{ kips}$$

Step 3F: Determine ε_{si}. The strain in the reinforcement ε_{si} at the various layers is determined by Eq. (5.43) where compression strains are positive:

- Layer 1 ($d_1 = 2.0$ in):

$$\varepsilon_{s1} = \frac{0.0030(76.8 - 2.0)}{76.8} = 0.0029$$

- Layer 2 ($d_2 = 20.0$ in):

$$\varepsilon_{s2} = \frac{0.0030(76.8 - 20.0)}{76.8} = 0.0022$$

- Layer 3 ($d_3 = 38.0$ in):

$$\varepsilon_{s3} = \frac{0.0030(76.8 - 38.0)}{76.8} = 0.0015$$

- Layer 4 ($d_4 = 56.0$ in):

$$\varepsilon_{s4} = \frac{0.0030(76.8 - 56.0)}{76.8} = 0.0008$$

- Layer 5 ($d_5 = 74.0$ in):

$$\varepsilon_{s5} = \frac{0.0030(76.8 - 74.0)}{76.8} = 0.0001$$

- Layer 6 ($d_6 = 92.0$ in):

$$\varepsilon_{s6} = \frac{0.0030(76.8 - 92.0)}{76.8} = -0.0006$$

- Layer 7 ($d_7 = 110.0$ in):

$$\varepsilon_{s7} = \frac{0.0030(76.8 - 110.0)}{76.8} = -0.0013$$

- Layer 8 ($d_8 = 128.0$ in):

$$\varepsilon_{s8} = \frac{0.0030(76.8 - 128.0)}{76.8} = -0.0020 \text{ (checks)}$$

It is evident that the top five layers of reinforcement are in compression and that the bottom three layers are in tension. Also, the two layers of reinforcement closest to the extreme compression fiber and the layer of reinforcement farthest from the extreme compression fiber yield.

Step 3G: Determine f_{si}. The stress in the reinforcement f_{si} at the various layers is determined by multiplying ε_{si} by the modulus of elasticity of the steel E_s:

- Layer 1: $f_{s1} = 0.0029 \times 29,000 = 84.1$ ksi > 60 ksi; use $f_{s1} = 60$ ksi
- Layer 2: $f_{s2} = 0.0022 \times 29,000 = 63.8$ ksi > 60 ksi; use $f_{s2} = 60$ ksi
- Layer 3: $f_{s3} = 0.0015 \times 29,000 = 43.5$ ksi
- Layer 4: $f_{s4} = 0.0008 \times 29,000 = 23.2$ ksi
- Layer 5: $f_{s5} = 0.0001 \times 29,000 = 2.9$ ksi
- Layer 6: $f_{s6} = -0.0006 \times 29,000 = -17.4$ ksi
- Layer 7: $f_{s7} = -0.0013 \times 29,000 = -37.7$ ksi
- Layer 8: $f_{s8} = -60$ ksi

Step 3H: Determine F_{si}. The force in the reinforcement F_{si} at the various layers is determined by Eq. (5.44) or (5.45), which depends on the location of the steel layer:

- Layer 1 ($d_1 = 2.0$ in $< a = 61.4$ in): $F_{s1} = [60 - (0.85 \times 5)] \times 2 \times 0.20 = 22.3$ kips
- Layer 2 ($d_2 = 20.0$ in $< a = 61.4$ in): $F_{s2} = [60 - (0.85 \times 5)] \times 2 \times 0.20 = 22.3$ kips
- Layer 3 ($d_3 = 38.0$ in $< a = 61.4$ in): $F_{s3} = [43.5 - (0.85 \times 5)] \times 2 \times 0.20 = 15.7$ kips
- Layer 4 ($d_4 = 56.0$ in $< a = 61.4$ in): $F_{s4} = [23.2 - (0.85 \times 5)] \times 2 \times 0.20 = 7.6$ kips
- Layer 5: $F_{s5} = 2.9 \times 2 \times 0.2 = 1.2$ kips
- Layer 6: $F_{s6} = -17.4 \times 2 \times 0.2 = -7.0$ kips
- Layer 7: $F_{s7} = -37.7 \times 2 \times 0.2 = -15.1$ kips
- Layer 8: $F_{s8} = -60.0 \times 2 \times 0.2 = -24.0$ kips

Note that the compression steel in the top four layers falls within the depth of the equivalent stress block; thus, Eq. (5.45) is used to determine the forces in the reinforcement in those layers.

Step 3I: Determine P_n and M_n. The nominal axial strength P_n and nominal flexural strength M_n of the section are determined by Eqs. (5.46) and (5.47), respectively:

$$P_n = C + \sum F_{si} = 2,609.5 + (22.3 + 22.3 + 15.7 + 7.6 + 1.2 - 7.0 - 15.1 - 24.0) = 2,632.5 \text{ kips}$$

$$\begin{aligned}
M_n &= 0.5C(h - a) + \sum F_{si}(0.5h - d_i) \\
&= [0.5 \times 2,609.5 \times (130 - 61.4)] + [22.3(65 - 2) + 22.3(65 - 20) \\
&\quad + 15.7(65 - 38) + 7.6(65 - 56) + 1.2(65 - 74) \\
&\quad + (-7.0)(65 - 92) + (-15.1)(65 - 110) + (-24.0)(65 - 128)] \\
&= 89,505.9 + 5,270.4 = 94,776.3 \text{ in kips} = 7,898.0 \text{ ft kips}
\end{aligned}$$

This section is compression-controlled because ε_t is equal to the compression-controlled strain limit of 0.0020 (see ACI 10.3.3). Thus, in accordance with ACI 9.3.2.2, the strength reduction factor ϕ is equal to 0.65 for a compression-controlled section with lateral reinforcement consisting of ties (or, equivalently, a compression-controlled section without spiral reinforcement conforming to ACI 10.9.3). Therefore, the design axial strength ϕP_n and the design flexural strength ϕM_n are

$$\phi P_n = 0.65 \times 2,632.5 = 1,711.1 \text{ kips}$$

$$\phi M_n = 0.65 \times 7,898.0 = 5,133.7 \text{ ft kips}$$

Additional points on the design strength interaction diagram, which is shown in Fig. 9.2, can be obtained in a similar manner.

Step 4: Check the adequacy of the section. As mentioned earlier, the design strength interaction diagram is shown in Fig. 9.2. The load combinations of Table 9.2 are also shown in the figure. It

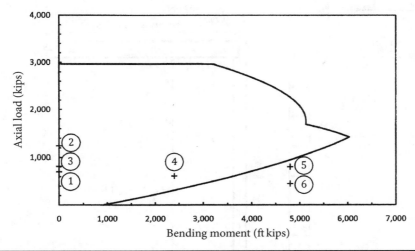

FIGURE 9.2   Design strength interaction diagram for the wall section of Example 9.1.

is evident that the wall section is not adequate because the points representing load combination numbers 5 and 6 fall outside of the design strength interaction curve.

It can be shown that two No. 6 bars spaced at 9 in on center are satisfactory for strength. Other bar sizes and spacing can be investigated as well.

Comments
The minimum horizontal reinforcement that is required in the section is equal to $0.0020 \times 10 \times 12 = 0.24 \ \text{in}^2/\text{ft}$, assuming Grade 60 deformed bars are not larger than No. 5 bars. This corresponds to two No. 4 bars spaced at 18 in on center (the provided area of steel is equal to $2 \times 0.20 \times 12/18 = 0.27 \ \text{in}^2/\text{ft}$). The adequacy of this reinforcement for shear is checked in Example 9.5.

Example 9.2   Check the adequacy of the 8-in reinforced concrete wall depicted in Fig. 9.3 for the following loads:

Dead load $= 120 \ \text{psf}$
Roof live load $= 20 \ \text{psf}$
Wind load $= 20 \ \text{psf}$

The wall is reinforced with a single layer of No. 3 bars spaced at 10 in on center and located at the middle of the wall. Assume that the gravity loads act through the centroid of the section and that the ends of the wall are pinned. Also assume normal-weight concrete with $f'_c = 4,000$ psi and Grade 60 reinforcement ($f_y = 60,000$ psi) and that the frame is nonsway.

Solution
Step 1: Determine the axial loads and bending moments on the wall. The axial loads and bending moments are determined for a 1-ft design strip:

$$P_D = 0.120 \times \frac{30}{2} = 1.8 \ \text{kips/ft}$$

$$P_{L_r} = 0.020 \times \frac{30}{2} = 0.3 \ \text{kips/ft}$$

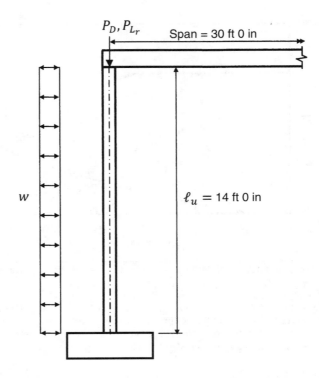

FIGURE 9.3
Reinforced
concrete wall of
Example 9.2.

The wind load does not cause any axial load on the wall, but it does cause the following bending moment that acts at the midheight section of the wall:

$$M_W = 0.020 \times \frac{14^2}{8} = 0.5 \text{ ft kips/ft} = 6.0 \text{ in kips/ft}$$

A summary of the factored load combinations is given in Table 9.3. The axial loads and bending moments are expressed in kips and inch-kips, respectively, per foot length of wall.

Step 2: Determine if slenderness effects need to be considered. Because the frame is a nonsway frame, slenderness effects need not be considered where Eq. (8.19) is satisfied:

$$\frac{k\ell_u}{r} \leq 34 - 12\left(\frac{M_1}{M_2}\right) \leq 40$$

For this 8-in wall, in the direction of analysis (i.e., bending about the minor axis) $r = 0.3 \times 8 = 2.4$ in.
Thus,

$$\frac{k\ell_u}{r} = \frac{1.0 \times 14 \times 12}{2.4} = 70 > 40$$

Thus, slenderness effects must be considered in the design of this wall in this direction.

Step 3: Determine the magnified moments in the column using the moment magnification procedure. In lieu of an elastic or a nonlinear second-order analysis, the moment magnification procedure of ACI 10.10.6 is used to determine the magnified factored moments due to second-order effects.

Load Case	Axial Load (kips)	Bending Moment (in kips)
Dead (D)	1.8	0
Roof live (L_r)	0.3	0
Wind (W)	0	±6.0
Load combination		
1 1.4D	2.5	0
2 1.2D + 0.5L_r	2.3	0
3 1.2D + 1.6L_r + 0.8W	2.6	4.8
4 1.2D + 1.6W + 0.5L_r	2.3	9.6
5 0.9D − 1.6W	1.6	−9.6

TABLE 9.3 Summary of Axial Loads and Bending Moments on the Wall Given in Example 9.2

Calculations are provided for load combination number 4 of Table 9.3.
From Eq. (8.22),

$$M_c = \delta_{ns} M_2$$

where

$$\delta_{ns} = \frac{C_m}{1 - (P_u/0.75P_c)} \geq 1.0$$

Because there are transverse loads between the supports of this wall, $C_m = 1.0$ (ACI 10.10.6.4)
The critical buckling load P_c is determined by Eq. (8.25):

$$P_c = \frac{\pi^2 EI}{(k\ell_u)^2}$$

Because the wall has one layer of vertical reinforcement, use Eq. (9.3) to determine the stiffness EI:

$$\frac{0.1 E_c I_g}{\beta} \leq EI = \frac{E_c I_g}{\beta} \left(0.5 - \frac{e}{h}\right) \leq \frac{0.4 E_c I_g}{\beta}$$

The modulus of elasticity of the concrete is determined by ACI 8.5.1:

$$E_c = w_c^{1.5} 33\sqrt{f_c'} = (150)^{1.5} \times 33\sqrt{4,000} = 3,834,254 \text{ psi}$$

For a 1-ft design strip,

$$I_g = \frac{12 \times 8^3}{12} = 512 \text{ in}^4$$

Because the dead load is the only sustained load, β_d is determined as follows:

$$\beta_d = \frac{1.2D}{1.2D + 1.6W + 0.5L_r} = 0.94$$

The eccentricity $e = M_u/P_u = 9.6/2.3 = 4.2$ in.
Therefore, the EI from Eq. (9.3) is

$$EI = \frac{E_c I_g}{\beta}\left(0.5 - \frac{e}{h}\right) = \frac{E_c I_g}{\beta}\left(0.5 - \frac{4.2}{8.0}\right) = -0.03\frac{E_c I_g}{\beta} < 0.1\frac{E_c I_g}{\beta}; \text{ use } 0.1\frac{E_c I_g}{\beta}$$

$$\rho = \frac{0.11 \times (12/10)}{8 \times 12} = 0.0014 > 0.0012$$

$$\beta = 0.9 + 0.5\beta_d^2 - 12\rho = 0.9 + (0.5 \times 0.94^2) - (12 \times 0.0014) = 1.33$$

Thus,

$$EI = 0.1\frac{E_c I_g}{\beta} = \frac{0.1 \times 3{,}834 \times 512}{1.33} = 147.6 \times 10^3 \text{ kip in}^2$$

Determine the critical buckling load P_c:

$$P_c = \frac{\pi^2 EI}{(k\ell_u)^2} = \frac{\pi^2 \times 147.6 \times 10^3}{(1.0 \times 14 \times 12)^2} = 51.6 \text{ kips}$$

The moment magnification factor δ_{ns} is equal to the following:

$$\delta_{ns} = \frac{C_m}{1 - (P_u/0.75P_c)} = \frac{1.0}{1 - [2.3/(0.75 \times 51.6)]} = 1.06 > 1.0$$

Thus,

$$M_c = \delta_{ns}M_2 = 1.06 \times 9.6 = 10.2 \text{ in kips}$$

Check the minimum factored bending moment requirement [see Eq. (8.29)]:

$$M_{2,min} = P_u(0.6 + 0.03h) = 2.3 \times [0.6 + (0.03 \times 8)] = 1.9 \text{ in kips} < M_c$$

Therefore, the factored axial load and moment in this load combination become $P_u = 2.3$ kips and $M_u = M_c = 10.2$ in kips.

Similar calculations can be performed for the other load combinations.

A summary of the results from the moment magnification procedure is given in Table 9.4.

Step 4: Check the adequacy of the section for combined axial load and bending. The design strength interaction diagram for a 1-ft-wide section of the wall reinforced with No. 3 bars at 10 in

Load Combination	P_u (kips)	M_u (in kips)	β_d	β	$EI \times 10^3$ (kip in^2)	P_c (kips)	δ_{ns}	M_c (in kips)
1	2.5	0	1.00	1.38	567.7	198.3	1.02	0
2	2.3	0	0.94	1.32	594.2	207.8	1.02	0
3	2.6	4.8	0.82	1.22	439.6	153.6	1.02	4.9
4	2.3	9.6	0.94	1.33	147.6	51.6	1.06	10.2
5	1.6	9.6	1.00	1.38	141.9	49.6	1.05	10.1

TABLE 9.4   Summary of the Results from the Moment Magnification Procedure Given in Example 9.2

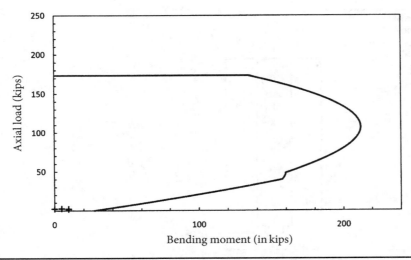

FIGURE 9.4   Design strength interaction diagram for the wall given in Example 9.2.

on center is shown in Fig. 9.4. Also shown in the figure are the factored axial loads and magnified bending moments for the five load combinations.

Because all five load combination points fall within the design strength interaction diagram, the wall is adequate.

Comments
It is clear from Fig. 9.4 that the section is tension-controlled for all load combinations. The amount of vertical steel that is provided in the wall is slightly greater than the minimum required in accordance with ACI 14.3.2.

9.2.3   Empirical Design Method

The provisions for the empirical design method are given in ACI 14.5. This method may be used for the design of walls where all of the following limitations are satisfied:

1. The wall has a solid, rectangular cross-section.

2. The resultant of all applicable factored loads falls within the middle third of the wall thickness.

3. The thickness of bearing walls is equal to or greater than the unsupported height or length of the wall, whichever is shorter, divided by 25, but not less than 4 in. Furthermore, the wall thickness is equal to or greater than 7.5 in for exterior basement walls and foundation walls.

Illustrated in Fig. 9.5 is a wall section subjected to an axial load acting at an eccentricity from the centroid of the section. In this case, the second limitation of this method is satisfied where $e \leq h/6$. In general, the total eccentricity caused by all applicable factored load effects, including those from lateral loads, must be compared with $h/6$.

FIGURE **9.5**
Eccentricity
limitations in the
empirical design
method.

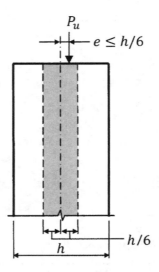

The design axial strength ϕP_n of a wall satisfying the limitations of this method is determined by ACI Eq. (14-1):

$$\phi P_n = 0.55\phi f'_c A_g \left[1 - \left(\frac{k\ell_c}{32h} \right)^2 \right] \tag{9.5}$$

In this equation, A_g is the gross area of the wall and k is the effective length factor prescribed in ACI 14.5.2.

For walls that are braced against lateral translation at both ends of the wall, k is defined as follows:

- $k = 0.8$ when the wall is restrained against rotation at one or both ends
- $k = 1.0$ when the wall is unrestrained against rotation at both ends

A k-value of 0.8 implies that the end of the wall is attached to a member that has a flexural stiffness that is at least equal to that of the wall in the direction of analysis. Members with lesser stiffnesses cannot adequately restrain the wall against rotation.

For walls that are not braced against lateral translation, $k = 2.0$. This would be applicable, for example, to freestanding (cantilever) walls or to walls that are connected to diaphragms that undergo significant deflections when subjected to lateral loads.

Equation (9.5) takes into consideration both load eccentricity and slenderness effects. The eccentricity factor 0.55 was originally selected to give strengths comparable to those determined by ACI Chap. 10 for members with an axial load applied at an eccentricity of $h/6$. The strength reduction factor ϕ corresponds to compression-controlled sections in accordance with ACI 9.3.2.2. Thus, ϕ is equal to 0.65 for wall sections designed by this method.

In order to satisfy strength requirements, the design strength ϕP_n determined by Eq. (9.5) must be equal to or greater than the factored axial load P_u.

The empirical design method is best suited for relatively short walls subjected to vertical loads. Because the total eccentricity must not exceed $h/6$, its application becomes extremely limited when lateral loads need to be considered.

Walls not meeting the limitations of this method must be designed as compression members subjected to axial load and bending by the provisions of Chap. 10 or, if applicable, by the alternative design method of ACI 14.8, which is covered later.

Example 9.3   A reinforced concrete wall with an unsupported length of 16 ft is subjected to the following service axial loads: $P_D = 15$ kips and $P_{L_r} = 8$ kips.

Assume that these loads act through the centroid of the wall and over a width of 12 in. Also assume normal-weight concrete with $f'_c = 4,000$ psi and Grade 60 reinforcement ($f_y = 60,000$ psi). The ends of the wall are braced against lateral translation and are unrestrained against rotation (i.e., the ends of the wall are pinned).

Design the wall in accordance with the empirical design method.

Solution

Step 1: Select a trial wall thickness. According to ACI 14.5.3.1, the minimum thickness of a wall that is designed by the empirical design method is

$$h = \frac{\ell_u}{25} \geq 4 \text{ in}$$

For a 16-ft wall height,

$$h = \frac{16 \times 12}{25} = 7.7 \text{ in}$$

Try $h = 8$ in.

Step 2: Determine the factored axial loads. The governing factored axial load is determined by ACI Eq. (9.3):

$$P_u = 1.2P_D + 1.6P_{L_r} = (1.2 \times 15) + (1.6 \times 8) = 31 \text{ kips}$$

Step 3: Determine the design strength of the wall. The design strength ϕP_n is determined by Eq. (9.5):

$$\phi P_n = 0.55\phi f'_c A_g \left[1 - \left(\frac{k\ell_c}{32h} \right)^2 \right]$$

$$= 0.55 \times 0.65 \times 4 \times (8 \times 12) \times \left[1 - \left(\frac{1.0 \times 16 \times 12}{32 \times 8} \right)^2 \right]$$

$$= 60 \text{ kips} > P_u = 31 \text{ kips}$$

Step 4: Provide minimum reinforcement in wall. Minimum requirements for vertical reinforcement and horizontal reinforcement are given in ACI 14.3.2 and 14.3.3, respectively.

Minimum vertical reinforcement $= 0.0012 \times 8 \times 12 = 0.12$ in^2.

Provide a single layer of No. 4 bars spaced 12 in on center ($A_s = 0.20$ in^2).

Minimum horizontal reinforcement $= 0.0020 \times 8 \times 12 = 0.19$ in^2.

As also for minimum vertical reinforcement, provide a single layer of No. 4 bars spaced 12 in on center ($A_s = 0.20$ in^2).

The 12-in spacing of the vertical and horizontal bars is less than the maximum allowable spacing of 18 in for this 8-in-thick wall (ACI 14.3.5).

Comments

Even though the loads in this example were not applied at an eccentricity, the design strength ϕP_n that was determined is applicable up to a total eccentricity of $h/6 = 1.3$ in.

9.2.4 Alternative Design of Slender Walls

Limitations

The design procedure given in ACI 14.8 is an alternative to the requirements of ACI 10.10 for the out-of-plane design of slender, reinforced concrete walls. The provisions are based on experimental research[2] and first appeared in the 1988 edition of the Uniform Building Code.[3]

According to ACI 14.8.1, the provisions of ACI 14.8 are considered to satisfy the slenderness provisions of ACI 10.10 when flexural tension controls the design of the wall. The following limitations apply to this method:

1. The wall panel must be simply supported, axially loaded, and subjected to an out-of-plane, uniform lateral load with the maximum moment and deflection occurring at the midheight of the wall.

2. The cross-section of the wall must be constant over the entire height.

3. The wall must be tension-controlled; that is, the net tensile strain in the extreme tension steel ε_t must be equal to or greater than 0.0050 when the concrete in compression reaches its assumed strain limit of 0.0030 (ACI 10.3.4).

4. The vertical reinforcement provided in the wall must provide a design strength ϕM_n that is equal to or greater than the cracking moment M_{cr}, which is determined by ACI Eq. (9-9) as $M_{cr} = f_r I_g / y_t$, where the modulus of rupture $f_r = 7.5\lambda\sqrt{f_c'}$.

5. Concentrated gravity loads applied to a wall above the design flexural section, which is located at the midheight of the wall, must be distributed over widths equal to those shown in Fig. 9.6 for loads near the edge and at the interior of the wall. In the figure, B is the bearing length at the top of the wall; S is the spacing between concentrated gravity loads; and E is the distance from the edge of bearing to the edge of the wall.

6. The axial stress P_u/A_g at the midheight must be equal to or less than 6% of the concrete compressive strength.

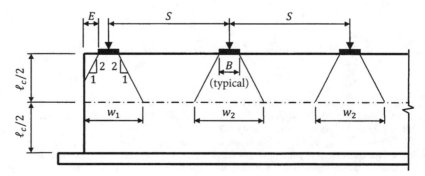

$$w_1 = B + (\ell_c/4) + E \leq S$$

$$w_2 = B + (\ell_c/2) \leq S$$

Figure 9.6 Distribution of concentrated loads in the alternative design method for walls.

When any of these six conditions is not satisfied, the wall must be designed as a compression member in accordance with the provisions of ACI 14.4. For example, it is common for walls to have relatively large window and door openings. Such walls are not considered to have a constant cross-section over their full height. Thus, the alternate design method is not applicable because the second limitation of this method is not satisfied. Instead, the wall must be designed in accordance with ACI 14.4.

Design Requirements

Flexure The following strength equation must be satisfied at the midheight of a wall:

$$\phi M_n \geq M_u \tag{9.6}$$

The factored moment M_u includes second-order effects and consists of two parts:

$$M_u = M_{ua} + P_u \Delta_u \tag{9.7}$$

The moment M_{ua} is the maximum factored moment at the midheight of the wall due to lateral loads (wind or seismic) and/or vertical factored loads P_u applied at an eccentricity from the centroid of the wall. The deflection Δ_u is the total deflection at the midheight of the wall due to the factored loads. This deflection is determined by ACI Eq. (14-5):

$$\Delta_u = \frac{5 M_u \ell_c^2}{(0.75)48 E_c I_{cr}} \tag{9.8}$$

In this equation, ℓ_c is the length of the wall, measured center-to-center of joints, and I_{cr} is the moment of inertia of the cracked wall section transformed into concrete. ACI Eq. (14-7) is used to determine I_{cr}:

$$I_{cr} = \frac{E_s}{E_c} \left(A_s + \frac{P_u h}{2 f_y d} \right)(d - c)^2 + \frac{\ell_w c^3}{3} \tag{9.9}$$

In this equation, E_s is the modulus of elasticity of the reinforcing steel; E_c is the modulus of elasticity of the concrete; A_s is the area of the vertical reinforcement in the wall; d is the distance from the extreme compression fiber to the centroid of A_s; c is the distance from the extreme compression fiber to the neutral axis; and ℓ_w is the length of the wall. The ratio E_s/E_c must be taken equal to or greater than 6 in this equation.

In the strength design method, the neutral axis depth c is related to the depth of the equivalent rectangular stress block a by $c = a/\beta_1$. In general, a is equal to the force in the tension reinforcement ($A_s f_y$) divided by the equivalent compressive stress ($0.85 f_c'$) times the width of the section (b) [see Eq. (5.9)]. In the alternative design method, an effective area of longitudinal reinforcement $A_{se,w}$ is used in the determination of a and c and is calculated by the following equation:

$$A_{se,w} = A_s + \frac{P_u h}{2 f_y d} \tag{9.10}$$

This effective area of longitudinal reinforcement is used in the determination of I_{cr} [see Eq. (9.9)].

Thus, the depth of the equivalent stress block for bending about the minor axis of the wall is

$$a = \frac{A_s f_y + (P_u h / 2d)}{0.85 f_c' \ell_w} \tag{9.11}$$

Also, the distance from the extreme compression fiber to the neutral axis is

$$c = \frac{A_s f_y + (P_u h / 2d)}{0.85 f_c' \ell_w \beta_1} \tag{9.12}$$

It is evident from Eq. (9.8) that the deflection Δ_u is a function of M_u. However, M_u is a function of Δ_u, as is apparent from Eq. (9.7). This clearly illustrates the iterative nature inherent to this method. Thus, M_u can be determined by assuming a value of Δ_u and then performing several calculation iterations until convergence occurs. Alternatively, it can be determined by ACI Eq. (14-6):

$$M_u = \frac{M_{ua}}{1 - \left[5 P_u \ell_c^2 / (0.75) 48 E_c I_{cr}\right]} \tag{9.13}$$

Consider the simply supported wall depicted in Fig. 9.7. The wall is subjected to a factored gravity load P_{ug} acting at an eccentricity e from the centroid of the section and a factored uniform lateral load w_u.

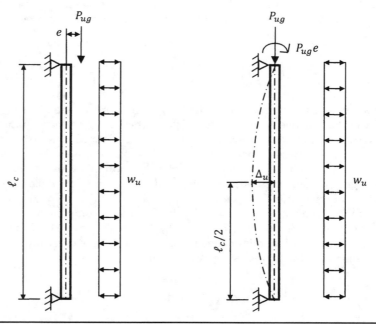

FIGURE 9.7 Wall analysis in accordance with ACI 14.8.

As noted earlier, the critical section occurs at the midheight. At that location, the total factored load P_u is equal to P_{ug} plus the weight of the wall P_{usw} from the top to the midheight.

The maximum factored moment at the midheight of the wall M_{ua} is equal to the moment due to the factored gravity load P_{ug} applied at an eccentricity e plus the moment due to w_u:

$$M_{ua} = \frac{P_{ug}e}{2} + \frac{w_u \ell_c^2}{8} \tag{9.14}$$

Assuming that the lateral load is due to wind and/or earthquakes, which means that it can act in either horizontal direction, the critical loading case occurs when the bending moments due to the gravity loads and the lateral loads are additive, which is reflected in Eq. (9.14).

Once M_{ua} has been calculated, M_u is determined by Eq. (9.13). The third, fourth, and sixth limitations of the procedure are subsequently checked. These calculations are performed for each load combination.

Deflection In addition to satisfying strength requirements, the deflection requirement of ACI 14.8.4 must be satisfied. The maximum deflection due to service loads Δ_s, which includes second-order effects, depends on the magnitude of the service load moment M_a at the midheight of the wall where M_a is equal to the following:

$$M_a = M_{sa} + P_s \Delta_s \tag{9.15}$$

The quantities M_{sa} and P_s are the first-order, service-level bending moment and axial load at the midheight of the wall, respectively.

It has been demonstrated that out-of-plane deflections increase rapidly when M_a exceeds two-thirds of the cracking moment M_{cr}.[2] Thus, Δ_s is determined by one of the following two equations:

- Case 1: $M_a \leq 2M_{cr}/3$

$$\Delta_s = \left(\frac{M_a}{M_{cr}}\right)\Delta_{cr} \tag{9.16}$$

- Case 2: $M_a > 2M_{cr}/3$

$$\Delta_s = \frac{2\Delta_{cr}}{3} + \left[\frac{M_a - (2M_{cr}/3)}{M_n - (2M_{cr}/3)}\right]\left(\Delta_n - \frac{2\Delta_{cr}}{3}\right) \tag{9.17}$$

In these equations, Δ_{cr} and Δ_n are the midheight deflections corresponding to the cracking moment M_{cr} and the nominal flexural strength M_n, respectively, and are determined as follows:

$$\Delta_{cr} = \frac{5M_{cr}\ell_c^2}{48E_cI_g} \tag{9.18}$$

$$\Delta_n = \frac{5M_n\ell_c^2}{48E_cI_{cr}} \tag{9.19}$$

Because Δ_s is a function of M_a [Eq. (9.16) or (9.17)] and M_a is a function of Δ_s [Eq. (9.15)], there is no closed-form solution for Δ_s. A value of Δ_s is obtained by iteration; that is, an initial value of Δ_s is assumed, and calculations are performed until convergence occurs.

In order to satisfy the deflection requirements of ACI 14.8.4, the service-level deflection Δ_s calculated by the method outlined earlier must not exceed $\ell_c/150$.

Example 9.4   A reinforced concrete wall with an unsupported length of 28 ft supports a roof system consisting of precast double-tee beams (see Fig. 9.8). The double tees weigh 720 lb per linear foot and have 3.75-in-wide webs that are spaced 5 ft on center. The webs bear on the top of the wall, and the span length of the double-tees is 48 ft 0 in.

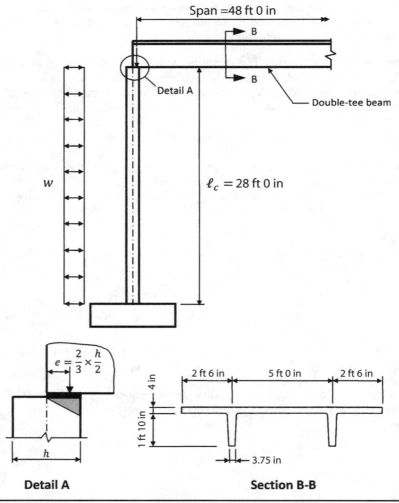

FIGURE 9.8   Reinforced concrete wall of Example 9.4.

The superimposed dead load on the roof is 10 psf; the roof live load is 25 psf; and the wind load is 25 psf.

Assume that the gravity loads act at the eccentricity indicated in Fig. 9.8. Also assume normal-weight concrete with $f'_c = 4{,}000$ psi and Grade 60 reinforcement ($f_y = 60{,}000$ psi). The ends of the wall are braced against lateral translation and are unrestrained against rotation.

Design the wall in accordance with the alternative design method of ACI 14.8.

Solution

Step 1: Select a trial wall thickness. Try an 8-in-thick wall, and assume one layer of No. 5 bars spaced at 8 in on center.

For a 1-ft-wide design strip, the vertical reinforcement ratio is equal to

$$\frac{0.31 \times (12/8)}{12 \times 8} = 0.0048 > 0.0012 \text{ [ACI 14.3.2(a)]}$$

Step 2: Determine axial loads and moments at the midheight of wall. The loads carried by the double-tee beams are distributed to the wall through their webs. The loads at the interior of the wall are distributed at the midheight over a width prescribed in ACI 14.8.2.5 (see Fig. 9.6):

$$w_2 = B + (\ell_c/2) = (3.75/12) + (28/2) = 14.3 \text{ ft} > \text{spacing} = 5 \text{ ft; use 5 ft}$$

Determine the axial loads per foot at the midheight of the wall:

$$\text{Weight of the wall from the top to the midheight} = \frac{8}{12} \times 0.150 \times \frac{28}{2} = 1.4 \text{ kips/ft}$$

$$\text{Roof dead load} = \frac{[(720/2 \text{ webs}) + (10 \times 5)] \times (48/2)}{5 \times 1{,}000} = 2.0 \text{ kips/ft}$$

$$\text{Roof live load} = \frac{(25 \times 5) \times (48/2)}{5 \times 1{,}000} = 0.6 \text{ kips/ft}$$

$$\text{Eccentricity of axial loads} = \frac{2}{3} \times 4 = 2.7 \text{ in}$$

A summary of the axial loads and bending moments per foot length of the wall at the midheight is given in Table 9.5. Included in the table are the applicable factored load combinations.

The service-level bending moments in the table are determined as follows (see Fig. 9.7):

$$\text{Service dead load moment} = \frac{(P_s)_D \times e}{2} = \frac{2.0 \times 2.7}{2} = 2.7 \text{ in kips (only the axial dead}$$

$$\text{load from the roof causes bending moment at the midheight)}$$

$$\text{Service roof live load moment} = \frac{(P_s)_{L_r} \times e}{2} = \frac{0.6 \times 2.7}{2} = 0.8 \text{ in kips}$$

$$\text{Service wind load moment} = \frac{w\ell_c^2}{8} = \frac{0.025 \times 28^2}{8} = 2.5 \text{ ft kips} = 29.4 \text{ in kips}$$

Step 3: Determine the factored axial loads and moments at the midheight of the wall, including slenderness effects. A summary of the factored axial loads and moments, which include slenderness effects, is given in Table 9.6.

Calculations are provided for load combination number 4.

The total moment M_u, which includes slenderness effects, is determined by Eq. (9.13):

$$M_u = \frac{M_{ua}}{1 - \left[5 P_u \ell_c^2 / (0.75) 48 E_c I_{cr}\right]}$$

Load Case	Axial Load P_s (kips)	Bending Moment (in kips)
Dead (D)	3.4	2.7
Roof live (L_r)	0.6	0.8
Wind (W)	0	±29.4

Load Combination		Axial Load P_u (kips)	Bending Moment M_{ua} (in kips)
1	$1.4D$	4.8	3.8
2	$1.2D + 0.5L_r$	4.4	3.6
3	$1.2D + 1.6L_r + 0.8W$	5.0	28.0
4	$1.2D + 1.6W + 0.5L_r$	4.4	50.7
5	$0.9D - 1.6W$	3.1	−44.6

TABLE 9.5 Summary of Axial Loads and Bending Moments for the Wall Given in Example 9.4

The only unknown in this equation is I_{cr}, which is determined by Eq. (9.9):

$$I_{cr} = \frac{E_s}{E_c}\left(A_s + \frac{P_u h}{2 f_y d}\right)(d - c)^2 + \frac{\ell_w c^3}{3} = \frac{E_s}{E_c} A_{se,w}(d - c)^2 + \frac{\ell_w c^3}{3} \quad \text{[see Eq. (9.10)]}$$

$$A_{se,w} = A_s + \frac{P_u h}{2 f_y d} = 0.31\left(\frac{12}{8}\right) + \frac{4.4 \times 8}{2 \times 60 \times 4} = 0.54 \text{ in}^2/\text{ft}$$

The neutral axis depth c is determined by Eq. (9.12):

$$c = \frac{A_s f_y + (P_u h/2d)}{0.85 f_c' \ell_w \beta_1} = \frac{A_{se,w} f_y}{0.85 f_c' \ell_w \beta_1} = \frac{0.54 \times 60}{0.85 \times 4 \times 12 \times 0.85} = 0.93 \text{ in}$$

Check if the section is tension-controlled:

$$\varepsilon_t = 0.0030\left(\frac{d_t}{c} - 1\right) = 0.0030\left(\frac{4}{0.93} - 1\right) = 0.0099 > 0.0050$$

Therefore, the section is tension-controlled. It can be determined that this wall section is tension-controlled for all load combinations.

Load Combination		P_u (kips)	M_{ua} (in kips)	$A_{se,w}$ (in²)	c (in)	I_{cr} (in⁴)	M_u (in kips)
1	$1.4D$	4.8	3.8	0.54	0.94	41.8	7.0
2	$1.2D + 0.5L_r$	4.4	3.6	0.54	0.93	41.9	6.3
3	$1.2D + 1.6L_r + 0.8W$	5.0	28.0	0.55	0.95	42.1	54.9
4	$1.2D + 1.6W + 0.5L_r$	4.4	50.7	0.54	0.93	41.9	88.9
5	$0.9D - 1.6W$	3.1	−44.6	0.52	0.89	40.5	71.5

TABLE 9.6 Summary of Factored Axial Loads and Bending Moments, Including Slenderness Effects, for the Wall Given in Example 9.4

Also,

$$E_c = 33w_c^{1.5}\sqrt{f_c'} = 33 \times (150)^{1.5} \times \sqrt{4{,}000} = 3{,}834{,}254 \text{ psi}$$

$$E_s/E_c = 29{,}000{,}000/3{,}834{,}254 = 7.6 > 6$$

Thus,

$$\begin{aligned}
I_{cr} &= \frac{E_s}{E_c} A_{se,w}(d-c)^2 + \frac{\ell_w c^3}{3} \\
&= 7.6 \times 0.54 \times (4-0.93)^2 + \frac{12 \times 0.93^3}{3} = 41.9 \text{ in}^4 \\
M_u &= \frac{M_{ua}}{\left[5P_u \ell_c^2/(0.75)48 E_c I_{cr}\right]} \\
&= \frac{50.7}{1 - [5 \times 4.4 \times (28 \times 12)^2/(0.75 \times 48 \times 3834 \times 41.9)]} = 88.9 \text{ in kips}
\end{aligned}$$

Similar calculations can be performed for the other load combinations.

Step 4: Determine the cracking moment M_{cr}. The cracking moment is determined by the following equation:

$$M_{cr} = \frac{f_r I_g}{y_t} = \frac{7.5\lambda\sqrt{f_c'}I_g}{y_t} = \frac{7.5 \times 1.0\sqrt{4{,}000} \times (12 \times 8^3/12)}{4 \times 1{,}000} = 60.7 \text{ in kips}$$

Step 5: Determine the design moment strength ϕM_n, and check the adequacy of the wall section. The design moment strength of the wall is determined by the following equation:

$$\phi M_n = \phi A_{se,w} f_y \left(d - \frac{a}{2}\right)$$

A summary of the design moment strengths for the applicable load combinations is given in Table 9.7. The strength reduction factor ϕ is equal to 0.9 because the section is tension-controlled for all load combinations. It is evident from the table that the design moment strength ϕM_n is greater than the required strength M_u and the cracking moment M_{cr}.

Step 6: Determine the maximum axial stress at the midheight section of the wall. The maximum stress at the midheight section of the wall is determined using the greatest axial load from the load combinations (see Table 9.6):

$$\frac{P_u}{A_g} = \frac{5{,}000}{8 \times 12} = 52.1 \text{ psi} < 0.06 f_c' = 240 \text{ psi}$$

	Load Combination	$A_{se,w}$ (in^2)	c (in)	a (in)	ϕM_n (in kips)	M_u (in kips)
1	$1.4D$	0.54	0.94	0.80	105.0	7.0
2	$1.2D + 0.5L_r$	0.54	0.93	0.79	105.1	6.3
3	$1.2D + 1.6L_r + 0.8W$	0.55	0.95	0.81	106.8	54.9
4	$1.2D + 1.6W + 0.5L_r$	0.54	0.93	0.79	105.1	88.9
5	$0.9D - 1.6W$	0.52	0.89	0.76	101.7	71.5

TABLE 9.7 Summary of Design Moment Strength for the Wall Given in Example 9.4

Step 7: Determine the service-level deflection at the midheight section of the wall. The maximum service-level deflection Δ_s occurs at the midheight section of the wall for the load combination that includes wind loads. Because there is no closed-form solution for Δ_s, assume that $M_a < 2M_{cr}/3$. Also assume the following value of Δ_s for the initial iteration:

$$\Delta_s = \left(\frac{M_{sa}}{M_{cr}}\right)\Delta_{cr}$$

The value of the maximum service-level bending moment M_{sa} at the midheight section of the wall is equal to the sum of the moments due to dead load, roof live load, and wind (see step 2 of this example):

$$M_{sa} = 2.7 + 0.8 + 29.4 = 32.9 \text{ in kips}$$

The deflection Δ_{cr} is determined by Eq. (9.18):

$$\Delta_{cr} = \frac{5M_{cr}\ell_c^2}{48E_cI_g}$$

$$= \frac{5 \times 60.7 \times (28 \times 12)^2}{48 \times 3{,}834 \times (12 \times 8^3)/12} = 0.36 \text{ in}$$

Thus,

$$\Delta_s = \left(\frac{32.9}{60.7}\right) \times 0.36 = 0.20 \text{ in}$$

Determine M_a from Eq. (9.15):

$$M_a = M_{sa} + P_s\Delta_s = 32.9 + [(3.4 + 0.6) \times 0.20] = 33.7 \text{ in kips}$$

Because it has been assumed that $M_a < 2M_{cr}/3$, determine Δ_s from Eq. (9.16):

$$\Delta_s = \left(\frac{M_a}{M_{cr}}\right)\Delta_{cr} = \left(\frac{33.7}{60.7}\right) \times 0.36 = 0.20 \text{ in}$$

This value of Δ_s is the same as the value that was initially assumed for Δ_s, so no additional iterations are required.

Check the initial assumption:

$$M_a = 33.7 \text{ in kips} < 2M_{cr}/3 = 40.5 \text{ in kips}$$

Check the deflection limit:

$$\Delta_s = 0.20 \text{ in} < \frac{\ell_c}{150} = \frac{28 \times 12}{150} = 2.2 \text{ in}$$

Therefore, the 8-in wall section is adequate with No. 5 bars spaced 8 in on center.

9.3 Design for Shear

9.3.1 Overview

Shear requirements for walls are given in ACI 11.9. Provisions are provided for shear forces perpendicular to the face of the wall and those in the plane of the wall.

According to ACI 11.9.1, design for horizontal shear forces acting perpendicular to the face of a wall shall be in accordance with the provisions for slabs in ACI 11.11. These design requirements are applicable to walls subjected to loads that act perpendicular to the face, such as wind loads. Both one-way and two-way shear must be investigated at the corresponding critical sections. Section 7.6 covers the requirements for both types of shear.

Design for horizontal in-plane forces shall be in accordance with ACI 11.9.2 through 11.9.9. In-plane shear forces are typically critical for walls with relatively small height-to-length ratios that resist the effects of lateral loads. Taller walls are usually governed by flexure rather than by shear.

It is permitted to design certain types of walls for shear using a strut-and-tie model in accordance with Appendix A of the Code. In particular, the walls must have a height-to-length ratio that is equal to or less than 2. The requirements of ACI 11.9.9.2 through 11.9.9.5 must also be satisfied.

The focus of this section is on the design of walls subjected to horizontal in-plane shear forces. Provided next are the shear strength design requirements given in ACI 11.9.2 through 11.11.9.

9.3.2 Design Shear Strength

The following equation must be satisfied for shear strength:

$$\phi V_n \geq V_u \tag{9.20}$$

In this equation, the strength reduction factor ϕ is equal to 0.75 in accordance with ACI 9.3.2.3. The nominal shear strength V_n consists of two parts: the nominal shear strength provided by the concrete V_c and the nominal shear strength provided by the shear reinforcement V_s.

The nominal strength V_n is limited to $10\sqrt{f_c'}hd$ at any horizontal section of a wall, where h is the thickness of the wall and d is the effective depth of the wall defined in ACI 11.9.4. The limiting shear stress of $10\sqrt{f_c'}$ is based on tests on walls with a thickness equal to the length of the wall divided by 25.[4]

The required strength V_u is determined from the analysis of the structure and the applicable load combinations of ACI 9.2.1.

9.3.3 Shear Strength Provided by Concrete

Two methods are given in the Code to determine V_c. The simpler of the two methods can be found in ACI 11.9.5. In this method, V_c is calculated by the following equation, which is applicable to walls subjected to axial compression:

$$V_c = 2\lambda\sqrt{f_c'}hd \tag{9.21}$$

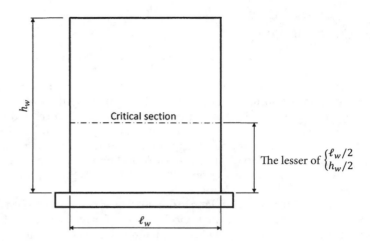

FIGURE 9.9 Critical section for shear.

ACI 11.9.4 permits d to be taken as 80% of the length of the wall. However, a larger value of d can be used in Eq. (9.21) if it is determined by a strain compatibility analysis.

For walls subjected to axial tension, ACI Eq. (11-8) may be used to determine V_c. This equation is also applicable to the design of columns subjected to axial tension (see Section 8.8).

In the second method, V_c is taken as the smaller of the values obtained by ACI Eqs. (11-27) and (11-28):

$$V_c = 3.3\lambda\sqrt{f_c'}hd + \frac{N_u d}{4\ell_w} \tag{9.22}$$

$$V_c = \left[0.6\lambda\sqrt{f_c'} + \frac{\ell_w\left(1.25\lambda\sqrt{f_c'} + 0.2N_u/\ell_w h\right)}{(M_u/V_u) - (\ell_w/2)}\right]hd \tag{9.23}$$

In these equations, the term N_u is positive for axial compression and negative for axial tension. Also, Eq. (9.23) is not applicable where $\ell_w/2 \geq M_u/V_u$.

Equation (9.22) corresponds to the occurrence of a principal tensile stress of approximately $4\lambda\sqrt{f_c'}$ at the centroid of the wall cross-section. Similarly, Eq. (9.23) corresponds to the occurrence of a flexural tensile stress of approximately $6\lambda\sqrt{f_c'}$ at a section $\ell_w/2$ above the section being investigated.

The location of the critical section for shear is given in ACI 11.9.7. The governing value of V_c at a section located a distance of 50% of the wall length or wall height, whichever is smaller, above the base of the wall applies not only to that section but also to all sections between that section and the base (see Fig. 9.9). However, as noted previously, the maximum factored shear force V_u at any section, including the base of the wall, is limited to ϕV_n in accordance with ACI 11.9.3.

9.3.4   Shear Strength Provided by Shear Reinforcement

Both horizontal shear reinforcement and vertical shear reinforcement are required for all walls. The required amount of shear reinforcement depends on the magnitude of the maximum factored shear force V_u.

- $V_u < 0.5\phi V_c$

    When V_u is less than 50% of ϕV_c, reinforcement in accordance with ACI 11.9.9 or ACI Chap. 14 must be provided. Table 9.1 contains a summary of the minimum reinforcement ratios for the vertical and horizontal shear reinforcement for Grade 60 reinforcement according to ACI 14.3.2 and 14.3.3, respectively. Note that these reinforcement ratios are based on the gross concrete area.

- $0.5\phi V_c \leq V_u \leq \phi V_c$

    In cases where V_u is equal to or between 50% and 100% of ϕV_c, both the horizontal and vertical reinforcement ratios must be at least equal to 0.0025 (ACI 11.9.9).

- $V_u > \phi V_c$

    When V_u exceeds ϕV_c, horizontal shear reinforcement must be provided to satisfy ACI Eqs. (11-1) and (11-2). The nominal shear strength provided by the horizontal shear reinforcement V_s is calculated by ACI Eq. (11-29):

$$V_s = \frac{A_v f_y d}{s} \tag{9.24}$$

In this equation, A_v is the area of the horizontal shear reinforcement within spacing s and d is determined in accordance with ACI 11.9.4.

The minimum ratio of horizontal shear reinforcement area to gross concrete area ρ_t is 0.0025. Also, the maximum spacing of the horizontal reinforcement is the smallest of $\ell_w/5$, $3h$ and 18 in.

The minimum ratio of vertical shear reinforcement area to gross concrete area ρ_ℓ is the larger of that determined by ACI Eq. (11-30) and 0.0025:

$$\rho_\ell = 0.0025 + 0.5 \left(2.5 - \frac{h_w}{\ell_w} \right) (\rho_t - 0.0025) \geq 0.0025 \tag{9.25}$$

The value of ρ_ℓ calculated by Eq. (9.25) need not be taken greater than ρ_t determined by ACI 11.9.9.1. For low-rise walls, tests indicate that horizontal shear reinforcement becomes less effective and vertical reinforcement becomes more effective in resisting the effects from shear.[5] This change in reinforcement effectiveness is recognized in Eq. (9.25). In cases where h_w/ℓ_w is less than 0.5, the amount of vertical reinforcement is equal to the amount of horizontal reinforcement. However, if h_w/ℓ_w is greater than 2.5, only a minimum amount of vertical reinforcement, which is equal to $0.0025sh$, is required.

The maximum spacing of the vertical reinforcement is the smallest of $\ell_w/3$, $3h$, and 18 in.

Example 9.5 Design the wall of Example 9.1 for shear. Figure 9.10 shows the service-level wind loads that act over the height of the wall.

Solution

   Step 1: Determine the factored load combinations. Table 9.8 contains a summary of the service-level loads at the base of the wall (see Table 9.2). The shear forces on the wall due to gravity load effects are negligible. Also included in the table are the applicable factored load combinations.

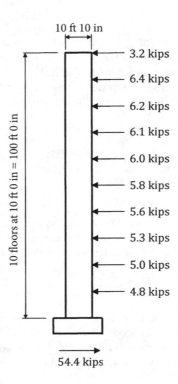

Figure 9.10 Wind load distribution over the height of the wall given in Example 9.5.

10 ft 10 in

3.2 kips
6.4 kips
6.2 kips
6.1 kips
6.0 kips
5.8 kips
5.6 kips
5.3 kips
5.0 kips
4.8 kips

10 floors at 10 ft 0 in = 100 ft 0 in

54.4 kips

Load Case	Axial Load (kips)	Bending Moment (ft kips)	Shear Force (kips)
Dead (D)	500	0	0
Roof live (L_r)	8	0	0
Live (L)	400	0	0
Wind (W)	0	±3,004	54.4
Load Combination			
1 $1.4D$	700	0	0
2 $1.2D + 1.6L + 0.5L_r$	1,244	0	0
3 $1.2D + 1.6L_r + 0.5L$	813	0	0
4 $1.2D + 1.6L_r + 0.8W$	613	2,403	43.5
5 $1.2D + 1.6W + 0.5L + 0.5L_r$	804	4,806	87.0
6 $0.9D - 1.6W$	450	−4,806	−87.0

Table 9.8 Summary of Axial Loads, Bending Moments, and Shear Forces on the Wall Given in Example 9.5

Step 2: Check maximum shear strength requirements. The design shear strength ϕV_n is limited to the following at any horizontal section of the wall (ACI 11.9.3):

$$\phi V_n = \phi 10\sqrt{f_c'}hd = 0.75 \times 10 \times \sqrt{5,000} \times 10 \times (0.8 \times 130)/1,000 = 551.5 \text{ kips}$$

Note that d was taken as 80% of the length of the wall in this equation (ACI 11.9.4).

Because $\phi V_n > V_u = 87.0$ kips, maximum shear strength requirements are satisfied.

Step 3: Determine the shear strength provided by the concrete. The nominal shear strength provided by the concrete V_c is determined in this example using both of the methods permitted in the Code. The critical section for shear is located at

$$\ell_w/2 = 10.83/2 = 5.4 \text{ ft (governs)}$$
$$h_w/2 = 100/2 = 50 \text{ ft}$$

This critical section occurs in the first story where the shear force is equal to that at the base of the wall.

Method 1:

$$V_c = 2\lambda\sqrt{f_c'}hd = 2 \times 1.0\sqrt{5,000} \times 10 \times (0.8 \times 130)/1,000 = 147.1 \text{ kips}$$

Method 2: The nominal shear strength V_c is equal to the smaller of the values obtained by Eqs. (9.22) and (9.23). The minimum value of V_c is obtained using the factored axial load in load combination number 6 because it is smaller than the axial load in load combination number 5.

Equation (9.22):

$$V_c = 3.3\lambda\sqrt{f_c'}hd + \frac{N_u d}{4\ell_w}$$

$$= \frac{3.3 \times 1.0\sqrt{5,000} \times 10 \times (0.8 \times 130)}{1,000} + \frac{450 \times (0.8 \times 130)}{4 \times 130} = 332.7 \text{ kips}$$

Equation (9.23):

$$V_c = \left[0.6\lambda\sqrt{f_c'} + \frac{\ell_w\left(1.25\lambda\sqrt{f_c'} + 0.2N_u/\ell_w h\right)}{(M_u/V_u) - (\ell_w/2)}\right]hd$$

At the base of the wall for load combination numbers 5 and 6,

$$\frac{M_u}{V_u} - \frac{\ell_w}{2} = \frac{4,806}{87.2} - \frac{10.83}{2} = 49.7 \text{ ft} > 0$$

The ratio M_u/V_u is the same for load combination number 4 as well.
Thus, for load combination number 6,

$$V_c = \left\{\frac{0.6 \times 1.0\sqrt{5,000}}{1,000} + \left[130 \times \left(\frac{1.25 \times 1.0\sqrt{5,000}}{1,000} + \frac{0.2 \times 450}{130 \times 10}\right)\right]\Big/(49.7 \times 12)\right\}$$
$$\times 10 \times (0.8 \times 130) = 79.9 \text{ kips}$$

Therefore, according to Method 2, $V_c = 79.9$ kips.

The value of V_c determined by Method 2 is used throughout the remainder of this example.

Step 4: Determine the shear strength provided by the shear reinforcement. Because $V_u = 87.0$ kips $> \phi V_c = 59.9$ kips, the nominal shear strength provided by the horizontal shear reinforcement V_s is calculated by Eq. (9.24):

$$V_s = \frac{A_v f_y d}{s}$$

Assuming two No. 4 horizontal bars spaced at 18 in on center (see Example 9.1), the provided horizontal shear reinforcement ratio is

$$\rho_t = \frac{2 \times 0.20 \times 12/18}{10 \times 12} = 0.0022 < 0.0025$$

Thus, additional horizontal reinforcement must be provided to satisfy the minimum requirement in accordance with ACI 11.9.9.2. The horizontal reinforcement ratio for two No. 5 bars spaced at 18 in is 0.0034, which is greater than the minimum ratio of 0.0025.

The maximum spacing of the horizontal shear reinforcement is the smallest of the following:

- $\ell_w/5 = 130/5 = 26.0$ in
- $3h = 30$ in
- 18 in (governs)

The required vertical shear reinforcement is determined by Eq. (9.25):

$$\rho_\ell = 0.0025 + 0.5\left(2.5 - \frac{h_w}{\ell_w}\right)(\rho_t - 0.0025)$$

$$= 0.0025 + 0.5\left(2.5 - \frac{100}{10.83}\right)(0.0034 - 0.0025) < 0$$

Thus, $\rho_\ell = 0.0025$.

The provided vertical shear reinforcement ratio for two No. 6 bars spaced at 9 in on center is (see Example 9.1)

$$\rho_t = \frac{2 \times 0.44 \times 12/9}{10 \times 12} = 0.0098 > 0.0025$$

The maximum spacing of the vertical shear reinforcement is the smallest of the following:

- $\ell_w/3 = 130/3 = 43.3$ in
- $3h = 30$ in
- 18 in (governs)

Check shear strength requirement where V_s is determined by Eq. (9.24):

$$\phi V_n = \phi(V_c + V_s)$$

$$= 0.75\left[79.9 + \frac{(2 \times 0.31) \times 60 \times (0.8 \times 130)}{18}\right] = 221.1 \text{ kips} > V_u = 87.0 \text{ kips}$$

Use two No. 5 bars at 18-in horizontal bars and two No. 6 bars at 9-in vertical bars at the base of the wall.

Comments

The reinforcement determined in this example is adequate for shear and combined axial load and bending. The amount of reinforcement can be decreased over the height of the wall, but it must not be less than the minimum prescribed in the Code.

9.4 Design Procedure

The following design procedure can be used in the design of walls. Included is the information presented in the previous sections on analysis, design, and detailing.

Step 1: Determine a preliminary wall thickness. The first step in the design procedure is to determine a preliminary wall thickness. From a practical standpoint, a minimum thickness of 6 in is required for a wall with a single layer of reinforcement and 10 in for a wall with a double layer. In the case of low-rise walls, shear requirements usually govern, so a preliminary thickness can be determined on the basis of shear. In high-rise structures, a preliminary wall thickness is not as obvious. In such structures, the thickness can vary a number of times over the height of the structure, and an initial value is usually determined from experience. Although fire resistance requirements seldom govern wall thickness, the governing building code requirements should not be overlooked.

More often than not, the size of openings required for stairwells and elevators dictates the minimum wall plan layouts. Thus, the lengths of walls are usually controlled by architectural considerations.

Step 2: Determine the design method (Section 9.2). The next step is to determine which of the three available methods can be used to design a wall. As discussed previously, designing walls as compression members in accordance with Chap. 10 of the Code is permitted in all cases.

Both the empirical design method and the alternative design method have limitations. The empirical design method is fairly limited in application and is best suited for relatively short walls with vertical loads. Wall sections designed by this method are compression-controlled. The alternative design method has a number of limitations that must be satisfied, and wall sections designed by this method must be tension-controlled.

Step 3: Determine the required reinforcement (Sections 9.2 and 9.3). Vertical and horizontal reinforcements must be determined to satisfy the strength requirements for axial load, bending, and shear. In low-rise walls, which are typically governed by shear requirements, the amount of vertical and horizontal reinforcements is initially determined on the basis of the shear provisions of ACI 11.9. The axial load and bending requirements of the appropriate design method are then checked on the basis of the reinforcement for shear. It is not uncommon for low-rise walls to have minimum amounts of reinforcement over their entire height.

In the case of high-rise walls, wall sections at the base of the structure are usually governed by the requirements for axial load and bending. Once the amount of reinforcement is determined on the basis of those requirements, the shear requirements of ACI 11.9 are checked. For walls subjected to relatively large bending demands, larger amounts of vertical reinforcement are sometimes concentrated at the ends of a wall to increase its flexural capacity. The amounts of vertical and horizontal reinforcements are typically varied over the height of high-rise walls. In no case shall the provided areas of reinforcement be less than the minimum values prescribed in the Code.

References

1. MacGregor, J. G. 1974. Design and safety of reinforced concrete compression Members. Paper presented at the International Association for Bridge and Structural Engineering Symposium, Quebec.

2. Athley, J. W. (ed.) 1982. *Test Report on Slender Walls*. Southern California Chapter of the American Concrete Institute and Structural Engineers Association of Southern California, Los Angeles, CA, 129 pp.

3. International Conference of Building Officials (ICBO). 1997. *Uniform Building Code*, Vol. 2. ICBO, Whittier, CA, 492 pp.

4. Cardenas, A. E., Hanson, J. M., Corley, W. G., and Hognestad, E. 1973. Design provisions for shear walls. *ACI Journal* 70(3):221-230.

5. Barda, F., Hanson, J. M., and Corley, W. G. 1977. Shear strength of low-rise walls with boundary elements. In: *Reinforced Concrete Structures in Seismic Zones*, SP-53. American Concrete Institute, Farmington Hills, MI, pp. 149-202.

Problems

9.1. Determine the minimum thickness of a load-bearing concrete wall that has a height of 12 ft 0 in and a length of 18 ft 6 in. It has been determined that the empirical design method can be used.

9.2. Determine the design axial strength ϕP_n for the wall in Problem 9.1, assuming a wall thickness of 7 in and normal-weight concrete with $f_c' = 4,000$ psi. Also assume that the wall is unrestrained against rotation at both ends.

9.3. Double-tee beams that have 60-ft spans are supported by a reinforced concrete wall that has a height of 16 ft. The double-tees have webs that are 4.75 in wide and are spaced 6 ft on center. Determine the midheight distribution width of interior concentrated loads. Use the provisions in ACI 14.8.

9.4. For the wall system described in Problem 9.3, determine the minimum area of vertical reinforcement for a 7-in-thick wall, assuming Grade 60 reinforcement and deformed bars not larger than No. 5 bars.

9.5. Determine the factored axial load P_u at the midheight of the 7-in-thick wall described in Problem 9.3 given the following loads:

> Dead load of double-tee beams = 1,000 plf
> Superimposed dead load = 20 psf
> Roof live load = 30 psf
> Wind load = 30 psf

Assume that the gravity loads from the double-tee beams are applied to the wall at an eccentricity of 2.3 in from the centroid of the wall. Also assume normal-weight concrete with $f_c' = 4,000$ psi.

9.6. For the wall system described in Problems 9.3 and 9.5, assume that the factored load due to the roof gravity loads is equal to 4.0 kips/ft. Determine the factored moment M_{sa} at the midheight of the wall.

9.7. Assume that the 7-in-thick wall described in Problem 9.3 is reinforced with one layer of No. 4 bars spaced at 10 in on center. Given that $M_{ua} = 30$ in kips and $P_u = 4$ kips/ft, determine the total factored moment M_u at the midheight of the wall.

9.8. For the wall system described in Problem 9.3, determine the midheight deflection Δ_s.

9.9. A 12-in-thick wall is 20 ft long and has a height of 12 ft. Assuming normal-weight concrete with $f_c' = 5,000$ psi, determine the maximum shear strength that is permitted.

9.10. The wall described in Problem 9.9 is subjected to a factored horizontal force of 250 kips applied at the top of the wall. Determine the nominal shear strength provided by the concrete at the critical section.

CHAPTER 10

Foundations

10.1 Introduction

10.1.1 Overview

The main function of a foundation is to transmit the loads from the structure above to the soil below. In buildings, the loads usually come directly or indirectly from columns and walls. Foundations must be located on a soil or rock stratum that has adequate strength to support the loads. The loads must be spread out over a sufficient area so that the resulting pressure is not greater than the allowable bearing capacity of the soil or rock. In addition to strength, total settlement of a structure and differential settlement between adjoining foundations must be limited to tolerable amounts in order to prevent possible damage to the structure. Any foundation system must also be safe against overturning, sliding, and excessive rotation. The overall stability of a building depends on the foundations performing as intended. Chapter 18 of the International Building Code (IBC) contains provisions for foundation systems used to support buildings.

There are numerous types of foundations, and this chapter focuses on those types that are commonly used to support building structures. Design requirements are presented for both shallow and deep foundations. Methods are provided on how to size the members and how to design and detail the required reinforcement.

10.1.2 Shallow Foundations

A *shallow foundation* transfers the load from the superstructure to a soil stratum that is relatively close to the ground surface. A number of factors need to be considered when locating the depth of a shallow foundation. As noted previously, a soil stratum that has adequate bearing capacity needs to be identified. Also, shallow foundations must be placed below the frost line to avoid possible frost heave. Local building codes contain approximate frost-depth contours. Topsoil, organic materials, and unconsolidated materials are just a few of the soil types that shallow foundations should not bear on. A geotechnical report provides guidance on the appropriate depth for particular site conditions.

Footings and mats are two types of shallow foundations. In general, a *spread footing* supports one or more vertical elements. An *isolated spread footing* carries a single column (Fig. 10.1*a*). Its function is to spread the column load to the soil so that the stress intensity is equal to or less than the allowable bearing capacity of the soil. Flexural reinforcement is provided in two orthogonal directions at the bottom of the footing. A *wall footing* is similar to a spread footing and is usually continuous under the length of the wall

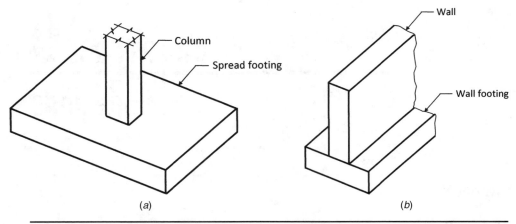

Figure 10.1   Types of footings: (a) spread footing and (b) wall footing.

(Fig. 10.1b). Flexural reinforcement is placed at the bottom of the footing, perpendicular to the face of the wall, and temperature and shrinkage reinforcement is provided parallel to the length of the wall.

A *combined footing* is a special type of spread footing that supports multiple columns or walls on the same footing (Fig. 10.2). They are commonly used where the space between adjoining isolated footings is small or where a building is close to a property line. Additional information on the analysis and design of combined footings can be found in Ref. 1.

If it is not possible to locate a column at the center of a spread footing or if moments from a column or wall are transferred to a footing, the resulting soil pressure is nonuniform. Excessive rotation or differential settlements can occur in such cases

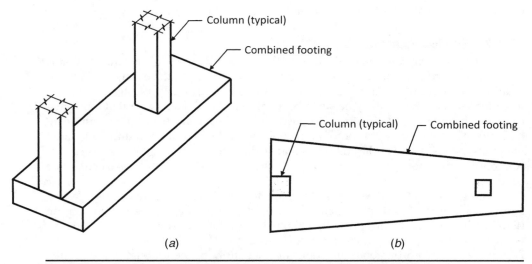

Figure 10.2   Combined footing: (a) rectangular footing and (b) trapezoidal footing.

because of relatively large eccentric loads or bending moments. It is possible to control these detrimental effects by placing two or more columns in a line on a combined footing. Reinforcement for flexure is required at the bottom of the section in both directions for negative bending moments. One layer of reinforcement is also required at the top of the section for positive bending moments.

The combined rectangular footing shown in Fig. 10.2a is commonly used where the load at the interior column is greater than the load at the edge column. The footing can be sized so that the pressure at the base of the footing is uniform. Where an exterior column has a larger load than an adjacent interior column, the combined footing can be constructed trapezoidal in shape to achieve a uniform distribution of soil pressure beneath the footing (Fig. 10.2b).

A *strap footing* or *cantilever footing* consists of an eccentrically loaded footing connected to an adjacent footing by a strap beam (Fig. 10.3). The purpose of the strap beam is to transmit the moment from the eccentrically loaded footing to the adjacent footing so that a uniform soil pressure is achieved under both footings. Strap beams perform essentially the same role as the interior portion of a combined footing but are narrower to save on materials. To simplify the formwork, the width is usually the same as that of the largest column. The strap beam must be rigid to avoid rotation of the exterior footing. Providing a strap beam that has a moment of inertia of at least two times that of the footing will usually achieve this.

A *mat foundation* is a large concrete slab that supports some or all of the columns and/or walls in a building (Fig. 10.4). They are usually used in cases where individual spread footings would cover over 50% of the required area. This can occur where the bearing capacity of the soil is relatively low.

One of the benefits of a mat foundation is that total and differential settlements are smaller than those from individual spread footings. The relatively stiff concrete mat

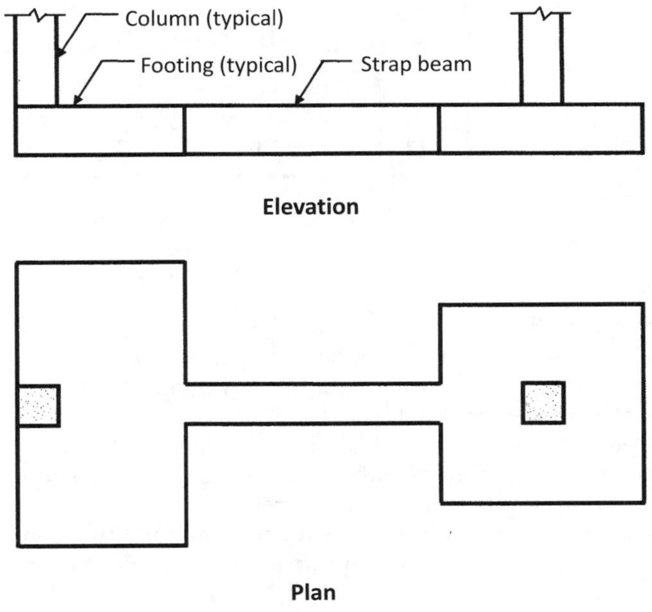

FIGURE 10.3 Strap footings.

Column (typical)

Footing (typical)   Strap beam

Elevation

Plan

FIGURE 10.4 Mat foundation.

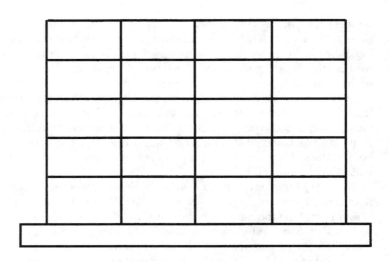

Elevation

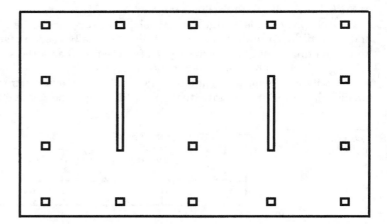

Plan

tends to spread the concentrated loads from the columns and walls over larger areas, thereby reducing the local effects caused by these loads. Settlement is also reduced at pockets of weaker soils; the concrete mat essentially bridges over these weak pockets and equalizes differential settlements.

For purposes of analysis and design, a mat foundation can be thought of as a concrete slab where the columns and walls are supports and the soil pressure is the applied load. Similar to elevated reinforced concrete slabs in a building, both positive reinforcement and negative reinforcement is required in both directions of a mat foundation. Reference 1 provides additional information on the analysis and design of this type of foundation.

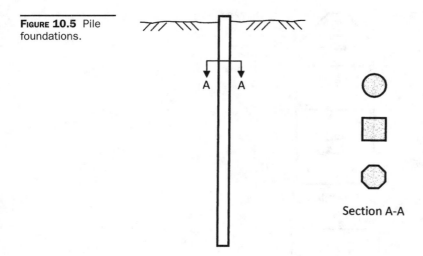

Figure 10.5 Pile foundations.

Section A-A

10.1.3 Deep Foundations

A *deep foundation* is generally any type of foundation that extends below strata of poor soil to a level where the soil is adequate to support the loads. The most common types of deep foundations utilized under buildings are *piles* and *drilled piers*.

Piles are foundation members that have relatively small cross-sectional dimensions compared with their length (Fig. 10.5). They are available in various materials and shapes, and most concrete piles are circular, square, or octagonal in cross-section. Piles are typically placed vertically into the soil but can be installed at a slight inclination to help resist lateral loads.

Concrete piles may be cast-in-place or precast. A cast-in-place pile is formed either by driving a casing or by drilling a hole into the ground. In the former, the casing is typically removed after concrete has been deposited into the hole. In the latter, concrete is poured directly into the hole. Many other procedures are used in the construction of cast-in-place concrete piles, most of which depend on the type of proprietary system that is utilized.

Precast piles are constructed in a precast casting yard and are transported to the site, where they are driven into the ground by a pile driver. Either mild reinforcing bars or prestressed tendons are used as reinforcement. Additional information on the design, manufacture, and installation of concrete piles can be found in Ref. 2.

More than one pile is usually used beneath a column or wall to support the loads. A *pile cap* is a reinforced concrete element that ties the tops of the piles together and distributes the loads to the individual piles in the group. In the case of symmetrically loaded pile caps with a symmetric arrangement of piles, it is assumed that the load supported by the pile cap is shared equally by each pile. Piles are embedded into the cap a minimum distance depending on the type of pile. Figure 10.6 illustrates just a few pile cap layouts for individual columns.

The capacity of an individual pile depends on a number of factors, including cross-sectional dimensions, material strength, and reinforcement type and arrangement. Capacity is typically reported by the manufacturer in tons, and the type and number of piles is chosen on the basis of the applied service loads.

FIGURE 10.6 Pile cap layouts for individual columns.

Three piles

Five piles

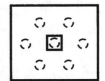

Seven piles

FIGURE 10.7 Typical belled concrete drilled pier without a cap.

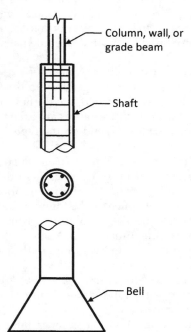

Column, wall, or grade beam

Shaft

Bell

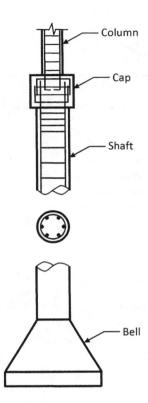

FIGURE 10.8 Typical belled concrete drilled pier with a cap.

— Column

— Cap

— Shaft

— Bell

A *drilled pier* is similar to a cast-in-place pile in that it is a shaft that is drilled into the soil. The shaft may be lined with a casing that may or may not be extracted as the shaft is filled with concrete. A permanent casing is generally required where unstable soil conditions are encountered, which could lead to the soil caving into the drilled shaft. A drilled pier is also referred to as a *pier* or a *caisson*. Reference 3 provides additional information on construction methods.

Drilled piers have a circular cross-section, and a circular bell is usually provided at the bottom of the shaft to distribute the load over a greater area of soil or rock (Fig. 10.7). It is common for a single column to be supported by a single drilled pier, whereas multiple piers are provided beneath walls. A column, wall, or grade beam can be supported directly on the top of the shaft as shown in Fig. 10.7. A *drilled pier cap* is proved under individual columns to increase bearing area (Fig. 10.8).

Under certain conditions, a socketed drilled shaft is provided, which consists of a permanent pipe or tube casing that extends to the top of a rock layer and an uncased socket that is drilled a predetermined depth into the rock. The socket helps to ensure that full bearing occurs between the shaft and the rock.

A *grade beam* is a reinforced concrete member that directly supports columns and walls from the superstructure and transfers the loads from these members to drilled piers. The loads from the supported members cause bending, shear, and possibly torsion in the grade beams. As such, the grade beams are designed for these load effects, using the general principles of the strength design method. As an example where grade

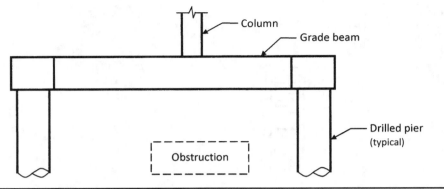

FIGURE 10.9   Example where a grade beam is utilized.

beams are utilized, consider the situation depicted in Fig. 10.9. The building column must be located as shown. Because there is an existing obstruction below the surface, it is not possible to locate a drilled pier immediately below this column. The grade beam transfers the column load to piers located on either side of the obstruction.

10.2   Footings

10.2.1   Overview

Chapter 15 of the Code contains the design requirements for isolated footings. Most of the provisions are also applicable to the design of combined footings and mats.

Methods are presented in the following sections on the following:

1. Sizing the base area and thickness
2. Determining the required amount of reinforcement
3. Detailing the reinforcement

10.2.2   Loads and Reactions

As in the design of any reinforced concrete member, footings must be proportioned to resist the effects from the governing factored loads determined in accordance with ACI 9.2. This includes axial loads, bending moments, and shear forces. The base area of a footing is determined using unfactored (service) loads and the allowable soil-bearing capacity that is provided in the geotechnical report or any other referenced document (ACI 15.2.2). When determining base dimensions, the minimum moment requirement for slenderness considerations of ACI 10.10.6.5 need not be considered; only the computed end moment that exists at the base of a column or pedestal needs to be transferred to the footing.

The thickness of a footing and the required area of flexural reinforcement are determined using the strength design method, which utilizes factored load effects. Shear requirements must be satisfied using factored shear forces and design shear strength.

In cases where piles are used in conjunction with footings (i.e., piles are embedded into the bottom of footings; these elements are commonly referred to as pile caps), the number and arrangement of the piles is determined using unfactored loads (ACI 15.2.2). In such cases, ACI 15.2.3 permits the moments and shears in the footing or pile cap to be determined assuming that the reaction from any pile is concentrated at the center of the pile.

10.2.3   Sizing the Base Area

Allowable Bearing Capacity

As noted earlier, the base dimensions of a footing are determined using unfactored loads and allowable soil bearing capacities. The bearing capacity of soil or rock can be obtained from soil borings and tests performed by a geotechnical engineer. These values along with other important information and data are typically summarized in a geotechnical report. In the absence of site-specific data, allowable bearing pressures may be available from local building authorities. Section 1806 of the IBC contains presumptive load-bearing values for a variety of different soil types.[4] It is important to check with the local building authority to ensure that these presumptive values are permitted to be used in the design of the foundation.

Soil Pressure Distribution

Once the bearing capacity of the soil has been established, the next step is to determine the pressure distribution at the base of the footing. Elastic analyses and observations reveal that the stress distribution beneath a symmetrically loaded footing is not uniform. The actual stress distribution is highly indeterminate and depends on the rigidity (or flexibility) of the footing and the type of soil beneath the footing. For footings on coarse-grained soil, like loose sand, soil near the edge of the footing tends to displace laterally, whereas soil located in the interior region is relatively confined (see Fig. 10.10a).

The pressure distribution for cohesive soils, like clay, is depicted in Fig. 10.10b. The high stresses at the edges of the footing are a result of shears that occur before settlement takes place. Because these soil types generally have low rupture strength, it is very likely that these stresses do not last long.

Figure 10.10 Soil pressure distribution beneath a footing bearing on (a) coarse-grained soils and (b) cohesive soils.

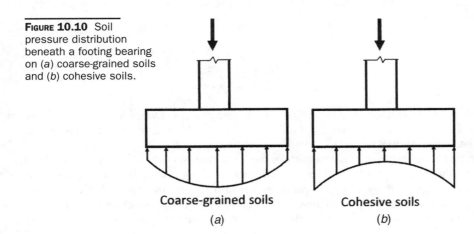

Coarse-grained soils
(a)

Cohesive soils
(b)

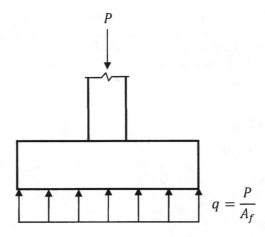

$$q = \frac{P}{A_f}$$

It is common practice to disregard nonuniform pressure distributions in design and to use a linear pressure distribution, because of the following: (1) the magnitude of the nonuniform pressure is uncertain and highly variable, and (2) the influence of the nonuniform stress on the magnitudes of bending moments and shear forces in the footing is relatively small.

Consider the footing depicted in Fig. 10.11 subjected to an applied service load P that acts through the centroid of the footing plan area A_f. For purposes of design, the footing is assumed to be rigid, and the resulting soil pressure q at the base of the footing is assumed to be uniform.

For footings subjected to an axial load and bending moment or, equivalently, to an axial load acting at an eccentricity e from the centroid of the footing area, the total combined stress at the base of the footing is equal to the sum of the stress due to the axial load P (axial load/footing area) and the bending moment M (bending moment/section modulus of footing). The pressure is assumed to vary linearly, as shown in Fig. 10.12a. This distribution is valid where the axial load falls within the kern of the footing area, that is, where the eccentricity e is less than $L/6$. The following equation for q can be used to determine the minimum and maximum pressures at the extreme edges of the footing:

$$q = \frac{P}{A_f} \pm \frac{6M}{BL^2} \tag{10.1}$$

When the eccentricity e is equal to $L/6$, the minimum pressure along one edge of the footing is equal to zero [see Eq. (10.1) and Fig. 10.12b]. The maximum pressure at the other edge is equal to $2P/A_f$.

When the eccentricity e falls outside of the kern, that is, where e is greater than $L/6$, the combined stress determined by Eq. (10.1) gives a negative value for the pressure along one edge of the footing. Because no tension can be transmitted between the footing and the soil at the contact area, Eq. (10.1) is no longer applicable, and the bearing pressure is distributed as shown in Fig. 10.12c. The maximum pressure is determined

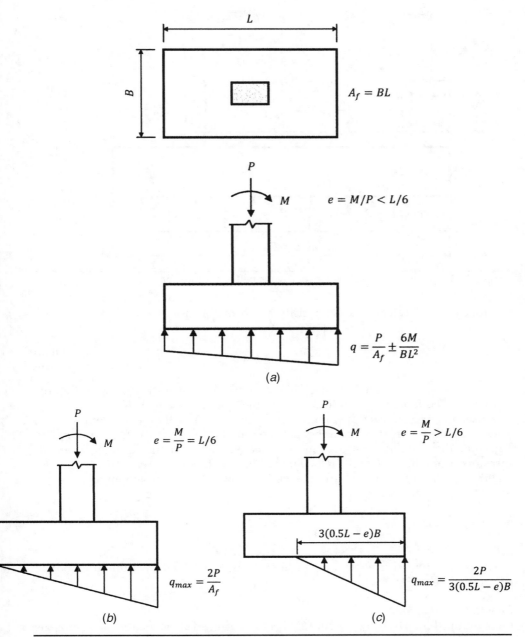

$$A_f = BL$$

$$e = M/P < L/6$$

$$q = \frac{P}{A_f} \pm \frac{6M}{BL^2}$$

(a)

$$e = \frac{M}{P} = L/6$$

$$q_{max} = \frac{2P}{A_f}$$

(b)

$$e = \frac{M}{P} > L/6$$

$$3(0.5L - e)B$$

$$q_{max} = \frac{2P}{3(0.5L - e)B}$$

(c)

Figure 10.12   Soil pressure distribution for footing subjected to axial load and bending moment: (a) $e < L/6$, (b) $e = L/6$, and (c) $e > L/6$.

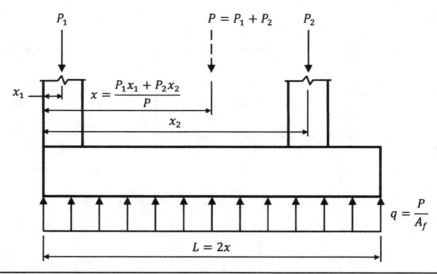

FIGURE 10.13 Soil pressure distribution for a combined rectangular footing.

as a function of the axial load, eccentricity, and plan dimensions of the footing:

$$q_{max} = \frac{2P}{3(0.5L - e)B} \tag{10.2}$$

In the case of combined footings, it is desirable to design the footing so that the centroid of the footing area coincides with the resultant of the column loads. This produces uniform bearing pressure over the entire area of the footing and helps in preventing the footing from rotating. For the combined rectangular footing shown in Fig. 10.13, the columns are supporting axial loads P_1 and P_2 at the locations x_1 and x_2 measured from the edge of the left column. The resultant force P is equal to $P_1 + P_2$, and the distance x from the edge of the footing to P is obtained by summing moments about this point and solving for x:

$$x = \frac{P_1 x_1 + P_2 x_2}{P} \tag{10.3}$$

A uniform pressure at the base of the footing is obtained by setting the footing dimension L equal to $2x$.

Required Footing Area

Once the maximum bearing pressure has been determined at the base of the footing, the required footing area can be determined using the net permissible soil pressure q_p. By definition, q_p is equal to the allowable bearing capacity of the soil q_a minus the weight of the surcharge above the footing. The weight of the surcharge typically consists of the weight of the soil and concrete above the base of the footing plus any additional service surcharge applied at the surface. Because the thickness of the footing is not known at this stage, an estimate of the concrete weight must be made. In general, a footing area is determined so that the maximum computed bearing pressure q_{max} is equal to or less than the permissible soil pressure q_p.

Isolated Spread Footings   For a concentrically loaded isolated spread footing, the required area of the footing A_f is determined by dividing the total service load P by the permissible soil pressure q_p:

$$A_f = BL = \frac{P}{q_p} \tag{10.4}$$

There are obviously an infinite number of solutions to this equation. When one of the plan dimensions is fixed, the other dimension can be easily computed by Eq. (10.4). In the case of square footings, the length or width is equal to $\sqrt{P/q_p}$.

For spread footings subjected to an axial load P and a moment M or, equivalently, to an axial load P at an eccentricity e where e is equal to or less than $L/6$, the footing area is found by trial and error, using the condition that the maximum combined pressure q_{max} is equal to or less than q_p:

$$q_{max} = \frac{P}{A_f} + \frac{6M}{BL^2} \le q_p \tag{10.5}$$

The required footing size must also be determined by trial and error in cases where the eccentricity is greater than $L/6$, on the basis of the maximum pressure along the edge of the footing:

$$q_{max} = \frac{2P}{3(0.5L - e)B} \le q_p \tag{10.6}$$

Combined Footings   For combined rectangular footings, the length L of the footing is equal to $2x$, where x is determined such that uniform pressure is obtained at the base of the footing [see Eq. (10.3)]. In this case, the width B of the footing is obtained as follows:

$$B = \frac{P}{Lq_p} \tag{10.7}$$

The dimensions of a combined trapezoidal footing can be determined using similar methods. Consider the trapezoidal footing shown in Fig. 10.14. Assuming that the length L is established on the basis of column size and spacing, the dimensions B_1 and B_2 can be determined so that a uniform soil pressure occurs at the base of the footing.

Defining $\bar{x}$ as the distance from the center of the heavier-loaded column on the left to the location where the centroid of the footing area coincides with the resultant of the column loads, it is evident that a trapezoidal footing is not possible if $\bar{x} + (c_1/2) < L/3$; in such cases, the result is a triangular footing with the column on the right, not fully supported on the foundation. A combined rectangular footing occurs where $\bar{x} + (c_1/2) = L/2$.

On the basis of the limits established earlier, a combined trapezoidal footing solution exists where $L/3 < \bar{x} + (c_1/2) < L/2$. Defining $x = \bar{x} + (c_1/2)$, the following equation

Figure 10.14
Dimensions of a
combined
trapezoidal footing.

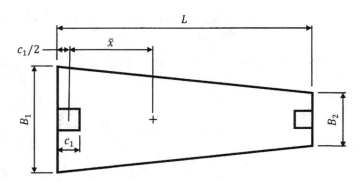

locates the centroid of the footing with respect to the left edge:

$$x = \frac{L}{3}\left(\frac{2B_2 + B_1}{B_1 + B_2}\right) \tag{10.8}$$

Like in the case of a combined rectangular footing, the resultant force P is equal to $P_1 + P_2$, where P_1 and P_2 are the service axial loads on the left and right columns, respectively. The distance $\bar{x}$ from the center of the left column to P is obtained by summing moments about this point and solving for $\bar{x}$:

$$\bar{x} = \frac{P_2 L'}{P} = x - \frac{c_1}{2} = \frac{L}{3}\left(\frac{2B_2 + B_1}{B_1 + B_2}\right) - \frac{c_1}{2} \tag{10.9}$$

In this equation, L' is the distance between P_1 and P_2, that is, the center-to-center distance between the columns.

Also, the pressure at the base of the footing must not exceed the permissible soil pressure:

$$q_p = \frac{P}{A_f} = \frac{2P}{L(B_1 + B_2)} \tag{10.10}$$

Equations (10.9) and (10.10) can be solved for the two unknowns B_1 and B_2.

Example 10.1   Determine the required area of the concentrically loaded spread footing depicted in Fig. 10.15, given an allowable soil-bearing capacity $q_a = 4{,}000$ psf. Assume that the combined weight of soil and concrete above the base of the footing is equal to 120 pcf.

Solution   Because the footing is concentrically loaded, Eq. (10.4) can be used to determine the required area.
Determine the permissible soil pressure q_p:

$$q_p = 4{,}000 - (120 \times 6) = 3{,}280 \text{ psf}$$

Therefore,

$$A_f = \frac{P}{q_p} = \frac{300{,}000 + 150{,}000}{3{,}280} = 137.2 \text{ ft}^2$$

Use an 11-ft 9-in square footing (provided area $= 138.1$ ft^2).

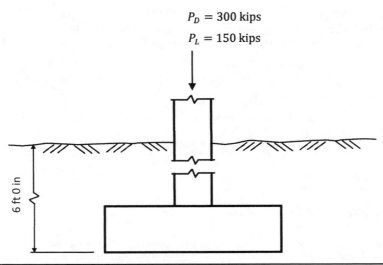

$P_D = 300$ kips

$P_L = 150$ kips

6 ft 0 in

FIGURE 10.15   The spread footing given in Example 10.1.

Example 10.2   Determine the required area of an isolated spread footing subjected to the following load effects:

$$P_D = 200 \text{ kips, } M_D = 8 \text{ ft kips}$$
$$P_L = 150 \text{ kips, } M_L = 4 \text{ ft kips}$$
$$P_W = \pm5 \text{ kips, } M_W = \pm225 \text{ ft kips, } V_W = \pm23 \text{ kips}$$

Assume a permissible soil pressure of 3,280 psf.

Solution   Because bending moments act on the footing in addition to axial loads, the area of the footing must be determined by trial and error.

Check the combined soil pressure, using a 12-ft 0-in square footing:

Pressure due to axial loads: $\dfrac{P}{A_f} = \dfrac{200 + 150 + 5}{12^2} = 2.47$ ksf

Pressure due to moments: $\dfrac{6M}{BL^2} = \dfrac{6 \times (8 + 4 + 225)}{12^3} = 0.82$ ksf

Total pressure $= 2.47 + 0.82 = 3.29$ ksf > 3.28 ksf

Increase the footing size to 12 ft 6 in, and check the maximum pressure:

Maximum pressure $= \dfrac{200 + 150 + 5}{12.5^2} + \dfrac{6 \times (8 + 4 + 225)}{12.5^3} = 2.27 + 0.73 = 3.00$ ksf < 3.28 ksf

Check if the resultant axial load is within the kern:

Eccentricity $e = \dfrac{M}{P} = \dfrac{8 + 4 + 225}{200 + 150 + 5} = 0.67$ ft $< \dfrac{L}{6} = \dfrac{12.5}{6} = 2.1$ ft

Use a 12-ft 6-in square footing.

Example 10.3   Determine the width of a wall footing like the one shown in Fig. 10.2, which supports a 10-in-thick normal-weight concrete wall that is 12 ft in height. The wall carries concentric service dead and live loads equal to 2 kips/ft and 1 kip/ft, respectively. Assume a permissible soil pressure of 2,000 psf.

Solution   The width of the footing is determined using loads per foot width of wall.

$$\text{Weight of the wall} = \frac{10}{12} \times 12 \times 150 = 1{,}500 \text{ plf}$$

$$\text{Total load} = 1{,}500 + 2{,}000 + 1{,}000 = 4{,}500 \text{ plf}$$

$$A_f = \frac{P}{q_p} = \frac{4{,}500}{2{,}000} = 2.3 \text{ ft}^2/\text{ft}$$

Use wall footing that is 2 ft 6 in wide.

Example 10.4   Determine the required area of a combined rectangular footing supporting the two columns shown in Fig. 10.16. Assume a permissible soil pressure of 3,000 psf.

Solution   The base dimension L of the footing is determined so that the pressure at the base of the footing is uniform. This is achieved by having the resultant of the axial loads fall at $L/2$. Determine the distance x from the edge of the footing to the resultant force P by Eq. (10.3):

$$x = \frac{P_1 x_1 + P_2 x_2}{P} = \frac{(415 \times 0.75) + (725 \times 22.25)}{1{,}140} = 14.4 \text{ ft}$$

Therefore, $L = 2x = 28.8$ ft. Use $L = 29$ ft 0 in.

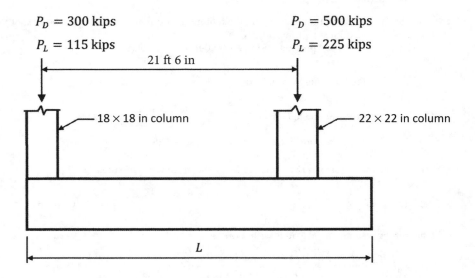

Figure 10.16   The combined footing given in Example 10.4.

The dimension B of the footing is determined by Eq. (10.7):

$$B = \frac{P}{Lq_p} = \frac{1,140}{29 \times 3} = 13.1 \text{ ft}$$

Use $B = 13$ ft 6 in.

Example 10.5   Determine the required area of a combined trapezoidal footing supporting the two columns shown in Fig. 10.17. Assume a permissible soil pressure of 3,500 psf.

Solution   The base dimensions B_1 and B_2 of the footing are determined so that the pressure at the base is uniform:

$$L = \frac{24}{2 \times 12} + 20 + \frac{20}{2 \times 12} = 21.83 \text{ ft}$$

The distance $\bar{x}$ from the center of the left column to P is obtained by summing moments about this point and solving for $\bar{x}$:

$$\bar{x} = \frac{P_2 L'}{P} = \frac{350 \times 20}{550 + 350} = \frac{7,000}{900} = 7.8 \text{ ft}$$

Check if a combined trapezoidal solution exists:

$$\frac{L}{3} = 7.3 \text{ ft} < \bar{x} + \frac{c_1}{2} = 8.8 \text{ ft} < \frac{L}{2} = 10.9 \text{ ft}$$

From Eq. (10.9),

$$\bar{x} = \frac{P_2 L'}{P} = \frac{L}{3}\left(\frac{2B_2 + B_1}{B_1 + B_2}\right) - \frac{c_1}{2}$$

$$7.8 = 7.3\left(\frac{2B_2 + B_1}{B_1 + B_2}\right) - 1$$

$$1.5B_1 - 5.8B_2 = 0$$

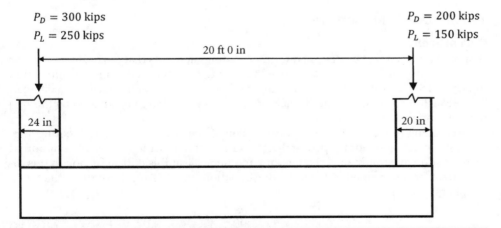

$P_D = 300$ kips
$P_L = 250$ kips

$P_D = 200$ kips
$P_L = 150$ kips

20 ft 0 in

24 in

20 in

FIGURE 10.17   The combined footing given in Example 10.5.

From Eq. (10.10),

$$q_p = \frac{P}{A_f} = \frac{2P}{L(B_1 + B_2)}$$

$$3.5 = \frac{2 \times 900}{21.83(B_1 + B_2)}$$

$$76.41 B_1 + 76.41 B_2 = 1{,}800$$

Solving Eqs. (10.9) and (10.10) for B_1 and B_2 results in

$$B_1 = 18.7 \text{ ft}$$
$$B_2 = 4.8 \text{ ft}$$

Use $L = 21$ ft 10 in.
Use $B_1 = 18$ ft 9 in.
Use $B_2 = 4$ ft 10 in.

10.2.4   Sizing the Thickness

Once the required area of the footing has been established on the basis of the service loads and the allowable bearing capacity of the soil, the thickness h of a footing must be determined considering both flexure and shear.

In general, a spread footing must be designed for the bending moments that are induced because of the pressure developed at the base of the footing from the factored loads. Requirements for both one- and two-way shear must also be satisfied. Methods to determine the thickness are provided in the following sections.

ACI 15.7 requires a minimum footing depth of 6 in above the bottom reinforcement for footings on soil and 12 in for footings on piles. According to ACI 7.7.1, the minimum concrete cover to the reinforcement is equal to 3 in for concrete cast against and permanently exposed to earth. Therefore, for footings on soil, the minimum overall thickness is equal to approximately 10 in. Similarly, the minimum overall thickness of footings on piles is 16 in.

10.2.5   Design for Flexure

Critical Section

A spread footing must be designed for the bending moments that are induced because of the pressure developed at the base of the footing from the factored loads. Illustrated in Fig. 10.18 is an isolated spread footing subjected to a concentric factored axial load P_u. The factored pressure q_u at the base of the footing is equal to P_u divided by the area of the footing A_f.

According to ACI 15.4.2, the critical section for flexure for an isolated footing supporting a concrete column, pedestal, or wall is located at the face of the supported member. The maximum factored bending moment M_u at this critical section in this direction can be determined by the following equation, which is applicable to cantilevered members:

$$M_u = \frac{q_u c^2}{2} \tag{10.11}$$

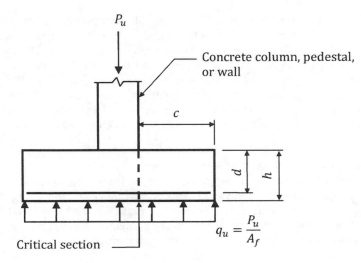

FIGURE 10.18
Critical section for
an isolated footing
supporting a
column, pedestal,
or wall.

In this equation, c is the distance from the critical section to the edge of the footing (i.e., c is the length of the cantilevered portion of the footing).

If the footing were subjected to a moment or load acting at an eccentricity, the resulting factored pressure would be nonuniform. In such cases, the bending moment at the critical section can be obtained from statics.

ACI 15.4.2 also contains critical section locations for two other cases. For footings supporting masonry walls, the critical section is located halfway between the middle and the edge of the wall (Fig. 10.19a), whereas for footings supporting columns with a steel base plate, the critical section is located halfway between the face of the column

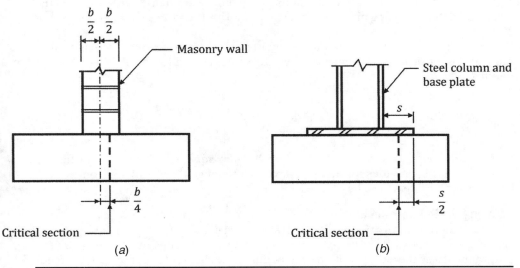

FIGURE 10.19 Critical section locations for (a) footings supporting masonry walls and (b) footings supporting columns with a steel base plate.

and the edge of the base plate (Fig. 10.19*b*). In the case of combined footings, the critical section for negative moments is taken at the face of the supports. A maximum positive moment occurs near the midspan between columns. The Direct Design Method of analysis given in ACI 13.6 is not permitted to be used to determine factored bending moments in combined footings or mat foundations (ACI 15.10.2).

ACI 15.3 permits circular or regular polygon-shaped columns or pedestals to be replaced by an equivalent square member with the same area as the original shape for location of critical sections for moments, shear, and development of flexural reinforcement.

Determining the Required Reinforcement

Once the maximum factored moment M_u at the critical section has been determined, the required area of reinforcing steel A_s can be calculated using the strength design requirements of Chap. 5. The following equation must be satisfied for a concentrically loaded isolated footing where the nominal flexural strength M_n is given by Eq. (5.10) for a rectangular section with tension reinforcement:

$$M_u = \frac{q_u c^2}{2} \le \phi M_n = \phi A_s f_y \left(d - \frac{a}{2}\right) \tag{10.12}$$

The required strength M_u must be equal to or less than the design strength ϕM_n. An efficient design for footings would be one where the section is tension-controlled. Thus, the strength reduction factor ϕ is equal to 0.9 in accordance with ACI 9.3.2.1. Similar equations can be derived for other pressure distributions.

According to ACI 10.5.4, the minimum area of flexural reinforcement $A_{s,min}$ for footings of uniform cross-section is equal to the required shrinkage and temperature reinforcement prescribed in ACI 7.12.2.1. For footings with Grade 60 reinforcement, the minimum area of steel is equal to 0.18% of the gross area of the footing, which is equal to the overall thickness h times footing plan dimension B or L, depending on the direction of analysis. The maximum spacing of flexural reinforcement is the lesser of $3h$ or 18 in.

Assuming that a square column is supported by a square footing that has a minimum area of flexural reinforcement, Eq. (10.12) can be solved for the required effective depth d:

$$d = 2.2c\sqrt{\frac{P_u}{A_f}} \tag{10.13}$$

In this equation, c is in feet; P_u is in kips; A_f is in square feet; and d is in inches. Equation (10.13) provides an initial estimate of footing thickness based on flexure. Shear provisions must also be checked.

Detailing the Reinforcement

Requirements for the distribution of flexural reinforcement in footings are given in ACI 15.4. For one-way (wall) and two-way square footings, reinforcement is to be distributed uniformly across the entire width of the footing. In square footings, the reinforcement in both orthogonal layers is the same because the maximum factored bending moments at the critical sections are the same. Uniform distribution of the reinforcement is shown

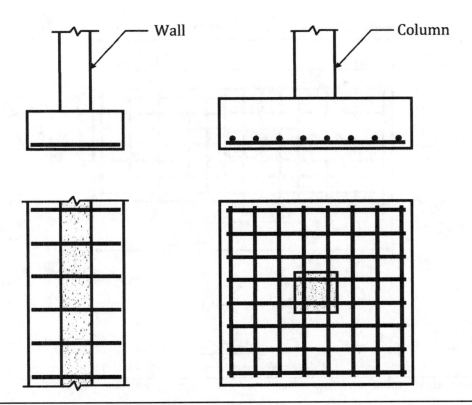

Figure 10.20 Distribution of flexural reinforcement in one- and two-way square footings.

in Fig. 10.20 for a one-way spread footing supporting a wall and for an isolated square spread footing supporting a column. Other reinforcement is not shown for clarity.

Flexural reinforcement in two-way rectangular footings must be distributed in accordance with ACI 15.4.4. Reinforcement in the long direction is uniformly distributed across the entire width of the footing. In the short direction, a portion of the total reinforcement $\gamma_s A_s$ must be uniformly distributed over a band width centered on the column or pedestal that is equal to the length of the short side of the footing. The term γ_s is determined by ACI Eq. (15-1):

$$\gamma_s = \frac{2}{\beta + 1} \tag{10.14}$$

In this equation, β is the ratio of the long side to the short side of the footing.

The remainder of the reinforcement outside of the center band must be uniformly distributed. This distribution reflects the fact that the moment is largest immediately under the column and decreases with increasing distance from the column.

Figure 10.21 illustrates the provisions of ACI 15.4.4. In the long direction, the required area of steel A_{sB} is uniformly distributed at a spacing of s_3 over the width B. In the short direction, the portion A_{s1} of the total required area of reinforcement A_{sL} is uniformly distributed at a spacing of s_1 over a width equal to B centered on the

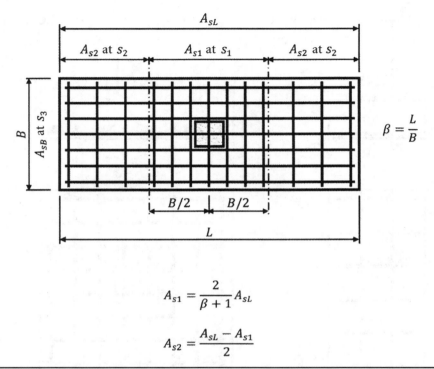

$$A_{s1} = \frac{2}{\beta + 1} A_{sL}$$

$$A_{s2} = \frac{A_{sL} - A_{s1}}{2}$$

FIGURE 10.21 Distribution of flexural reinforcement in a rectangular footing.

column. Note that A_{s1} is equal to $\gamma_s A_{sL}$. The remaining reinforcement A_{s2} is uniformly distributed at a spacing of s_2 on either side of the center band. The minimum reinforcement and maximum spacing requirements of the Code must be satisfied in all portions of the footing.

Development of Reinforcement

Flexural reinforcement in footings must be fully developed in accordance with the applicable provisions of ACI Chap. 12. The bars must extend at least a tension development length ℓ_d beyond the critical section defined in ACI 15.4.2 (see ACI 15.6 and Figs. 10.18 and 10.19 of this book).

For a concrete column supported by an isolated footing, the required development length ℓ_d must be equal to or less than the available development length:

$$\ell_d \leq \frac{L - c_1}{2} - 3 \text{ in} \tag{10.15}$$

In this equation, L and c_1 are the lengths of the footing and the column in the direction of analysis, respectively. As noted previously, a minimum cover of 3 in is required for concrete cast against and permanently exposed to earth.

Section 6.2 contains information on how to determine ℓ_d for flexural reinforcement. The provided reinforcing bar size and spacing in the footing must satisfy Eq. (10.15) so that the bars can be fully developed.

Bar Size (No.)	Development Length ℓ_d (in)
4	19
5	24
6	29
7	42
8	48
9	54
10	60
11	67

TABLE 10.1 Minimum Tension Development Length for Flexural Reinforcement in Footings

In typical cases, the clear spacing and cover requirements listed in the first row of the table given in ACI 12.2.2 are satisfied for flexural reinforcement in footings, and the required ℓ_d can be calculated by the appropriate equations given in the table:

- For No. 6 and smaller bars, $\ell_d = \left(\dfrac{f_y \Psi_t \Psi_e}{25 \lambda \sqrt{f_c'}} \right) d_b$

- For No. 7 and larger bars, $\ell_d = \left(\dfrac{f_y \Psi_t \Psi_e}{20 \lambda \sqrt{f_c'}} \right) d_b$

The factors Ψ_t and Ψ_e are the horizontal reinforcement factor and the coating factor, respectively [see ACI 12.2.4(a) and 12.2.4(b)]. Table 10.1 contains the minimum development lengths ℓ_d for normal-weight concrete that has a compressive strength of 4,000 psi and Grade 60 reinforcement that is uncoated and is placed at the bottom of the footing (i.e., not top bars). The values have been rounded up to the next whole number.

The provisions of ACI 12.2.2 usually produce conservative development lengths compared with those obtained by the provisions of ACI 12.2.3. ACI Eq. (12-1) may be used to determine ℓ_d in any case and must be used where the spacing and/or cover requirements of ACI 12.2.2 are not satisfied.

10.2.6 Design for Shear

Overview
Provisions for shear strength in footings are the same as those required for slabs and are given in ACI 11.11. Requirements for both one- and two-way shear must be satisfied. The required footing thickness for shear is based on the severer of these two conditions.

The critical section for shear is measured from the face of a column, pedestal, or wall for footings supporting such elements (ACI 15.5.2). For footings supporting columns or pedestals with steel base plates, the critical section is measured from the position halfway between the face of the column or pedestal and the edge of the steel base plate.

One-Way Shear
The factored shear force V_u at the critical section, based on the factored pressure at the base of the footing within the tributary area, must be equal to or less than the design

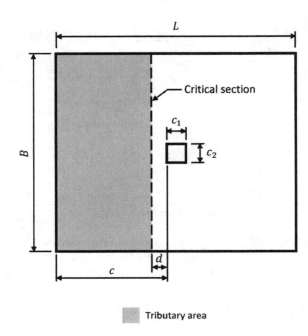

Figure 10.22
Critical section
for one-way
shear in a footing.

Tributary area

shear strength ϕV_c determined in accordance with ACI 11.2.1.1. Figure 10.22 illustrates the tributary area that is to be used in the calculation of V_u for a column supported by an isolated rectangular footing. The following equation must be satisfied at the critical section, which in this case is located a distance d from the face of the column (see ACI 11.11.1.1 and 15.5.2):

$$V_u = q_u B(c - d) \leq \phi V_c = \phi 2\lambda\sqrt{f'_c}Bd \qquad (10.16)$$

According to ACI 9.3.2.3, the strength reduction factor ϕ is equal to 0.75 for shear. One-way shear needs to be checked in the other direction as well.

Equation (10.16) can be used to determine the minimum effective depth d that satisfies one-way shear requirements:

$$d = \frac{q_u c}{q_u + \phi 2\sqrt{f'_c}} \qquad (10.17)$$

This equation is applicable where a concrete column is supported by a square footing.

Even though one-way shear requirements rarely control the design of footings, they must still be checked. Providing an effective depth d that is equal to or greater than that obtained by Eq. (10.17) ensures that one-way shear strength requirements are satisfied.

Two-Way Shear

The factored shear force V_u at the critical section, based on the factored pressure at the base of the footing within the tributary area, must be equal to or less than the design shear strength ϕV_c determined in accordance with ACI 11.11.2.1. The tributary area that is to be used in the calculation of V_u for a column supported by an isolated rectangular

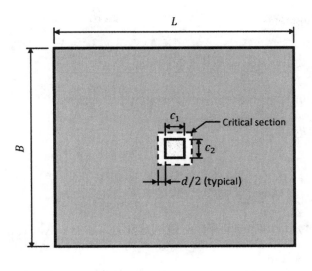

FIGURE 10.23
Critical section
for two-way shear
in a footing.

Tributary area

footing is illustrated in Fig. 10.23. The following equation must be satisfied at the critical section, which in this case is located a distance $d/2$ from the face of the column (see ACI 11.11.1.2 and 15.5.2):

$$V_u = q_u \left[BL - (c_1 + d)(c_2 + d) \right] \le \phi V_c \qquad (10.18)$$

In this equation, the design shear strength of the concrete ϕV_c is the smallest of the values defined in ACI Eqs. (11-31) through (11-33):

$$\phi V_c = \begin{cases} \phi \left(2 + \dfrac{4}{\beta} \right) \lambda \sqrt{f_c'} b_o d \\[2mm] \phi \left(\dfrac{\alpha_s d}{b_o} + 2 \right) \lambda \sqrt{f_c'} b_o d \\[2mm] \phi 4 \lambda \sqrt{f_c'} b_o d \end{cases} \qquad (10.19)$$

Illustrated in Fig. 10.23 is the perimeter b_o of a four-sided critical section.

ACI Eq. (11-31) accounts for the effect of β, which is equal to ratio of the long side to the short side of the column, concentrated load, or reaction area. As β increases, the design shear strength decreases. ACI Eq. (11-32) accounts for the effect of b_o/d. Also included in this equation is α_s, which is equal to 40 for critical sections with four sides, 30 for critical sections with three sides, and 20 for critical sections with two sides. ACI Eq. (11-32) yields the smallest design strength where $d/c_1 \le 0.25$, which rarely occurs. For footings supporting square columns, ACI Eq. (11-33) usually governs.

In cases where ACI Eq. (11-33) results in the smallest design shear strength, the following equation can be used to determine the minimum effective depth d that is required to satisfy two-way shear requirements for a square column supported by a square footing:

$$d = c_1 \left[\frac{-a + \sqrt{a^2 + q_u bc}}{2b} \right] \qquad (10.20)$$

where $a = \dfrac{q_u}{2} + \phi v_c$

$b = \dfrac{q_u}{4} + \phi v_c$

$c = \dfrac{A_f}{c_1^2} - 1$

$v_c = 4\lambda\sqrt{f_c'}$

Generally, two-way shear is more critical than one-way shear. Because shear reinforcement is not economical, the depth of the footing must be increased where shear capacity is not sufficient.

The largest d computed by Eqs. (10.13), (10.17), and (10.20) is to be used in determining the overall thickness of a footing.

10.2.7   Force Transfer at Base of Supported Members

Overview
The interface between the supported and the supporting member must be designed to adequately transfer vertical and horizontal forces between the members. ACI 15.8 contains design requirements for force transfer from a column, wall, or pedestal to a pedestal or footing.

Vertical compressive loads are transferred by bearing on the concrete or by a combination of bearing and reinforcement. Tensile loads must be resisted entirely by reinforcement, which may consist of extended longitudinal bars, dowels, anchor bolts, or mechanical connectors. Provisions are given in ACI 15.8.1 for both cast-in-place and precast members supported by footings. Lateral loads are transferred using the shear-friction provisions of ACI 11.6 or other appropriate methods. Interface reinforcement is designed to resist lateral loads.

Vertical Transfer
Bearing Stress   According to ACI 15.8.1.1, the bearing strength requirements of ACI 10.14 must be satisfied for both the supported and the supporting member. For bearing on the supported member, the factored axial load P_u must be equal to or less than the design-bearing strength ϕP_{nb}:

$$P_u \leq \phi P_{nb} = \phi 0.85 f_c' A_1 \qquad (10.21)$$

In this equation, A_1 is the area of the column, wall, or pedestal that is supported by the footing and the strength reduction factor ϕ for bearing is equal to 0.65 (ACI 9.3.2.4).

The following equation must be satisfied for bearing on a footing or pedestal:

$$P_u \leq \phi P_{nb} = \phi 0.85 f_c' A_1 \sqrt{A_2/A_1} \leq 2\phi 0.85 f_c' A_1 \qquad (10.22)$$

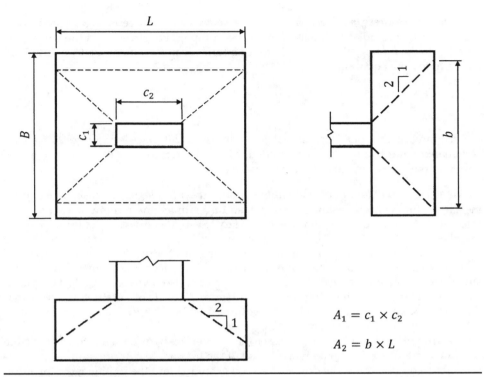

$$A_1 = c_1 \times c_2$$

$$A_2 = b \times L$$

Figure 10.24   Determination of area A_2.

The term A_1 is the area of the supported member and A_2 is defined as the area of the lower base of the largest frustrum of a pyramid, cone, or tapered wedge contained wholly within the footing and having for its upper base the loaded area A_1 and having side slopes of 1 vertical to 2 horizontal.

When the supporting area is wider than the loaded area on all sides, the surrounding concrete confines the bearing area, resulting in an increase in bearing strength. This increase is reflected in Eq. (10.22), where the basic design bearing strength is multiplied by $\sqrt{A_2/A_1}$. An upper limit of two times the basic design bearing strength is required regardless of the magnitude of the adjustment factor.

The determination of A_2 where the lower base falls within and outside of the base dimensions of a footing is illustrated in Fig. 10.24. In this case, A_2 is equal to the projected length b, which is less than the footing width B times L.

Reinforcement across the Interface   The amount of reinforcement that is required between the supported and the supporting member depends on the type of stress in the bars of the supported member under all applicable load combinations. Minimum embedment lengths into both members also depend on this stress.

Dowels are commonly used as interface reinforcement between columns or walls and footings. The dowel bars are set in the footing prior to casting the footing concrete and are subsequently spliced to the column or wall bars.

Compressive stress in the bars of the supported member   Where $P_u > \phi P_{nb}$, reinforcement must be provided to transfer the excess compressive stress from the supported member to the footing. The required area of interface reinforcement can be determined by the following equation:

$$A_s = \frac{P_u - \phi P_{nb}}{\phi f_y} \geq A_{s,min} \tag{10.23}$$

A minimum area of reinforcement $A_{s,min}$ across the interface is required even where concrete-bearing strength is not exceeded. Minimum reinforcement requirements are based on the type of member that is supported.

Cast-in-place columns or pedestals supported on footings.   In cases where cast-in-place columns or pedestals are supported on footings, the minimum area of reinforcement $A_{s,min}$ across the interface is equal to 0.5% of the gross area of the supported member A_1 (ACI 15.8.2.1). This reinforcement can consist of (1) extended reinforcing bars from the column or pedestal into the footing or (2) dowels emanating from the footing (ACI 15.8.2). Providing a minimum area of interface reinforcement helps to ensure ductile behavior between the two members and also helps to provide a degree of structural integrity during the construction stage and the life of the structure. Note that the Code does not require that all of the column bars be extended and anchored into a footing when the stress in them is compressive.

Illustrated in Fig. 10.25 are dowels across the interface between a reinforced concrete column and footing. In situations where all of the bars are in compression in all applicable load combinations, the dowels must be extended into the footing a compression development length ℓ_{dc} determined in accordance with ACI 12.3.2:

$$\ell_{dc} = \begin{cases} (0.02 f_y/\lambda \sqrt{f_c'})d_b \\ (0.0003 f_y)d_b \end{cases} \geq 8 \text{ in} \tag{10.24}$$

The value of ℓ_{dc} determined by Eq. (10.24) may be reduced by the applicable factors given in ACI 12.3.3.

The dowel bars are usually hooked and extend to the level of the flexural reinforcement in the footing. The hooked portion of the dowels cannot be considered effective for developing the dowels in compression (ACI 12.5.5). The following equation must be satisfied to ensure adequate development of the dowels into the footing:

$$h \geq \ell_{dc} + r + (d_b)_{dowel} + 2(d_b)_f + 3 \text{ in} \tag{10.25}$$

In this equation, r is the radius of the dowel bar bend, which is defined in ACI Table 7.2. The bar diameters $(d_b)_{dowel}$ and $(d_b)_f$ correspond to the dowels and flexural reinforcement, respectively. In cases where Eq. (10.25) is not satisfied, either a greater number of smaller dowel bars can be used, or the depth of the footing must be increased.

Dowels must also be fully developed in the column. These bars are typically lap spliced to the column bars. In cases where the dowel bars are the same size as the column bars, the minimum compression lap splice length is equal to $(0.0005 f_y)d_b = 30d_b$ or 12 in for Grade 60 reinforcement. ACI 12.16.1 contains additional requirements for other grades of reinforcement and concrete strengths less than 3,000 psi.

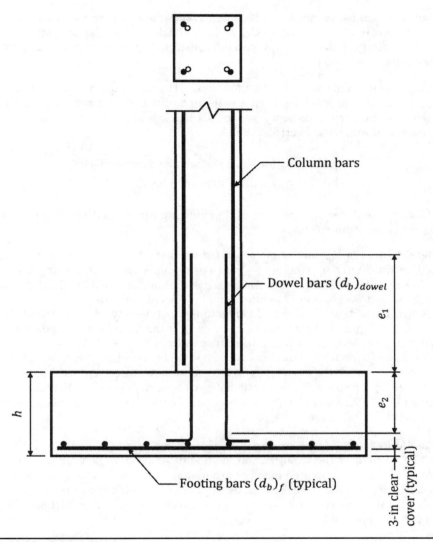

FIGURE 10.25 Footing dowels.

Where the dowel bars are smaller in diameter than the column bars, the compression lap splice length must be equal to or greater than the larger of the following (ACI 12.16.2):

1. The development length in compression ℓ_{dc} of the larger bar, which is determined in accordance with ACI 12.3

2. The compression lap splice length of the smaller bar, which is determined in accordance with ACI 12.16.1

According to ACI 15.8.2.3, it is permitted to lap splice Nos. 14 and 18 column bars that are in compression to dowel bars that are No. 11 or smaller. The required lap splice length is determined in accordance with the requirements outlined earlier. This provision is an exception to ACI 12.14.2.1, which prohibits lap splicing of Nos. 14 and 18 bars and is based on many years of satisfactory performance of these types of connections.

Cast-in-place walls supported on footings. The minimum area of reinforcement $A_{s,min}$ across the interface between a cast-in-place wall and footing is equal to the minimum vertical reinforcement given in ACI 14.3.2, which depends on the size of the bars passing through the interface:

$$A_{s,min} = \begin{cases} 0.0012 A_g \text{ for No. 5 and smaller deformed bars} \\ 0.0015 A_g \text{ for other deformed bars} \end{cases} \qquad (10.26)$$

Like cast-in-place columns, extended reinforcing bars from the wall or dowels can be used to satisfy this requirement.

Tensile stress in the bars of the supported member Tensile forces—either applied directly or transferred by a moment—from a supported element to a footing or pedestal must be resisted entirely by reinforcement across the interface (ACI 15.8.1.2 and 15.8.1.3). In such cases, dowel bars are provided for all of the column bars.

Tensile anchorage of the dowel bars into the footing is typically accomplished by providing 90-degree standard hooks at the ends of the dowel bars. The development length of the hooked bars is determined in accordance with ACI 12.5.

A tension lap splice or a mechanical connection in accordance with ACI 12.17 must be provided between the dowel bars and the column reinforcement (ACI 15.8.1.3). The type of tension lap splice (Class A or B) that must be used depends on the conditions set forth in ACI 12.17.2.2 and 12.17.2.3 (see Section 8.7 of this book for a detailed discussion on lap splices in columns).

Horizontal Transfer

ACI 15.8.1.4 permits the shear-friction method of ACI 11.6 to be used for transfer of lateral loads from a supported member to a footing or pedestal. The reinforcement A_{vf} provided across the interface between the supported and the supporting member must satisfy the following equation, which is applicable where the shear-friction reinforcement is perpendicular to the interface:

$$A_{vf} \geq \frac{V_u}{\phi f_y \mu} \qquad (10.27)$$

In this equation, V_u is the maximum factored shear force due to the lateral load effects obtained by the applicable load combinations given in ACI 9.2.1 and μ is the coefficient of friction that is determined by ACI 11.6.4.3. The strength reduction factor ϕ is equal to 0.75.

It is commonly assumed that the concrete of the supported member is placed against the hardened concrete of the footing or pedestal without intentionally roughening the surface of the supporting member. In such cases, the effects of the lateral loads are primarily resisted by dowel action of the reinforcement across the interface, and tests have shown that μ can be taken as 0.6λ.[5] The modification factor λ is equal to 1.0 for normal-weight concrete and 0.75 for all lightweight concrete. Note that λ can also be determined on the basis of the volumetric proportions of the lightweight and normal-weight aggregates in the concrete mixture, but it must not exceed 0.85 (ACI 11.6.4.3).

In cases where no intentional roughening is provided, the following strength equation must also be satisfied (ACI 11.6.5):

$$V_u \leq \phi V_n = \phi 0.2 f_c' A_c \leq \phi 800 A_c \tag{10.28}$$

The term A_c is the area of the concrete section that resists V_u. For example, A_c is equal to the area of a column that is supported by a footing. Where the supported and supporting concretes have different compressive strengths, the smaller of the two strengths must be used. Equation (10.28) provides an upper limit on shear-friction strength, and it should be checked prior to design.

Because shear-friction reinforcement acts in tension, full tension anchorage must be provided into the footing or pedestal and the supported member. The lengths of these dowel bars are determined in the same way as those for vertical transfer where tension forces or bending moments are present.

Typically, the area of the dowel bars is initially determined on the basis of the requirements for vertical transfer. That area is then compared with the area determined by Eq. (10.27) for horizontal transfer and the larger of the two areas is provided at the interface.

10.2.8 Design Procedure

The following design procedure can be used in the design of footings. Included is the information presented in the previous sections on analysis, design, and detailing.

Step 1: Determine the area of the footing. The base dimensions of a footing are calculated using service loads and the permissible bearing capacity of the soil.

For a concentrically loaded isolated footing, the maximum pressure at the base of the footing is uniform, and A_f can be obtained by dividing the service axial load by the permissible bearing capacity. For isolated footings subjected to an axial load and bending moment, A_f is determined by trial and error. Closed-form solutions are available for combined rectangular and trapezoidal footings where the dimensions are determined so that uniform pressure occurs at the base of the footing.

Step 2: Determine the thickness of the footing. The thickness of the footing h must satisfy flexural requirements and the one- and two-way shear requirements of ACI 15.5. Factored loads are used to compute the maximum bending moments and shear forces at the critical sections.

According to ACI 15.7, the depth of the footing above the bottom reinforcement layer must be 6 in for footings supported on soil and 12 in for footings on piles. The minimum cover to reinforcement in concrete elements cast against and permanently exposed to earth is 3 in in accordance with ACI 7.7.1.

Step 3: Determine the required flexural reinforcement. Once a footing thickness has been established, the flexural reinforcement can be determined using the general requirements of the strength design method. The location of the critical section for the maximum factored moment in a footing is given in ACI 15.4 for three cases (see Figs. 10.18 and 10.19).

The required area of reinforcement is calculated at the critical section, assuming a tension-controlled section. The provided area of reinforcement must be equal to or greater than the minimum reinforcement prescribed in ACI 10.5.4.

Step 4: Distribute the flexural reinforcement. The flexural reinforcement is distributed in the footing according to the provisions of ACI 15.4. For one-way footings and two-way square footings, reinforcement is uniformly distributed across the entire width of the footing. In two-way rectangular footings, reinforcement in the long direction is uniformly distributed across the entire width of the footing. In the short direction, distribution must satisfy the requirements of ACI 15.4.4.2 (see Figs. 10.20 and 10.21).

Step 5: Develop the flexural reinforcement. The flexural reinforcement is developed according to the provisions of ACI Chap. 12. The location of the critical sections for development of reinforcement are the same as those defined in ACI 15.4.2 for maximum factored moments.

Step 6: Check transfer of forces at the base of the supported member. Both vertical and horizontal force transfer must be checked at the interface between the supported and the supporting member. Vertical compression forces are transferred by bearing on the concrete and by reinforcement, if required. Provisions for bearing are given in ACI 10.14. Tensile forces must be resisted entirely by reinforcement.

A minimum area of reinforcement is required across the interface of the supported member and the footing even where concrete bearing strength is not exceeded. Horizontal force transfer must satisfy the shear-friction requirements of ACI 11.7.

Example 10.6 Determine the thickness and reinforcement for the footing given in Example 10.1. The column that is supported by the footing is 16×16 in, has normal-weight concrete with a compressive strength of 5,000 psi, and is reinforced with eight No. 7 bars. Assume that the footing has normal-weight concrete with a compressive strength of 4,000 psi and Grade 60 reinforcement.

Solution

Step 1: Determine the area of the footing. The area of the footing has been determined in Example 10.1 on the basis of the permissible soil pressure. The base dimensions are 11 ft 9 in $\times$ 11 ft 9 in.

Step 2: Determine the thickness of the footing. The thickness of the footing must be determined using the factored pressure at the base of the footing, considering both flexure and shear. The maximum factored pressure q_u is calculated by dividing the factored load P_u by the area of the footing A_f:

$$q_u = \frac{P_u}{A_f} = \frac{(1.2 \times 300) + (1.6 \times 150)}{11.75^2} = \frac{600}{138.1} = 4.35 \text{ ksf}$$

- *Design for flexure.* The critical section for flexure for a footing supporting a concrete column is at the face of the column (see Fig. 10.18). Using Eq. (10.13), the required effective depth d of the footing for flexure assuming minimum flexural reinforcement is

$$d = 2.2c\sqrt{\frac{P_u}{A_f}} = 2.2 \times \left(\frac{11.75}{2} - \frac{16}{2 \times 12}\right) \times \sqrt{4.35} = 23.9 \text{ in}$$

The required flexural reinforcement is determined after the effective depth is established considering both flexure and shear strength.

- *Design for shear.* The minimum effective depth d that is required to satisfy one-way shear requirements is determined by Eq. (10.17):

$$d = \frac{q_u c}{q_u + \phi 2\sqrt{f_c'}} = \frac{(4{,}350/144) \times \{(11.75/2) - [16/(2 \times 12)]\} \times 12}{(4{,}350/144) + (0.75 \times 2 \times \sqrt{4{,}000})} = 15.1 \text{ in}$$

Equation (10.20) is used to determine the minimum effective depth d to satisfy two-way shear requirements:

$$d = c_1 \left[\frac{-a + \sqrt{a^2 + q_u bc}}{2b} \right]$$

$$= 16 \left\{ \frac{-204.9 + \sqrt{204.9^2 + [(4{,}350/144) \times 197.3 \times 76.7]}}{2 \times 197.3} \right\} = 20.3 \text{ in}$$

where

$$a = \frac{q_u}{2} + \phi v_c = \frac{4{,}350}{2 \times 144} + (0.75 \times 253.0) = 204.9 \text{ psi}$$

$$b = \frac{q_u}{4} + \phi v_c = \frac{4{,}350}{4 \times 144} + (0.75 \times 253.0) = 197.3 \text{ psi}$$

$$c = \frac{A_f}{c_1^2} - 1 = \frac{11.75^2}{(16/12)^2} - 1 = 76.7$$

$$v_c = 4\lambda\sqrt{f_c'} = 4 \times 1.0\sqrt{4{,}000} = 253.0 \text{ psi}$$

Therefore, a $23.9 + 4 = 27.9$ in footing is adequate for flexure and shear.

Try a 28-in thick footing ($d = 24$ in).

Step 3: Determine the required flexural reinforcement. The maximum factored moment M_u at the critical section is (see Fig. 10.18)

$$M_u = \frac{q_u c^2}{2} = \frac{4.35 \times \{(11.75/2) - [16/(2 \times 12)]\}^2}{2} = 59.0 \text{ ft kips/ft}$$

The flowchart shown in Fig. 6.4 is used to determine A_s. It is modified to account for the differences applicable to footing design.

Step 3A: Assume tension-controlled section. Footings should be designed as tension-controlled sections whenever possible. Thus, assume that the strength reduction factor $\phi = 0.9$.

Step 3B: Determine the nominal strength coefficient of resistance R_n. For a rectangular section, R_n is determined by Eq. (6.5), which is a function of the factored bending moment M_u:

$$R_n = \frac{M_u}{\phi b_w d^2} = \frac{59.0 \times 12{,}000}{0.9 \times 12 \times 24^2} = 113.8 \text{ psi}$$

Step 3C: Determine the required reinforcement ratio ρ. The reinforcement ratio ρ is determined by Eq. (6.7):

$$\rho = \frac{0.85 f_c'}{f_y} \left[1 - \sqrt{1 - \frac{2 R_n}{0.85 f_c'}} \right] = \frac{0.85 \times 4}{60} \left[1 - \sqrt{1 - \frac{2 \times 113.8}{0.85 \times 4{,}000}} \right] = 0.0019$$

Step 3D: Determine the required area of tension reinforcement A_s.

$$A_s = \rho b d = 0.0019 \times 12 \times 24 = 0.55 \text{ in}^2/\text{ft}$$

Step 3E: Determine the minimum required area of reinforcement $A_{s,min}$. The minimum amount of reinforcement is determined by ACI 10.5.4:

$$A_{s,min} = 0.0018bh = 0.0018 \times 12 \times 28 = 0.61 \text{ in}^2/\text{ft} > 0.55 \text{ in}^2/\text{ft}$$

Use $A_s = 0.61 \text{ in}^2/\text{ft}$.

Step 3F: Determine the depth of the equivalent rectangular stress block a.

$$a = \frac{A_s f_y}{0.85 f'_c b} = \frac{0.61 \times 60,000}{0.85 \times 4,000 \times 12} = 0.9 \text{ in}$$

Step 3G: Determine β_1. According to ACI 10.2.73, $\beta_1 = 0.85$ for $f'_c = 4,000$ psi (see Section 5.2).

Step 3H: Determine the neutral axis depth c.

$$c = \frac{a}{\beta_1} = \frac{0.9}{0.85} = 1.1 \text{ in}$$

Step 3I: Determine ε_t.

$$\varepsilon_t = 0.0030 \left(\frac{d_t}{c} - 1 \right) = 0.0030 \left(\frac{24}{1.1} - 1 \right) = 0.0625 > 0.0040$$

Because $\varepsilon_t > 0.0040$, the maximum reinforcement requirement of ACI 10.3.5 is satisfied. Also, the section is tension-controlled because $\varepsilon_t > 0.0050$ (ACI 10.3.4), and the initial assumption that the section is tension-controlled is correct.

Step 3J: Choose the size and spacing of the reinforcing bars.

$$\text{Total area of steel required} = 0.61 \times 11.75 = 7.17 \text{ in}^2$$

Try 12 No. 7 bars (provided $A_s = 7.20 \text{ in}^2$) spaced 12 in on center.

Step 4: Distribute the flexural reinforcement. The provided spacing of 12 in is less than the maximum spacing of $3h = 84$ in or 18 in (governs).

Because the footing is square, the bars are uniformly distributed in both orthogonal directions.

Step 5: Develop the flexural reinforcement. Check if the tension development length ℓ_d of the No. 7 bars can be determined by ACI 12.2.2:

$$\text{Clear spacing of bars being developed} = 12 - 0.875 = 11.1 \text{ in} > 2d_b = 1.8 \text{ in}$$

$$\text{Clear cover} = 3.0 \text{ in} > d_b = 0.875 \text{ in}$$

Thus, ℓ_d may be determined by the following equation that can be found in the first row of the table given in ACI 12.2.2, which is applicable to No. 7 bars and larger:

$$\ell_d = \left(\frac{f_y \Psi_t \Psi_e}{20\lambda \sqrt{f'_c}} \right) d_b = \left(\frac{60,000 \times 1.0 \times 1.0}{20 \times 1.0\sqrt{4,000}} \right) \times 0.875 = 41.5 \text{ in}$$

where $\psi_t = 1.0$ for bars other than top bars
$\psi_e = 1.0$ for uncoated reinforcement

In lieu of this equation, ℓ_d can be calculated by Eq. (6.28):

$$\ell_d = \left[\frac{3}{40} \frac{f_y}{\lambda \sqrt{f'_c}} \frac{\psi_t \psi_e \psi_s}{(c_b + K_{tr})/d_b} \right] d_b \geq 12 \text{ in}$$

where $\quad \lambda = 1.0$ for normal-weight concrete

$\psi_t = 1.0$ for bars other than top bars

$\psi_e = 1.0$ for uncoated reinforcement

$\psi_s = 1.0$ for No. 7 bars

$$c_b = 3.0 + \frac{0.875}{2} = 3.4 \text{ in (governs)}$$

$$= \frac{12}{2} = 6.0 \text{ in}$$

$$K_{tr} = 0$$

$$\frac{c_b + K_{tr}}{d_b} = \frac{3.4 + 0}{0.875} = 3.9 > 2.5; \text{ use } 2.5$$

Therefore,

$$\ell_d = \left(\frac{3}{40} \frac{60{,}000}{1.0\sqrt{4{,}000}} \frac{1.0 \times 1.0 \times 1.0}{2.5} \right) \times 0.875 = 24.9 \text{ in}$$

As expected, Eq. (6.28) results in a development length that is less than that determined by ACI 12.2.2.

The available development length is determined by Eq. (10.15):

$$\frac{L - c_1}{2} - 3 = \frac{(11.75 \times 12) - 16}{2} - 3 = 59.5 \text{ in}$$

Because this available length is greater than the required development length, the No. 7 bars are fully developed for flexure.

Use 12 No. 7 bars at 12 in, which are 11 ft long each way.

Step 6: Check force transfer at the base of the column.

1. Check bearing stress on the concrete column and footing.

The bearing strength of the column is determined by Eq. (10.21):

$$\phi P_{nb} = \phi 0.85 f_c' A_1 = 0.65 \times 0.85 \times 5 \times 16^2 = 707 \text{ kips} > P_u = 600 \text{ kips}$$

The bearing strength of the footing is determined by Eq. (10.22):

$$\phi P_{nb} = \phi 0.85 f_c' A_1 \sqrt{A_2/A_1} \le 2\phi 0.85 f_c' A_1$$

Using Fig. 10.24, the area A_2 is determined as follows:

Thickness of footing $h = 28$ in

Horizontal projection for a 1:2 slope $= 2 \times 28 = 56$ in

Projected length $b = 56 + 16 + 56 = 128$ in $< 11.75 \times 12 = 141$ in

$$A_2 = b^2 = 128^2 = 16{,}384 \text{ in}^2$$

Thus,

$$\sqrt{\frac{A_2}{A_1}} = \sqrt{\frac{16{,}384}{16^2}} = 8.0 > 2.0; \text{ use } 2.0$$

$$\phi P_{nb} = 2\phi 0.85 f_c' A_1 = 2 \times 0.65 \times 0.85 \times 4 \times 16^2 = 1{,}132 \text{ kips} > P_u = 600 \text{ kips}$$

2. Determine the required interface reinforcement.

Because the design-bearing strength is adequate for the column and footing, provide the minimum area of reinforcement across the interface in accordance with ACI 15.8.2.1:

$$A_{s,min} = 0.005 A_g = 0.005 \times 16^2 = 1.28 \text{ in}^2$$

Provide four No. 6 dowel bars ($A_s = 1.76 \text{ in}^2$).

3. The dowel bars in compression are developed as follows.

- *Development of the dowel bars into the footing.* The dowel bars must be extended into the footing a compression development length ℓ_{dc} determined by Eq. (10.24):

$$\ell_{dc} = \begin{cases} (0.02 f_y / \lambda \sqrt{f_c'}) d_b = [(0.02 \times 60{,}000)/(1.0\sqrt{4{,}000})] \times 0.75 = 14.2 \text{ in (governs)} \\ (0.0003 f_y) d_b = 0.0003 \times 60{,}000 \times 0.75 = 13.5 \text{ in} \end{cases}$$

The minimum footing thickness for development of the dowel bars is determined by Eq. (10.25):

$$\ell_{dc} + r + (d_b)_{dowel} + 2(d_b)_f + 3 \text{ in} = 14.2 + (6 \times 0.75) + 0.75 + (2 \times 0.875) + 3 = 24.2 \text{ in}$$

Because the provided footing thickness $h = 28$ in is greater than 24.2 in, the hooked dowel bars can be fully developed in the footing.

- *Development of the dowel bars into the column.* The dowel bars must be lap spliced to the column bars. Because the dowel bars are smaller in diameter than the column bars, the compression lap splice length must be equal to or greater than the larger of the following:

(a) The development length in compression ℓ_{dc} of the larger bar, which is determined in accordance with ACI 12.3.

The development length ℓ_{dc} in compression of the No. 7 column bars is

$$\ell_{dc} = \begin{cases} (0.02 f_y / \lambda \sqrt{f_c'}) d_b = [(0.02 \times 60{,}000)/(1.0\sqrt{5{,}000})] \times 0.875 = 14.9 \text{ in} \\ (0.0003 f_y) d_b = 0.0003 \times 60{,}000 \times 0.875 = 15.8 \text{ in (governs)} \end{cases}$$

This length can be reduced in accordance with ACI 12.3.3(a), which takes into account excess reinforcement in the section; no reduction is taken in this example.

(b) The compression lap splice length of the smaller bar.

The compression lap splice length of the No. 6 bars is determined in accordance with ACI 12.16.1:

$$\text{Compression lap splice length} = \begin{cases} (0.0005 f_y) d_b = 30 d_b = 30 \times 0.75 = 22.5 \text{ in (governs)} \\ 12 \text{ in} \end{cases}$$

Therefore, provide a lap splice length equal to 2 ft 0 in.

No horizontal forces are transferred from the column to the footing, so horizontal force transfer is not investigated.

Reinforcement details for this column and footing are similar to those shown in Fig. 10.25.

Example 10.7 Determine the thickness and reinforcement for the footing given in Example 10.2. The column that is supported by the footing is 20 × 20 in, has normal-weight concrete with a compressive strength of 5,000 psi, and is reinforced with 12 No. 7 bars. Assume that the footing has normal-weight concrete with a compressive strength of 4,000 psi and Grade 60 reinforcement.

Solution

   Step 1: Determine the area of the footing. The area of the footing has been determined in Example 10.2 on the basis of the permissible soil pressure. The base dimensions are 12 ft 6 in × 12 ft 6 in.

   Step 2: Determine the thickness of the footing. The thickness of the footing must be determined using the factored pressure at the base of the footing, considering both flexure and shear. The total factored pressure q_u at the edge of the footing is equal to the stress due to the factored load P_u plus the stress due to the factored moment M_u, considering the applicable load combinations given in ACI 9.2.1. A summary of the total pressure is given in Table 10.2 for the governing load combinations.

   It is evident from Table 10.2 that the third load combination yields the maximum pressure. A preliminary footing thickness will be determined on the basis of two-way shear requirements.

1. *Two-way shear*: A preliminary effective depth d to satisfy two-way shear requirements can be obtained from Eq. (10.20), which is based on a uniform pressure distribution; the total shear stress considering both axial load and bending moment will be checked later. Conservatively assume that the maximum factored pressure is uniformly distributed over the entire area of the footing.

$$d = c_1 \left[\frac{-a + \sqrt{a^2 + q_u bc}}{2b} \right]$$

$$= 20 \left\{ \frac{-200.9 + \sqrt{200.9^2 + [(3{,}210/144) \times 195.3 \times 55.3]}}{2 \times 195.3} \right\} = 16.9 \text{ in}$$

where $a = \dfrac{q_u}{2} + \phi v_c = \dfrac{3{,}210}{2 \times 144} + (0.75 \times 253.0) = 200.9 \text{ psi}$

$b = \dfrac{q_u}{4} + \phi v_c = \dfrac{3{,}210}{4 \times 144} + (0.75 \times 253.0) = 195.3 \text{ psi}$

$c = \dfrac{A_f}{c_1^2} - 1 = \dfrac{12.5^2}{(20/12)^2} - 1 = 55.3$

$v_c = 4\lambda\sqrt{f_c'} = 4 \times 1.0\sqrt{4{,}000} = 253.0 \text{ psi}$

   Because the bending moment in this example is relatively large, the shear stress due to this moment will have a relatively large impact on the total shear stress. Thus, try a 24-in-thick footing ($d = 20$ in).

   The critical section around the 20-in column occurs at a distance $d/2 = 10$ in from the face of the column (see Fig. 10.23). The total factored shear stresses due to the axial load and bending moment must be determined.

Load Combination	Axial Load P_u (kips)	Bending Moment M_u (ft kips)	Pressure Due to Axial Load P_u/A_f (ksf)	Pressure Due to Bending Moment $6M_u/BL^2$, ksf	Total Pressure (ksf)
$1.4D$	280	11.2	1.79	0.03	1.82
$1.2D + 1.6L$	480	16.0	3.07	0.05	3.12
$1.2D + 0.5L + 1.6W$	323	371.6	2.07	1.14	3.21
$0.9D - 1.6W$	172	−352.8	1.10	1.08	2.18

TABLE 10.2 Summary of Total Pressure at the Base of the Footing Given in Example 10.6

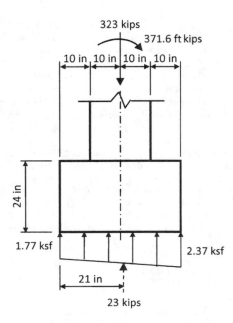

FIGURE 10.26
Free-body diagram
of the critical
section given in
Example 10.7.

The stress distribution in this example is similar to that depicted in Fig. 10.12a, where the maximum factored pressure is 3.21 ksf and the minimum factored pressure is $2.07 - 1.14 = 0.93$ ksf (see Table 10.2). The magnitude of the factored pressure along the length of the footing can be determined by the following equation:

$$\text{Factored pressure} = \frac{3.21 - 0.93}{12.5}x + 0.93 = 0.1824x + 0.93$$

The distance x is measured from the edge of the footing with minimum factored pressure.

The free-body diagram of the critical section illustrated in Fig. 10.26 will assist in the determination of V_u and M_u at this location.

The boundaries of the critical section are at the following locations from the edge of the footing with minimum factored pressure:

$$x_1 = \frac{12.5}{2} - \frac{20}{2 \times 12} - \frac{10}{12} = 4.58 \text{ ft}$$

$$x_2 = \frac{12.5}{2} + \frac{20}{2 \times 12} + \frac{10}{12} = 7.92 \text{ ft}$$

The corresponding pressures at these locations are (see Fig. 10.26)

$$q_u(x_1) = (0.1824 \times 4.58) + 0.93 = 1.77 \text{ ksf}$$

$$q_u(x_2) = (0.1824 \times 7.92) + 0.93 = 2.37 \text{ ksf}$$

The factored shear force at the critical section is equal to the factored column load minus the factored soil pressure in the area bounded by the critical section:

$$V_u = 323 - \left[1.77 \times \left(\frac{40}{12}\right)^2\right] - \left[\frac{1}{2}(2.37 - 1.77) \times \left(\frac{40}{12}\right)^2\right] = 323 - 23 = 300 \text{ kips}$$

The unbalanced moment M_u is obtained by summing the moments due to the load from the column and the load from the soil about the centroid of the critical section:

$$M_u = 371.6 - \left(23 \times \frac{1}{12}\right) = 369.7 \text{ ft kips}$$

Determine the shear factor γ_v by Eq. (7.38):

$$\gamma_f = \frac{1}{1 + (2/3)\sqrt{b_1/b_2}} = \frac{1}{1 + (2/3)\sqrt{40/40}} = 0.6$$

$$\gamma_v = 1 - \gamma_f = 1 - 0.6 = 0.4$$

The section properties of the critical section are determined using Fig. 7.53 for an interior column:

$$c_{AB} = b_1/2 = 40/2 = 20.0 \text{ in}$$

$$f_1 = 2\left[\left(1 + \frac{c_2}{c_1}\right)\left(\frac{c_1}{d}\right) + 2\right] = 2\left[(1 + 1)\left(\frac{20}{20}\right) + 2\right] = 8.0$$

$$A_c = f_1 d^2 = 8.0 \times 20^2 = 3{,}200 \text{ in}^2$$

$$f_2 = \frac{1}{6}\left[\left(1 + \frac{3c_2}{c_1}\right)\left(\frac{c_1}{d}\right)^2 + \left(5 + \frac{3c_2}{c_1}\right)\left(\frac{c_1}{d}\right) + 5\right]$$

$$= \frac{1}{6}\left[(1 + 3)\left(\frac{20}{20}\right)^2 + (5 + 3)\left(\frac{20}{20}\right) + 5\right] = 2.83$$

$$J_c/c_{AB} = 2 f_2 d^3 = 2 \times 2.83 \times 20^3 = 45{,}280 \text{ in}^3$$

These section properties can also be obtained using the tables given in Appendix B. The total factored shear stress is determined by Eq. (7.39):

$$v_{u(AB)} = \frac{V_u}{A_c} + \frac{\gamma_v M_u c_{AB}}{J_c}$$

$$= \frac{300{,}000}{3{,}200} + \frac{0.4 \times 369.7 \times 12{,}000}{45{,}280} = 93.8 + 39.2 = 133.0 \text{ psi}$$

The allowable stress for a square column is obtained by Eq. (10.19):

$$\phi v_c = \frac{\phi V_c}{b_o d} = \phi 4\lambda\sqrt{f_c'} = 0.75 \times 4 \times 1.0\sqrt{4{,}000} = 189.7 \text{ psi} > 133.0 \text{ psi}$$

Therefore, try a 24-in-thick footing ($d = 20$ in).

2. *One-way shear.* The critical section for one-way shear is located a distance $d = 20$ in from the face of the column (see Fig. 10.22).

Conservatively assuming a uniform factored pressure of 3.21 ksf over the entire footing area, the maximum factored shear force at the critical section is [see Eq. (10.16)]

$$V_u = q_u B(c - d) = 3.21 \times 12.5 \times \left(\frac{12.5}{2} - \frac{20}{2 \times 12} - \frac{20}{12}\right) = 150.5 \text{ kips}$$

Design shear strength is also given in Eq. (10.16):

$$\phi V_c = \phi 2\lambda\sqrt{f_c'}\,Bd = 0.75 \times 2 \times 1.0\sqrt{4,000} \times (12.5 \times 12) \times 20/1,000 = 284.6 \text{ kips}$$

Because $V_u < \phi V_c$, one-way shear strength requirements are satisfied.
Use $h = 24.0$ in. ($d = 20.0$ in.).

Step 3: Determine the required flexural reinforcement. The maximum factored moment M_u at the critical section is determined considering the nonuniform distribution of pressure at the base of the footing.

The distance x is measured from the edge of the footing with minimum factored pressure. At the critical section,

$$x = \frac{12.5}{2} + \frac{20}{2 \times 12} = 7.08 \text{ ft}$$

Therefore, the pressure at the critical section = $(0.1824 \times 7.08) + 0.93 = 2.22$ ksf.

The factored bending moment at the critical section can be obtained by summing moments about the critical section for flexure. The trapezoidal pressure is divided into rectangular and triangular portions:

$$M_u = \frac{2.22 \times [(12.5/2) - 20/(2 \times 12)]^2}{2} + \frac{1}{2} \times (3.21 - 2.22) \times \frac{2}{3} \times \left(\frac{12.5}{2} - \frac{20}{2 \times 12}\right)^2$$
$$= 42.3 \text{ ft kips/ft}$$

For comparison purposes, conservatively calculate the factored moment at the critical section, assuming that the maximum pressure is uniformly distributed over the entire area of the footing:

$$M_u = \frac{3.21 \times [(12.5/2) - 20/(2 \times 12)]^2}{2} = 47.1 \text{ ft kips}$$

This bending moment is approximately 11% greater than the bending moment obtained from the actual pressure distribution. The factored bending moment of 42.3 ft kips/ft is used throughout the remainder of this example.

The flowchart shown in Fig. 6.4 is used to determine A_s. It is modified to account for the differences applicable to footing design.

Step 3A: Assume tension-controlled section. Footings should be designed as tension-controlled sections whenever possible. Thus, assume that the strength reduction factor $\phi = 0.9$.

Step 3B: Determine the nominal strength coefficient of resistance R_n. For a rectangular section, R_n is determined by Eq. (6.5), which is a function of the factored bending moment M_u:

$$R_n = \frac{M_u}{\phi b_w d^2} = \frac{42.3 \times 12,000}{0.9 \times 12 \times 20^2} = 117.5 \text{ psi}$$

Step 3C: Determine the required reinforcement ratio ρ. The reinforcement ratio ρ is determined by Eq. (6.7):

$$\rho = \frac{0.85 f_c'}{f_y}\left[1 - \sqrt{1 - \frac{2R_n}{0.85 f_c'}}\right] = \frac{0.85 \times 4}{60}\left[1 - \sqrt{1 - \frac{2 \times 117.5}{0.85 \times 4,000}}\right] = 0.0020$$

Step 3D: Determine the required area of tension reinforcement A_s.

$$A_s = \rho bd = 0.0020 \times 12 \times 20 = 0.48 \text{ in}^2/\text{ft}$$

Step 3E: Determine the minimum required area of reinforcement $A_{s,min}$. The minimum amount of reinforcement is determined by ACI 10.5.4:

$$A_{s,min} = 0.0018bh = 0.0018 \times 12 \times 24 = 0.52 \text{ in}^2/\text{ft} > 0.48 \text{ in}^2/\text{ft}$$

Use $A_s = 0.52 \text{ in}^2/\text{ft}$.

Step 3F: Determine the depth of the equivalent rectangular stress block a.

$$a = \frac{A_s f_y}{0.85 f'_c b} = \frac{0.52 \times 60{,}000}{0.85 \times 4{,}000 \times 12} = 0.8 \text{ in}$$

Step 3G: Determine β_1. According to ACI 10.2.73, $\beta_1 = 0.85$ for $f'_c = 4{,}000$ psi (see Section 5.2).

Step 3H: Determine the neutral axis depth c.

$$c = \frac{a}{\beta_1} = \frac{0.8}{0.85} = 0.9 \text{ in}$$

Step 3I: Determine ε_t.

$$\varepsilon_t = 0.0030 \left(\frac{d_t}{c} - 1 \right) = 0.0030 \left(\frac{20}{0.9} - 1 \right) = 0.0637 > 0.0040$$

Because $\varepsilon_t > 0.0040$, the maximum reinforcement requirement of ACI 10.3.5 is satisfied. Also, the section is tension-controlled because $\varepsilon_t > 0.0050$ (ACI 10.3.4), and the initial assumption that the section is tension-controlled is correct.

Step 3J: Choose the size and spacing of the reinforcing bars.

$$\text{Total area of steel required} = 0.52 \times 12.5 = 6.50 \text{ in}^2$$

Try 15 No. 6 bars (provided $A_s = 6.60 \text{ in}^2$) spaced 10 in on center.

Step 4: Distribute the flexural reinforcement. The provided spacing of 10 in is less than the maximum spacing of $3h = 72$ in or 18 in (governs).

Because the footing is square, the bars are uniformly distributed in both orthogonal directions.

Step 5: Develop the flexural reinforcement. Determine ℓ_d by Eq. (6.28):

$$\ell_d = \left[\frac{3}{40} \frac{f_y}{\lambda \sqrt{f'_c}} \frac{\psi_t \psi_e \psi_s}{(c_b + K_{tr})/d_b} \right] d_b \geq 12 \text{ in}$$

where
$\lambda = 1.0$ for normal-weight concrete
$\psi_t = 1.0$ for bars other than top bars
$\psi_e = 1.0$ for uncoated reinforcement
$\psi_s = 0.8$ for No. 6 bars
$$c_b = 3.0 + \frac{0.75}{2} = 3.4 \text{ in (governs)}$$
$$= \frac{10}{2} = 5.0 \text{ in}$$
$$K_{tr} = 0$$
$$\frac{c_b + K_{tr}}{d_b} = \frac{3.4 + 0}{0.75} = 4.5 > 2.5; \text{ use } 2.5$$

Therefore,

$$\ell_d = \left(\frac{3}{40} \frac{60{,}000}{1.0\sqrt{4{,}000}} \frac{1.0 \times 1.0 \times 0.8}{2.5} \right) \times 0.75 = 17.1 \text{ in}$$

The available development length is determined by Eq. (10.12):

$$\frac{L - c_1}{2} - 3 = \frac{(12.5 \times 12) - 20}{2} - 3 = 62.0 \text{ in}$$

Because this available length is greater than the required development length, the No. 6 bars can be fully developed for flexure.

Use 15 No. 6 bars at 10 in, which are 12 ft long each way.

Step 6: Check force transfer at the base of the column.

1. Check bearing stress on the concrete column and footing.

   The third load combination produces the largest stresses on the column:

   $$\text{Axial compressive stress} = \frac{P_u}{c_1^2} = \frac{323{,}000}{20^2} = 808 \text{ psi}$$

   $$\text{Bending stress} = \frac{6M_u}{c_1^3} = \frac{6 \times 371.6 \times 12{,}000}{20^3} = 3{,}344 \text{ psi (compression and tension)}$$

   Total maximum compressive stress $p_u = 808 + 3{,}344 = 4{,}152$ psi

   Allowable bearing stress is obtained by Eq. (10.21):

   $$\phi p_{nb} = \phi 0.85 f_c' = 0.65 \times 0.85 \times 5{,}000 = 2{,}763 \text{ psi} < p_u = 4{,}152 \text{ psi}$$

   Therefore, reinforcement must be provided across the interface to resist the additional compressive stress.

   Note that tensile stresses are also present because the net stress at the other face is equal to $p_u = 808 - 3{,}344 = -2{,}536$ psi. Interface reinforcement must be provided to resist the entire tensile force.

   There is no need at this stage to check the bearing strength of the footing, because interface reinforcement must be provided on the basis of the bearing strength of the column.

2. Determine the required interface reinforcement.

   For simplicity, provide 12 No. 7 dowel bars. This reinforcement matches the reinforcement in the column and ensures that both the compressive and tensile forces will be adequately transferred through the interface.

   Check the minimum area of reinforcement across the interface in accordance with ACI 15.8.2.1:

   $$A_{s,\,min} = 0.005 A_g = 0.005 \times 20^2 = 2.00 \text{ in}^2$$

   The provided interface reinforcement $A_s = 7.20$ in^2 > 2.00 in^2.

3. The dowel bars in tension are developed as follows:

   - *Development of the dowel bars into the footing.* Because some of the dowel bars are in tension, a standard 90-degree hook will be provided at the ends of all of the dowel bars. The tension development length for deformed bars terminating in a standard hook ℓ_{dh} is determined in accordance with ACI 12.5.2:

$$\ell_{dh} = \frac{0.02 \Psi_e f_y d_b}{\lambda \sqrt{f_c'}} = \frac{0.02 \times 1.0 \times 60{,}000 \times 0.875}{1.0 \sqrt{4{,}000}} = 16.6 \text{ in} > 8d_b = 7.0 \text{ in and 6 in}$$

Because the hooked portion of the dowel bars can be developed in tension, the minimum thickness of the footing for development of the dowel bars is determined by the following:

$$\ell_{dh} + 2(d_b)_f + 3 \text{ in} = 16.6 + (2 \times 0.75) + 3 = 21.1 \text{ in}$$

Because the provided footing thickness $h = 24$ in is greater than 21.1 in, the hooked dowel bars can be fully developed in tension into the footing.

- *Development of the dowel bars into the column.* The dowel bars must be lap spliced to the column bars. The design strength interaction diagram for the column is shown in Fig. 10.27. Also shown in the figure are the factored load combinations from Table 10.2.

It is evident that the third and fourth load combinations fall within the region of the interaction diagram where the tensile stress in the column bars is greater than 50% of the yield stress of the reinforcement. Thus, the column and dowel bars must be spliced with a Class B tension splice in accordance with ACI 12.15.1:

$$\text{Class B splice length} = 1.3\ell_d \geq 12 \text{ in}$$

The tension development length ℓ_d is determined by Eq. (6.28):

$$\ell_d = \left[\frac{3}{40} \frac{f_y}{\lambda \sqrt{f_c'}} \frac{\psi_t \psi_e \psi_s}{(c_b + K_{tr})/d_b} \right] d_b \geq 12 \text{ in}$$

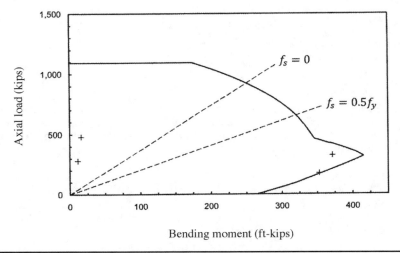

FIGURE 10.27 Design strength interaction diagram for the column given in Example 10.7.

where $\lambda = 1.0$ for normal-weight concrete
$\psi_t = 1.0$ for bars other than top bars
$\psi_e = 1.0$ for uncoated reinforcement
$\psi_s = 1.0$ for No. 7 bars

Assuming No. 3 ties in the column,

$$c_b = 1.5 + 0.375 + \frac{0.875}{2} = 2.3 \text{ in (governs)}$$

$$= \frac{20 - 2(1.5 + 0.375) - 0.875}{2 \times 3} = 2.6 \text{ in}$$

The required spacing of the No. 3 ties is the smallest of the following:

- 16 (longitudinal bar diameters) $= 16 \times 0.875 = 14.0$ in (governs)
- 48 (tie bar diameters) $= 48 \times 0.375 = 18.0$ in
- Least column dimension $= 20$ in

Use No. 3 ties spaced 14.0 in on center.

Clear space between the longitudinal bars $= \dfrac{20 - 2(1.5 + 0.375) - 0.875}{3} - 0.875 = 4.3 \text{ in}$

Because the clear space between the bars is less than 6 in, one of the longitudinal bars on each face does not need lateral support. However, for symmetry, provide a No. 3 crosstie on each of the interior bars (i.e., provide two crossties in each direction).

$$K_{tr} = \frac{40 A_{tr}}{sn} = \frac{40 \times 4 \times 0.11}{14 \times 4} = 0.3$$

$$\frac{c_b + K_{tr}}{d_b} = \frac{2.3 + 0.3}{0.875} = 3.0 > 2.5; \text{ use } 2.5$$

Therefore,

$$\ell_d = \left(\frac{3}{40} \frac{60,000}{1.0\sqrt{5,000}} \frac{1.0 \times 1.0 \times 1.0}{2.5} \right) \times 0.875 = 22.3 \text{ in} = 1.9 \text{ ft}$$

Class B splice length $= 1.3 \times 1.9 = 2.5$ ft

Use a splice length of 2 ft 6 in.

- *Horizontal transfer.* From Example 10.2, a service shear force of 23 kips is transferred between the column and the footing. Use Eq. (10.27) to compute the required area of shear-friction reinforcement, assuming that the surface between the column and the footing has not been intentionally roughened:

$$A_{vf} = \frac{V_u}{\phi f_y \mu} = \frac{1.6 \times 23}{0.75 \times 60 \times 0.6 \times 1.0} = 1.36 \text{ in}^2$$

This reinforcement is less than the area of the dowel bars that are provided. Check the upper shear limit using Eq. (10.28):

$$V_u = 1.6 \times 23 = 36.8 \text{ kips} < \phi 0.2 f'_c A_c = 0.75 \times 0.2 \times 4 \times 20^2 = 240.0 \text{ kips}$$

Thus, horizontal transfer requirements are satisfied using the dowel bars determined previously.

Reinforcement details for this column and footing are similar to those shown in Fig. 10.25.

Example 10.8 Determine the thickness and reinforcement for the footing given in Example 10.3. The 10-in-thick wall that is supported by the footing has normal-weight concrete with a compressive strength of 4,000 psi and is reinforced with two No. 4 vertical bars spaced at 12 in on center. Assume that the footing has normal-weight concrete with a compressive strength of 4,000 psi and Grade 60 reinforcement.

Solution
Step 1: Determine the width of the footing. The width of the footing has been determined in Example 10.3 on the basis of the permissible soil pressure and is equal to 2 ft 6 in.
Step 2: Determine the thickness of the footing. The thickness of the footing must be determined using the factored pressure at the base of the footing, considering both flexure and shear. The maximum factored pressure q_u is determined by dividing the factored load P_u by the area of the footing A_f:

$$q_u = \frac{P_u}{A_f} = \frac{[1.2 \times (1.5 + 2.0)] + (1.6 \times 1.0)}{2.5 \times 1} = 2.32 \text{ ksf}$$

- *Design for flexure.* The critical section for flexure for a footing supporting a concrete wall is at the face of the wall (see Fig. 10.18). Using Eq. (10.13), the required effective depth d of the footing for flexure, assuming minimum flexural reinforcement, is

$$d = 2.2c\sqrt{\frac{P_u}{A_f}} = 2.2 \times \left(\frac{2.5}{2} - \frac{10}{2 \times 12}\right) \times \sqrt{2.32} = 2.8 \text{ in}$$

The required flexural reinforcement is determined after the required effective depth is established considering both flexure and shear strength.

- *Design for shear.* The minimum effective depth d to satisfy one-way shear requirements is determined by Eq. (10.17):

$$d = \frac{q_u c}{q_u + \phi 2\sqrt{f_c'}} = \frac{(2,320/144) \times [(2.5/2) - 10/(2 \times 12)] \times 12}{(2,320/144) + (0.75 \times 2 \times \sqrt{4,000})} = 1.5 \text{ in}$$

Two-way shear requirements are not applicable to one-way footings.
It is clear that the minimum thickness requirements of ACI 15.7 govern in this situation.

Try a 12-in-thick footing ($d = 8$ in).
Step 3: Determine the required flexural reinforcement. The maximum factored moment M_u at the critical section is (see Fig. 10.18)

$$M_u = \frac{q_u c^2}{2} = \frac{2.32 \times [(2.5/2) - 10/(2 \times 12)]^2}{2} = 0.8 \text{ ft kips/ft}$$

The required area of reinforcement for this factored moment is equal to 0.02 in^2/ft.
The minimum amount of reinforcement is determined by ACI 10.5.4:

$$A_{s,min} = 0.0018bh = 0.0018 \times 12 \times 12 = 0.26 \text{ in}^2/\text{ft} > 0.02 \text{ in}^2/\text{ft}$$

Use $A_s = 0.26$ in^2/ft.

It can be determined that the maximum reinforcement requirement of ACI 10.3.5 is satisfied and that the section is tension-controlled.

Try No. 5 bars spaced 12 in on center (provided $A_s = 0.31$ in^2/ft).

Step 4: Distribute the flexural reinforcement. The provided spacing of 12 in is less than the maximum spacing of $3h = 36$ in or 18 in (governs).

Step 5: Develop the flexural reinforcement. Determine the development length ℓ_d by Eq. (6.28):

$$\ell_d = \left[\frac{3}{40} \frac{f_y}{\lambda \sqrt{f_c'}} \frac{\psi_t \psi_e \psi_s}{(c_b + K_{tr})/d_b} \right] d_b \geq 12 \text{ in}$$

where

$$\lambda = 1.0 \text{ for normal-weight concrete}$$
$$\psi_t = 1.0 \text{ for bars other than top bars}$$
$$\psi_e = 1.0 \text{ for uncoated reinforcement}$$
$$\psi_s = 0.8 \text{ for No. 5 bars}$$
$$c_b = 3.0 + \frac{0.625}{2} = 3.3 \text{ in (governs)}$$
$$ = \frac{12}{2} = 6.0 \text{ in}$$
$$K_{tr} = 0$$
$$\frac{c_b + K_{tr}}{d_b} = \frac{3.3 + 0}{0.625} = 5.3 > 2.5; \text{ use } 2.5$$

Therefore,

$$\ell_d = \left(\frac{3}{40} \frac{60{,}000}{1.0\sqrt{4{,}000}} \frac{1.0 \times 1.0 \times 0.8}{2.5} \right) \times 0.625 = 14.2 \text{ in}$$

The available development length is determined by Eq. (10.12):

$$\frac{L - c_1}{2} - 3 = \frac{(2.5 \times 12) - 10}{2} - 3 = 7.0 \text{ in}$$

Because this available length is less than the required development length determined by Eq. (6.28), the No. 5 bars cannot be fully developed.

Try a footing that is 3 ft 6 in wide, reinforced with No. 4 bars spaced at 9 in on center (provided $A_s = 0.27$ in^2/ft).

$$\ell_d = \left(\frac{3}{40} \frac{60{,}000}{1.0\sqrt{4{,}000}} \frac{1.0 \times 1.0 \times 0.8}{2.5} \right) \times 0.5 = 11.4 \text{ in} < 12.0 \text{ in; use } 12.0 \text{ in}$$

$$\frac{L - c_1}{2} - 3 = \frac{(3.5 \times 12) - 10}{2} - 3 = 13.0 \text{ in} > \ell_d = 12.0 \text{ in}$$

Use 3-ft-long No. 4 bars at 9 in.

The required temperature and shrinkage reinforcement is determined in accordance with ACI 7.12.2.1 and is placed perpendicular to the main flexural reinforcement:

$$A_s = 0.0018bh = 0.0018 \times 3.5 \times 12 \times 12 = 0.91 \text{ in}^2$$

Provide four No. 5 bars perpendicular to the main reinforcement for temperature and shrinkage.

Step 6: Check force transfer at the base of the wall.

1. Check bearing stress on the concrete wall and footing.

   The bearing strength of the wall is determined by Eq. (10.21):

   $$\phi P_{nb} = \phi 0.85 f_c' A_1 = 0.65 \times 0.85 \times 4 \times 10 \times 12 = 265 \text{ kips} > P_u = 6 \text{ kips}$$

   The bearing strength of the footing is determined by Eq. (10.22):

   $$\phi P_{nb} = \phi 0.85 f_c' A_1 \sqrt{A_2/A_1} \leq 2\phi 0.85 f_c' A_1$$

   Using Fig. 10.24, the area A_2 is determined as follows:

   Thickness of footing $h = 12$ in
   Horizontal projection for a 1:2 slope $= 2 \times 12 = 24$ in
   Projected length $b = 24 + 10 + 24 = 58$ in > 42 in
   $A_2 = 12 \times 12 = 144$ in^2

   Thus,

   $$\sqrt{\frac{A_2}{A_1}} = \sqrt{\frac{144}{10 \times 12}} = 1.1 < 2.0$$

   $$\phi P_{nb} = \phi 0.85 f_c' A_1 \sqrt{A_2/A_1}$$
   $$= 0.65 \times 0.85 \times 4 \times 10 \times 12 \times 1.1 = 292 \text{ kips} > P_u = 6 \text{ kips}$$

2. Determine the required interface reinforcement.

   Because the design-bearing strength is adequate for the wall and footing, provide the minimum area of reinforcement across the interface in accordance with ACI 15.8.2.2, which is equal to the minimum vertical reinforcement given in ACI 14.3.2:

   $$A_{s,min} = 0.0012 A_g = 0.0012 \times 10 \times 12 = 0.14 \text{ in}^2/\text{ft}$$

   Try two No. 4 dowel bars spaced at 12 in on center. This matches the vertical reinforcement in the wall and provides 0.40 in^2/ft interface reinforcement. The size of the dowel bars will be confirmed after the requirements for development have been checked.

3. The dowel bars in compression are developed as follows:
   - *Development of the dowel bars into the footing.* The dowel bars must be extended into the footing a compression development length ℓ_{dc}, which is determined by Eq. (10.24):

   $$\ell_{dc} = \begin{cases} (0.02 f_y/\lambda\sqrt{f_c'})d_b = [(0.02 \times 60{,}000)/(1.0\sqrt{4{,}000})] \times 0.50 = 9.5 \text{ in (governs)} \\ (0.0003 f_y)d_b = 0.0003 \times 60{,}000 \times 0.50 = 9.0 \text{ in} \end{cases}$$

   The minimum required thickness of the footing for development of the dowel bars is

   $$\ell_{dc} + r + (d_b)_{dowel} + 2(d_b)_f + 3 \text{ in} = 9.5 + (6 \times 0.5) + 0.5 + (2 \times 0.5) + 3 = 17.0 \text{ in}$$

Because the provided footing thickness $h = 12$ in is less than 17.0 in, the hooked dowel bars cannot be fully developed in the footing.

In lieu of increasing the footing depth to 17 in, change the size of the dowel bars. Providing No. 3 dowel bars satisfies the minimum interface reinforcement requirements and results in a minimum footing thickness of 13.2 in for development.

Use a 14-in-thick footing with two No. 3 dowel bars spaced at 12 in on center.

The flexural and temperature reinforcement provided for the 12-in-thick footing is also adequate for the 14-in-thick footing.

- *Development of the dowel bars into the wall.* The dowel bars must be lap spliced to the wall bars. Because the dowel bars are smaller in diameter than the wall bars, the compression lap splice length must be equal to or greater than the larger of the following:

(a) The development length in compression ℓ_{dc} of the larger bar, which is determined in accordance with ACI 12.3.

The development length ℓ_{dc} in compression of the No. 4 wall bars is

$$\ell_{dc} = \begin{cases} (0.02 f_y/\lambda\sqrt{f'_c})d_b = [(0.02 \times 60{,}000)/(1.0\sqrt{4{,}000})] \times 0.5 = 9.5 \text{ in (governs)} \\ (0.0003 f_y)d_b = 0.0003 \times 60{,}000 \times 0.5 = 9.0 \text{ in} \end{cases}$$

This length can be reduced to account for excess reinforcement in accordance with ACI 12.3.3(a); no reduction is taken in this example.

(b) The compression lap splice length of the smaller bar.

The compression lap splice length of the No. 3 bars is determined in accordance with ACI 12.16.1:

$$\text{Compression lap splice length} = \begin{cases} (0.0005 f_y)d_b = 30 d_b = 30 \times 0.375 = 11.3 \text{ in} \\ 12 \text{ in (governs)} \end{cases}$$

Therefore, provide a lap splice length equal to 1 ft 0 in.

No horizontal forces are transferred from the wall to the footing, so horizontal force transfer is not investigated.

Reinforcement details for this wall footing are given in Fig. 10.28.

Example 10.9 Determine the thickness and reinforcement for the footing given in Example 10.4. The 18-in square column has normal-weight concrete with a compressive strength of 5,000 psi and is reinforced with eight No. 8 bars. The 22-in square column also has normal-weight concrete with a compressive strength of 5,000 psi and is reinforced with eight No. 10 bars. Assume that the footing has normal-weight concrete with a compressive strength of 4,000 psi and Grade 60 reinforcement.

Solution

Step 1: Determine the area of the footing. The area of the footing has been determined in Example 10.4 on the basis of the permissible soil pressure. The base dimensions are $L = 29$ ft 0 in and $B = 13$ ft 6 in.

Step 2: Determine the thickness of the footing. The thickness of the footing must be determined using the factored pressure at the base of the footing, considering both flexure and shear.

The factored uniformly distributed load along the length of the footing is obtained by dividing the total factored axial load by the length of the footing L:

$$q_u = \frac{P_u}{L} = \frac{[1.2 \times (300 + 500)] + [(1.6 \times (115 + 225)]}{29.0} = \frac{1{,}504}{29.0} = 51.86 \text{ kips/ft}$$

Shear and moment diagrams for the footing are shown in Fig. 10.29.

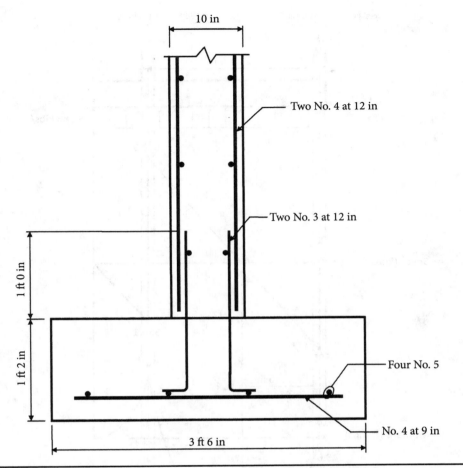

10 in

Two No. 4 at 12 in

Two No. 3 at 12 in

Four No. 5

No. 4 at 9 in

1 ft 0 in

1 ft 2 in

3 ft 6 in

FIGURE 10.28   Reinforcement details for the wall footing given in Example 10.8.

1. *One-way shear*: The effective depth d will be determined on the basis of one-way shear requirements. Two-way shear requirements will be checked later on the basis of this d.

   The maximum factored shear force occurs at the interior column (see Fig. 10.29). The factored shear force at a distance d from the face of the column is

   $$V_u = 609.9 - 51.86 \left(\frac{22}{2 \times 12} + d \right) = 562.4 - 51.86d$$

   The design one-way shear strength is given in Eq. (10.16):

   $$\phi V_c = \phi 2 \lambda \sqrt{f_c'} B d = 0.75 \times 2 \times 1.0 \sqrt{4{,}000} \times 13.5 \times 12 \times d / 1{,}000 = 15.37d \text{ kips}$$

   Set the required strength equal to the design strength and solve for d:

   $$562.4 - (51.86/12)d = 15.37d$$
   $$d = 28.6 \text{ in}$$

   Thus, an effective depth of 29 in is adequate for one-way shear.

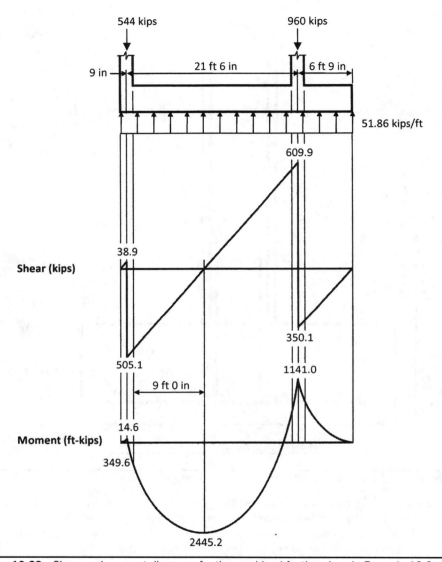

FIGURE 10.29   Shear and moment diagrams for the combined footing given in Example 10.9.

2. *Two-way shear*: Preliminary calculations indicate that an effective depth of 32 in is required for two-way shear. Check two-way shear requirements at both columns using $d = 32$ in.

  • *Edge column*: The edge column has a three-sided critical section, which is located at a distance $d/2$ from the face of the column in both directions.

    The maximum factored shear force at the critical section is equal to the factored column load minus the factored soil pressure in the area bounded by the critical section:

$$V_u = 544 - \frac{1,504}{29 \times 13.5 \times 144}\left[(18+32) \times \left(18 + \frac{32}{2}\right)\right] = 544.0 - 45.4 = 498.6 \text{ kips}$$

Unlike two-way slabs, the unbalanced moment at the face of the column and that at the centroid of the critical section in a combined footing typically vary significantly because of the relatively large critical area and the variation in moment along the span. Therefore, determine the unbalanced moment at the centroid of the critical section of the edge column because this moment will be greater than that at the face of the column.

A free-body diagram of the edge column is shown in Fig. 10.30. Figure 7.55 is used to determine the location of the centroid of the critical section from the edge of the column:

FIGURE 10.30
Free-body diagram of critical section at the edge column given in Example 10.9.

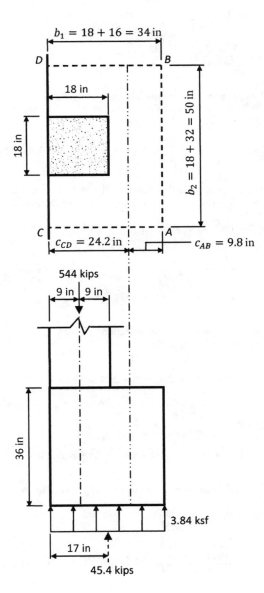

$$c_{CD} = \left(\frac{f_2}{f_2 + f_3} \right) \left(c_1 + \frac{d}{2} \right)$$

$$f_2 = \frac{[(c_1/d) + 1/2]^2 \, [(c_1/d)(1 + 2c_2/c_1) + 5/2] + [(c_1/d)(1 + c_2/2c_1) + 1]}{6\,[(c_1/d) + 1/2]}$$

$$= \frac{[(18/32) + 1/2]^2 \, [(18/32)(1 + 2) + 5/2] + [(18/32)(1 + 1/2) + 1]}{6\,[(18/32) + 1/2]} = 1.03$$

$$f_3 = \frac{[(c_1/d) + 1/2]^2 \, [(c_1/d)(1 + 2c_2/c_1) + 5/2] + [(c_1/d)(1 + c_2/2c_1) + 1]}{6\,[(c_1/d)(1 + c_2/c_1) + 3/2]}$$

$$= \frac{[(18/32) + 1/2]^2 \, [(18/32)(1 + 2) + 5/2] + [(18/32)(1 + 1/2) + 1]}{6\,[(18/32)(1 + 1) + 1/2]} = 0.42$$

$$c_{CD} = \left(\frac{1.03}{1.03 + 0.42} \right) \left(18 + \frac{32}{2} \right) = 24.2 \text{ in}$$

The unbalanced moment is obtained by summing the moments due to the load from the column and the load from the soil about the centroid of the critical section (see Fig. 10.30):

$$M_u = \left(544 \times \frac{24.2 - 9}{12} \right) - \left(45.4 \times \frac{24.2 - 17}{12} \right) = 689.1 - 27.2 = 661.9 \text{ ft kips}$$

For comparison purposes, the moment at the face of the edge column is equal to 349.6 ft kips, which is slightly greater than one-half of the moment at the centroid of the critical section.

Determine the shear factor γ_v by Eq. (7.38):

$$\gamma_f = \frac{1}{1 + (2/3)\sqrt{b_1/b_2}} = \frac{1}{1 + (2/3)\sqrt{(18 + 16)/(18 + 32)}} = 0.65$$

$$\gamma_v = 1 - \gamma_f = 1 - 0.65 = 0.35$$

The section properties of the critical section are determined using Fig. 7.55 for an edge column bending perpendicular to the edge:

$$c_{AB} = b_1 - c_{CD} = 34 - 24.2 = 9.8 \text{ in}$$

$$f_1 = 2 + \frac{c_1}{d} \left(2 + \frac{c_2}{c_1} \right) = 2 + \frac{18}{32}(2 + 1) = 3.69$$

$$A_c = f_1 d^2 = 3.69 \times 32^2 = 3{,}779 \text{ in}^2$$

$$J_c/c_{AB} = 2 f_2 d^3 = 2 \times 1.03 \times 32^3 = 67{,}502 \text{ in}^3$$

The total factored shear stress is determined by Eq. (7.39):

$$v_{u(AB)} = \frac{V_u}{A_c} + \frac{\gamma_v M_u c_{AB}}{J_c}$$

$$= \frac{498{,}600}{3{,}779} + \frac{0.35 \times 661.9 \times 12{,}000}{67{,}502} = 131.9 + 41.2 = 173.1 \text{ psi}$$

The allowable stress for a square column is obtained by Eq. (10.19):

$$\phi v_c = \frac{\phi V_c}{b_o d} = \phi 4\lambda\sqrt{f_c'} = 0.75 \times 4 \times 1.0\sqrt{4{,}000} = 189.7 \text{ psi} > 173.1 \text{ psi}$$

- *Interior column*: The interior column has a four-sided critical section (see Fig. 10.23), which is located a distance $d/2$ from the face of the column in both directions.

  The maximum factored shear force at the critical section is equal to the factored column load minus the factored soil pressure in the area bounded by the critical section:

$$V_u = 960 - \frac{1{,}504}{29 \times 13.5 \times 144}\,[(22+32) \times (22+32)] = 882.2 \text{ kips}$$

The load from the column and that from the soil act through the centroid of the critical section; thus, no unbalanced moment occurs at this location.

The section properties of the critical section are determined using Fig. 7.53 for an interior column:

$$c_{AB} = c_{CD} = b_1/2 = (22+32)/2 = 27.0 \text{ in}$$

$$f_1 = 2\left[\left(1+\frac{c_2}{c_1}\right)\left(\frac{c_1}{d}\right)+2\right] = 2\left[(1+1)\left(\frac{22}{32}\right)+2\right] = 6.75$$

$$A_c = f_1 d^2 = 6.75 \times 32^2 = 6{,}912 \text{ in}^2$$

The total factored shear stress is

$$v_u = \frac{V_u}{A_c} = \frac{882{,}200}{6{,}912} = 127.6 \text{ psi} < \phi v_c = 189.7 \text{ psi}$$

Therefore, a 32-in effective depth is adequate for one- and two-way shear.
Use a 36-in-thick footing ($d = 32$ in).
Step 3: Determine the required flexural reinforcement.

- *Moment in the longitudinal direction near the midspan*: The maximum moment near the midspan is equal to 2,445.2 ft kips (see Fig. 10.29). The flexural reinforcement needed to resist this moment must be placed at the top of the footing section; it is analogous to positive reinforcement.

The flowchart shown in Fig. 6.4 is used to determine A_s. It is modified to account for the differences applicable to footing design.

Step 3A: Assume tension-controlled section. Footings should be designed as tension-controlled sections whenever possible. Thus, assume that the strength reduction factor $\phi = 0.9$.

Step 3B: Determine the nominal strength coefficient of resistance R_n. For a rectangular section, R_n is determined by Eq. (6.5), which is a function of the factored bending moment M_u:

$$R_n = \frac{M_u}{\phi b_w d^2} = \frac{2{,}445.2 \times 12{,}000}{0.9 \times (13.5 \times 12) \times 32^2} = 196.5 \text{ psi}$$

Step 3C: Determine the required reinforcement ratio ρ. The reinforcement ratio ρ is determined by Eq. (6.7):

$$\rho = \frac{0.85 f_c'}{f_y}\left[1 - \sqrt{1 - \frac{2R_n}{0.85 f_c'}}\right] = \frac{0.85 \times 4}{60}\left[1 - \sqrt{1 - \frac{2 \times 196.5}{0.85 \times 4{,}000}}\right] = 0.0034$$

Step 3D: Determine the required area of tension reinforcement A_s.

$$A_s = \rho b d = 0.0034 \times (13.5 \times 12) \times 32 = 17.6 \text{ in}^2$$

Step 3E: Determine the minimum required area of reinforcement $A_{s,min}$. The minimum amount of reinforcement is determined by ACI 10.5.4:

$$A_{s,min} = 0.0018bh = 0.0018 \times (13.5 \times 12) \times 36 = 10.5 \text{ in}^2 < 17.6 \text{ in}^2$$

Use $A_s = 17.6 \text{ in}^2$.

Step 3F: Determine the depth of the equivalent rectangular stress block a.

$$a = \frac{A_s f_y}{0.85 f_c' b} = \frac{17.6 \times 60,000}{0.85 \times 4,000 \times (13.5 \times 12)} = 1.9 \text{ in}$$

Step 3G: Determine β_1. According to ACI 10.2.73, $\beta_1 = 0.85$ for $f_c' = 4,000$ psi (see Section 5.2).

Step 3H: Determine the neutral axis depth c.

$$c = \frac{a}{\beta_1} = \frac{1.9}{0.85} = 2.2 \text{ in}$$

Step 3I: Determine ε_t.

$$\varepsilon_t = 0.0030 \left(\frac{d_t}{c} - 1 \right) = 0.0030 \left(\frac{32}{2.2} - 1 \right) = 0.0406 > 0.0040$$

Because $\varepsilon_t > 0.0040$, the maximum reinforcement requirement of ACI 10.3.5 is satisfied. Also, the section is tension-controlled because $\varepsilon_t > 0.0050$ (ACI 10.3.4), and the initial assumption that the section is tension-controlled is correct.

Step 3J: Choose the size and spacing of the reinforcing bars. Try 18 No. 9 top bars (provided $A_s = 18.0 \text{ in}^2$).

- *Moment in the longitudinal direction at the interior column*: The maximum moment at the interior column is equal to 1,141.0 ft kips (see Fig. 10.29). Because it is anticipated that minimum flexural reinforcement will govern at this critical section, determine the required flexural reinforcement on the basis of this moment rather than the moment at the face of the column. The flexural reinforcement needed to resist this moment must be placed at the bottom of the footing section; it is analogous to negative reinforcement.

  The flowchart shown in Fig. 6.4 is used to determine A_s. Following the steps similar to those shown earlier for the moment near the midspan, the required reinforcement at the interior column is $A_s = 7.8 \text{ in}^2$, which is less than $A_{s,min} = 10.5 \text{ in}^2$.

  Try 11 No. 9 bars (provided $A_s = 11.0 \text{ in}^2$).

- *Moment in the longitudinal direction at the face of the edge column*: The maximum moment at the face of the edge column is equal to 349.6 ft kips. Minimum reinforcement requirements govern at this critical section also.

  Try 11 No. 9 bars (provided $A_s = 11.0 \text{ in}^2$).

  Check that the provided flexural reinforcement at the edge column is adequate to satisfy the moment transfer requirements of ACI 13.5.3.

  The total unbalanced moment at this slab–column connection is equal to 661.9 ft kips, which was determined in Step 2.

  A fraction of this moment $\gamma_f M_u$ must be transferred over an effective width equal to $c_2 + 3h = 18 + (3 \times 36) = 126$ in.

It was determined in step 2 that $\gamma_f = 0.65$. For edge columns bending perpendicular to the edge, the value of γ_f may be increased to 1.0 provided that $V_u \leq 0.75\phi V_c$ [ACI 13.5.3.3(a)]. No adjustment to γ_f is made in this example.

Unbalanced moment transferred by flexure = $\gamma_f M_u = 0.65 \times 661.9 = 430.2$ ft kips. The required area of steel to resist this moment in the 126-in-wide strip is $A_s = 2.82$ in², which is equivalent to three No. 9 bars.

Check minimum reinforcement requirements:

$$A_{s,min} = 0.0018bh = 0.0018 \times 126 \times 36 = 8.2 \text{ in}^2 > A_s$$

Thus, nine No. 9 bars are required within the 126-in-wide strip.

Uniformly spacing 12 No. 9 bars at 14 in within the 162-in-wide footing provides a sufficient amount of reinforcement within the 126-in-wide strip centered on the column for moment transfer; that is, nine No. 9 bars are provided within the 126-in-wide strip.

- *Moment in the transverse direction at the interior column*: The bending moment in the transverse direction is determined by assuming that the factored load from the column is distributed over a width on each side of the column. The actual width is not important at this stage because the moments are independent of it.

The factored distributed load at the interior column in the transverse direction is (see Fig. 10.31)

$$q_u = \frac{960}{13.5} = 71.1 \text{ kips/ft}$$

Thus, the factored moment at the critical section is

$$M_u = \frac{71.1 \times 5.83^2}{2} = 1{,}208.3 \text{ ft kips}$$

Assuming that the effective width of the transverse member is equal to the width of the column plus the effective depth d (i.e., the width extends $d/2$ on each side of the column), the required area of flexural reinforcement is $A_s = 0.0051 \times 54 \times 32 = 8.8$ in².

FIGURE 10.31
Bending moment in the transverse direction at the interior column of Example 10.9.

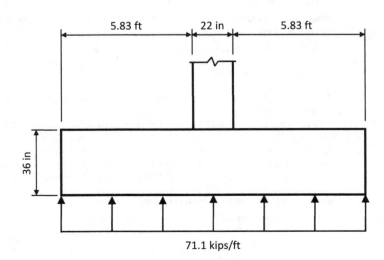

5.83 ft   22 in   5.83 ft

36 in

71.1 kips/ft

Check minimum reinforcement requirements:

$$A_{s,min} = 0.0018bh = 0.0018 \times 54 \times 36 = 3.5 \text{ in}^2 < A_s$$

Try nine No. 9 (provided $A_s = 9.0 \text{ in}^2$) bottom bars.

- *Moment in the transverse direction at the edge column*: Similar calculations can be performed for the required transverse reinforcement at the edge column. At this location, the assumed effective width of the transverse members is $18 + (32/2) = 34$ in.

$$M_u = \frac{(544/13.5) \times 6.0^2}{2} = 725.3 \text{ ft kips}$$

$$A_s = 0.0048 \times 34 \times 32 = 5.2 \text{ in}^2 > A_{s,min} = 2.2 \text{ in}^2$$

Try six No. 9 bars (provided $A_s = 6.0 \text{ in}^2$) bars.

Step 4: Distribute the flexural reinforcement.

- *Reinforcement in the longitudinal direction near the midspan*: Providing a uniform spacing of 9 in is less than the maximum spacing of $3h = 108$ in or 18 in (governs).
- *Reinforcement in the longitudinal direction at the interior and edge columns*: Providing a uniform spacing of 14 in is less than the maximum spacing of $3h = 108$ in or 18 in (governs).
- *Reinforcement in the transverse direction at the interior and edge columns*: Uniformly distributing the reinforcement in the respective widths satisfies maximum spacing requirements.

Step 5: Develop flexural reinforcement.

- *Reinforcement in the longitudinal direction near the midspan*: The 18 No. 9 top bars must be developed on either side of the critical section, which is located 9 ft 0 in from the face of the edge column (see Fig. 10.29).

  It is evident from the moment diagram shown in Fig. 10.29 that the points of inflection occur within and very close to the edge and interior columns, respectively. Consequently, the top bars will be extended over the entire footing length.

  Use 18 No. 9 top bars at 9 in, which are 28 ft long.

- *Reinforcement in the longitudinal direction at the columns*: Bottom bars are provided over the entire width of the footing for simpler detailing.

  Use 12 No. 9 bottom bars at 14 in, which are 28 ft long.

- *Reinforcement in the transverse direction*: The transverse reinforcement is developed by providing hooks at both ends of the bars.

  Use nine No. 9 bottom bars 13 ft long, uniformly distributed within a 4-ft 6-in width below the interior column.

  Use six No. 9 bottom bars 13 ft long, uniformly distributed within a 3-ft 0-in width below the edge column.

  Use No. 9 bottom bars spaced at 12 in on center in the remainder of the footing.

  Step 6: Check force transfer at the base of the columns. Vertical force transfer is illustrated for the edge column. Similar calculations can be performed for the interior column.

1. Check bearing stress on the concrete column and footing.

   The bearing strength of the column is determined by Eq. (10.21):

$$\phi P_{nb} = \phi 0.85 f_c' A_1 = 0.65 \times 0.85 \times 5 \times 18^2 = 895 \text{ kips} > P_u = 544 \text{ kips}$$

The bearing strength of the footing is determined by Eq. (10.22). Because the supporting area is not on all sides, assume the following bearing strength:

$$\phi P_{nb} = \phi 0.85 f'_c A_1 = 0.65 \times 0.85 \times 4 \times 18^2 = 716 \text{ kips} > P_u = 544 \text{ kips}$$

2. Determine the required interface reinforcement.

Provide the minimum area of reinforcement across the interface in accordance with ACI 15.8.2.1:

$$A_{s,min} = 0.005 A_g = 0.005 \times 18^2 = 1.62 \text{ in}^2$$

Provide four No. 6 dowel bars ($A_s = 1.76 \text{ in}^2$).

3. The dowel bars in compression are developed as follows:

   • *Development of the dowel bars into the footing*: The dowel bars must be extended into the footing a compression development length ℓ_{dc}, which is determined by Eq. (10.24):

$$\ell_{dc} = \begin{cases} (0.02 f_y / \lambda \sqrt{f'_c}) d_b = [(0.02 \times 60{,}000)/(1.0 \sqrt{4{,}000})] \times 0.75 = 14.2 \text{ in (governs)} \\ (0.0003 f_y) d_b = 0.0003 \times 60{,}000 \times 0.75 = 13.5 \text{ in} \end{cases}$$

The minimum required thickness of the footing for development of the dowel bars is determined by Eq. (10.25):

$$\ell_{dc} + r + (d_b)_{dowel} + 2(d_b)_f + 3 \text{ in} = 14.2 + (6 \times 0.75) + 0.75 + (2 \times 1.128) + 3 = 24.7 \text{ in}$$

Because the provided footing thickness $h = 36$ in is greater than 24.7 in, the hooked dowel bars can be fully developed in the footing.

   • *Development of the dowel bars into the column*: The dowel bars must be lap spliced to the column bars. Because the dowel bars are smaller in diameter than the column bars, the compression lap splice length must be equal to or greater than the larger of the following:

   (a) The development length in compression ℓ_{dc} of the larger bar, which is determined in accordance with ACI 12.3.

   The development length ℓ_{dc} in compression of the No. 8 column bars is

$$\ell_{dc} = \begin{cases} (0.02 f_y / \lambda \sqrt{f'_c}) d_b = [(0.02 \times 60{,}000)/(1.0 \sqrt{5{,}000})] \times 1.0 = 17.0 \text{ in} \\ (0.0003 f_y) d_b = 0.0003 \times 60{,}000 \times 1.0 = 18.0 \text{ in (governs)} \end{cases}$$

This length can be reduced to account for excess reinforcement in accordance with ACI 12.3.3(a); no reduction is taken in this example.

   (b) The compression lap splice length of the smaller bar.

   The compression lap splice length of the No. 6 bars is determined in accordance with ACI 12.16.1:

$$\text{Compression lap splice length} = \begin{cases} (0.0005 f_y) d_b = 30 d_b = 30 \times 0.75 = 22.5 \text{ in (governs)} \\ 12 \text{ in} \end{cases}$$

Therefore, provide a lap splice length equal to 2 ft 0 in.

No horizontal forces are transferred from the column to the footing, so horizontal force transfer is not investigated.

Reinforcement details for this column and footing are shown in Fig. 10.32.

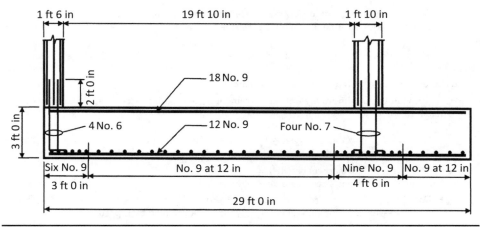

Figure 10.32   Reinforcement details for the combined footing given in Example 10.9.

10.3 Mat Foundations

10.3.1 Overview

Mat foundations are used to support all or a portion of the vertical elements of a building (see Fig. 10.4). They are commonly specified where erratic or relatively weak soil strata are encountered or where a large number of spread footings would be needed to support the loads and bending moments from the structure above.

Mats are commonly designed and analyzed as either rigid bodies or flexible plates supported by an elastic foundation, which is the soil. Each of these methods is discussed in the sections that follow.

The base dimensions of the mat are proportioned so that the maximum pressure due to service loads is equal to or less than the permissible soil pressure. The thickness is typically based on shear strength requirements. Both one- and two-way shear must be investigated at the critical sections around the vertical elements supported by the mat. Like footings, it is common practice not to use shear reinforcement in mats for overall economy.

10.3.2 Analysis Methods

Rigid Mats

A wide variety of analysis methods can be used in the design of mat foundations. The most appropriate method typically depends on the rigidity of the mat. For purposes of analysis, a mat is considered to be rigid when all of the following criteria are met[1]:

- Variation in adjacent column loads is equal to or less than 20%.
- Columns are regularly spaced where the distance between adjacent columns does not differ by more than 20%.

- Column spacing is less than $1.75/\lambda$, where λ is the stiffness evaluation factor defined as follows:

$$\lambda = \left(\frac{K_s}{4E_cI}\right)^{0.25} \tag{10.29}$$

In this equation, K_s is the spring constant of the soil, which is related to the coefficient (or modulus) of the subgrade reaction k_s as follows:

$$K_s = k_sB' \tag{10.30}$$

The terms B' and I are the width and moment of inertia of the design strip, respectively (it will become evident shortly why design strips are used in the analysis of rigid mats). The value of k_s is determined by the geotechnical engineer and is typically provided in the geotechnical report that is prepared for the site. The stiffness evaluation factor given in Eq. (10.29) is used in the solution of the basic differential equation for a beam on an elastic foundation.

Once it has been established that a mat is rigid, the soil pressures beneath the mat can be determined using the service loads from the structure above. Because the mat is rigid, the pressure diagram is linear and the magnitude of the pressure at any point can be determined by the following equation:

$$q = \sum P\left(\frac{1}{BL} \pm \frac{e_xx}{I_y} \pm \frac{e_yy}{I_x}\right) \tag{10.31}$$

The term $\sum P$ is the resultant of the axial loads from the columns. Eccentricities e_x and e_y are measured from the centroid of the mat to the location of this resultant force and include the effects from any moments transferred from the columns. The variables x and y locate the point where the soil pressure is computed, and I_y and I_x are the moments of inertia of the mat with respect to the y-axis and x-axis, respectively.

Because the plan dimensions B and L of the mat are usually dictated by the overall geometry of the building, it is common to initially assume values of B and L and then to calculate pressures at various locations by Eq. (10.31). The maximum pressure at the base of the mat must be equal to or less than permissible soil pressure q_p. Depending on q_p and other factors, B and L may need to be adjusted accordingly.

Like in the case of footings and flat-plate slabs, the depth of a mat is usually controlled by shear requirements. Factored one- and two-way shear stresses need to be checked at the critical locations around the interior, edge, and corner columns. It is common practice not to use shear reinforcement in mats; this results in a thicker and more rigid mat, which increases the reliability of this analysis method.

The maximum bending moments and corresponding flexural reinforcement are determined using factored loads. For purposes of determining these moments, rigid mats can be divided into design strips in both directions, similar to two-way slabs (see Fig. 10.33). Each design strip is assumed to act independently and is analyzed as a combined footing subjected to known bearing pressures and column loads.

The design strips do not actually act independently; there is some shear transfer between adjoining strips. Thus, vertical equilibrium may not be satisfied on any given

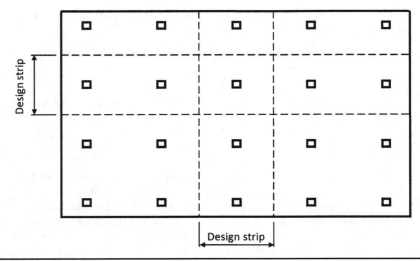

FIGURE 10.33 Design strips in a mat foundation.

design strip. It may be found that the resultant of the column loads and the centroid of the bearing pressure are not equal and do not act at the same point. Reference 6 presents a method of analysis that can be used so that equilibrium is satisfied in such cases.

It is important to note that the Direct Design Method of analysis is not permitted to be used in the design of mat foundations (ACI 15.10.2).

Nonrigid Mats

If a mat does not meet the rigidity requirements outlined earlier, it must be designed as a flexible plate. Approximate methods and computer analyses can be utilized.

An approximate flexible method was introduced in 1966 by ACI Committee 436.[7] In this method, the required mat depth is computed on the basis of shear requirements, and bending moments, shear forces, and deflections are determined using charts. Additional information on the details of this method can be found in Refs. 8 and 9.

A computer analysis is typically based on an approximation where the mat is divided into a number of discrete or finite elements. The finite difference method, the finite grid method, and the finite element method are three common methods of solution. Most commercial computer programs are based on the finite element method, which idealizes the mat as a mesh of rectangular and/or triangular elements that are connected at the nodes. The soil is modeled as a set of isolated springs. The details of these methods are beyond the scope of this book; more information can be found in Ref. 1 and other publications devoted to finite element analysis.

10.3.3 Design Procedure

The following design procedure can be used in the design of mat foundations. Included is the information presented in the previous section on analysis and additional information on design and detailing.

Step 1: Determine the preliminary plan dimensions of the mat. The plan dimensions of a mat are typically dictated by the geometry of the building and the layout of the vertical elements supported by the mat. Property lines and other factors may also influence plan dimensions. The assumed plan dimensions can be adjusted later as required.

Step 2: Determine the soil pressure distribution and refine the plan dimensions of the mat. Assuming that the mat is rigid, soil pressures at various locations beneath the mat can be determined by Eq. (10.31). The maximum pressure determined using the preliminary plan dimensions must be equal to or less than the permissible soil pressure. If this is not the case, the dimensions of the mat must be adjusted accordingly.

The third of the three conditions of a rigid mat is checked once a preliminary thickness is determined. It is assumed at this stage that the first two conditions related to column load and spacing have been satisfied.

A computer analysis of the mat gives the soil pressure contours based on the applied loads. The maximum pressure determined from this method must be equal to or less than the permissible soil pressure.

Step 3: Determine the thickness of the mat. Both one- and two-way shear must be investigated at the critical sections around the vertical elements supported by the mat in accordance with the requirements of ACI 11.11. Factored loads are used to compute the shear forces at the critical sections. Like footings, it is common practice not to use shear reinforcement in mats.

In cases where a rigid mat has been assumed, the third of the three conditions should be checked at this stage on the basis of the mat thickness and the width of the design strip, using Eqs. (10.29) and (10.30). The thickness of the mat may need to be increased or a computer analysis of the foundation system can be performed if this condition is not satisfied.

Step 4: Determine the factored moments. Maximum bending moments are determined at the critical sections in the mat, using factored loads. In the case of rigid mats, the mat can be divided into a series of design strips (see Fig. 10.33), and the bending moments are determined from statics. Because of the simplifying assumptions of this method, adjustments to the design strips may be needed to ensure that vertical equilibrium is satisfied. A computer analysis provides moment contours of the mat based on the factored loads.

Step 5: Check shear requirements. Once the factored bending moments have been determined, shear strength requirements need to be checked at the critical sections of the interior, edge, and corner columns. Like in two-way slabs, shear stress due to direct shear and the portion of the unbalanced moment transferred by shear is usually critical at the edge and corner columns.

Step 6: Determine the required flexural reinforcement. The required flexural reinforcement is determined using the general requirements of the strength design method. Like footings, mat foundations should be designed as tension-controlled sections. The provided area of reinforcement must be equal to or greater than the minimum reinforcement prescribed in ACI 10.5.4.

Step 7: Develop the flexural reinforcement. Flexural reinforcement must be developed at the critical sections in the mat. The requirements of ACI Chap. 12 are used to determine the required lengths and cutoff points of the reinforcing bars.

Step 8: Check transfer of forces at the base of the supported members. Transfer of forces between the supported elements and the mat foundation needs to be checked. It is important to ensure that both the vertical and lateral loads are adequately transferred across the interface by satisfying the appropriate Code requirements, which are the same as those for footings.

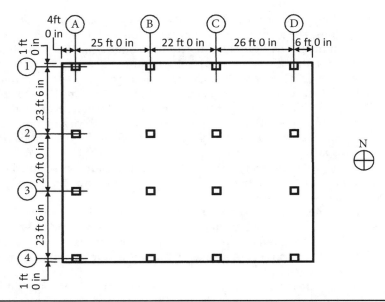

Figure 10.34   The mat foundation given in Example 10.10.

Example 10.10   Design the mat foundation shown in Fig. 10.34. Determine the required reinforcement in the end span of the interior design strip along line C. The interior columns are 24 × 24 in, and the edge columns are 20 × 20 in. The column loads are given in Table 10.3. Assume that the mat has normal-weight concrete with a compressive strength of 4,000 psi and Grade 60 reinforcement. Also assume that the soil has a permissible bearing capacity of 2,000 psf and a modulus of subgrade reaction of 150 kips/ft^3.

Solution   The analysis of the mat will proceed assuming that the mat is rigid, even though adjoining axial loads differ by more than 20% (see Table 10.3). Columns are regularly spaced, and the distance between adjacent columns does not differ by more than 20%.

Step 1: Determine the preliminary plan dimensions of mat. The plan dimensions of the mat shown in Fig. 10.34 are used and may be adjusted later in the analysis.

Column	Service Axial Load (kips)	Factored Axial Load (kips)
A1, A4	224	302
A2, A3	381	515
B1, B4	318	430
B2, B3	542	733
C1, C4	325	439
C2, C3	554	749
D1, D4	258	348
D2, D3	439	593

Table 10.3   Service and Factored Axial Loads on the Mat Foundation Given in Example 10.10

Step 2: Determine soil pressure distribution and refine the plan dimensions of the mat.
Equation (10.31) is used to determine the soil pressure on the basis of the service axial loads given
in Table 10.3.

The location of the resultant force in the east-west direction is determined by summing moments
about the west edge of the mat:

$$\bar{x} = \frac{2\{[(224 + 381) \times 4] + [(318 + 542) \times 29] + [(325 + 554) \times 51] + [(258 + 439) \times 77]\}}{2(224 + 381 + 318 + 542 + 325 + 554 + 258 + 439)}$$

$$= \frac{251{,}716}{6{,}082} = 41.4 \text{ ft}$$

Similarly, the location of the resultant force in the north-south direction is determined by summing moments about the north edge of the mat. Because of the symmetric load distribution,
$\bar{y} = 34.5$ ft.

The eccentricities e_x and e_y are measured from the geometric center of the mat:

$$e_x = 41.5 - 41.4 = 0.1 \text{ ft (west of center)}$$
$$e_y = 0$$

The pressure q at any point on the mat can be determined by Eq. (10.31):

$$q = \sum P \left(\frac{1}{BL} \pm \frac{e_x x}{I_y} \pm \frac{e_y y}{I_x} \right)$$

$$= \frac{6{,}082{,}000}{69 \times 83} \pm \frac{(6{,}082{,}000 \times 0.1)x}{(69 \times 83^3)/12} \pm 0 = 1{,}062 \pm 0.2x$$

The pressure is greatest along the west edge of the mat ($x = -41.5$ ft):

$$q = 1{,}062 + (0.2 \times 41.5) = 1{,}062 + 8 = 1{,}070 \text{ psf} < 2{,}000 \text{ psf}$$

It is evident from this calculation that the soil pressure is essentially uniform beneath the mat.
Use a 69-ft 0-in × 83-ft 0-in mat.

Step 3: Determine the thickness of the mat. The thickness of the mat is determined on the
basis of two-way shear strength requirements at the interior column C2 (or C3), based on factored
soil pressure (see Table 10.3):

$$\text{Total factored axial load } P_u = 8{,}218 \text{ kips}$$

As shown in step 2, the soil pressure is essentially uniform; thus,

$$q_u = \frac{8{,}218}{69 \times 83} = 1.44 \text{ ksf}$$

For column C2, the critical section for two-way shear occurs at a distance of $d/2$ from the face
of the column. The factored shear force at the critical section is

$$V_u = \frac{1.44 \times 1{,}000}{144}[(24 \times 21.75 \times 144) - (24 + d)^2] = 745{,}920 - 10d^2 - 480d$$

For an interior square column, the design shear strength is given by ACI Eq. (11-33):

$$\phi V_c = \phi 4\lambda \sqrt{f_c'} b_o d = 0.75 \times 4 \times 1.0\sqrt{4{,}000} \times [4 \times (24 + d)] \times d = 18{,}215d + 7{,}59d^2$$

Equating the required and design strengths results in the following equation that can be solved for d:

$$769d^2 + 18{,}695d - 745{,}920 = 0$$
$$d = 21.3 \text{ in}$$

Try $d = 22$ in.

Step 4: Determine the factored moments. Check the third condition for a rigid mat, assuming that the width of the design strip B' is 24 ft and that the thickness of the mat is 26 in:

$$I = \frac{24 \times (26/12)^3}{12} = 20.3 \text{ ft}^4$$

The subgrade spring constant is determined by Eq. (10.30):

$$K_s = k_s B' = 150 \times 24 = 3{,}600 \text{ ksf}$$

The modulus of elasticity of the concrete is determined in accordance with ACI 8.5.1:

$$E_c = 57{,}000\sqrt{f_c'} = 57{,}000 \times \sqrt{4{,}000}/1{,}000 = 3{,}605 \text{ ksi} = 519{,}120 \text{ ksf}$$

The stiffness evaluation factor is determined by Eq. (10.29):

$$\lambda = \left(\frac{K_s}{4E_c I}\right)^{0.25} = \left(\frac{3{,}600}{4 \times 519{,}120 \times 20.3}\right)^{0.25} = 0.096$$

$$\text{Maximum column spacing} = \frac{1.75}{\lambda} = \frac{1.75}{0.096} = 18.2 \text{ ft} < 23.6 \text{ ft}$$

In lieu of performing a computer analysis of the mat, increase the thickness of the mat.

A thickness of approximately 36 in results in a rigid mat based on the requirement of the third condition.

Check vertical equilibrium on the 24-ft-wide design strip:

$$\text{Soil reaction} = q_u B' B = 1.44 \times 24 \times 69 = 2{,}385 \text{ kips}$$
$$\text{Column loads} = 2(439 + 749) = 2{,}376 \text{ kips}$$

The soil pressure and the column loads will be averaged so that the strip is in equilibrium. Note that the resultant of the soil pressure and column loads acts at the center of the design strip, so a uniform soil distribution can be used.

$$\text{Average load in design strip} = \frac{2{,}385 + 2{,}376}{2} = 2{,}381 \text{ kips}$$

$$\text{Soil pressure for this average load} = 1.44 \times \frac{2{,}381}{2{,}385} = 1.438 \text{ ksf}$$

The column loads must be increased by the following factor: $2{,}381/2{,}376 = 1.002$. The column loads are multiplied by this factor so that vertical equilibrium is satisfied in this design strip.

A free-body diagram of the design strip and the shear and moment diagrams are shown in Fig. 10.35.

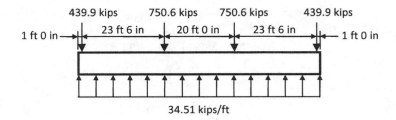

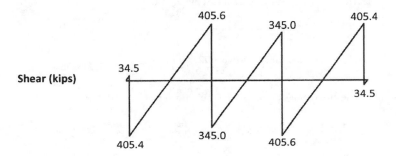

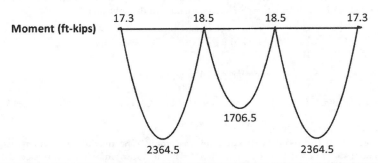

FIGURE 10.35 Shear and moment diagrams for the design strip given in Example 10.10.

Step 5: Check shear requirements.

1. *Two-way shear*: The edge column has a three-sided critical section, which is located at a distance $d/2$ from the face of the column in both directions.

   The maximum factored shear force at the critical section is equal to the factored column load minus the factored soil pressure in the area bounded by the critical section:

   $$V_u = 439.9 - \frac{1.438}{144}\left[(20+22) \times \left(20 + \frac{22}{2}\right)\right] = 439.9 - 13.0 = 426.9 \text{ kips}$$

   The unbalanced moment M_u is determined at the centroid of the critical section, similar to that shown in step 2 of Example 10.9; in this example, $M_u = 425.8$ ft kips.

The section properties of the critical section are determined using Fig. 7.55 for an edge column bending perpendicular to the edge:

$$c_{AB} = b_1 - c_{CD} = 31 - 21.8 = 9.2 \text{ in}$$

$$f_1 = 2 + \frac{c_1}{d}\left(2 + \frac{c_2}{c_1}\right) = 2 + \frac{20}{22}(2 + 1) = 4.73$$

$$A_c = f_1 d^2 = 4.73 \times 22^2 = 2{,}289 \text{ in}^2$$

$$J_c/c_{AB} = 2f_2 d^3 = 2 \times 1.51 \times 22^3 = 32{,}157 \text{ in}^3$$

The total factored shear stress is determined by Eq. (7.39):

$$
\begin{aligned}
v_{u(AB)} &= \frac{V_u}{A_c} + \frac{\gamma_v M_u c_{AB}}{J_c} \\
&= \frac{426{,}900}{2{,}289} + \frac{0.36 \times 425.8 \times 12{,}000}{32{,}157} = 186.5 + 57.2 = 243.7 \text{ psi}
\end{aligned}
$$

The allowable stress for a square column is obtained from Eq. (10.19):

$$\phi v_c = \frac{\phi V_c}{b_o d} = \phi 4\lambda\sqrt{f_c'} = 0.75 \times 4 \times 1.0\sqrt{4{,}000} = 189.7 \text{ psi} < 243.7 \text{ psi}$$

It can be determined that an effective depth of 28 in satisfies two-way shear requirements at the edge column and at an interior column.

2. *One-way shear*: The maximum factored shear force occurs at the interior column (see Fig. 10.35). The factored shear force at a distance of d from the face of the column is

$$V_u = 405.6 - 34.51\left(\frac{24}{2 \times 12} + \frac{28}{12}\right) = 290.6 \text{ kips}$$

The design one-way shear strength is given in Eq. (10.16):

$$\phi V_c = \phi 2\lambda\sqrt{f_c'}B'd = 0.75 \times 2 \times 1.0\sqrt{4{,}000} \times 24 \times 12 \times 28/1{,}000 = 765.0 \text{ kips} > 290.6 \text{ kips}$$

Use a 32-in-thick mat ($d = 28$ in).

Step 6: Determine the required flexural reinforcement. The maximum moment near the midspan is equal to 2,364.5 ft kips (see Fig. 10.35). The flexural reinforcement needed to resist this moment must be placed at the top of the mat; it is analogous to positive reinforcement.

The flowchart shown in Fig. 6.4 is used to determine A_s. It can be determined that the required area of flexural reinforcement is $A_s = 19.2 \text{ in}^2 > A_{s,min} = 16.6 \text{ in}^2$

The bar size and spacing must also be selected considering the moment transfer requirements of ACI 13.5.3.

Minimum flexural reinforcement must be provided at the bottom of the footing on the basis of the factored moments shown in Fig. 10.35.

Step 7: Develop the flexural reinforcement. The flexural reinforcement must be developed on either side of the critical sections for flexure in accordance with ACI Chap. 12.

Continuous bars are provided at the top of the mat. Bottom bars can be cut off where they are no longer required. In certain situations, it may be advantageous to provide continuous bars at both the top and the bottom of the mat in both directions.

Step 8: Check transfer of forces at the base of the supported members. Transfer of forces between the supported elements and the mat foundation needs to be checked. In this example, only vertical force transfer needs to be investigated.

Calculations for vertical force transfer have been illustrated in previous examples.

Comments
To complete the design of this mat, other strips need to be designed in both directions. It is usually advantageous to provide the same amount of flexural reinforcement wherever possible; this facilitates bar placement in the field.

10.4 Pile Caps

10.4.1 Overview

A pile cap is a reinforced concrete element that distributes the loads from the super-structure to the individual piles in a pile group below the supported member. These elements are sometimes referred to as footings on piles.

The plan dimensions of a pile cap depend on the number of piles that are needed to support the load. Pile spacing is generally a function of pile type and capacity; a common spacing is 3 ft on center. The number and arrangement of the piles is determined using unfactored loads.

The thickness of a pile cap depends primarily on shear. One- and two-way shear requirements must be satisfied at critical sections around both the supported member and the piles, using factored loads. Once a thickness has been established, the required flexural reinforcement is calculated using the general principles of the strength design method. Reinforcing bars are located at the bottom of the pile cap and must be fully developed in accordance with ACI Chap. 12.

Vertical and horizontal transfer between the supported member and the pile cap must be checked similar to footings. Vertical compression forces are transferred by bearing on the concrete and by reinforcement (if required), whereas vertical tension forces must be resisted entirely by reinforcement. Shear-friction requirements must be satisfied for horizontal transfer.

10.4.2 Design for Shear

Shear Requirements

General shear strength requirements for pile caps are given in ACI 15.5. For purposes of analysis, it is assumed that the reaction from any pile is a concentrated load that acts at the center of the pile (ACI 15.2.3).

Geometric properties dictate the shear requirements that need to be satisfied. According to ACI 15.5.3, the one- and two-way shear strength requirements for slabs and footings given in ACI 11.11 and the additional requirements of ACI 15.5.4 must be satisfied for pile caps where the distance x between the axis of any pile to the axis of the supported column is greater than two times the distance y between the top of the pile cap and the top of the pile (see Fig. 10.36). For pile caps that do not meet this condition, the pile cap is assumed to be "deep," and the requirements of either ACI Appendix A or ACI 11.11 and 15.5.4 must be satisfied.

Critical Sections—Supported Members

Similar to footings, critical sections for one- and two-way shear occur at d and $d/2$ from the face of the supported member, respectively (see Figs. 10.22 and 10.23). The magnitude of the factored shear force that needs to be considered at these critical sections

FIGURE 10.36
Geometric
properties of
a pile cap.

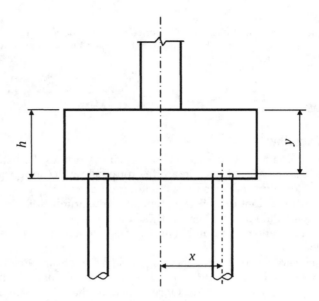

depends on the location of the center of a pile with respect to these sections (ACI 15.5.4). Consider the three piles illustrated in Fig. 10.37. The discussion that follows is applicable to the critical section for two-way shear; a similar discussion is applicable to the critical section for one-way shear.

The reaction from Pile A, which is assumed to act at the center of the pile, is located at a distance that is greater than 50% of the pile diameter d_{pile} inside the critical section for two-way shear; thus, the reaction from this pile does not produce shear force on that section (ACI 15.5.4.2). The center of Pile C is located more than $d_{pile}/2$ outside

FIGURE 10.37
Consideration of
shear forces on
pile cap.

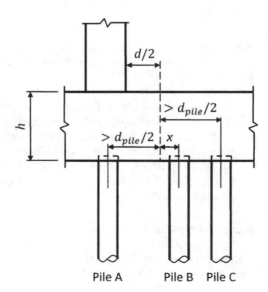

Pile A   Pile B   Pile C

of the critical section, so the full reaction from this pile must be considered in shear calculations (ACI 15.5.4.1). Because the center of Pile B is located outside of the critical section, all or part of the reaction will contribute to the total shear at that section. If $x \geq d_{pile}/2$, then the full reaction of the pile must be considered. However, if $x < d_{pile}/2$, then only a part of the reaction needs to be included; ACI 15.5.4.3 permits a reduced reaction to be determined by linear interpolation between the full value of the reaction at $d_{pile}/2$ outside of the section and zero at $d_{pile}/2$ inside the section.

For relatively deep pile caps, it is possible that all or most of the pile reactions can be excluded in shear calculations based on the requirements of ACI 15.5.4.2; in other words, the length of the critical section is large enough so that most or all of the piles fall inside the critical section by more than $d_{pile}/2$. In extreme cases, the critical section may fall outside of the pile cap. Reference 10 contains two special investigation methods that can be used to determine the required shear force in such cases.

Critical Sections—Piles

In addition to critical sections around supported members, shear requirements need to be investigated at critical sections around piles. Figure 10.38 illustrates four possible conditions that may need to be investigated.

- *Condition 1: Two-way shear at an interior pile.* Two-way shear requirements must be investigated at the critical section of an interior pile. The factored shear force at the critical section is equal to the factored axial load in the pile. The design shear strength is the smallest of the values obtained by ACI Eqs. (11-31) through (11-33). For the usual case of square or round piles, ACI Eq. (11-33) governs.

- *Condition 2: Two-way shear at interior piles with overlapping critical perimeters.* In cases where the shear perimeters of adjacent piles overlap, a modified shear

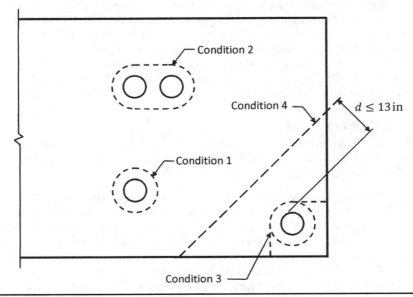

FIGURE 10.38 Critical sections around piles.

critical perimeter b_o must be used. The perimeter b_o should be taken as that portion of the smallest envelope of individual shear perimeter for the piles under consideration. ACI Fig. R15.5 illustrates the modified critical perimeter for two piles with overlapping critical perimeters.

- *Condition 3: Two-way shear at corner piles.* For piles located at the corners of pile caps, the shear around two critical sections needs to be investigated, as shown in Fig. 10.38. In particular, the smaller of the two perimeters b_o must be used in determining the design shear strength. This is similar to the investigation of two-way shear in two-way slabs where a cantilevered portion of slab is adjacent to an edge column (see Fig. 7.46).

- *Condition 4: One-way shear at corner piles.* One-way shear requirements need to be checked at a distance d, but no more than 13 in, from the face of a corner pile. The critical section occurs at a 45-degree angle with respect to the edges of the pile cap.

10.4.3 Design for Flexure

Once a thickness has been established on the basis of shear requirements, design of the pile cap for flexure follows the general requirements of the strength design method.

The maximum factored moment at the critical section for flexure, which is defined in ACI 15.4.2, is determined by multiplying the factored pile loads by the respective distances from the center of the pile to the critical section. Required flexural reinforcement is calculated on the basis of this moment. Because the thickness of a pile cap is typically established on the basis of shear requirements, the minimum flexural reinforcement requirements of ACI 10.5.4 usually govern.

Like isolated spread footings, flexural reinforcement is placed at the bottom of a pile cap. A minimum 3-in clear cover is typically provided from the top of the embedded piles to the reinforcing steel. Concrete piles are typically embedded a minimum distance of 4 in into the pile cap, whereas steel piles are usually embedded a minimum of 6 in.

For square pile caps, reinforcing bars are uniformly distributed in both orthogonal directions, whereas for rectangular pile caps, the spacing requirements of ACI 15.4.4.2 must be satisfied. The bars must be developed in accordance with the applicable provisions of ACI Chap. 12. Because of the relatively short length that is available for development of the bars in tension, hooks are frequently provided at both ends of the bars.

10.4.4 Design Procedure

The following design procedure can be used in the design of pile caps. Included is the information presented in the previous section on analysis and additional information on design and detailing.

Step 1: Determine the plan dimensions of the pile cap. The number of piles that are needed to support the load from the superstructure dictates the plan dimensions of a pile cap. Piles are typically spaced 3 ft on center, and the distance from the center of an edge pile to the edge of the pile cap is usually 1 ft 3 in. Reference 10 contains plan dimensions for a variety of pile cap layouts.

Step 2: Determine the thickness of the pile cap. Both one- and two-way shear must be investigated at the critical sections around the vertical elements supported by the pile cap in accordance with the requirements of ACI 11.11 and 15.5.4. Factored loads are used to compute the shear forces at the critical sections. Like footings, it is common practice not to use shear reinforcement in pile caps.

A preliminary thickness can be determined on the basis of two-way shear requirements at the critical section around the supported column, considering the requirements of ACI 15.5.4. One-way shear requirements are subsequently checked at the critical section of the column, and one- and two-way shear requirements are checked at the piles. Adjustments are made to the thickness, if required.

Step 3: Determine the factored moments. Maximum bending moments are determined at the critical sections in the pile cap, using factored loads. In particular, M_u is determined by multiplying the factored pile loads by the respective distances from the center of the pile to the critical section.

Step 4: Determine the required flexural reinforcement. The required flexural reinforcement is determined using the general requirements of the strength design method. Like footings, pile caps should be designed as tension-controlled sections. The provided area of reinforcement must be equal or greater than the minimum reinforcement prescribed in ACI 10.5.4.

The reinforcing bars are uniformly distributed in both orthogonal directions in square pile caps. For rectangular pile caps, the reinforcement in the short direction must be spaced in accordance with ACI 15.4.4.2.

Step 5: Develop the flexural reinforcement. The flexural reinforcement must be developed at the critical sections in the pile cap. The requirements of ACI Chap. 12 are used to determine the required lengths of the reinforcing bars. Hooks are frequently provided at both ends of the bars to achieve the required development.

Step 6: Check transfer of forces at the base of the supported members. Transfer of forces between the supported elements and the pile cap needs to be checked. It is important to ensure that both vertical and lateral loads are adequately transferred across the interface by satisfying the appropriate Code requirements, which are the same as those for footings.

Example 10.11 Design the pile cap illustrated in Fig. 10.39. The piles have a diameter of 12 in and a service load capacity of 50 tons each. Assume that the pile cap has normal-weight concrete with a compressive strength of 4,000 psi and Grade 60 reinforcement. Also assume that the piles are embedded 4 in into the pile cap. The axial loads on the column are due to dead and live loads and are equal to 425 and 250 kips, respectively.

Solution

Step 1: Determine the plan dimensions of the pile cap. The plan dimensions of the pile cap are given in Fig. 10.39. A 3-ft spacing is provided between the piles (including the piles closest to the column and the adjacent piles, which occur on the diagonal), and the edge distance is 1 ft 3 in.

Step 2: Determine the thickness of the pile cap. The thickness of the pile cap is initially determined on the basis of two-way shear requirements at the column. One-way shear requirements will be checked around the column, and one- and two-way shear requirements will be checked around the piles based on this thickness.

The total factored axial load on the column is

$$P_u = (1.2 \times 425) + (1.6 \times 250) = 910 \text{ kips}$$

Load per pile = $910/8 = 114$ kips

Assuming that the centers of the two piles closest to the column are located $d_{pile}/2 = 6$ in or more inside the perimeter of the critical section for two-way shear, their contribution to the total

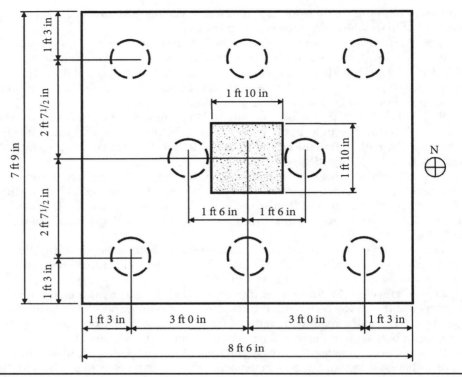

Figure 10.39   The pile cap given in Example 10.11.

shear can be neglected (ACI 15.5.4.2). This assumption will be checked later. Therefore, the factored shear force at the critical section is $V_u = 6 \times 114 = 684$ kips.

For an interior square column, the design shear strength is given by ACI Eq. (11-33):

$$\phi V_c = \phi 4\lambda\sqrt{f_c'}b_o d = 0.75 \times 4 \times 1.0\sqrt{4,000} \times [4 \times (22 + d)] \times d = 16,697d + 759d^2$$

Equating the required and design shear strengths results in the following equation that can be solved for d:

$$759d^2 + 16,697d - 684,000 = 0$$
$$d = 21.0 \text{ in}$$

On the basis of $d = 21.0$ in, the center of the two piles that are closest to the column is located $(43/2) - 18 = 3.5$ in inside the section, which is less than $d_{pile}/2 = 6$ in (see Fig. 10.40). Thus, the initial assumption is not valid; that is, part of the reactions from these two piles must be considered in the shear force at the critical section. By increasing d to 26 in, the center of the piles is located at $d_{pile}/2$ inside the critical section, and these reactions can be neglected.

Try $d = 26.0$ in

1. Check shear requirements at the column.

   • *Two-way shear*: Because the provided effective depth d is greater than that determined for the actual factored shear force, two-way shear requirements are automatically satisfied. Note that

Figure 10.40
Critical section
for two-way
shear at column.

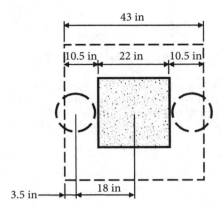

with $d = 26.0$ in, the center of the six piles on the north and south sides of the column is 7.5 in outside of the critical section, which means that the full pile reactions must be considered.

- *One-way shear*: For the critical section located a distance $d = 26$ in from the north or south face of the column, the center of the three piles is located $26 - (31.5 - 11) = 5.5$ in inside of the critical section, which is less than $d_{pile}/2 = 6$ in (see Fig. 10.41).

ACI 15.5.4.3 permits the magnitude of the pile reaction to be determined by linear interpolation in this case. At a distance of $26 - (d_{pile}/2) = 20$ in from the face of the column, the reactions from the piles are zero. At $26 + (d_{pile}/2) = 32$ in from the face of the column, the reaction from each pile is 114 kips. Because the center of the piles is located at 20.5 in from the face of the column, the reaction in each pile is

$$114 \times \left(\frac{20.5}{32 - 20} - 1.667 \right) = 5 \text{ kips}$$

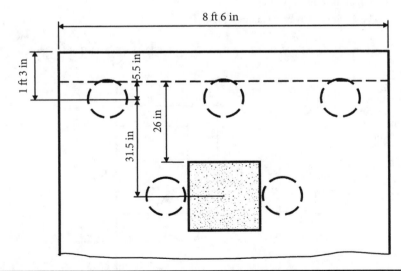

Figure 10.41 Critical section for one-way shear north (or south) of column.

Therefore, $V_u = 3 \times 5 = 15$ kips.
The design one-way shear strength is

$$\phi V_c = \phi 2\lambda\sqrt{f_c'}Ld = 0.75 \times 2 \times 1.0\sqrt{4{,}000} \times 8.5 \times 12 \times 26/1{,}000 = 252 \text{ kips} > 15 \text{ kips}$$

For the critical section located a distance $d = 26$ in from the east or west face of the column, the center of the two piles closet to the column is located $26 - 1 - 6 = 19$ in inside the critical section; thus, the reactions from these two piles are not considered. Two piles are located $26 - (36 - 11) = 1$ in inside of the critical section, which is less than $d_{pile}/2 = 6$ in. The reaction in each pile is

$$114 \times \left(\frac{25}{32 - 20} - 1.667\right) = 48 \text{ kips}$$

Therefore, $V_u = 2 \times 48 = 96$ kips.
The design one-way shear strength is

$$\phi V_c = \phi 2\lambda\sqrt{f_c'}Bd = 0.75 \times 2 \times 1.0\sqrt{4{,}000} \times 7.75 \times 12 \times 26/1{,}000 = 229 \text{ kips} > 96 \text{ kips}$$

2. Check shear requirements at the corner pile.

  • *Two-way shear*: The factored shear force at the critical section is equal to the reaction of the pile, which is 114 kips (see Fig. 10.42 for the critical shear perimeter).

    The design two-way shear strength is

    $$\phi V_c = \phi 4\lambda\sqrt{f_c'}b_o d = 0.75 \times 4 \times 1.0\sqrt{4{,}000} \times \left[\frac{\pi}{4} \times (12 + 26) + (2 \times 15)\right] \times 26/1{,}000$$
    $$= 295 \text{ kips} > 114 \text{ kips}$$

  • *One-way shear*: As in the case for two-way shear, the factored shear force at the critical section for one-way shear is equal to the reaction of the pile, which is 114 kips (see Fig. 10.43 for the location of the critical section).

    The design one-way shear strength is

    $$\phi V_c = \phi 2\lambda\sqrt{f_c'}bd = 0.75 \times 2 \times 1.0\sqrt{4{,}000} \times (2 \times 40.2) \times 26/1{,}000 = 198 \text{ kips} > 114 \text{ kips}$$

Figure 10.42
Critical perimeter for two-way shear at the corner pile given in Example 10.11.

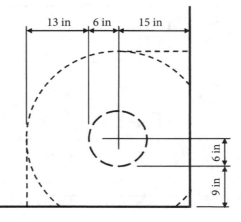

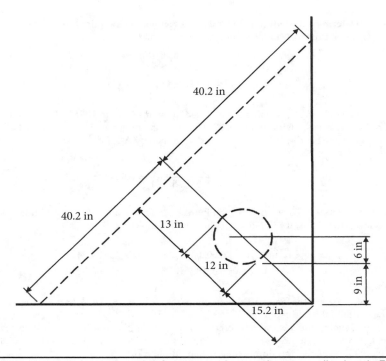

FIGURE 10.43 Location of critical section for one-way shear at the corner pile given in Example 10.11.

3. Check shear requirements at other piles.

The critical sections for two-way shear of the piles closest to the column overlap. Thus, the design two-way shear strength is

$$\phi V_c = \phi 4\lambda\sqrt{f_c'}b_o d = 0.75 \times 4 \times 1.0\sqrt{4{,}000} \times 2 \times \left[\frac{\pi}{2} \times (12 + 26) + 36\right] \times 26/1{,}000$$

$$= 944 \text{ kips}$$

$$> V_u = 2 \times 114 = 228 \text{ kips}$$

The critical sections for two-way shear of the edge piles overlap and are incomplete (i.e., they fall outside of the pile cap). Thus, shear requirements do not control for these piles.

The minimum overall thickness for footings on piles (pile caps) is 16 in (ACI 15.7). Use a pile cap that is 2 ft 9 in thick ($d = 26$ inches).

Step 3: Determine the factored moments. In the short direction, the factored moment at the critical section (i.e., at the face of the column) is obtained by multiplying the reactions from three piles by the distance from the center of the piles to the critical section:

$$M_u = (3 \times 114) \times \left(\frac{7.75}{2} - 1.25 - \frac{22}{2 \times 12}\right) = 584.3 \text{ ft kips}$$

Similarly, in the long direction,

$$M_u = 114 \times \left(1.5 - \frac{22}{2 \times 12}\right) + (2 \times 114) \times \left(3 - \frac{22}{2 \times 12}\right) = 541.5 \text{ ft kips}$$

Step 4: Determine the required flexural reinforcement. The required area of steel is determined using the general principles of the strength design method.

$$\text{Short direction: } A_s = 0.0019 \times (8.5 \times 12) \times 26 = 5.04 \text{ in}^2$$
$$< A_{s,min} = 0.0018 \times (8.5 \times 12) \times 33 = 6.06 \text{ in}^2$$

Providing 11 No. 7 bars satisfies strength requirements ($A_s = 6.60 \text{ in}^2$). These short bars must be spaced in accordance with ACI 15.4.4.2. The fraction of the total bars that must be distributed uniformly over the band width of 7 ft 9 in is determined by ACI Eq. (15-1):

$$\gamma_s = \frac{2}{\beta + 1} = \frac{2}{(8.5/7.75) + 1} = 0.95$$

Provide 10 bars spaced at 9 in on center in the center band, and add one additional bar outside of the center band. Thus, try 12 No. 7 bars in the short direction.

$$\text{Long direction: } A_s = 0.0020 \times (7.75 \times 12) \times 26 = 4.84 \text{ in}^2$$
$$< A_{s,min} = 0.0018 \times (7.75 \times 12) \times 33 = 5.52 \text{ in}^2$$

Try 10 No. 7 bars in the long direction (provided $A_s = 6.00 \text{ in}^2$). These bars are spaced uniformly over the width of the pile cap.

Step 5: Develop the flexural reinforcement. The flexural reinforcement must be developed in tension a minimum distance of ℓ_d past the critical section. Using ACI Eq. (12-1), $\ell_d = 2.1$ ft for the No. 7 bars. The available development length in the short direction is

$$\frac{7.75}{2} - \frac{22}{24} - \frac{3}{12} = 2.7 \text{ ft} > 2.1 \text{ ft}$$

Thus, the bars in the short direction can be fully developed. Because the available development length is greater in the long direction, the bars in that direction can be fully developed as well.

Use 12 No 7 bars in the short direction, which are 7 ft 0 in long.

Use 10 No. 7 bars in the long direction, which are 8 ft 0 in long.

Step 6: Check transfer of forces at the base of the column. Transfer of forces between the column and the pile cap needs to be checked. In this example, only vertical force transfer needs to be investigated.

Calculations for vertical force transfer have been illustrated in previous examples.

10.5 Drilled Piers

10.5.1 Overview

A drilled pier, which is sometimes referred to as a pier or caisson, transfers the loads from the superstructure to a soil or rock stratum that is usually well below the ground surface. Often, the bottom of the shaft is belled out to provide a larger end-bearing area (see Figs. 10.7 and 10.8). Concrete is deposited into the shaft after the required reinforcing bars have been set into place.

The loads from the supported member are transferred to the shaft by bearing. Skin resistance (or friction), point bearing, and a combination of the two are ways in which the load is transferred to the soil surrounding and below the shaft or bell. Drilled piers are categorized on the basis of the manner in which the loads are transferred to the soil or rock.

10.5.2 Determining the Shaft and Bell Sizes

The diameter of the shaft and bell are determined using service loads and the allowable bearing capacity of the concrete and soil or rock, respectively.[11] The current edition of the Code does not contain design provisions based on allowable stresses; the working stress method last appeared in Appendix A of the 1999 Code and was identified in that document as the Alternate Design Method.

Shaft Diameter

Table 1810.3.2.6 of Ref. 4 contains allowable stresses for materials used in deep foundation elements. For cast-in-place concrete with a permanent casing that satisfies the requirements of IBC 1810.3.2.7, the allowable stress in compression is $0.4 f_c'$. When a permanent casing is not provided, the allowable stress is $0.3 f_c'$, which is the same as that provided for bearing in the working stress method (see, e.g., Appendix A of the 1999 edition of the Code).

The diameter of the shaft d_{shaft} can be determined by the following equation, which is valid for drilled piers without permanent casing subjected to a total service axial dead and live load P:

$$d_{shaft} = \left[\frac{4P}{\pi (0.3 f_c')} \right]^{1/2} \tag{10.32}$$

The diameter is specified in multiples of 6 in. A similar equation can be derived for drilled piers with permanent casing.

Equation (10.32) is based on the assumption that full lateral support is provided over the entire length of the shaft; that is, the surrounding soil provides sufficient lateral resistance to prevent buckling of the piers. Guidelines on the soil properties that may be considered to provide lateral support are given in Ref. 11. If available, the geotechnical report for the site should provide information on this as well. In situations where soil cannot provide lateral resistance or where piers extend above the soil surface or through subsurface layers of air or water, the piers should be designed as columns. This includes checking if the shaft must be designed for the effects of slenderness (see Section 8.5).

Where significant bending moments and shear forces are transferred to the shaft in combination with axial loads, the shaft is typically designed by approximate methods. Reference 11 provides a number of such methods, including ways to calculate the lateral deflection at the top of the pier.

Reference 11 also provides guidelines for piers that resist loads by skin friction or by a combination of end bearing and shear friction. Allowable frictional resistance can be found in Ref. 4.

IBC 1810.3.5.2 gives minimum shaft dimensions for both uncased and cased drill piers:

- Cased: $d_{shaft} \geq 8$ in
- Uncased: $d_{shaft} \geq 12$ in

For uncased shafts, the length of the pier is limited to 30 times the diameter of the shaft; the length is permitted to exceed this limiting value where the design and installation are under the direct supervision of the registered design professional.

Bell Diameter

For the common case of end-bearing drilled piers, the diameter of the bell d_{bell} is determined on the basis of the total service axial dead and live load P and the allowable bearing capacity of the soil or rock q_a:

$$d_{bell} = \left[\frac{4P}{\pi q_a} \right]^{1/2} \tag{10.33}$$

Like the diameter of the shaft, the bell diameter is specified in multiples of 6 in.

The height of the bell depends on the diameter of the shaft and the slope of the bell (see Figs. 10.7 and 10.8). The thickness of the edge of the bell (i.e., the lower portion of the bell that is not sloped) is typically 1 ft, although a thickness not less than 6 in has been recommended in various sources. The angle of the sloped portion of the bell is usually 60 degrees or more so that effects of vertical shear do not need to be considered in the design (IBC 1810.3.9.5).

Reference 10 contains tables that facilitate the selection of shaft and bell diameters for a variety of cases.

10.5.3   Reinforcement Details

The recommended reinforcement details for drilled piers subjected to axial compressive loads are given in Fig. 10.44.[10] A minimum longitudinal reinforcement ratio of 0.005 is used, which corresponds to compression members with cross-sections that are larger than required for the applied loads (ACI 10.8.4). The minimum embedment length of the longitudinal bars is three times the diameter of the shaft or 10 ft, whichever is greater.

Ties are provided over the length of the longitudinal bars and must conform to the requirements for lateral reinforcement for compression members (ACI 7.10.5).

The recommended number, size, and spacing of the longitudinal bars and tie bars in drilled piers subjected to axial compression loads can be found in Ref. 10.

In cases where drilled piers are subjected to uplift or to bending moments that exceed the cracking moment of the shaft, reinforcement must be provided to resist these tension effects (IBC 1810.3.9.2). The shaft is designed for the effects of factored loads, using the general principles of the strength design method. Longitudinal reinforcement must extend into the shaft a sufficient distance to fully develop the bars in tension.

Vertical and horizontal loads must be transferred from the supported member to the top of the shaft or cap. See Section 10.2 for more information on force transfer.

Example 10.12   Design a drilled pier subjected to the following compressive axial loads: $P_D = 1,000$ kips and $P_L = 575$ kips. A bell will be provided at the bottom of the pier and will bear on a rock stratum with an allowable bearing capacity of 12,000 psf. Assume normal-weight concrete with a compressive strength of 4,000 psi and Grade 60 reinforcement. Also assume that the pier will not be permanently cased and that the surrounding soil provides adequate lateral support for the shaft.

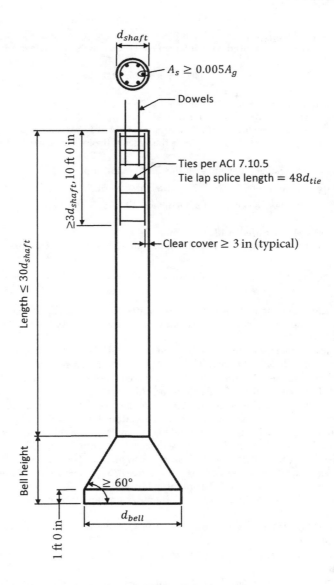

FIGURE 10.44
Reinforcement details for drilled piers subjected to axial compression loads.

Solution   The diameter of the shaft is determined by Eq. (10.32):

$$d_{shaft} = \left[\frac{4P}{\pi(0.3f_c')} \right]^{1/2} = \left[\frac{4 \times 1,575}{\pi(0.3 \times 4)} \right]^{1/2} = 40.9 \text{ in}$$

Use a shaft with a diameter of 3 ft 6 in.
The diameter of the bell is determined by Eq. (10.33):

$$d_{bell} = \left[\frac{4P}{\pi q_a} \right]^{1/2} = \left[\frac{4 \times 1,575}{\pi \times 12} \right]^{1/2} = 12.9 \text{ ft}$$

Use a bell with a diameter of 13 ft 0 in.
The area of the longitudinal reinforcement is equal to the following:

$$A_s = 0.005 \times \pi \times (3.5 \times 12/2)^2 = 6.92 \text{ in}^2$$

The minimum extension into the shaft $= 3d = 10.5$ ft (governs) or 10 ft.
Use seven No. 9 bars, which are 10 ft 6 in long.
No. 3 ties are used for longitudinal bars that are No. 10 bars and smaller (ACI 7.10.5.1), and the maximum spacing of the circular ties is the lesser of the following (ACI 7.10.5.2):

- 16 (longitudinal bar diameters) $= 16 \times 1.128 = 18.1$ in
- 48 (tie bar diameters) $= 48 \times 0.375 = 18.0$ in (governs)
- Least dimension of compression member $= 42.0$ in

Use No. 3 ties spaced at 18 in on center.
Reinforcement details for this drilled pier are similar to those shown in Fig. 10.44.

References

1. American Concrete Institute (ACI), Committee 336. 1988 (Reapproved 2002). *Suggested Analysis and Design Procedures for Combined Footings and Mats*, ACI 336.2. ACI, Farmington Hills, MI.
2. American Concrete Institute (ACI), Committee 543. 2000 (Reapproved 2005). *Design, Manufacture, and Installation of Concrete Piles*, ACI 543. ACI, Farmington Hills, MI.
3. American Concrete Institute (ACI), Committee 336. 2001. *Specification for the Construction of Drilled Piers*, ACI 336.1. ACI, Farmington Hills, MI.
4. International Code Council (ICC). 2009. *International Building Code*. ICC, Washington, DC.
5. Mattock, A. H. 1977. Discussion of "Considerations for the design of precast concrete bearing wall buildings to withstand abnormal loads," by PCI Committee on Precast Concrete Bearing Wall Buildings. *Journal of the Prestressed Concrete Institute* 22(3):105–106.
6. Bowles, J. E. 1995. *Foundation Analysis and Design*, 5th ed. McGraw-Hill, New York.
7. American Concrete Institute (ACI), Committee 336. 1966. *Suggested Design Procedures for Combined Footings and Mats*. ACI, Farmington Hills, MI.
8. Shukla, S. N. 1984. A simplified method for design of mats on elastic foundations. *ACI Journal* 81(5):469-475.
9. Hetenyi, M. 1946. *Beams on Elastic Foundation*, University of Michigan Press, Ann Arbor, MI, pp. 100-106.
10. Concrete Reinforcing Steel Institute (CRSI). 2008. *CRSI Design Handbook*, 10th ed. CRSI, Schaumburg, IL.
11. American Concrete Institute (ACI), Committee 336. 2001. *Design and Construction of Drilled Piers*, ACI 336.3. ACI, Farmington Hills, MI.

Problems

10.1.   Determine the net permissible soil pressure beneath a footing, given the following information: (1) allowable bearing capacity of soil $= 4{,}000$ psf; (2) base of footing is located 4 ft below ground level; and (3) service surcharge at ground level $= 100$ psf. Assume that the weight of the soil and concrete above the footing base is 130 pcf.

10.2.   Determine the required size of the footing in Problem 10.1, assuming a net permissible soil pressure of 4,000 psf. The applied axial service dead and live loads are 450 kips and 200 kips, respectively.

10.3. Determine the factored one-way shear force at the critical section for the footing described in Problems 10.1 and 10.2, assuming the following: (1) footing thickness = 30 in; (2) plan dimensions of footing = 13 ft 0 in × 13 ft 0 in, (3) normal-weight concrete with a compressive strength of 3,000 psi, (4) 18 × 24-in column centered on the footing.

10.4. For the design conditions of Problem 10.3, determine the design one-way shear strength at the critical section.

10.5. For the design conditions of Problem 10.3, determine the factored two-way shear force at the critical section.

10.6. For the design conditions of Problem 10.3, determine the design two-way shear strength at the critical section.

10.7. For the design conditions of Problems 10.2 and 10.3, determine the required area of flexural reinforcement at the critical section.

10.8. A combined footing supports two 20 × 20 in columns. The centerline of the column on the left is located 12 in from the edge of the footing, and the centerline of the column on the right is located 12 ft from the centerline of the column on the left. The column on the left supports service axial dead and live loads of 125 and 55 kips, respectively, and the column on the right supports service axial dead and live loads of 200 and 125 kips, respectively. Determine the length of the footing that results in uniform pressure at the base of the footing.

10.9. Determine the bearing strength of the columns in Problem 10.8, assuming normal-weight concrete with a compressive strength of 6,000 psi for the columns and 3,500 psi for the footing.

10.10. A 12-ft square footing supports a 24-in square column that transmits a total service axial load of 700 kips and a service bending moment of 500 ft kips. Determine the maximum soil pressure at the base of the footing.

10.11. Design the combined footing in Example 10.9, assuming that the 22-in square column and accompanying loads occur at the edge and that the 18-in square column and accompanying loads occur at the interior of the footing.

10.12. Design the edge strip in the east-west direction for the mat foundation depicted in Fig. 10.34, given the design data of Example 10.10.

10.13. Design the pile cap for the five-pile arrangement depicted in Fig. 10.6. The piles are spaced 3 ft 0 in on center and have an edge distance of 1 ft 3 in. Assume that the pile cap supports a 16-in square column with axial dead and live loads of 250 and 120 kips, respectively. Also assume 3,000 psi concrete and Grade 60 reinforcement.

10.14. Design a drilled pier for service axial dead and live loads of 2,000 and 1,275 kips, respectively. Assume that the shaft is uncased, that the soil provides lateral support to the shaft over the entire length, and that a bell is provided, which will bear on a rock stratum with an allowable bearing capacity of 15,000 psf. Also assume 4,000 psi concrete and Grade 60 reinforcement.

APPENDIX A

Steel Reinforcement Information

A.1 ASTM Standard Reinforcing Bars

Bar Size Number	Nominal Diameter (in)	Nominal Area (in^2)	Nominal Weight (lb/ft)
3	0.375	0.11	0.376
4	0.500	0.20	0.668
5	0.625	0.31	1.043
6	0.750	0.44	1.502
7	0.875	0.60	2.044
8	1.000	0.79	2.670
9	1.128	1.00	3.400
10	1.270	1.27	4.303
11	1.410	1.56	5.313
14	1.693	2.25	7.650
18	2.257	4.00	13.600

TABLE A.1 Information for ASTM Standard Reinforcing Bars

A.2 Wire Reinforcement Institute (WRI) Standard Wire Reinforcement

W & D Size		Nominal Diameter (in)	Nominal Area (in²)	Nominal Weight (lb/ft)	Area (in²/ft)						
					Center-to-Center Spacing (in)						
Plain	Deformed				2	3	4	6	8	10	12
W31	D31	0.628	0.310	1.054	1.86	1.24	0.93	0.62	0.46	0.37	0.31
W30	D30	0.618	0.300	1.020	1.80	1.20	0.90	0.60	0.45	0.36	0.30
W28	D28	0.597	0.280	0.952	1.68	1.12	0.84	0.56	0.42	0.33	0.28
W26	D26	0.575	0.260	0.934	1.56	1.04	0.78	0.52	0.39	0.31	0.26
W24	D24	0.553	0.240	0.816	1.44	0.96	0.72	0.48	0.36	0.28	0.24
W22	D22	0.529	0.220	0.748	1.32	0.88	0.66	0.44	0.33	0.26	0.22
W20	D20	0.504	0.200	0.680	1.20	0.80	0.60	0.40	0.30	0.24	0.20
W18	D18	0.478	0.180	0.612	1.08	0.72	0.54	0.36	0.27	0.21	0.18
W16	D16	0.451	0.160	0.544	0.96	0.64	0.48	0.32	0.24	0.19	0.16
W14	D14	0.422	0.140	0.476	0.84	0.56	0.42	0.28	0.21	0.16	0.14
W12	D12	0.390	0.120	0.408	0.72	0.48	0.36	0.24	0.18	0.14	0.12
W11	D11	0.374	0.110	0.374	0.66	0.44	0.33	0.22	0.16	0.13	0.11
W10.5	—	0.366	0.105	0.357	0.63	0.42	0.315	0.21	0.15	0.12	0.105
W10	D10	0.356	0.100	0.340	0.60	0.40	0.30	0.20	0.15	0.12	0.10
W9.5	—	0.348	0.095	0.323	0.57	0.38	0.285	0.19	0.14	0.11	0.095
W9	D9	0.338	0.090	0.306	0.54	0.36	0.27	0.18	0.13	0.10	0.09
W8.5	—	0.329	0.085	0.289	0.51	0.34	0.255	0.17	0.12	0.10	0.085
W8	D8	0.319	0.080	0.272	0.48	0.32	0.24	0.16	0.12	0.09	0.08
W7.5	—	0.309	0.075	0.255	0.45	0.30	0.225	0.15	0.11	0.09	0.075
W7	D7	0.298	0.070	0.238	0.42	0.28	0.21	0.14	0.10	0.08	0.07
W6.5	—	0.288	0.065	0.221	0.39	0.26	0.195	0.13	0.09	0.07	0.065
W6	D6	0.276	0.060	0.204	0.36	0.24	0.18	0.12	0.09	0.07	0.06
W5.5	—	0.264	0.055	0.187	0.33	0.22	0.165	0.11	0.08	0.06	0.055
W5	D5	0.252	0.050	0.170	0.30	0.20	0.15	0.10	0.07	0.06	0.05
W4.5	—	0.240	0.045	0.153	0.27	0.18	0.135	0.09	0.06	0.05	0.045
W4	D4	0.225	0.040	0.136	0.24	0.16	0.12	0.08	0.06	0.04	0.04
W3.5	—	0.211	0.035	0.119	0.21	0.14	0.105	0.07	0.05	0.04	0.035
W3	—	0.195	0.030	0.102	0.18	0.12	0.09	0.06	0.04	0.03	0.03
W2.9	—	0.192	0.029	0.098	0.174	0.116	0.087	0.058	0.04	0.03	0.029
W2.5	—	0.178	0.025	0.085	0.15	0.10	0.075	0.05	0.03	0.03	0.025
W2	—	0.159	0.020	0.068	0.12	0.08	0.06	0.04	0.03	0.02	0.02
W1.4	—	0.135	0.014	0.049	0.084	0.056	0.042	0.028	0.02	0.01	0.014

TABLE A.2 Information for WRI Standard Wire Reinforcement

Critical Section Properties for Two-Way Shear

B.1 Derivation of Critical Section Properties

B.1.1 Overview

Equations to determine the properties of critical sections for use in two-way shear calculations are given in Figs. 7.53 through 7.57 for interior, edge, and corner rectangular columns and for circular interior columns. This section contains derivations of the critical section properties for a variety of support conditions.

B.1.2 Interior Rectangular Column

Properties are derived for the critical section of an interior rectangular column. Refer to Fig. B.1.

Perimeter of Critical Section

$$b_o = 2(c_1 + d) + 2(c_2 + d) = 2(b_1 + b_2)$$

Area of Critical Section

$$A_c = b_o d = 2(b_1 + b_2)d = 2c_1 d + 2c_2 d + 4d^2$$

$$\frac{A_c}{d^2} = 2\left(\frac{c_1}{d}\right) + 2\left(\frac{c_2}{d}\right) + 4$$

$$= 2\left[\left(1 + \frac{c_2}{c_1}\right)\left(\frac{c_1}{d}\right) + 2\right]$$

Define

$$f_1 = 2\left[\left(1 + \frac{c_2}{c_1}\right)\left(\frac{c_1}{d}\right) + 2\right]$$

Therefore,

$$A_c = f_1 d^2$$

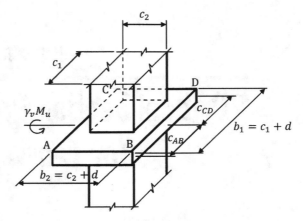

Interior rectangular column

$c_{AB} = c_{CD} = b_1/2$

$A_c = f_1 d^2$

$J_c/c_{AB} = J_c/c_{CD} = 2f_2 d^3$

$f_1 = 2\left[\left(1 + \frac{c_2}{c_1}\right)\left(\frac{c_1}{d}\right) + 2\right]$

$f_2 = \frac{1}{6}\left[\left(1 + \frac{3c_2}{c_1}\right)\left(\frac{c_1}{d}\right)^2 + \left(5 + \frac{3c_2}{c_1}\right)\left(\frac{c_1}{d}\right) + 5\right]$

Center of Gravity

From symmetry,

$$c_{AB} = c_{CD} = b_1/2 = (c_1 + d)/2$$

Polar Moment of Inertia

For faces $c_1 + d$,

$$(J_c)_1 = I_{xx} + I_{zz} = 2 \times \left[\frac{d(c_1 + d)^3}{12} + \frac{(c_1 + d)d^3}{12}\right] = \frac{b_1^3 d + b_1 d^3}{6}$$

For faces $c_2 + d$:

$$(J_c)_2 = I_{xx} = 2 \times d(c_2 + d)\left(\frac{c_1 + d}{2}\right)^2 = \frac{b_1^2 b_2 d}{2}$$

Thus,

$$J_c = (J_c)_1 + (J_c)_2 = \frac{b_1^3 d + b_1 d^3}{6} + \frac{b_1^2 b_2 d}{2}$$

$$\frac{J_c}{c_{AB}} = \frac{J_c}{c_{CD}} = \frac{b_1^2 d + d^3}{3} + b_1 b_2 d = \frac{1}{3}\left[\left(1 + \frac{3c_2}{c_1}\right)\frac{c_1^2}{d^2} + \left(5 + \frac{3c_2}{c_1}\right)\frac{c_1}{d} + 5\right]d^3$$

Define

$$f_2 = \frac{1}{6}\left[\left(1 + \frac{3c_2}{c_1}\right)\left(\frac{c_1}{d}\right)^2 + \left(5 + \frac{3c_2}{c_1}\right)\left(\frac{c_1}{d}\right) + 5\right]$$

Therefore,

$$\frac{J_c}{c_{AB}} = \frac{J_c}{c_{CD}} = 2f_2 d^3$$

B.1.3 Edge Rectangular Column Bending Parallel to the Edge

Properties are derived for the critical section of an edge rectangular column that is bending parallel to the edge of the slab. Refer to Fig. B.2.

Perimeter of Critical Section

$$b_o = (c_1 + d) + 2\left(c_2 + \frac{d}{2}\right) = b_1 + 2b_2$$

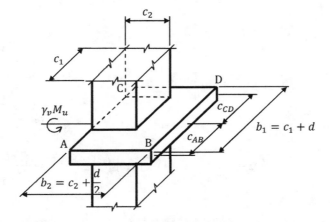

FIGURE B.2 Edge rectangular column bending parallel to the edge.

Edge rectangular column bending parallel to the edge

$$c_{AB} = c_{CD} = b_1/2$$

$$A_c = f_1 d^2$$

$$J_c/c_{AB} = J_c/c_{CD} = 2f_2 d^3$$

$$f_1 = \left(1 + \frac{2c_2}{c_1}\right)\left(\frac{c_1}{d}\right) + 2$$

$$f_2 = \frac{1}{12}\left[\left(1 + \frac{6c_2}{c_1}\right)\left(\frac{c_1}{d}\right)^2 + \left(5 + \frac{6c_2}{c_1}\right)\left(\frac{c_1}{d}\right) + 5\right]$$

Area of Critical Section

$$A_c = b_o d = (b_1 + 2b_2)d = (c_1 + d)d + (2c_2 + d)d$$

$$\frac{A_c}{d^2} = \left(\frac{c_1 + d}{d}\right) + \left(\frac{2c_2 + d}{d}\right)$$

$$= \left(1 + \frac{2c_2}{c_1}\right)\left(\frac{c_1}{d}\right) + 2$$

Define

$$f_1 = \left(1 + \frac{2c_2}{c_1}\right)\left(\frac{c_1}{d}\right) + 2$$

Therefore,

$$A_c = f_1 d^2$$

Center of Gravity
From symmetry,

$$c_{AB} = c_{CD} = b_1/2 = (c_1 + d)/2$$

Polar Moment of Inertia
For face $c_1 + d$,

$$(J_c)_1 = I_{xx} + I_{zz} = \left[\frac{d(c_1 + d)^3}{12} + \frac{(c_1 + d)d^3}{12}\right] = \frac{b_1^3 d + b_1 d^3}{12}$$

For faces $c_2 + (d/2)$,

$$(J_c)_2 = I_{xx} = 2 \times d\left(c_2 + \frac{d}{2}\right)\left(\frac{c_1 + d}{2}\right)^2 = \frac{b_1^2 b_2 d}{2}$$

Thus,

$$J_c = (J_c)_1 + (J_c)_2 = \frac{b_1^3 d + b_1 d^3}{12} + \frac{b_1^2 b_2 d}{2}$$

$$\frac{J_c}{c_{AB}} = \frac{J_c}{c_{CD}} = \frac{b_1^2 d + d^3}{6} + b_1 b_2 d = \frac{1}{6}\left[\left(1 + \frac{6c_2}{c_1}\right)\frac{c_1^2}{d^2} + \left(5 + \frac{6c_2}{c_1}\right)\frac{c_1}{d} + 5\right]d^3$$

Define

$$f_2 = \frac{1}{12}\left[\left(1 + \frac{6c_2}{c_1}\right)\left(\frac{c_1}{d}\right)^2 + \left(5 + \frac{6c_2}{c_1}\right)\left(\frac{c_1}{d}\right) + 5\right]$$

Therefore,

$$\frac{J_c}{c_{AB}} = \frac{J_c}{c_{CD}} = 2f_2 d^3$$

B.1.4   Edge Rectangular Column Bending Perpendicular to the Edge

Properties are derived for the critical section of an edge rectangular column that is bending perpendicular to the edge of the slab. Refer to Fig. B.3.

Perimeter of Critical Section

$$b_o = 2\left(c_1 + \frac{d}{2}\right) + (c_2 + d) = 2b_1 + b_2$$

FIGURE B.3 Edge rectangular column bending perpendicular to the edge.

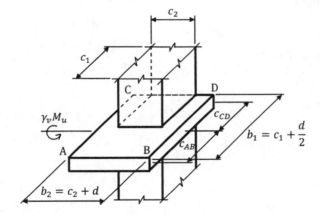

Edge rectangular column bending perpendicular to the edge

$c_{AB} = [f_3/(f_2 + f_3)][c_1 + (d/2)]$

$c_{CD} = [f_2/(f_2 + f_3)][c_1 + (d/2)]$

$A_c = f_1 d^2$

$J_c/c_{AB} = 2f_2 d^3$

$J_c/c_{CD} = 2f_3 d^3$

$f_1 = 2 + \dfrac{c_1}{d}\left(2 + \dfrac{c_2}{c_1}\right)$

$$f_2 = \frac{\left(\frac{c_1}{d} + \frac{1}{2}\right)^2 \left[\frac{c_1}{d}\left(1 + \frac{2c_2}{c_1}\right) + \frac{5}{2}\right] + \left[\frac{c_1}{d}\left(1 + \frac{c_2}{2c_1}\right) + 1\right]}{6\left(\frac{c_1}{d} + \frac{1}{2}\right)}$$

$$f_3 = \frac{\left(\frac{c_1}{d} + \frac{1}{2}\right)^2 \left[\frac{c_1}{d}\left(1 + \frac{2c_2}{c_1}\right) + \frac{5}{2}\right] + \left[\frac{c_1}{d}\left(1 + \frac{c_2}{2c_1}\right) + 1\right]}{6\left[\frac{c_1}{d}\left(1 + \frac{c_2}{c_1}\right) + \frac{3}{2}\right]}$$

Area of Critical Section

$$A_c = b_o d = (2b_1 + b_2)d = 2\left(c_1 + \frac{d}{2}\right)d + (c_2 + d)d$$

$$\frac{A_c}{d^2} = \left(\frac{2c_1 + d}{d}\right) + \left(\frac{c_2 + d}{d}\right)$$

$$= 2 + \frac{c_1}{d}\left(2 + \frac{c_2}{c_1}\right)$$

Define

$$f_1 = 2 + \frac{c_1}{d}\left(2 + \frac{c_2}{c_1}\right)$$

Therefore,

$$A_c = f_1 d^2$$

Center of Gravity
Summing moments about face $c_2 + d$,

$$c_{AB} = \frac{2 \times b_1 d \times (b_1/2)}{A_c} = \frac{b_1^2 d}{(2b_1 + b_2)d} = \frac{b_1^2}{2b_1 + b_2} = \frac{[c_1 + (d/2)]^2}{2[c_1 + (d/2)] + (c_2 + d)}$$

$$= \frac{[c_1 + (d/2)]^2}{[2 + (c_2/c_1)]c_1 + 2d}$$

Define

$$f_2 = \frac{[(c_1/d) + (1/2)]^2\,\{(c_1/d)[1 + (2c_2/c_1)] + (5/2)\} + \{(c_1/d)[1 + (c_2/2c_1)] + 1\}}{6[(c_1/d) + (1/2)]}$$

$$f_3 = \frac{[(c_1/d) + (1/2)]^2\,\{(c_1/d)[1 + (2c_2/c_1)] + (5/2)\} + \{(c_1/d)[1 + (c_2/2c_1)] + 1\}}{6\{(c_1/d)[1 + (c_2/c_1)] + (3/2)\}}$$

Therefore,

$$c_{AB} = \left(\frac{f_3}{f_2 + f_3}\right)\left(c_1 + \frac{d}{2}\right)$$

$$c_{CD} = b_1 - c_{AB} = \left(\frac{f_2}{f_2 + f_3}\right)\left(c_1 + \frac{d}{2}\right)$$

Polar Moment of Inertia
For faces $c_1 + (d/2)$,

$$(J_c)_1 = I_{xx} + I_{zz} = \frac{2 \times d\,[c_1 + (d/2)]^3}{12} + 2d\left(c_1 + \frac{d}{2}\right)\left[\frac{c_1 + (d/2)}{2} - c_{AB}\right]^2$$

$$+ \frac{2 \times [c_1 + (d/2)]\,d^3}{12}$$

For face $c_2 + d$,

$$(J_c)_2 = I_{xx} = (c_2 + d)dc_{AB}^2$$

Thus,

$$J_c = (J_c)_1 + (J_c)_2 = \frac{b_1 d^3}{6} + \frac{2d}{3}\left(c_{AB}^3 + c_{CD}^3\right) + db_2 c_{AB}^2$$

$$\frac{J_c}{c_{AB}} = \frac{b_1 d^3}{6c_{AB}} + \frac{2d}{3}\left(c_{AB}^2 + \frac{c_{CD}^3}{c_{AB}}\right) + db_2 c_{AB}^2 = f_2 d^3$$

$$\frac{J_c}{c_{CD}} = \frac{b_1 d^3}{6c_{CD}} + \frac{2d}{3}\left(\frac{c_{AB}^3}{c_{CD}} + c_{CD}^2\right) + db_2\left(\frac{c_{AB}^2}{c_{CD}}\right) = 2f_3 d^3$$

B.1.5 Corner Rectangular Column Bending Perpendicular to the Edge

Properties are derived for the critical section of a corner rectangular column that is bending perpendicular to the edge of the slab. Refer to Fig. B.4.

Perimeter of Critical Section

$$b_o = \left(c_1 + \frac{d}{2}\right) + \left(c_2 + \frac{d}{2}\right) = b_1 + b_2$$

Area of Critical Section

$$A_c = b_o d = (b_1 + b_2)d = \left(c_1 + \frac{d}{2}\right)d + \left(c_2 + \frac{d}{2}\right)d$$

$$\frac{A_c}{d^2} = \left(\frac{c_1}{d} + \frac{1}{2}\right) + \left(\frac{c_2}{d} + \frac{1}{2}\right)$$

$$= 1 + \frac{c_1}{d}\left(1 + \frac{c_2}{c_1}\right)$$

Define

$$f_1 = 1 + \frac{c_1}{d}\left(1 + \frac{c_2}{c_1}\right)$$

Therefore,

$$A_c = f_1 d^2$$

Center of Gravity

Summing moments about face $c_2 + (d/2)$,

$$c_{AB} = \frac{b_1 d \times (b_1/2)}{A_c} = \frac{b_1^2 d}{2(b_1 + b_2)d} = \frac{b_1^2}{2(b_1 + b_2)} = \frac{[c_1 + (d/2)]^2}{2\{[c_1 + (d/2)] + [c_2 + (d/2)]\}}$$

$$= \frac{[c_1 + (d/2)]^2}{2\{c_1[1 + (c_2/c_1)] + d\}}$$

FIGURE B.4 Corner rectangular column bending perpendicular to the edge.

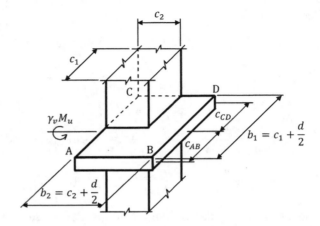

Corner rectangular column bending perpendicular to the edge

$$c_{AB} = [f_3/(f_2 + f_3)][c_1 + (d/2)]$$

$$c_{CD} = [f_2/(f_2 + f_3)][c_1 + (d/2)]$$

$$A_c = f_1 d^2$$

$$J_c/c_{AB} = 2f_2 d^3$$

$$J_c/c_{CD} = 2f_3 d^3$$

$$f_1 = 1 + \frac{c_1}{d}\left(1 + \frac{c_2}{c_1}\right)$$

$$f_2 = \frac{\left(\frac{c_1}{d} + \frac{1}{2}\right)^2 \left[\frac{c_1}{d}\left(1 + \frac{4c_2}{c_1}\right) + \frac{5}{2}\right] + \left[\frac{c_1}{d}\left(1 + \frac{c_2}{c_1}\right) + 1\right]}{12\left(\frac{c_1}{d} + \frac{1}{2}\right)}$$

$$f_3 = \frac{\left(\frac{c_1}{d} + \frac{1}{2}\right)^2 \left[\frac{c_1}{d}\left(1 + \frac{4c_2}{c_1}\right) + \frac{5}{2}\right] + \left[\frac{c_1}{d}\left(1 + \frac{c_2}{c_1}\right) + 1\right]}{12\left[\frac{c_1}{d}\left(1 + \frac{2c_2}{c_1}\right) + \frac{3}{2}\right]}$$

Define

$$f_2 = \frac{[(c_1/d) + (1/2)]^2 \{(c_1/d)[1 + (4c_2/c_1) + (5/2)]\} + \{(c_1/d)[1 + (c_2/c_1)] + 1\}}{12[(c_1/d) + (1/2)]}$$

$$f_3 = \frac{[(c_1/d) + (1/2)]^2 \{(c_1/d)[1 + (4c_2/c_1) + (5/2)]\} + \{(c_1/d)[1 + (c_2/c_1)] + 1\}}{12\{(c_1/d)[1 + (2c_2/c_1)] + (3/2)\}}$$

Therefore,

$$c_{AB} = \left(\frac{f_3}{f_2 + f_3}\right)\left(c_1 + \frac{d}{2}\right)$$

$$c_{CD} = b_1 - c_{AB} = \left(\frac{f_2}{f_2 + f_3}\right)\left(c_1 + \frac{d}{2}\right)$$

Polar Moment of Inertia

For face $c_1 + (d/2)$,

$$(J_c)_1 = I_{xx} + I_{zz} = \frac{d[c_1 + (d/2)]^3}{12} + d\left(c_1 + \frac{d}{2}\right)\left[\frac{c_1 + (d/2)}{2} - c_{AB}\right]^2 + \frac{[c_1 + (d/2)]d^3}{12}$$

For face $c_2 + d$,

$$(J_c)_2 = I_{xx} = \left(c_2 + \frac{d}{2}\right)dc_{AB}^2$$

Thus,

$$J_c = (J_c)_1 + (J_c)_2 = \frac{b_1 d^3}{12} + \frac{d}{3}\left(c_{AB}^3 + c_{CD}^3\right) + db_2 c_{AB}^2$$

$$\frac{J_c}{c_{AB}} = \frac{b_1 d^3}{12c_{AB}} + \frac{d}{3}\left(c_{AB}^2 + \frac{c_{CD}^3}{c_{AB}}\right) + db_2 c_{AB}^2 = f_2 d^3$$

$$\frac{J_c}{c_{CD}} = \frac{b_1 d^3}{12c_{CD}} + \frac{d}{3}\left(\frac{c_{AB}^3}{c_{CD}} + c_{CD}^2\right) + db_2\left(\frac{c_{AB}^2}{c_{CD}}\right) = 2f_3 d^3$$

B.1.6   Circular Interior Column

Properties are derived for the critical section of a circular interior. Refer to Fig. B.5.

Perimeter of Critical Section

$$b_o = \pi(D + d)$$

Area of Critical Section

$$A_c = b_o d = \pi(D + d)d$$

$$\frac{A_c}{d^2} = \pi\left(\frac{D}{d} + 1\right)$$

Define

$$f_1 = \pi\left(\frac{D}{d} + 1\right)$$

Therefore,

$$A_c = f_1 d^2$$

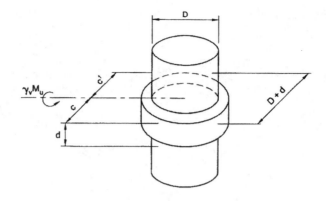

Circular interior column

$$c = c' = (D + d)/2$$

$$A_c = f_1 d^2$$

$$J_c/c = J_c/c' = 2f_2 d^3$$

$$f_1 = \pi \left(\frac{D}{d} + 1 \right)$$

$$f_2 = \frac{\pi}{8} \left(\frac{D}{d} + 1 \right)^2 + \frac{1}{6}$$

Center of Gravity

From symmetry,

$$c = c' = (D + d)/2$$

Polar Moment of Inertia

Moments of inertia are obtained for projections on two orthogonal planes:

$$I_{xx} = d \int_0^{2\pi} y^2 ds$$

$$= d \int_0^{2\pi} \left(\frac{D+d}{2} \cos\theta \right)^2 \left(\frac{D+d}{2} \right) d\theta$$

$$= d \left(\frac{D+d}{2} \right)^3 \int_0^{2\pi} (\cos\theta)^2 d\theta$$

$$= d \left(\frac{D+d}{2} \right)^3 \left(\frac{\theta}{2} + \frac{\sin 2\theta}{4} \right) \Big|_0^{2\pi}$$

$$= \pi d \left(\frac{D+d}{2} \right)^3$$

$$I_{zz} = 2 \times 2 \left(\frac{D+d}{2} \right) \int_{-d/2}^{d/2} z^2 dz$$

$$= 4 \left(\frac{D+d}{2} \right) \left(\frac{z^3}{3} \right) \Big|_{-d/2}^{d/2}$$

$$= \left(\frac{D+d}{6} \right) d^3$$

Thus,

$$J_c = I_{xx} + I_{yy} = \pi d \left(\frac{D+d}{2} \right)^3 + \left(\frac{D+d}{6} \right) d^3$$

$$\frac{J_c}{c} = \frac{J_c}{c'} = \pi \left(\frac{D+d}{2} \right)^2 + \frac{d^3}{3}$$

$$= \left[\frac{\pi}{4} \left(\frac{D}{d} + 1 \right)^2 + \frac{1}{3} \right] d^3$$

Define

$$f_2 = \frac{\pi}{8} \left(\frac{D}{d} + 1 \right)^2 + \frac{1}{6}$$

Therefore,

$$\frac{J_c}{c} = \frac{J_c}{c'} = 2 f_2 d^3$$

B.2 Tabulated Values of Critical Shear Constants

B.2.1 Overview

This section contains tabulated values of the constants f_1, f_2, and f_3 that are used to facilitate the calculation of the critical section properties for two-way shear. Included are constants for the following:

- Interior rectangular columns (Table B.1; Fig. 7.53)
- Edge rectangular columns bending parallel to the edge (Table B.2; Fig. 7.54)
- Edge rectangular columns bending perpendicular to the edge (Table B.3; Fig. 7.55)
- Corner rectangular columns bending perpendicular to the edge (Table B.4; Fig. 7.56)
- Circular interior columns (Table B.5; Fig. 7.57)

Linear interpolation may be used to obtain constants that are not listed in the tables.

B.2.2 Interior Rectangular Column

| | f_1 | | | | | | | f_2 | | | | | |
| | c_2/c_1 | | | | | | | c_2/c_1 | | | | | |
c_1/d	0.50	0.75	1.00	1.25	1.50	1.75	2.00	0.50	0.75	1.00	1.25	1.50	1.75	2.00
1.00	7.00	7.50	8.00	8.50	9.00	9.50	10.00	2.33	2.58	2.83	3.08	3.33	3.58	3.83
1.50	8.50	9.25	10.00	10.75	11.50	12.25	13.00	3.40	3.86	4.33	4.80	5.27	5.74	6.21
2.00	10.00	11.00	12.00	13.00	14.00	15.00	16.00	4.67	5.42	6.17	6.92	7.67	8.42	9.17
2.50	11.50	12.75	14.00	15.25	16.50	17.75	19.00	6.15	7.24	8.33	9.43	10.52	11.61	12.71
3.00	13.00	14.50	16.00	17.50	19.00	20.50	22.00	7.83	9.33	10.83	12.33	13.83	15.33	16.83
3.50	14.50	16.25	18.00	19.75	21.50	23.25	25.00	9.73	11.70	13.67	15.64	17.60	19.57	21.54
4.00	16.00	18.00	20.00	22.00	24.00	26.00	28.00	11.83	14.33	16.83	19.33	21.83	24.33	26.83
4.50	17.50	19.75	22.00	24.25	26.50	28.75	31.00	14.15	17.24	20.33	23.43	26.52	29.61	32.71
5.00	19.00	21.50	24.00	26.50	29.00	31.50	34.00	16.67	20.42	24.17	27.92	31.67	35.42	39.17
5.50	20.50	23.25	26.00	28.75	31.50	34.25	37.00	19.40	23.86	28.33	32.80	37.27	41.74	46.21
6.00	22.00	25.00	28.00	31.00	34.00	37.00	40.00	22.33	27.58	32.83	38.08	43.33	48.58	53.83
6.50	23.50	26.75	30.00	33.25	36.50	39.75	43.00	25.48	31.57	37.67	43.76	49.85	55.95	62.04
7.00	25.00	28.50	32.00	35.50	39.00	42.50	46.00	28.83	35.83	42.83	49.83	56.83	63.83	70.83
7.50	26.50	30.25	34.00	37.75	41.50	45.25	49.00	32.40	40.36	48.33	56.30	64.27	72.24	80.21
8.00	28.00	32.00	36.00	40.00	44.00	48.00	52.00	36.17	45.17	54.17	63.17	72.17	81.17	90.17
8.50	29.50	33.75	38.00	42.25	46.50	50.75	55.00	40.15	50.24	60.33	70.43	80.52	90.61	100.71
9.00	31.00	35.50	40.00	44.50	49.00	53.50	58.00	44.33	55.58	66.83	78.08	89.33	100.58	111.83
9.50	32.50	37.25	42.00	46.75	51.50	56.25	61.00	48.73	61.20	73.67	86.14	98.60	111.07	123.54
10.00	34.00	39.00	44.00	49.00	54.00	59.00	64.00	53.33	67.08	80.83	94.58	108.33	122.08	135.83

TABLE B.1 Properties of the Critical Section—Interior Rectangular Column

B.2.3   Edge Rectangular Column Bending Parallel to the Edge

	f_1							f_2						
	c_2/c_1							c_2/c_1						
c_1/d	0.50	0.75	1.00	1.25	1.50	1.75	2.00	0.50	0.75	1.00	1.25	1.50	1.75	2.00
1.00	4.00	4.50	5.00	5.50	6.00	6.50	7.00	1.42	1.67	1.92	2.17	2.42	2.67	2.92
1.50	5.00	5.75	6.50	7.25	8.00	8.75	9.50	2.17	2.64	3.10	3.57	4.04	4.51	4.98
2.00	6.00	7.00	8.00	9.00	10.00	11.00	12.00	3.08	3.83	4.58	5.33	6.08	6.83	7.58
2.50	7.00	8.25	9.50	10.75	12.00	13.25	14.50	4.17	5.26	6.35	7.45	8.54	9.64	10.73
3.00	8.00	9.50	11.00	12.50	14.00	15.50	17.00	5.42	6.92	8.42	9.92	11.42	12.92	14.42
3.50	9.00	10.75	12.50	14.25	16.00	17.75	19.50	6.83	8.80	10.77	12.74	14.71	16.68	18.65
4.00	10.00	12.00	14.00	16.00	18.00	20.00	22.00	8.42	10.92	13.42	15.92	18.42	20.92	23.42
4.50	11.00	13.25	15.50	17.75	20.00	22.25	24.50	10.17	13.26	16.35	19.45	22.54	25.64	28.73
5.00	12.00	14.50	17.00	19.50	22.00	24.50	27.00	12.08	15.83	19.58	23.33	27.08	30.83	34.58
5.50	13.00	15.75	18.50	21.25	24.00	26.75	29.50	14.17	18.64	23.10	27.57	32.04	36.51	40.98
6.00	14.00	17.00	20.00	23.00	26.00	29.00	32.00	16.42	21.67	26.92	32.17	37.42	42.67	47.92
6.50	15.00	18.25	21.50	24.75	28.00	31.25	34.50	18.83	24.93	31.02	37.11	43.21	49.30	55.40
7.00	16.00	19.50	23.00	26.50	30.00	33.50	37.00	21.42	28.42	35.42	42.42	49.42	56.42	63.42
7.50	17.00	20.75	24.50	28.25	32.00	35.75	39.50	24.17	32.14	40.10	48.07	56.04	64.01	71.98
8.00	18.00	22.00	26.00	30.00	34.00	38.00	42.00	27.08	36.08	45.08	54.08	63.08	72.08	81.08
8.50	19.00	23.25	27.50	31.75	36.00	40.25	44.50	30.17	40.26	50.35	60.45	70.54	80.64	90.73
9.00	20.00	24.50	29.00	33.50	38.00	42.50	47.00	33.42	44.67	55.92	67.17	78.42	89.67	100.92
9.50	21.00	25.75	30.50	35.25	40.00	44.75	49.50	36.83	49.30	61.77	74.24	86.71	99.18	111.65
10.00	22.00	27.00	32.00	37.00	42.00	47.00	52.00	40.42	54.17	67.92	81.67	95.42	109.17	122.92

TABLE B.2   Properties of the Critical Section—Edge Rectangular Column Bending Parallel to Edge

B.2.4 Edge Rectangular Column Bending Perpendicular to the Edge

| | f_1 | | | | | | | f_2 | | | | | | | f_3 | | | | | | |
| | c_2/c_1 | | | | | | | c_2/c_1 | | | | | | | c_2/c_1 | | | | | | |
c_1/d	0.50	0.75	1.00	1.25	1.50	1.75	2.00	0.50	0.75	1.00	1.25	1.50	1.75	2.00	0.50	0.75	1.00	1.25	1.50	1.75	2.00
1.00	4.50	4.75	5.00	5.25	5.50	5.75	6.00	1.38	1.51	1.65	1.79	1.93	2.07	2.21	0.69	0.70	0.71	0.72	0.72	0.73	0.74
1.50	5.75	6.13	6.50	6.88	7.25	7.63	8.00	2.07	2.34	2.60	2.87	3.14	3.40	3.67	1.11	1.13	1.16	1.18	1.19	1.21	1.22
2.00	7.00	7.50	8.00	8.50	9.00	9.50	10.00	2.94	3.38	3.81	4.24	4.68	5.11	5.54	1.63	1.69	1.73	1.77	1.80	1.82	1.85
2.50	8.25	8.87	9.50	10.13	10.75	11.38	12.00	3.98	4.62	5.26	5.91	6.55	7.19	7.83	2.27	2.36	2.43	2.49	2.53	2.58	2.61
3.00	9.50	10.25	11.00	11.75	12.50	13.25	14.00	5.18	6.08	6.97	7.86	8.76	9.65	10.54	3.02	3.15	3.25	3.34	3.41	3.46	3.51
3.50	10.75	11.62	12.50	13.37	14.25	15.12	16.00	6.56	7.74	8.93	10.11	11.30	12.48	13.67	3.89	4.06	4.20	4.31	4.41	4.49	4.56
4.00	12.00	13.00	14.00	15.00	16.00	17.00	18.00	8.10	9.62	11.13	12.65	14.17	15.69	17.21	4.86	5.09	5.27	5.42	5.55	5.65	5.74
4.50	13.25	14.37	15.50	16.62	17.75	18.87	20.00	9.80	11.70	13.59	15.49	17.38	19.27	21.17	5.94	6.24	6.47	6.66	6.82	6.95	7.06
5.00	14.50	15.75	17.00	18.25	19.50	20.75	22.00	11.68	13.99	16.30	18.61	20.92	23.23	25.54	7.14	7.51	7.80	8.03	8.22	8.38	8.51
5.50	15.75	17.12	18.50	19.87	21.25	22.62	24.00	13.72	16.49	19.26	22.03	24.80	27.56	30.33	8.44	8.89	9.24	9.52	9.76	9.95	10.11
6.00	17.00	18.50	20.00	21.50	23.00	24.50	26.00	15.93	19.20	22.46	25.73	29.00	32.27	35.54	9.86	10.40	10.82	11.15	11.43	11.65	11.85
6.50	18.25	19.87	21.50	23.12	24.75	26.37	28.00	18.30	22.11	25.92	29.73	33.54	37.36	41.17	11.39	12.02	12.51	12.91	13.23	13.50	13.72
7.00	19.50	21.25	23.00	24.75	26.50	28.25	30.00	20.84	25.24	29.63	34.02	38.42	42.81	47.21	13.03	13.77	14.34	14.79	15.17	15.47	15.74
7.50	20.75	22.62	24.50	26.37	28.25	30.12	32.00	23.55	28.57	33.59	38.61	43.63	48.65	53.67	14.78	15.63	16.29	16.81	17.24	17.59	17.89
8.00	22.00	24.00	26.00	28.00	30.00	32.00	34.00	26.42	32.11	37.80	43.48	49.17	54.86	60.54	16.64	17.61	18.36	18.95	19.44	19.84	20.18
8.50	23.25	25.37	27.50	29.62	31.75	33.87	36.00	29.47	35.86	42.25	48.65	55.04	61.44	67.83	18.61	19.71	20.56	21.23	21.78	22.23	22.61
9.00	24.50	26.75	29.00	31.25	33.50	35.75	38.00	32.67	39.82	46.96	54.11	61.25	68.40	75.54	20.69	21.93	22.88	23.63	24.25	24.75	25.18
9.50	25.75	28.12	30.50	32.87	35.25	37.62	40.00	36.05	43.98	51.92	59.86	67.79	75.73	83.67	22.89	24.27	25.33	26.17	26.85	27.41	27.89
10.00	27.00	29.50	32.00	34.50	37.00	39.50	42.00	39.59	48.36	57.13	65.90	74.67	83.44	92.21	25.19	26.72	27.90	28.83	29.59	30.21	30.74

TABLE B.3   Properties of the Critical Section—Edge Rectangular Column Bending Perpendicular to the Edge

B.2.5 Corner Rectangular Column Bending Perpendicular to the Edge

	f_1							f_2							f_3						
	c_2/c_1							c_2/c_1							c_2/c_1						
c_1/d	0.50	0.75	1.00	1.25	1.50	1.75	2.00	0.50	0.75	1.00	1.25	1.50	1.75	2.00	0.50	0.75	1.00	1.25	1.50	1.75	2.00
1.00	2.50	2.75	3.00	3.25	3.50	3.75	4.00	0.83	0.97	1.10	1.24	1.38	1.52	1.66	0.35	0.36	0.37	0.37	0.38	0.38	0.38
1.50	3.25	3.63	4.00	4.38	4.75	5.13	5.50	1.30	1.57	1.83	2.10	2.36	2.63	2.90	0.58	0.60	0.61	0.62	0.63	0.64	0.64
2.00	4.00	4.50	5.00	5.50	6.00	6.50	7.00	1.90	2.34	2.77	3.20	3.64	4.07	4.50	0.87	0.90	0.92	0.94	0.96	0.97	0.98
2.50	4.75	5.37	6.00	6.62	7.25	7.87	8.50	2.63	3.27	3.92	4.56	5.20	5.84	6.49	1.21	1.27	1.31	1.33	1.36	1.37	1.39
3.00	5.50	6.25	7.00	7.75	8.50	9.25	10.00	3.49	4.38	5.27	6.16	7.06	7.95	8.84	1.63	1.70	1.76	1.80	1.83	1.85	1.88
3.50	6.25	7.12	8.00	8.87	9.75	10.62	11.50	4.46	5.65	6.83	8.02	9.20	10.39	11.57	2.10	2.20	2.28	2.33	2.37	2.41	2.44
4.00	7.00	8.00	9.00	10.00	11.00	12.00	13.00	5.57	7.09	8.60	10.12	11.64	13.16	14.68	2.64	2.77	2.87	2.94	2.99	3.04	3.07
4.50	7.75	8.87	10.00	11.12	12.25	13.37	14.50	6.80	8.69	10.58	12.48	14.37	16.26	18.16	3.24	3.41	3.53	3.62	3.68	3.74	3.78
5.00	8.50	9.75	11.00	12.25	13.50	14.75	16.00	8.15	10.46	12.77	15.08	17.39	19.70	22.01	3.90	4.11	4.26	4.37	4.45	4.52	4.57
5.50	9.25	10.62	12.00	13.37	14.75	16.12	17.50	9.63	12.40	15.17	17.94	20.70	23.47	26.24	4.62	4.88	5.06	5.19	5.29	5.37	5.43
6.00	10.00	11.50	13.00	14.50	16.00	17.50	19.00	11.23	14.50	17.77	21.04	24.31	27.58	30.85	5.41	5.71	5.92	6.08	6.20	6.29	6.37
6.50	10.75	12.37	14.00	15.62	17.25	18.87	20.50	12.96	16.77	20.58	24.39	28.21	32.02	35.83	6.26	6.61	6.86	7.04	7.18	7.29	7.38
7.00	11.50	13.25	15.00	16.75	18.50	20.25	22.00	14.82	19.21	23.60	28.00	32.39	36.79	41.18	7.17	7.58	7.87	8.08	8.24	8.36	8.46
7.50	12.25	14.12	16.00	17.87	19.75	21.62	23.50	16.79	21.81	26.83	31.85	36.87	41.89	46.91	8.14	8.62	8.94	9.18	9.36	9.51	9.62
8.00	13.00	15.00	17.00	19.00	21.00	23.00	25.00	18.90	24.58	30.27	35.96	41.64	47.33	53.02	9.18	9.72	10.09	10.36	10.57	10.73	10.86
8.50	13.75	15.87	18.00	20.12	22.25	24.37	26.50	21.13	27.52	33.92	40.31	46.71	53.10	59.50	10.28	10.89	11.31	11.61	11.84	12.02	12.17
9.00	14.50	16.75	19.00	21.25	23.50	25.75	28.00	23.48	30.63	37.77	44.92	52.06	59.21	66.35	11.44	12.12	12.59	12.93	13.19	13.39	13.56
9.50	15.25	17.62	20.00	22.37	24.75	27.12	29.50	25.96	33.90	41.83	49.77	57.71	65.64	73.58	12.66	13.42	13.94	14.32	14.61	14.83	15.02
10.00	16.00	18.50	21.00	23.50	26.00	28.50	31.00	28.56	37.33	46.10	54.87	63.64	72.41	81.18	13.95	14.79	15.37	15.79	16.10	16.35	16.55

TABLE B.4 Properties of the Critical Section—Corner Rectangular Column Bending Perpendicular to the Edge

B.2.6 Circular Interior Column

D/d	f_1	f_2
1.00	6.28	1.74
1.50	7.85	2.62
2.00	9.42	3.70
2.50	11.00	4.98
3.00	12.57	6.45
3.50	14.14	8.12
4.00	15.71	9.98
4.50	17.28	12.05
5.00	18.85	14.30
5.50	20.42	16.76
6.00	21.99	19.41
6.50	23.56	22.26
7.00	25.13	25.30
7.50	26.70	28.54
8.00	28.27	31.98
8.50	29.85	35.61
9.00	31.42	39.44
9.50	32.99	43.46
10.00	34.56	47.68

TABLE B.5 Properties of the Critical Section—Circular Interior Column

Index

Note: Page numbers followed by *t* and *f* indicate tables and figures, respectively.